SYMMETRIC SPACES

Short Courses Presented at Washington University

PURE AND APPLIED MATHEMATICS

A Series of Monographs and Textbooks

COORDINATOR OF THE EDITORIAL BOARD

S. Kobayashi

UNIVERSITY OF CALIFORNIA AT BERKELEY

1. Kentaro Yano. Integral Formulas in Riemannian Geometry (1970)
2. S. Kobayashi. Hyperbolic Manifolds and Holomorphic Mappings (1970)
3. V. S. Vladimirov. Equations of Mathematical Physics (A. Jeffrey, editor; A. Littlewood, translator) (1970)
4. B. N. Pshenichnyi. Necessary Conditions for an Extremum (L. Neustadt, translation editor; K. Makowski, translator) (1971)
5. L. Narici, E. Beckenstein, and G. Bachman. Functional Analysis and Valuation Theory (1971)
6. D. S. Passman. Infinite Group Rings (1971)
7. L. Dornhoff. Group Representation Theory (in two parts). Part A: Ordinary Representation Theory. Part B: Modular Representation Theory (1971, 1972)
8. W. Boothby and G. L. Weiss (eds.). Symmetric Spaces: Short Courses Presented at Washington University (1972)

In Preparation:

Y. Matsushima. Differentiable Manifolds (E. T. Kobayashi, translator)
Ararat Babakhanian. Cohomological Methods in Group Theory
Robert Gilmer. Multiplicative Ideal Theory
L. E. Ward, Jr. Topology: An Outline for a First Course
Jerry M. Mendel. Discrete Techniques of Parameter Estimation: The Equation Error Formulation

SYMMETRIC SPACES

SHORT COURSES PRESENTED AT WASHINGTON UNIVERSITY

Edited by William M. Boothby and Guido L. Weiss

DEPARTMENT OF MATHEMATICS
WASHINGTON UNIVERSITY
ST. LOUIS, MISSOURI

MARCEL DEKKER, INC. New York 1972

MARCEL DEKKER, INC.
95 Madison Avenue, New York, New York 10016

LIBRARY OF CONGRESS CATALOG CARD NUMBER: 74–182213

ISBN : 0–8247–1047–9

PRINTED IN THE UNITED STATES OF AMERICA

PREFACE

During the academic year 1969–1970 the Department of Mathematics of Washington University held a special year devoted to various topics touching on symmetric spaces. A number of experts in various aspects of this field were invited to give short courses of one to three weeks duration designed in each case to lead anyone with a rudimentary knowledge of Lie groups and symmetric spaces to some understanding of current research in the area of special interest of the lecturer. This volume consists of the texts of these courses, prepared by the lecturers themselves. Except for the courses of B. Kostant and J. Cheeger, which are being published elsewhere as research papers, all of the short courses given during the special year are included. Beyond this there were a number of one or two hour colloquia given by various visitors. Although they contributed substantially to the program, none are included in the present volume which, it was decided, should be limited to the courses only.

As is to be expected, the articles vary considerably in style and content. Some contain rather detailed preliminary exposition, some give a rather broad survey of problems and results in the field, and some concentrate on research in a more circumscribed area. There are a number of new results or improved proofs of hitherto published results. Inevitably there is a certain amount of overlapping and duplication; however, we have felt that any attempt to homogenize the material, standardize the notation, etc., would be quite beyond the competence of the editors and would greatly delay publication without materially adding to the value of the book. We have contented ourselves with grouping the papers in a fashion which, we hope, has a certain rough logic. A description of the five groups in order of their appearance now follows.

The course of N. Wallach, which happened to be the first one given, stands somewhat apart from the others in that it deals most directly with differential geometry. The author gives a rather complete exposition of the theory of minimal immersions of Riemannian manifolds and then treats in detail such immersions of homogeneous, especially symmetric, spaces into spheres. It is interesting to note the appearance of certain tools used in other papers on quite different topics.

The next group consists of the courses given by R. Gangolli (two separate courses given during a six week period), K. Gross and R. Kunze (presented by R. Kunze), S. Helgason, R. Hermann, and A. Knapp. These, roughly speaking, deal with representations of semisimple Lie groups and with the closely related topics of harmonic and special functions on symmetric spaces of these groups. Several papers, for example the first of Gangolli or the course of Knapp, have rather complete expositions of large parts of this theory.

In large areas of this subject the noncompact Hermitian symmetric spaces play a major role. These are complex analytic manifolds which generalize the unit disk or upper half-plane of the complex plane. They may be considered abstractly—as quotient spaces of semisimple Lie groups, or in various realizations—as in the case of the (holomorphically equivalent) half-plane and unit disk. They admit harmonic functions, have a natural Riemannian geometry, and allow totally discontinuous groups of automorphisms—thus leading to automorphic function theory. The article of S. Kobayashi deals with generalizations of the classical Schwarz Lemma to even more general complex manifolds than the Hermitian symmetric ones by making use of intrinsic metric properties. Y. Matsushima discusses closely related but more general complex domains (generalized Siegel domains) than those which are realizations of Hermitian symmetric spaces. The last paper of this group, that of J. A. Wolf, gives a very complete account of noncompact Hermitian symmetric spaces, of their realizations, and of an important topic not mentioned until now—the various natural ways in which a boundary can be attached to such spaces.

The next three courses, those of H. Furstenberg, A. Koranyi, and N. Weiss, deal with precisely the definition of the boundary and the closely related questions of the behavior of harmonic functions at the boundary. Taken together with the lectures of Wolf, this group of papers gives considerable insight into an active area of recent research. Its importance is clear from its relation to similar problems involving the disk and upper-half plane, e.g., Fatou's lemma.

The last two series of lectures, those of S. Rallis and H.-C. Wang deal in quite different ways with totally discontinuous groups, which of course is fundamental to the theory of automorphic functions in the classical case of the unit disk. As is well known, this leads to the study of discrete subgroups of the corresponding semisimple Lie groups acting on the symmetric space, the case where the quotient is compact or of bounded volume being of special importance. The second of Gangolli's courses could have been placed with this group, and it is also interesting to note that H. Furstenberg applies his methods to problems involving such discrete subgroups.

For a much better idea of the contents of this volume the reader is invited to glance through the articles themselves—they quite defy any easy classification or brief description. Taken in their totality they represent, it is hoped, a useful survey of many areas of active interest in symmetric spaces and their applications together with enough preliminary material to help the reader bridge any gap between his own general knowledge and the frontiers of current research.

The editors would like to express in warmest terms their gratitude to the many visitors who lectured at Washington University during our special year, to the authors who submitted material for this volume, to Washington University which paid the stipends of the lecturers, and to the Office of Naval Research which met the other expenses of the program and enabled G. Jensen and S. Kaneyuki to participate as postdoctoral fellows.

St. Louis, Missouri WILLIAM M. BOOTHBY
January, 1972 GUIDO WEISS

CONTRIBUTORS

1. Harry Furstenberg,* Hebrew University, Jerusalem, Israel
2. Ramesh Gangolli, University of Washington, Seattle, Washington
3. K. I. Gross, Dartmouth College, Hanover, New Hampshire
4. Sigurdur Helgason, Department of Mathematics, Massachusetts Institute of Technology, Cambridge, Massachusetts
5. Robert Hermann, Department of Mathematics, Rutgers University, New Brunswick, New Jersey
6. A. W. Knapp, Department of Mathematics, Cornell University, Ithaca, New York
7. Shoshichi Kobayashi, Department of Mathematics, University of California, Berkeley, California
8. Adam Koranyi, Department of Mathematics, Belfer Graduate School of Science, Yeshiva University, New York, New York
9. R. A. Kunze, University of California, Irvine, California
10. Yozo Matsushima, Department of Mathematics, University of Notre Dame, Notre Dame, Indiana
11. Stephen J. Rallis, Department of Mathematics, State University of New York, Stony Brook, New York

**Present Address:* Department of Mathematics, University of California, Berkeley, California

12. Nolan R. Wallach, Department of Mathematics, Rutgers–The State University, New Brunswick, New Jersey

13. Hsien-Chung Wang, Department of Mathematics, Cornell University, Ithaca, New York

14. Norman J. Weiss, Department of Mathematics, Columbia University, New York, New York

15. Joseph A. Wolf,* Department of Mathematics, Rutgers University, New Brunswick, New Jersey

**Present address:* Department of Mathematics, University of California, Berkeley, California 94720.

CONTENTS

MINIMAL IMMERSIONS OF SYMMETRIC SPACES INTO SPHERES

NOLAN R. WALLACH

Department of Mathematics
Rutgers—The State University
New Brunswick, New Jersey

Dedicated to Dana Kathleen

1. Introduction

The purpose of this article is to introduce the reader to a rapidly growing part of differential geometry: the study of minimal immersions into Riemannian manifolds. In this exposition, we concentrate on minimal immersions of homogeneous spaces (especially symmetric spaces) into spheres. Our reason for this emphasis is two-fold. The first (and more important) is that we can give many interesting concrete examples of minimal immersions of homogeneous spaces into spheres and that the analysis of these immersions leads to a nontrivial interaction between representation theory and differential geometry. Our second reason for this emphasis is that this is the only direction in the subject of minimal submanifolds that the author can write on with authority. This emphasis allows this article to be an introduction to the field of minimal immersions but (unfortunately) leaves out the very exciting directions taken by the subject that have little connection with Lie group theory (for example, the work of Chern, Osserman, and Lawson, to mention a few, on minimal immersions of surfaces into Euclidean space and spheres). Our emphasis does, however, allow the exposition to be self-contained modulo the standard material in a first course in Riemannian geometry.

We now summarize the material of the article to give the reader an idea of the flow of the exposition. In the second section, we study the Laplace–Beltrami operator of a Riemannian manifold. The results in this section are canonical. In the third section, we define the second fundamental form of an immersion of a manifold into a Riemannian manifold. The Gauss–Codazzi equations are derived. In the fourth section, the notion of minimal immersion of a manifold into a Riemannian manifold is defined. We derive the first variation formula for area and prove that n-dimensional minimal submanifolds are stationary points for n-dimensional area. As an application of the first variation formula (and the Mostow–Palais theorem), we prove that every orientable compact homogeneous space can be minimally, equivariantly immersed in a sphere (a result announced by Hsiang in [9]). In Section 5 we relate minimal, isometric immersions of Riemannian manifolds into Euclidean space and spheres with the Laplace–Beltrami operator. We also derive estimates on the scalar curvature of a minimal submanifold of a Riemannian manifold. In Section 6, we give a description of all minimal immersions of connected, simply connected (but not necessarily complete) spaces of constant positive curvature. In Section 7, several results on the Laplace–Beltrami operator on a naturally reductive homogeneous space are proved. We use the formalism of O'Neil [*13*] to derive our results and in the process put the fundamental equations of [*13*] into the formalism of moving frames. In Section 8, we give an exposition of the theory of class 1 representations of symmetric spaces of compact type. We give explicit descriptions (with proofs) for spheres and complex projective spaces. In Section 9, we reduce the problem of finding minimal isometric immersions of irreducible symmetric spaces of compact type into spheres to a problem in group theory. We also give examples in the nonirreducible case. The problem in group theory is the problem of rigidity. The last three sections are concerned with this problem. We point out that several of the results and ideas in these sections are new and have bearing on problems outside of minimal submanifolds. The notion of linear rigidity of Section 11 is an outgrowth of the work of doCarmo–Wallach [*8*]. Proposition 11.1 is a new generalization of a result in [*8*] and its Corollaries 11.1 and 11.3 are new. The higher fundamental forms are essentially due to Cartan. The use of the fundamental forms to decompose representations relative to isotropy groups deserves further attention.

2. The Laplace–Beltrami Operator

Let E^n be n-dimensional Euclidean space. Let $x_1, \ldots, x_n$ be the standard coordinates of E^n. The differential operator $\Delta = \sum \partial^2/\partial x_i^{\,2}$ is called the Laplacian of E^n. To give the reader an idea of how powerful a geometric object Δ is, we derive an interesting property of it.

Let $\phi : E^n \to E^n$ be a diffeomorphism. We define $\Delta^\phi f = (\Delta(f \circ \phi)) \circ \phi^{-1}$ for $f \in C^\infty(E^n)$ ($C^\infty(E^n)$ is the space of all real-valued C^∞ functions on E^n). Suppose $\Delta^\phi = \Delta$. Then we have for each $f \in C^\infty(E^n)$, $\Delta(f \circ \phi) = (\Delta f) \circ \phi$. Set $\phi_i = x_i \circ \phi$. Then clearly $\phi = (\phi_1, \ldots, \phi_n)$. We have $\Delta\phi_i = (\Delta x_i) \circ \phi = 0$. We also note that $\Delta((x_i \circ \phi)(x_j \circ \phi)) = (\Delta(x_i x_j)) \circ \phi = 2\delta_{ij}$. Combining the first observation with the second, we find

$$\sum_k \frac{\partial \phi_i}{\partial x_k} \frac{\partial \phi_j}{\partial x_k} = \delta_{ij}.$$

But this immediately implies that at each $p \in E^n$, the Jacobian matrix of ϕ is an orthogonal transformation. We have shown

Lemma 2.1. If $\phi : E^n \to E^n$ is a diffeomorphism such that $\Delta^\phi = \Delta$, then ϕ is an isometry.

Thus the Laplacian is one isometry invariant differential operator that determines whether or not a diffeomorphism is an isometry.

We now give three equivalent definitions of the Laplace–Beltrami operator on a Riemannian manifold. Let $(M, \langle,\rangle)$ be a Riemannian manifold. Let $U \subset M$ be open and let $x_1, \ldots, x_n$ be local coordinates on U. Set $g_{ij} = \langle(\partial/\partial x_i), (\partial/\partial x_j)\rangle$ on U and let $(g^{ij}) = (g_{ij})^{-1}$. Then on U we set $g = \det(g_{ij})$ and

$$\Delta = \frac{1}{\sqrt{g}} \sum_k \frac{\partial}{\partial x_k}\left(\sum_i g^{ik}\sqrt{g}\frac{\partial}{\partial x_i}\right).$$

It is not hard to check that Δ is independent of the choice of local coordinates and thus defines a differential operator on M. We note that if $(M, \langle,\rangle) = E^n$, then Δ is just the ordinary Laplacian. It is not hard to prove that if $\phi : M \to M$ is a diffeomorphism such that $\Delta^\phi = \Delta$, then ϕ is an isometry.

We now give a second definition which is more useful in practice. Let $(M, \langle,\rangle)$ be a Riemannian manifold. A Riemannian vector bundle with base M is a vector bundle $E \overset{\Pi}{\to} M$ and a correspondence $p \to \langle,\rangle_p$ where $\langle,\rangle_p$ is an inner product on $\Pi^{-1}(p)$ so that if X, Y are C^∞ cross sections of E, then $\langle X, Y\rangle \in C^\infty(M)$. An invariant connection on $(E, \langle,\rangle)$ is a real bilinear pairing of vector fields and C^∞ cross sections to C^∞ cross sections $X, \psi \to \nabla_X\psi$, subject to

(i) $\nabla_X F\psi = (XF)\psi + F\nabla_X\psi$ for ψ a cross section of E, X a vector field, and $F \in C^\infty(M)$.

(ii) $X\cdot\langle\phi,\psi\rangle = \langle\nabla_X\phi,\psi\rangle + \langle\phi,\nabla_X\psi\rangle$ for ϕ, ψ cross sections and X a vector field.

We will be mainly using the case $E = M \times E^n$ with the constant Riemannian structure $\langle(m,x), (m,y)\rangle_m = \langle x,y\rangle$. The invariant connection that we take is $\nabla_X\psi = \psi_*(X)$ where ψ is looked upon as a C^∞ mapping of M into E^n, ψ_* is the differential of ψ, and we identify the tangent space at each point of E^n with E^n. In particular, if $n = 1$, then $\nabla_X F = X\cdot F$.

If E and F are Riemannian vector bundles with invariant connections ∇, $\overline{\nabla}$, respectively, then $\operatorname{Hom}(E, F)$ is a vector bundle over M and we give it a Riemannian structure $\langle A, B\rangle = \operatorname{tr} {}^tBA$. Equivalently, it is given by $\langle A,B\rangle_p = \sum_{i=1}^n\langle A(e_i), B(e_i)\rangle_p$ where $e_1, \ldots, e_n$ is an orthonormal basis of the fiber of E at p. We define an invariant connection on $\operatorname{Hom}(E, F)$ as follows: $(\nabla_X A)(\psi) = \overline{\nabla}_X(A\psi) - A(\nabla_X\psi)$. By $\operatorname{Hom}((E, \langle,\rangle, \nabla), (F, \langle,\rangle, \overline{\nabla}))$, we mean $\operatorname{Hom}(E, F)$ with the above Riemannian structure and connection.

Let $T(M)$ be the tangent bundle of M. Then $\langle,\rangle$ induces the Riemannian connection on $T(M)$, ∇. Let now $(E, \langle,\rangle, \nabla)$ be an affine connected Riemannian vector bundle over M. If ψ is a cross section of E, we define $\nabla\psi$ a cross section of $\mathrm{Hom}(T(M), E)$ via

$$(\nabla\psi)(X) = \nabla_X\psi.$$

Now applying the affine connection on $\mathrm{Hom}(T(M), E)$, we define $\nabla^2\psi(X, Y) = (\nabla_X(\nabla\psi))(Y)$ for X, Y vector fields on M. We note that if $E = M \times R$ with the above defined affine connection and if $f \in C^\infty(M)$, then $\nabla^2 f$ is usually called the Hessian of f. Indeed, if $df_p = 0$ and if $x_1, \ldots, x_n$ are local coordinates at p, then $\nabla^2 f(\partial/\partial x_{i_p}, \partial/\partial x_{j_p}) = (\partial^2/\partial x_{i_p}\partial x_j)f$.

We can now give our second definition of Δ. $\Delta_p\psi = \sum_i \nabla^2\psi(e_i, e_i)$ where $e_1, \ldots, e_n$ is an orthonormal basis of $T_p(M)$.

We now give a third definition of the Laplace–Beltrami operator that is useful for integration. Let M be an orientable Riemannian manifold. Let for each $p \in M$, $e_1, \ldots, e_n$ be a positively oriented orthonormal frame. Let $\omega_1, \ldots, \omega_n$ be the corresponding coframe. Set $\omega_p = \omega_1 \wedge \cdots \wedge \omega_n$. Then ω defines an n-form on M, the volume form of M. We extend the Riemannian structure of M to $\Lambda^p T_q(M)^*$ in the usual manner. We define $* : \Lambda^p T_q(M)^* \to \Lambda^{n-p} T_q(M)^*$ via $\langle *\eta, \alpha\rangle\omega = \eta \wedge \alpha$, $\eta \in \Lambda^p T_q(M)^*$, $\alpha \in \Lambda^{n-p} T_q(M)^*$. Let $D^p(M)$ be the space of all C^∞-p-forms on M. We define $\delta\colon D^p \to D^{p-1}$ by

$$\delta D^0 = 0,$$

$$\delta\eta = (-1)^{np+n+1} * d * \eta \qquad \text{for} \quad \eta \in D^p, \quad p \neq 0.$$

On p-forms we now have the Hodge–Laplacian $d\delta + \delta d$. For $p = 0$, we set $\Delta f = -\delta df$. Then Δ is the Laplace–Beltrami operator of M. We note that if η, α are respectively a p-form and a $p + 1$-form, one of which has compact support, then

$$\int_M d\eta \wedge *\alpha = \int_M \eta \wedge *(\delta\alpha).$$

That is, δ is the formal adjoint to d. For functions, this says

$$\int_M (\Delta f) g * 1 = \int_M f \Delta g * 1.$$

Using this definition of Δ, we have

Proposition 2.1. (E. Hopf). Let M be a compact, connected Riemannian manifold. If $f \in C^\infty(M)$ and $\Delta f \geqq 0$, then f is constant. In particular, the only solutions to $\Delta f = 0$ on M are the constants.

Proof. We first note that $\int_M \Delta f * 1 = 0$ since $\int_M \Delta f * 1 = \int_M f \Delta 1 * 1 = 0$. But $\Delta f \geq 0$ now implies $\Delta f \equiv 0$. Now $0 = \int_M \Delta f f * 1 = \int_M \Delta f * f = \int_M df \wedge *df$. But $\int_M df \wedge *df = 0$ implies $df \equiv 0$. The result is thus proved.

Let D be the Laplace–Beltrami operator of E^{n+1}, Δ the Laplace–Beltrami operator of $S^n = \{p \in E^{n+1} \mid \|p\|^2 = 1\}$ with the induced metric. We find a formula for D in terms of Δ. Let $R^+ = \{t \in R \mid t > 0\}$. We look upon $E^{n+1} - \{0\}$ as $R^+ \cdot S^n$. Let $\psi : R^+ \times S^n \to E^{n+1} - \{0\}$ be defined by $\psi(t, p) = tp$. If $f \in C^\infty(E^{n+1})$, set $f_t(p) = f(tp)$. We pull back the Riemannian structure on E^{n+1} to $R^+ \times S^n$. Let $\omega = dx_1 \wedge dx_2 \cdots dx_{n+1}$. If ω_0 is the volume element of S^n, $\psi^*\omega = t^n dt \wedge \omega_0$. Now $df = (\partial f/\partial t)dt + df_t$. We note that $*dt = t^n \omega_0$ and $*df_t = -dt \wedge *_{tS^n} df_t$ (where tS^n is $\{p \in E^{n+1} \mid \|p\|^2 = t^2\}$ with the induced Riemannian structure). Thus,

$$*df = t^n \frac{\partial f}{\partial t} \omega_0 - dt \wedge *_{tS^n} df_t \tag{2.1}$$

$$d*df = nt^{n-1} \frac{\partial f}{\partial t} dt \wedge \omega_0 + t^n \frac{\partial^2 f}{\partial t^2} dt \wedge \omega_0 + dt \wedge d*_{tS^n} df_t. \tag{2.2}$$

Now let Δ_t be the Laplace–Beltrami operator of tS^n, then $\Delta_t = \dfrac{1}{t^2}\Delta$. Thus

$$*d*df(tp) = \frac{n}{t}\frac{\partial f}{\partial t}(tp) + \frac{\partial^2 f}{\partial t^2}(tp) + \dot{\Delta}_t f_t(p).$$

We have thus proved

Lemma 2.2.

$$D = \frac{n}{t}\frac{\partial}{\partial t} + \frac{\partial^2}{\partial t^2} + \frac{1}{t^2}\Delta.$$

3. The Second Fundamental Form

Let $(\overline{M}, \langle,\rangle)$ be a Riemannian manifold and let M be a C^∞-manifold. Let $\overline{\nabla}$ be the Riemannian connection on $\overline{M}$. Let $x : M \to \overline{M}$ be an immersion. Now x is locally an imbedding; thus we may identify M locally with its image in $\overline{M}$.

The normal bundle $N(M)$ of the immersion is the vector bundle over M with fiber at $p \in M$, $N_p = \{v \in T_p(\overline{M}) \mid \langle v, T_p(M)\rangle_p = 0\}$. $N(M)$ is naturally a Riemannian vector bundle.

x induces a Riemannian structure on M. Let ∇ be the Riemannian connection on M. If $p \in M$ and $v \in T_p(\overline{M})$, let v^T, v^N be such that $v = v^T + v^N$ and $v^T \in T_p(M)$, $v^N \in N_p$.

Lemma 3.1. Let $p \in M$, $v \in T_p(M)$, and let Y be a vector field on M. Extend Y in a small neighborhood U of p in $\overline{M}$ to a vector field on U. Then $\nabla_v Y = (\overline{\nabla}_v Y)^T$.

Proof. Let Z be another vector field on M extended to a vector field on U. Then $v \cdot \langle Y,Z \rangle = \langle \overline{\nabla}_v Y,Z \rangle + \langle Y,\overline{\nabla}_v Z \rangle = \langle (\overline{\nabla}_v Y)^T,Z \rangle + \langle Y,(\overline{\nabla}_v Z)^T \rangle$. Now if Y is a field on a neighborhood of U so that $Y|_{M \cap U} = 0$, then $v \cdot \langle Y,Z \rangle = 0$. Indeed, if σ is a curve in M so that $\dot{\sigma}(0) = v$, then $v \cdot \langle Y,Z \rangle = d/dt_{t=0} \langle Y_{\sigma(t)}, Z_{\sigma(t)} \rangle = 0$. But then $\langle \overline{\nabla}_v Y,Z \rangle_p = -\langle Y,\overline{\nabla}_v Z \rangle_p = 0$. Thus $(\overline{\nabla}_v Y)^T$ does not depend on the extension of Y. Furthermore $(\overline{\nabla}_{Y_p} Z)^T - (\overline{\nabla}_{Z_p} Y)^T = (\overline{\nabla}_{Y_p} Z - \overline{\nabla}_{Z_p} Y)^T = ([Y,Z]_p)^T = [Y,Z]_p$. Thus by the fundamental theorem of Riemannian geometry $(\overline{\nabla}_v Y)^T = \nabla_v Y$.

We can now define the second fundamental form of $x : M \to \overline{M}$. Let $p \in M$. Let X, Y be vector fields in a neighborhood of p in M, extend X, Y to a small neighborhood of p in $\overline{M}$. We set $B_p(X, Y) = \overline{\nabla}_{X_p} Y - \nabla_{X_p} Y$. We note that $B_p(X, Y) - B_p(Y, X) = \overline{\nabla}_{X_p} Y - \overline{\nabla}_{Y_p} X - \nabla_{X_p} Y + \nabla_{Y_p} X = [X, Y]_p - [X, Y]_p = 0$. Thus $B_p(X, Y) = B_p(Y, X)$. Hence $B_p(X, Y)$ depends only on X_p, Y_p. We have

Lemma 3.2. Let $p \in M$, $y_1, y_2 \in T_p(M)$, extend y_1, y_2 to vector fields, Y_1, Y_2, in a neighborhood of p in $\overline{M}$. Then $B_p(y_1, y_2) = \overline{\nabla}_{y_1} Y_2 - \nabla_{y_1} Y_2$ defines a symmetric, bilinear map $B_p : T_p(M) \times T_p(M) \to N_p$. B_p is called the second fundamental form of x at p.

We now show that B contains all of the curvature information of M and that B tells how M sits in $\overline{M}$.

Lemma 3.3. M is totally geodesic in $\overline{M}$ if and only if $B \equiv 0$.

Proof. If $B \equiv 0$ and if σ is a geodesic in M, then $0 = B(\dot{\sigma}(t), \dot{\sigma}(t)) = \overline{\nabla}_{\dot{\sigma}(t)}\dot{\sigma}(t) - \nabla_{\dot{\sigma}(t)}\dot{\sigma}(t) = \overline{\nabla}_{\dot{\sigma}(t)}\dot{\sigma}(t)$. Thus σ is a geodesic in $\overline{M}$. Conversely, suppose that M is totally geodesic in $\overline{M}$. Let $p \in M$ and let $y \in T_p(M)$. Let σ be the geodesic in $\overline{M}$ such that $\dot{\sigma}(0) = y$. Then $B_p(y, y) = \overline{\nabla}_{\dot{\sigma}(0)}\dot{\sigma} - \nabla_{\dot{\sigma}(0)}\dot{\sigma} = -\nabla_{\dot{\sigma}(0)}\dot{\sigma}$. But σ is also a geodesic in M, thus $B_p(y, y) = 0$. Hence $B_p \equiv 0$.

To show how B determines the curvature tensor of M, we derive the Gauss–Codazzi equation (for hypersurfaces in E^3 this is Gauss' "Theorema Egregium").

Theorem 3.1. Let R, $\bar{R}$ be respectively the curvature tensors of M and $\bar{M}$, let $p \in M$, and let $w, y \in T_p(M)$. Then $\langle \bar{R}(w, y)\, y, w\rangle = \langle R(w, y)\, y, w\rangle + \langle B(w, y), B(w, y)\rangle - \langle B(w, w), B(y, y)\rangle$.

Proof. Let X, Y be vector fields on a neighborhood of p in $\bar{M}$ extending w and y. Then $\langle \bar{R}(w, y)\, y, w\rangle = \langle \bar{\nabla}_w \bar{\nabla}_Y Y - \bar{\nabla}_y \bar{\nabla}_X Y - \bar{\nabla}_{[X,Y]_p} Y, w\rangle = \langle \bar{\nabla}_w(\nabla_Y Y + B(Y, Y)) - \bar{\nabla}_y(\nabla_X Y + B(X, Y)) - \nabla_{[X,Y]p} Y, w\rangle$ (since $\langle B([X, Y]_p, Y_p), w\rangle = 0) = \langle \nabla_w \nabla_Y Y, w\rangle + \langle \bar{\nabla}_w B(Y, Y), w\rangle - \langle \nabla_Y \nabla_X Y, w\rangle - \langle \bar{\nabla}_Y B(X, Y), w\rangle - \langle \nabla_{[X,Y]p} Y, w\rangle$. Now $\langle \bar{\nabla}_w B(Y, Y), w\rangle = w \cdot \langle B(Y, Y), X\rangle - \langle B(Y, Y), \bar{\nabla}_w X\rangle = -\langle B(Y, Y), B(X, X)\rangle$. Similarly $-\langle \bar{\nabla}_y B(X, Y), w\rangle = \langle B(X, Y), B(X, Y)\rangle$.

4. Minimal Submanifolds

Let $x: M \to \bar{M}$ be an immersion where M is a C^∞-manifold and $(\bar{M}, \langle,\rangle)$ is a Riemannian manifold. Let B be the second fundamental form of x. We define the mean curvature vector H of x as follows: Let $p \in M$. Let $e_1, \ldots, e_n$ be an orthonormal basis of $T_p(M)$. Then $H_p = (1/n)\sum_{i=1}^n B_p(e_i, e_i)$.

H defines a normal field on M.

Definition 4.1. $x : M \to \bar{M}$ is said to be *minimal* if $H \equiv 0$.

We note that if $\dim M = 1$, then $x(M)$ is locally a C^∞ curve in $\bar{M}$. We may thus reparametrize x, locally, by arc length and find that x is minimal if and only if this reparametrization of x makes x into a geodesic. Thus minimal submanifolds generalize the notion of geodesics.

We now show that the notion of minimality is a strict generalization of the notion of geodesic. That is, we show that if $x : M \to \bar{M}$ is a minimal submanifold of $\bar{M}$, then x is locally a critical point for $n(= \dim M)$-dimensional area.

We first need a notion of local variation. Let M be a compact orientable manifold possibly with boundary ∂M. Let $x : M \to \bar{M}$ be an immersion. A variation of x is a C^∞ mapping $F : (-\varepsilon, \varepsilon) \times M \to \bar{M}$ so that

(1) If $F_t(p) = F(t, p)$, then $F_t : M \to \overline{M}$ is an immersion and $F_0 = x$.

(2) If $p \in \partial M$, then $F_t(p) = x(p)$ for $t \in (-\varepsilon, \varepsilon)$.

We note that if $\sigma: [a, b] \to \overline{M}$ is a C^∞-curve, then a variation of σ in the above sense is exactly what is usually meant by a variation of σ.

The variation field of the variation F of x is defined by $E_p = F_{*(0,p)}(\partial/\partial t)$ for $p \in M$. We note that $E_p = 0$ for $p \in \partial M$. Let $A_F(t)$ be the Riemannian volume of M relative to $F_t^*\langle,\rangle$. We wish to compute explicitly $A_F'(0)$, for M, possibly "shrunk."

We are interested (as in the case of geodesics) in the case of local critical points for A_F. Thus we assume that $x : M \to \overline{M}$ is an imbedding. Thus E^T defines a vector field on M with compact support, vanishing on ∂M. Hence E^T defines a 1-parameter group of diffeomorphisms of M, ϕ_t. Let ψ : $(-\varepsilon\ \ \varepsilon) \times M \to (-\varepsilon\ \ \varepsilon) \times M$ be defined by $\psi(t, p) = (t, \phi_{-t}(p))$. Let $\tilde{F} = F \circ \psi$. Then $\tilde{F}$ is a variation of x and the variation field of $\tilde{F}$ is E^N. Furthermore $A_{\tilde{F}}(t) = A_F(t)$. We may thus assume $E = E^N$, $F = \tilde{F}$.

Set $g_t = F_t^*\langle,\rangle$. We compute g_t to second order in t. Let $\mathring{M} = M - \partial M$. Let $U, V \in T_p(M)$. Then $g_{t_p}(U, V) = \langle F_{t_{*p}}U, F_{t_{*p}}V\rangle$. Set $\sigma(t) = F_t(p)$ and let D/dt be covariant differentiation along σ in $\overline{M}$. Set $X(t) = F_{t_{*p}}U$, $Y(t) = F_{t_{*p}}V$. By Taylor's theorem,

$$g_{t_p}(U, V) = g_{0_p}(U, V) + t(d/dt)_{t=0}\langle X(t), Y(t)\rangle + O(t^2).$$

But

$$\frac{d}{dt_{t=0}}\langle X(t), Y(t)\rangle = \langle \frac{DX}{dt}(0), V\rangle + \langle U, \frac{DY}{dt}(0)\rangle.$$

Now if we extend X, Y, E to a small neighborhood of p in $\overline{M}$, we have

$$\frac{D}{dt_{t=0}}X = \bar{\nabla}_U E + [E, X]_p\ (\dot{\sigma}(t) = E_{\sigma(t)}).$$

But by the definition of E, we have $[E, X]_p = [E, Y]_p = 0$. ($E = F_*(\partial/\partial t)$ and U, V are tangent to M.) Thus,

$$\frac{d}{dt_{t=0}}\langle X(t), Y(t)\rangle = \langle \bar{\nabla}_U E, Y\rangle + \langle X, \bar{\nabla}_V E\rangle.$$

Now if ω (resp. γ) is a curve in M such that $\dot{\omega}(0) = U$ (resp. $\dot{\gamma}(0) = V$), then $\langle E, Y\rangle_{\omega(t)} = 0$ (resp. $\langle E, X\rangle_{\gamma(t)} = 0$). Thus $U\cdot\langle E, Y\rangle = U\langle E, X\rangle = 0$. Hence $\langle \bar{\nabla}_U E, Y\rangle + \langle X, \bar{\nabla}_V E\rangle = -\langle E, \bar{\nabla}_U Y\rangle - \langle E, \bar{\nabla}_V X\rangle = -2\langle E, B(U, V)\rangle$.

We thus find that

$$g_{t_p}(U, V) = g_{0_p}(U, V) - 2t\langle E, B(U, V)\rangle + O(t^2).$$

Let $e_1, e_2, \ldots, e_n$ be a positive frame in a neighborhood of p for M. Then

$$g_t(e_i, e_j) = \delta_{ij} - 2t \langle B(e_i, e_j), E\rangle + O(t^2).$$

Let ω_t be the volume element of g_t on M. Then $\omega_t = ([\det(g_t(e_i, e_j))]^{1/2})\omega_0 = (1 - nt \langle H, E\rangle)\omega_0 + O(t^2)$.

Thus, we find that

$$A_F(t) = A(0) - tn \int_M \langle H, E\rangle w_0 + O(t^2).$$

We thus have the first variation formula

Proposition 4.1. Let $x : M \to \overline{M}$ be an imbedding of M, a compact, orientable manifold with boundary into $(\overline{M}, \langle,\rangle)$, a Riemannian manifold. If F is a variation of M with variation vector E and if $A_F(t)$ is the volume of M induced by F_t, then

$$A_F'(0) = -n \int_M \langle H, E\rangle w_0.$$

Corollary 4.1. Let $x : M \to \overline{M}$ be an immersion. Then x is minimal if and only if M is locally a critical point for n ($=\dim M$) volume.

Proof. If $H \equiv 0$, then M is clearly locally a critical point for n-dimensional volume. If M is locally a critical point for n-dimensional volume, let $p \in U \subset M$, U open in M, be such that $x|_U$ is a critical point for any variation of $x|_U$. Let $\phi \in C^\infty(U)$ be a nonnegative function having compact support and such that $\phi(p) = 1$, $\phi = 0$ on ∂U. Set $F_t(q) = \mathrm{Exp}_{x(q)}\,(t\phi(q)H_q)$. Then the variation vector of F is ϕH. Thus $A_F'(0) = -n\int_M \phi\langle H, H\rangle\omega_0$. Thus $A_F'(0) = 0$ implies $H_p = 0$.

Using Corollary 4.1, we give a proof of a theorem of Hsiang [9]. We first need the notion of equivariant immersion. Let M be a manifold, G a compact Lie group acting on M. Let $\overline{M}$ be a Riemannian manifold and let $I(\overline{M})$ be the group of isometries of $\overline{M}$. An immersion $x : M \to \overline{M}$ is said to be *G-equivariant* if there is a continuous group homomorphism $\rho : G \to I(\overline{M})$ such that $x(gp) = \rho(g)x(p)$ for $g \in G$, $p \in M$.

Theorem 4.1. (Hsiang [9]). Let G be a compact Lie group and let K

be a closed subgroup of G. Let $M = G/K$. Then there is a G-equivariant minimal immersion of M into S^n for n large enough.

Proof. The Mostow–Palais theorem [*12*, *14*] says that there is an equivariant embedding $x : G/K \to S_1^n \subset E^{n+1}$. Let $x(gz) = \rho(g)x(z)$, $g \in G$, $z \in G/K$. Let $X = \{z \in S_1^n | \rho(k)z = z \text{ for all } k \in K\}$. Then X is a compact subset of S_1^n and $X \neq \emptyset$. Let for each $z \in X$, $f_z : G/K \to S_1^n$ be defined by $f_z(gK) = \rho(g)z$. Then f_z is a C^∞-mapping for each $z \in X$. If f_z is regular, define $B(z)$ to be the volume of G/K_0 induced by the Riemannian structure pulled back by f_z. K_0 is the connected open subgroup of K. If f_z is not regular, define $B(z)$ to be 0. We assert that the function $B : X \to R$ so defined is continuous. Indeed, let $\mathfrak{g}$ be the Lie algebra of G, $\mathfrak{k}$ the Lie algebra of K. Let $\langle , \rangle$ be an $\mathrm{Ad}(G)$ invariant inner product on $\mathfrak{g}$. Let $\mathfrak{p}$ be the orthogonal complement to $\mathfrak{k}$ in $\mathfrak{g}$. Let $x_1, \ldots, x_n$ be an orthonormal basis for $\mathfrak{p}$, relative to $\langle , \rangle$. Let $\omega_1, \ldots, \omega_m$ be the dual basis. Then $\omega = \omega_1 \wedge \cdots \wedge \omega_m$ induces a G-invariant volume element $\bar{\omega}$ on G/K_0. We may assume that $\int_{G/K} \bar{\omega} = 1$. Now if $z \in X$ is an element so that f_z is regular, then the induced volume element on G/K_0 is $c(z)\bar{\omega}$, $c(z)$ a positive constant. We compute $c(z)$. Let $x(gw) = \rho(g)x(w)$, $w \in G/K$, $g \in G$. $\rho : G \to O(n + 1)$ is a representation. Let us denote also by ρ the representation of $\mathfrak{g}$ on E^{n+1}. Then $c(z) = [\det(\langle \rho(x_i)z, \ \rho(x_j)z \rangle)]^{1/2}$. Thus $B(z) = [\det(\langle \rho(x_i)z, \rho(x_j)z \rangle)]^{1/2}$ and is clearly continuous.

Since X is compact, B takes a maximum, say at z_0. $B(z_0) \neq 0$; thus $f_{z_0} : G/K \to S_1^n$ is an equivariant immersion. Let H be the mean curvature vector of f_{z_0}. Then $\rho(g)H_p = H_{gp}$ for $g \in G$, $p \in G/K$. Let $F_t(p) = \mathrm{Exp}_{f_{z_0}(p)}(t\,H_p)$. Then $F_t(g \cdot p) = \mathrm{Exp}_{f_{z_0}(g \cdot p)}(t\,H_{g \cdot p}) = \mathrm{Exp}_{\rho(g) f_{z_0}(p)}(t\,\rho(g)H_p) = \rho(g)F_t(p)$. Thus $F_t = f_{F_t(eK)}$. The variation vector of F_t is H. Let $A(t)$ be the induced area given by F_t. Then $A(t) = B(F_t(eK))$. Thus, since $A(0) \geqq A(t)$, $t \neq 0$, we have $A'(0) = 0$. But then $0 = A'(0) = -m\int_{G/K} \langle H, H \rangle \, A(0)\bar{\omega}$. Thus, $H = 0$. Q.E.D.

5. Minimal Immersions into Euclidean Space and into Spheres

Let M be a C^∞ manifold and let $x : M \to E^n$ be an immersion. We compute the mean curvature vector of x. Since the mean curvature vector is locally defined, we may assume that x is an imbedding. We thus assume that M is a submanifold of E^n.

Let $p \in M$. Let $E_1, \ldots, E_m$ be a (local) moving frame on M in a neighborhood of p (that is, $E_1, \ldots, E_m$ are mutually orthogonal relative to the induced Riemannian structure). We may extend $E_1, \ldots, E_m$ to a local moving frame $E_1, \ldots, E_n$ on a sufficiently small neighborhood of p in E^n.

We assume that $\nabla_{E_{i_p}} E_j = 0$, $1 \leq i, j \leq m$, ∇ is the induced connection on M. We look at x as a cross section of the trivial bundle $M \times E^n$. Using the second definition of the Laplace–Beltrami operator, we find

$$\Delta x(p) = \sum_{i=1}^{m} E_{i_p} E_i x = \sum_{i=1}^{m} \overline{\nabla}_{E_{i_p}} E_i$$

where $\overline{\nabla}$ is the connection on E^n. But $\nabla_{E_{i_p}} E_i = 0$, $i = 1, \ldots, m$. Thus $\overline{\nabla}_{E_{i_p}} E_i = (\overline{\nabla}_{E_{i_p}} E_i)^N$. Hence $\Delta x(p) = \sum_{i=1}^{m} (\overline{\nabla}_{E_{i_p}} E_i)^N = mH_p$. We have proved

Lemma 5.1. Let $x : M \to E^n$ be an immersion; then $\Delta x = mH$.

Corollary 5.1. There are no minimal immersions of compact manifolds into Euclidean space.

Proof. If M is compact, $x : M \to E^n$ minimal, then $\Delta x = 0$. Thus if $x = (x_1, \ldots, x_n)$, $\Delta x_i = 0$. Hence x_i is constant on components. Thus, x cannot be an immersion.

Corollary 5.2. (Takahashi [*16*]). Let M be a connected m-dimensional Riemannian manifold. Let $x : M \to E^n$ be an isometric immersion. Suppose that $\Delta x = -\lambda x$, $\lambda \neq 0$. (Here we look at x as a cross section of $M \times E^n$. In any event, it is the componentwise Laplacian.) Then

(a) $\lambda > 0$,
(b) $x(M) \subset r \cdot S^{n-1}$ (the sphere of radius r),
(c) $r = (m/\lambda)^{1/2}$, and
(d) $x : M \to rS^{n-1}$ is minimal.

Furthermore, if $x : M \to rS^{n-1}$ is minimal, then $\Delta x = (-m/r^2)x$.

Proof. By Lemma 5.1, $H = -(\lambda/m)x$. Thus, since $\lambda \neq 0$, we have for $p \in M$ and $v \in T_p(M)$, $\langle x(p), x_{*p}(v)\rangle = 0$. Thus $p \to \langle x(p), x(p)\rangle$ is constant; this proves (b). Let $E_1, \ldots, E_m$ be a moving frame on M in a neighborhood of p. Let $E_1, \ldots, E_{n-1}$ be an extension of $E_1, \ldots, E_m$ to a moving frame in a neighborhood of p in rS^{n-1}. Set $E_{n_q} = (1/r)x(q)$. Let N_1 be a normal projection relative to x in rS^{n-1}. The mean curvature vector of x in rS^{n-1} is given by

$$\frac{1}{m} \sum_{i=1}^{m} (\overline{\nabla}_{E_i} E_i)^{N1} = H - \frac{1}{r^2} \langle H, x\rangle x = 0.$$

Thus (d) is proved. Now

$$mH_p = \sum_{i=1}^{m} \left\langle \bar{\nabla}_{E_i} E_i, \frac{p}{r} \right\rangle \frac{p}{r} = -\frac{1}{r^2} \sum_{i=1}^{m} \langle E_i, E_i \rangle p = -\frac{m}{r^2} p.$$

This proves (a), (c), and the last statement of the Corollary.

We now derive the relationship between the square of the norm of the second fundamental form of a minimal immersion and scalar curvature. First, we recall some definitions. Let $(M, \langle , \rangle)$ be a Riemannian manifold. Let R be the curvature tensor of M. If $p \in M$ and $x, y \in T_p(M)$, we define

$$S_p(x, y) = \operatorname{tr}(z \to R(z, x)y).$$

S defines a tensor field on M called the Ricci tensor. We note that $S(x, y) = S(y, x)$. Using the Ricci tensor, we define the scalar curvature

$$\tau_M(p) = \sum_{i=1}^{m} S_p(e_i, e_i)$$

where $e_1, \ldots, e_n$ is an orthonormal frame at p. Using the frame $e_1, \ldots, e_n$, we note that $\tau_M(p) = \sum_{i=1}^{m} \sum_{j \neq i} \langle R(e_j, e_i)e_i, e_j \rangle$.

Now suppose that $x : M \to \overline{M}$ is a minimal immersion where $(\overline{M}, \langle , \rangle)$ is a Riemannian manifold. Let $\overline{R}$ be the curvature tensor of $\overline{M}$, and let R be the induced curvature tensor of M. Then by Theorem 3.1, $\langle \overline{R}(e_j, e_i)e_i, e_j \rangle = \langle R(e_j, e_i)e_i, e_j \rangle + \langle B(e_i, e_j), B(e_i, e_j) \rangle - \langle B(e_i, e_i), B(e_j, e_j) \rangle$.

From this, we see that if $x : M \to \overline{M}$ is minimal, then

$$\sum_{j=1}^{m} \langle \overline{R}(e_j, e_i)e_i, e_j \rangle = \sum_{j=1}^{m} \langle R(e_j, e_i)e_i, e_j \rangle + \sum_{j=1}^{m} \langle B(e_i, e_j), B(e_i, e_j) \rangle.$$

This implies immediately

Lemma 5.2. If all of the sectional curvatures of $\overline{M}$ are ≤ 0, then the Ricci tensor of M is negative semidefinite.

We recall that the sectional curvature of a plane P in $T_p(M)$ spanned by u, v, an orthonormal basis, is $K(P) = \langle R(u, v)u, v \rangle$.

Suppose that all of the plane sectional curvatures of $\overline{M}$ are bounded above by a constant $\overline{K}$. Then we have by the above computation

$$S_p(e_i, e_j) \leqq (n-1)\overline{K} - \sum_{j=1}^{n} \langle B(e_j, e_i), B(e_j, e_i) \rangle.$$

Now $\|B_p\|^2 = \sum_{i,j=1}^{n} \langle B(e_i, e_j), B(e_i, e_j) \rangle$. Thus, we have

$$\tau_M(p) \leq n(n-1)\overline{K} - \|B_p\|^2.$$

Lemma 5.3. Let $x : M \to \overline{M}$ be a minimal immersion. Suppose that all of the sectional curvatures of $\overline{M}$ are bounded above by $\overline{K}$; then

$$\tau_M \leq n(n-1)\overline{K} - \|B\|^2.$$

Corollary 5.3. Let M be a Riemannian manifold whose scalar curvature is nonnegative. Let $x : M \to E^n$ be a minimal isometric immersion. Then M is flat and $x : M \to E^n$ is totally geodesic.

Proof. $0 \leq \tau_M = -\|B\|^2 \leq 0$. Thus $\|B\|^2 = 0$. Hence $B \equiv 0$.

6. Minimal Immersions of Spheres into Spheres

Let S_k^n be the unit sphere of R^{n+1} with the Riemannian metric of constant curvature k. Let $U \subset S_k^n$ be a connected open subset so that if $x \in U$, then $-x \notin U$, and let

$$f : U \to S_1^p$$

be a minimal isometric immersion. We analyze this situation in this section.

Let $f = (f_1, \ldots, f_{p+1}), f_i : U \to R$. Then by Takahashi's result (Corollary 5.2), if $\tilde{\Delta}$ is the Laplace–Beltrami operator on S_k^n, $\tilde{\Delta} f_i = -nf_i$. Let Δ be the Laplace–Beltrami operator of S_1^n. Then $\tilde{\Delta} = k\Delta$. Thus $\Delta f_i = -(n/k)f_i$.

We need an auxiliary notion.

Definition 6.1. Let $CU = \{tx | \; t \in R,\ t \geq 0,\ x \in U\} = R^+U$. Let $\lambda \in R$. Then a fractional spherical harmonic of degree λ on CU is a C^∞ function,

$$h : CU \to R$$

such that

(1) $h(tx) = t^\lambda h(x)$ for $x \in CU$, $t \in R$ so that $tx \in CU$.

(2) $\displaystyle\sum_{i=1}^{n+1} \frac{\partial^2}{\partial x_i^2} h = 0.$

Set as usual $D = \sum \partial^2/\partial x_i{}^2$.

Now using Lemma 2.2, we see that if h is a fractional spherical harmonic of degree λ on CU, then $\Delta(h|_U) = -\lambda(\lambda + n - 1)(h|_U)$. Using Lemma 2.2 we also easily derive a converse.

Lemma 6.1. Let $h : U \to R$ be a C^∞ function so that $\Delta h = -\mu h$, $\mu > 0$. If $\lambda = (1 - n + [(n-1)^2 + 4\mu]^{1/2})/2$, then the function $\tilde{h} : CU \to R$ defined by $\tilde{h}(tp) = t^\lambda h(p)$, $t \in R$, $t > 0$, $p \in U$ is a fractional spherical harmonic of degree λ.

Let $f_1, \ldots, f_{p+1}$ be as above. Then we extend $f_1, \ldots, f_{p+1}$ to CU as in Lemma 6.1 as fractional spherical harmonics of degree determined by the appropriate $\lambda > 0$. Thus f induces a mapping $f : CU \to R^{p+1}$.

Let g be the constant curvature 1 metric on S^n, and let cg be the constant curvature k metric. ($c = 1/k$.) The isometry condition on f says that if $i : U \to CU$ is the canonical injection, then

$$i^*\sum df_i \otimes df_i = cg.$$

Let $x_1, \ldots, x_{n+1}$ be the standard coordinates on R^{n+1}, then

$$g = i^*\sum dx_i \otimes dx_i.$$

We use these formulas to derive a formula for $\sum df_i \otimes df_i$.

Let $\Phi : R^+ \times U \to CU$ be defined by $\Phi(t, x) = tx$. Explicit computations using the differential Φ_* of Φ yield

$$\Phi^*(\sum dx_i \otimes dx_i)_{t,x} = t^2 g_x + (dt \otimes dt)_t. \qquad (6.1)$$

$$\Phi^*(\sum df_i \otimes df_i)_{t,x} = \lambda^2 t^{2\lambda-2}(dt \otimes dt) + (t^{2\lambda}) i^*(\sum df_i \otimes df_i)_x. \qquad (6.2)$$

Thus we have

$$\begin{aligned} \Phi^*(\sum_{i=1}^{p+1} df_i \otimes df_i)_{(t,x)} &= \lambda^2 t^{2\lambda-2}(dt \otimes dt)_t + ct^{2\lambda} g_x \\ &= t^{2\lambda-2}(\lambda^2 - c)(dt \otimes dt)_t + ct^{2\lambda-2}(dt \otimes dt_t + t^2 g_x). \qquad (6.3) \end{aligned}$$

Setting $r(x) = \sum x_i^2$, we have $\Phi^* r^{1/2} = t$. Thus $(\Phi^{-1})^*(dt)_x = r^{-1/2}\sum x_i dx_i$.

This combined with (6.3) yields

$$\sum df_i \otimes df_i = cr^{\lambda-1}\sum dx_i \otimes dx_i + r^{\lambda-2}(\lambda^2 - c)\sum x_i x_j \, dx_i \otimes dx_j. \qquad (6.4)$$

Equation (6.4) is equivalent to the system of partial differential equations

$$\sum_{k=1}^{p+1} \frac{\partial f_k}{\partial x_i} \frac{\partial f_k}{\partial x_j} = c\delta_{ij} r^{\lambda-1} + (\lambda^2 - c) x_i x_j r^{\lambda-2}. \qquad (6.5)$$

This motivates

Lemma 6.2. Let $f_1, \ldots, f_{p+1}$ be nonzero fractional spherical harmonics of degree $\lambda \geq 0$ on CU satisfying the system of differential equations

$$\sum_k \frac{\partial f_k}{\partial x_i} \frac{\partial f_k}{\partial x_j} = c_1 \delta_{ij} r^{\lambda-1} + c_2 x_i x_j r^{\lambda-2} \tag{6.6}$$

for $c_1, c_2 \in R$, $1 \leq i, j \leq n+1$. Then $c_1 \geq 0$. If $c_1 = 0$, then $\lambda = 0$ and $f_1, \ldots, f_{n+1}$ are constant on CU. If $c_1 > 0$, then $\lambda \geq 1$. If $\lambda = 0$, then $c_1 = 0$.

Proof. Equation (6.6) implies that $i^* \sum df_k \otimes df_k = c_1 g$; thus $c_1 \geq 0$. If $c_1 = 0$, then $i^* df_k = 0$, $k = 1, \ldots, p+1$. Thus, since U is connected, $f_i|_U$ is constant for $i = 1, \ldots, p+1$. But then $f_i = b_i r^{\lambda/2}$. But $Df_i = 0$ implies $\lambda = 0$, and $f_i = b_i$. If $c_1 > 0$ and $\lambda = 0$, then $i^* \sum df_k \otimes df_k = c_1 g$, and $\Delta f_i = 0$, $i = 1, \ldots, p+1$. Thus $(f_1, \ldots, f_{p+1}) = f|_U : U \to R^{p+1}$ is a minimal isometric immersion. But $\tau_U = n(n-1)/c_1 > 0$, contradicting Corollary 5.3. Thus $\lambda > 0$. Now, by Corollary 5.2, $f(U) \subset r_0 S_1^p$. Thus setting $y = (1/r_0)f$, we have $y : U \to S_1^p$ and $\Delta y|_U = -\lambda(\lambda + n - 1)(y|_U)$. Thus the induced curvature on U is $[n/\lambda(\lambda + n - 1)] \leq 1$ by Corollary 5.2; thus $\lambda \geq 1$. Finally, suppose $\lambda = 0$. If $c_1 > 0$, then $\lambda \geq 1$; hence $c_1 = 0$. Q.E.D.

Lemma 6.3. Let $f_1, \ldots, f_{p+1}$ be fractional spherical harmonics of degree λ, $\lambda \geq 1$, satisfying system (6.6). Let $h_1, \ldots, h_{(n+1)(p+1)}$ be the $\partial f_i/\partial x_j$ arranged in some order. Then $h_1, \ldots, h_{(n+1)(p+1)}$ are fractional spherical harmonics of degree $\lambda - 1$ satisfying the system (6.6) for appropriate c_1', c_2'.

Proof.

$$D \sum \frac{\partial f_k}{\partial x_i} \frac{\partial f_k}{\partial x_j} = 2 \sum_{m=1}^{n+1} \sum_{k=1}^{p+1} \frac{\partial^2 f_k}{\partial x_i \partial x_m} \frac{\partial^2 f_k}{\partial x_j \partial x_m} = \sum_{k=1}^{(n+1)(m+1)} \frac{\partial h_k}{\partial x_i} \frac{\partial h_k}{\partial x_j}$$

Now the result follows from $Dr^\gamma = 2\gamma(n + 2\gamma - 1) r^{\gamma-1}$, $Dx_i x_j = 2\delta_{ij}$, and $\sum_k (\partial/\partial x_k)(x_i x_j)(\partial/\partial x_k) r^\gamma = 4\gamma\, x_i x_j r^{\gamma-1}$.

We can now prove

Proposition 6.1. Let $f_1, \ldots, f_{p+1}$ be nonzero fractional spherical harmonics of degree $\lambda \geq 0$ satisfying on CU the system (6.6). Then λ is an integer and f_i extends to a harmonic homogeneous polynomial of degree λ in R^{n+1}. Furthermore, $f_1, \ldots, f_{p+1}$ satisfy (6.6) on R^{n+1}.

Proof. We show that λ must be an integer. If $\lambda = 0$, there is nothing to show. If $\lambda > 0$, then by Lemma 6.2, $\lambda \geq 1$. Let $[\lambda]$ be the greatest integer less than or equal to λ. Apply Lemma 6.3, $[\lambda]$ times and we have a solution to Eq. (6.6) by spherical harmonics of degree $0 \leq \lambda - [\lambda] < 1$. Hence $\lambda = [\lambda]$. Now Lemma 6.3 implies that the $\partial^\lambda f/\partial x_{i_1} \cdots \partial x_{i_\lambda}$ are constants. Thus, $f_1, \ldots, f_{p+1}$ extend to harmonic homogeneous polynomials on R^{n+1} of degree λ. Since (6.6) is a polynomial system for $\lambda \geq 1$, the extension of $f_1, \ldots, f_{p+1}$ satisfies (6.6) on R^{n+1}.

Corollary 6.1. Let U be an open connected Riemannian submanifold of S_k^n. Let $f : U \to S_1^p \subset E^{p+1}$ be a minimal isometric immersion, $f = (f_1, \ldots, f_{p+1})$. Then

(1) There is a positive integer m so that $k = [n/m(m + n - 1)]$.
(2) f extends to a minimal isometric immersion of S_k^n into S_1^p.
(3) f_i is a spherical harmonic of degree m for $i = 1, \ldots, p + 1$.

In Section 9 we prove an abstract (global) version of this result for arbitrary symmetric spaces.

7. The Laplace–Beltrami Operator of a Homogeneous Space

Let M, $\overline{M}$ be Riemannian manifolds. Let $f : M \to \overline{M}$ be a C^∞ mapping which is onto. Then f is said to be a *Riemannian submersion* if

(1) f is a submersion. That is, for each $p \in M$, $f_{*p} : T_p(M) \to T_{f(p)}(\overline{M})$ is onto.

(2) By (1), $f^{-1}(p)$ is a submanifold for each $p \in \overline{M}$. Let $p \in \overline{M}$, $p_1 \in f^{-1}(p)$. Then $f_{*p_1} : T_{p_1}(f^{-1}(p))^\perp \to T_p(\overline{M})$ is an isometry (where $T_{p_1}(f^{-1}(p))^\perp$ is the orthogonal complement of $T_{p_1}(f^{-1}(p))$ in $T_{p_1}(M)$).

Let $f : M \to \overline{M}$ be a Riemannian submersion. We derive the fundamental equation of f using moving frames. The notion of Riemannian submersion is due to O'Neil [*13*].

Let $p \in \overline{M}$, $p_1 \in f^{-1}(p)$. Let $E_1, \ldots, E_n$ be a moving frame in a neighborhood U of p_1 in M so that $f_*E_{m+j} = 0$, $j = 1, \ldots, n - m$, and $f_*E_i = \bar{E}_i$, $i = 1, \ldots, m$ defines a moving frame in $\overline{M}$ in a neighborhood of p. We let our indices run as follows:

$$1 \leq i, j, k \cdots \leq m,$$
$$m + 1 \leq \alpha, \beta, \gamma \leq n,$$
$$1 \leq A, B, C \leq n.$$

Let $\theta_1, \ldots, \theta_n$ be the coframe for $E_1, \ldots, E_n$ and let $\omega_1, \ldots, \omega_m$ be the coframe for $\bar{E}_1, \ldots, \bar{E}_m$. Then $f^*\omega_i = \theta_i$, $i = 1, \ldots, m$.

The connection forms for M relative to $E_1, \ldots, E_n$ are uniquely defined by

$$d\theta_A = \sum_{B=1}^{n} \theta_{AB} \wedge \theta_B, \qquad \theta_{AB} + \theta_{BA} = 0.$$

The connection forms for $\bar{M}$ relative to $\bar{E}_1, \ldots, \bar{E}_m$ are uniquely defined by

$$d\omega_i = \sum_{j=1}^{m} \omega_{ij} \wedge \omega_j, \qquad \omega_{ij} + \omega_{ji} = 0.$$

Now $f^*d\omega_i = \sum_{j=1}^m f^*\omega_{ij} \wedge \theta_j$, $i = 1, \ldots, m$. Thus $d\theta_i = \sum_{j=1}^m f^*\omega_{ij} \wedge \theta_j = \sum_{B=1}^n \theta_{iB} \wedge \theta_B$.

From this we find the fundamental equation of a submersion

$$\sum_{j=1}^{m} (\theta_{ij} - f^*\omega_{ij}) \wedge \theta_j = -\sum_{\alpha=m+1}^{n} \theta_{i\alpha} \wedge \theta_\alpha \qquad \text{for} \quad i = 1, \ldots, m. \tag{7.1}$$

We now derive some consequences of Eq. (7.1). Let $1 \leq r, s \leq m$; then evaluating both sides of (7.1) on (E_r, E_s), we have

$$\sum_{j=1}^{m} (\theta_{ij} - f^*\omega_{ij}) \wedge \theta_j(E_r, E_s) = 0.$$

Let for each $q \in U$, $\hat{T}_q$ be the linear span of $E_{1_q}, \ldots, E_{m_q}$. Then we have $\sum_{j=1}^m (\theta_{ij} - f^*\omega_{ij})|_{\hat{T}_q} \wedge \theta_j|_{\hat{T}_q} = 0$, $i = 1, \ldots, m$. Hence

$$(\theta_{ij} - f^*\omega_{ij})|_{\hat{T}_q} = \sum_{k=1}^{m} h_{ijk}\theta_k|_{\hat{T}_q}$$

and $h_{ijk} = h_{ikj}$, $h_{ijk} = -h_{jik}$. But this implies that $h_{ijk} = 0$. (In fact, $h_{ijk} = -h_{jik} = -h_{jki} = h_{kji} = h_{kij} = -h_{ikj} = -h_{ijk}$.) Thus, (7.1) implies

$$\theta_{ij}|_{\hat{T}_q} = f^*\omega_{ij}|_{\hat{T}_q}, \qquad q \in U. \tag{7.2}$$

We now evaluate both sides of (7.1) on (E_α, E_β), $m + 1 \leq \alpha, \beta \leq n$. Then

$$\sum_{\gamma=m+1}^{n} \theta_{i\gamma} \wedge \theta_\gamma(E_\alpha, E_\beta) = 0, \qquad i = 1, \ldots, m.$$

Let $\tilde{T}_q$ for $q \in U$ be the linear span of $E_{m+1_q}, \ldots, E_{n_q}$. Then as above we find

$$\theta_{i\gamma}\big|_{\tilde{T}_q} = \sum a_{i\gamma\delta}\theta_\delta \qquad \text{with} \quad a_{i\gamma\delta} = a_{i\delta\gamma} \tag{7.3}$$

Clearly the $a_{i\gamma\delta}$ are the entries of the second fundamental form of $f^{-1}(f(q))$ at q in the ith direction. Hence if $f^{-1}(p)$ is totally geodesic for each $p \in \overline{M}$ then we have

$$\theta_{i\gamma}\big|_{\tilde{T}_q} = 0 \tag{7.3'}$$

Putting Eqs. (7.1), (7.2), and (7.3′) together, we have

Proposition 7.1. Let $f : M \to \overline{M}$ be a Riemannian submersion. Suppose $f^{-1}(p)$ is totally geodesic in M for each p in $\overline{M}$. Let Δ, $\underline{\Delta}$ be respectively the Laplace–Beltrami operators of M, $\overline{M}$. Let $h \in C^\infty(\overline{M})$. Then

$$(\underline{\Delta} h) \circ f = \Delta(h \circ f).$$

Proof. Let $p \in M$, $p_1 \in f^{-1}(p)$, and let E_A, $\overline{E}_i$, θ_A, ω_i, θ_{AB}, ω_{ij} be as above. If $\phi \in C^\infty(M)$ (resp. $h \in C^\infty(\overline{M})$), then

$$\Delta\phi(p_1) = \sum_{A=1}^{n} (\nabla_{E_{A_{p_1}}} d\phi)(E_{A_{p_1}}) \ (\text{resp. } \underline{\Delta} h(p) = \sum(\overline{\nabla}_{\overline{E}_{i_p}} dh)\,(\overline{E}_{i_p})\,).$$

Now

$$d\phi = \sum(E_A\phi)\theta_A \quad (\text{resp. } dh = \sum(\overline{E}_i h)w_i).$$

Thus

$$\begin{aligned}
\Delta\phi(p_1) &= \textstyle\sum_{AB}\{E_{A_{p_1}}(E_B\phi)\}\theta_B(E_{A_{p_1}}) + \sum_{AB}(E_B\phi)\,(p_1)(\nabla_{E_{A_{p_1}}}\theta_B)(E_{A_{p_1}}) \\
&= \textstyle\sum_A E_{A_{p_1}} E_A\phi - \sum_{AB}(E_{B_{p_1}}\phi)\,\theta_B(\nabla_{E_{A_{p_1}}} E_A) \\
&= \textstyle\sum_A E_{A_{p_1}} E_A\phi - \sum_{AB}(E_B\phi)\,(p_1)\,\theta_{AB}(E_{A_{p_1}}) \\
&\quad (\text{resp. } (\underline{\Delta} h)(p) = \textstyle\sum_{i=1}^m \overline{E}_{i_p}\overline{E}_i h - \sum_{ij}(E_{j_p}h)\omega_{ij}(E_i)\,).
\end{aligned}$$

Now suppose that $\phi = h \circ f$. Then $E_\alpha\phi = 0$, $\alpha = m+1, \ldots, n$. Thus

$$\begin{aligned}
\Delta\phi(p_1) &= \sum_{i=1}^{m} E_{i_{p_1}} E_i\phi - \sum_A\sum_j (E_{j_{p_1}}\phi)\theta_{A_j}(E_A) \\
&= \sum_{i=1}^{m} E_{i_{p_1}} E_i\phi + \sum_A\sum_j (E_{j_{p_1}}\phi)\,(\theta_{jA})\,(E_{A_{p_1}}).
\end{aligned}$$

Now Eq. (7.3′) says that $\theta_{j\alpha}(E_\alpha) = 0$ for $m + 1 \leqq \alpha \leqq n$. Thus

$$(\Delta\phi)(p_1) = \sum E_{j_{p_1}} E_j\phi + \sum_{i,j} (E_{j_{p_1}}\phi)\theta_{ji}(E_{i_p})$$

$$= \sum_{i=1}^{m} \bar{E}_{i_p}\bar{E}_i h + \sum_{j,i} (E_{j_p}h)(f^*\omega_{ji})(E_{i_{p_1}}) \qquad \text{(by (7.2))}$$

$$= \sum_{k=1}^{m} \bar{E}_{i_q}\bar{E}_i h + \sum_{i,j} (E_{j_p}h)\,\omega_{ji}(\bar{E}_{i_p}) = (\Delta h)(f(p_1)). \qquad \text{Q.E.D.}$$

Let G be a compact Lie group and let K be a closed subgroup. Let $\langle,\rangle$ be a Riemannian structure on G that is invariant under the action of $G \times G$ on G given by $(g_1, g_2)\cdot g = g_1 g g_2^{-1}$. Let $M = G/K$ and let $\pi: G \to G/K$ be the canonical map. Let e be the identity of G. Identify $T_e(G)$ with $\mathfrak{g}$, the Lie algebra of G. Let $\mathfrak{k}$ be the Lie algebra of K and let $\mathfrak{p} = \mathfrak{k}^\perp$ relative to $\langle,\rangle_e$. Let $\langle,\rangle_{eK}$ be the inner product on $T_{eK}(M)$ which makes $\pi_{*e}|_\mathfrak{p}$ an isometry. Then the corresponding Riemannian structure on M is such that $\pi : G \to M$ is a Riemannian submersion so that $\pi^{-1}(p)$ is totally geodesic for each p in M. A Riemannian structure on M defined as above is called *naturally reductive*. By Proposition 7.1, in order to find the Laplace–Beltrami operator of $(M, \langle,\rangle)$ we need only find the Laplace–Beltrami operator of G, relative to $\langle,\rangle$.

Let $\mathfrak{g}$ be the Lie algebra of G looked upon as the space of right invariant vector fields. Let $E_1, \ldots, E_n$ be an orthonormal basis of $\mathfrak{g}$ relative to $\langle,\rangle$.

Lemma 7.1. $\Delta = \sum E_i^2$.

Proof. If $X, Y \in \mathfrak{g}$, we define $\nabla_X Y = \frac{1}{2}[X, Y]$. Then $\nabla_X Y - \nabla_Y X - [X, Y] = 0$. Furthermore, if $X, Y, Z \in \mathfrak{g}$, then $X\cdot\langle Y, Z\rangle = 0$. Thus ∇ is the Riemannian connection of $\langle,\rangle$. In particular, if $g \in G$, then $(\nabla_{E_i}E_i)_g = 0$. Now if $f \in C^\infty(G)$, $(\Delta f)(g) = \sum_{i=1}^n (\nabla_{E_{i_g}}(df))(E_{i_g}) = \sum_{i=1} E_{i_g}E_i f - \sum df(\nabla_{E_{i_g}}E_i) = \sum E_{i_g}E_i f$. Q.E.D.

Corollary 7.1. Let $G/K = M$ as above and let $\langle,\rangle$ be the Riemannian structure on M defined as above. If $X \in \mathfrak{g}$ and $f \in C^\infty(M)$, let $X_p^* f = d/dt_{t=0} f(\exp tXp)$. Then the Laplace–Beltrami operator on M, Δ_M, is given by $\sum_{i=1}^n (E_i^*)^2$.

Proof. Let $f \in C^\infty(M)$. Then $(\Delta_M f)\circ\pi = \Delta(f\circ\pi)(g) = \sum_{i=1}^n E_i^2(f\circ\pi) = \sum d/dt^2 f(\exp tE_i\pi(g)) = \sum (E_i^*)^2_{\pi(g)}(f)$. Q.E.D.

Corollary 7.2. Let M, Δ_M be as above. Let G act on $C^\infty(M)$ via $(g_0 f)(p) = f(g_0^{-1}p)$.

(a) If V is a G-invariant finite dimensional subspace of $C^\infty(M)$, then $\Delta_M(V) \subset V$.

(b) If V is irreducible and G-invariant, then $\Delta_M|_V = \lambda I$.

Proof. (a) Since V is G-invariant, V is $\mathfrak{g}$-invariant under the action $Xf(p) = d/dt_{t=0} f(\exp(-tX)p)$. Thus by Corollary 7.1, $\Delta_M(V) \subset V$.

(b) Let (,) be any G-invariant inner product on V. Then relative to (,), X^* is skew-symmetric for each $X \in \mathfrak{g}$. Hence $(X^*)^2$ is symmetric. Thus Δ_M is symmetric on V.

8. Class 1 Representations

Let G be a compact Lie group. Then an orthogonal G-module, V, is a real inner product space on which G acts linearly via orthogonal transformations. Let K be a closed subgroup of G. A class 1 representation of (G, K) is an irreducible orthogonal G-module V so that there is a unit vector $v \in V$ such that $k \cdot v = v$ for all $k \in K$.

Proposition 8.1. Let G be a compact Lie group and let K be a closed subgroup of G. Let $M = G/K$ and suppose that the isotropy action of K is irreducible. Let V be a nontrivial Class 1 representation of (G, K) with K-fixed unit vector v. Then the map $x : M \to S$ (the unit sphere of V) defined by $x(gK) = g \cdot v$ is a minimal isometric immersion of a multiple of the G-invariant metric of M.

Proof. Let $v = v_1, v_2, \ldots, v_{p+1}$ be an orthonormal basis of V. Define $f_i(gK) = \langle g \cdot v, v_i \rangle$. Then $f_i \in C^\infty(M)$, $i = 1, \ldots, p+1$. Let $\phi : V \to C^\infty(M)$ be defined by $\phi(w)(gK) = \langle gv, w \rangle$. Let G act on $C^\infty(M)$ via $(g_0 f)(g) = f(g_0^{-1}g)$. Then ϕ is a G-module isomorphism. In fact, $\phi(g_0 w)(gK) = \langle gv, g_0 w \rangle = \langle g_0^{-1}gv, w \rangle = \phi(w)(g_0^{-1}gK)$. Thus ϕ is a G-module homomorphism. But V is irreducible; thus ϕ is an isomorphism. Clearly $f_i \in \phi(V)$, $i = 1, \ldots, p+1$.

Let $\langle , \rangle$ be a G-invariant Riemannian structure on M. Let Δ be the Laplace–Beltrami operator of M. Since M is isotropy irreducible, M is naturally reductive as in Section 7. Thus, by Corollary 7.2, $\Delta|_{\phi(V)} = \lambda I$. Thus $\Delta f_i = \lambda f_i$. Since V is nontrivial, $\lambda \neq 0$. Now $\sum_{i=1}^{p+1} df_i \otimes df_i$ is the metric on M induced by x. $\sum_{i=1}^{p+1} df_i \otimes df_i$ is G-invariant, thus $\sum_i df_i \otimes df_i = c\langle,\rangle$, $c > 0$. Thus x is an immersion. Corollary 5.2 now implies that x is minimal. The proposition is now proved.

We now confine our attention to the case (G, K) is a symmetric pair of compact type. That is, G is a compact, connected Lie group with discrete center and there is an involutive automorphism $\sigma : G \to G$ so that if $k \in K$, $\sigma(k) = k$ and K is open in the fixed point set of σ. The classification of symmetric spaces says that G/K is the most general symmetric space having nonnegative curvature up to locally Euclidean factors. Also, every G-invariant Riemannian structure on G/K is naturally reductive as in Section 7.

We give a complete description of $C^\infty(G/K, C)$ through complex, class 1 representations of (G, K). A complex class 1 representation of (G, K) is an irreducible unitary representation of G with a K-fixed unit vector. Let $\{V_\alpha\}_{\alpha \in I}$ be a collection of complex class 1 representations of (G, K) so that if $\alpha \neq \beta$, V_α is not equivalent to V_β, and if V is a complex class 1 representation of (G, K), that V is equivalent to V_α for some $\alpha \in I$.

Let $\phi_\alpha : V_\alpha \to C^\infty(G/K, C)$ be defined by $\phi_\alpha(\omega)(gK) = \langle gv_\alpha, \omega\rangle$, where $v_\alpha \in V_\alpha$ is a K-fixed unit vector. Then ϕ_α is a G-module isomorphism.

Theorem 8.1. (E. Cartan [*3*]). $\sum_{\alpha\in I} \phi_\alpha(V_\alpha)$ is a dense subspace of $C^\infty(G/K, C)$ with respect to the L^2 topology relative to any G-invariant volume element on G/K.

Note. This says that every irreducible G-subrepresentation of $C^\infty(G/K, C)$ is class 1 for (G, K) and that each such subrepresentation appears exactly once in $C^\infty(G/K, C)$.

Proof. We prove the statement of the note since the note combined with the Peter–Weyl theorem implies the theorem. Suppose that V, W are irreducible subrepresentations of $C^\infty(G/K, C)$ such that as G-modules, V is equivalent to W. Let $e_1, \ldots, e_m$ (resp. $f_1, \ldots, f_m$) be an orthonormal basis of V (resp. W), so that $g \cdot e_i = \sum g_{ji} e_j$ (resp. $g \cdot f_i = \sum g_{ji} f_j$) for $g \in G$. This can be done since V and W are equivalent. Let $s_0 : G/K \to G/K$ be defined by $s_0(gK) = \sigma(g)K$. Set $Sf = f \circ s_0$ for $f \in C^\infty(G/K, C)$. Let $\pi: G \to G/K$ be the canonical map, and abusing notation, let $e_i = e_i \circ \pi$, $f_i = f_i \circ \pi$. Let $\mathfrak{g}$ be the Lie algebra of G, $\mathfrak{k}$ the Lie algebra of K, and let $\mathfrak{p} = \{x \in \mathfrak{g} | \sigma_*(X) = -X\}$ (where σ_* is the differential of σ). Then

$\mathfrak{g} = \mathfrak{k} \oplus \mathfrak{p}$. Let $P = \exp \mathfrak{p}$. Set $p_0 = eK$. Then $G/K = Pp_0$. Define $F(g_1, g_2) = \sum_{i=1}^{n} e_i(g_1)\overline{f_i(g_2)}$. Then if $g \in G$, $F(gg_1, gg_2) = F(g_1, g_2)$. Hence if $p \in P$, we have $F(p, e) = F(e, p^{-1})$. Now $s_0(p) = p^{-1}$. $F(p, e) = \sum \overline{f_i(e)} e_i(p)$, $F(e, p^{-1}) = \sum e_i(e)\overline{f}_i(p^{-1}) = \sum e_i(e)\overline{Sf_i}(p)$. Thus $V \cap S\overline{W} \neq (0)$. (Now letting $W = V$, we also have $W \cap S\overline{W} \neq 0$.) But W and V are irreducible and $S\overline{W}$ is G-invariant. Thus $V \subset S\overline{W}$ (resp. $W \subset S\overline{W}$). Thus since $\dim V = \dim W = \dim S\overline{W}$, we see $V = S\overline{W} = W$.

Now let V be an irreducible subrepresentation of $C^\infty(G/K, C)$. Let $f_1, \ldots, f_m$ be an orthonormal basis of V. Let $f: G/K \to C^m$ be defined by $f(gK) = (f_1(gK), \ldots, f_m(gK))$. Let $A: V \to C^m$ be defined by $A(\sum a_i f_i) = (a_i, \ldots, a_m)$.

If $g \in G$, let $g \cdot f_i = \sum g_{ji} f_j$. Set $\rho(g) = (g_{ij})$. Then $A(g \cdot h) = \rho(g)Ah$ for $h \in V$. Thus $f(g \cdot p) = \overline{\rho(g)f(p)}$. Set $v_0 = \overline{f(eK)}$. Then $A^{-1}(v_0) \in V$ and is K-fixed. The theorem is now proved.

We now prove a formal result known as the Frobenius reciprocity theorem. We first need the notion of an induced representation. Let G be a compact group and let K be a closed subgroup. Let W be a unitary (resp. orthogonal) K-module. Let $\Gamma(W)$ be the vector space of all continuous functions $f: G \to W$ so that $f(kg) = k \cdot f(g)$. Let dg be Haar measure on G and define for $f_1, f_2 \in \Gamma(W)$, $(f_1, f_2) = \int_G \langle f_1(g), f_2(g) \rangle \, dg$. Define for $f \in \Gamma(W)$, $(g_0 f)(g) = f(gg_0)$. Then $(\Gamma(W), \langle , \rangle)$ is a unitary (resp. orthogonal) G-module called the G-module *induced* by W.

If W_1, W_2 are unitary (resp. orthogonal) G-modules, then by $\operatorname{Hom}_G(W_1, W_2)$ we mean the space of all continuous linear maps $A: W_1 \to W_2$ so that $Agw_1 = gAw_1$.

Theorem 8.2. (Frobenius Reciprocity). Let G be a compact group, K a closed subgroup of G. Let M be a finite dimensional unitary (resp. orthogonal) K-module, let $\Gamma(M)$ be the induced unitary (resp. orthogonal) G-module. Let U be a finite dimensional unitary (resp. orthogonal) G-module. Then $\operatorname{Hom}_G(U, \Gamma(M))$ is canonically isomorphic with $\operatorname{Hom}_K(U, M)$.

Proof. Let $A \in \operatorname{Hom}_G(U, \Gamma(M))$. Define $\tilde{A}: U \to M$ by $\tilde{A}(u) = A(u)(e)$ for $u \in U$. Then $\tilde{A}(ku) = k\tilde{A}(u)$. Hence $\tilde{A} \in \operatorname{Hom}_K(U, M)$.

Now let $A \in \operatorname{Hom}_K(U, M)$. If $u \in U$, set $\hat{A}(u)(g) = A(gu)$. Then $\hat{A}(u)(kg) = A(kgu) = kA(gu) = k(\hat{A}(u)(g))$. Thus $\hat{A}(u) \in \Gamma(M)$. $\hat{A}(g_0 u)(g) = A(gg_0 u) = \hat{A}(u)(gg_0) = (g_0 \hat{A}(u))(g)$. Thus $\hat{A} \in \operatorname{Hom}_G(U, \Gamma(M))$. Clearly $\hat{\tilde{A}} = A$, $\tilde{\hat{A}} = A$. Q.E.D.

Corollary 8.1. (E. Cartan [*3*]). Let V be a class 1 unitary module for the symmetric pair (G, K). Then $V^K = \{v \in V \mid kv = v, k \in K\}$ is one-dimensional

Proof. Let C be the trivial K-module. Then. if $f \in C^\infty(G/K, C), \hat{f} \in \Gamma(C)$, where $\hat{f}(g) = f(\pi(g^{-1}))$. The map $f \to f \circ \pi$ maps $C^\infty(G/K, C) \to \Gamma(C)$ is a unitary G-module isomorphism. It is standard that the image of this isomorphism is dense in $\Gamma(C)$. The result now follows by Frobenius reciprocity.

Let (G, K) be a symmetric pair as above. Let V be an irreducible G-submodule of $C^\infty(G/K, C)$. Then since $\dim V^K = 1$, there is $\phi \in V$ such that $\phi(kgK) = \phi(gK)$ for all $k \in K$, $g \in G$. Furthermore, by the proof of the existence of ϕ, $\phi(eK) \neq 0$. Let $\phi_V \in V^K$ be the element such that $\phi_V(eK) = 1$. ϕ_V is called the *zonal spherical function* of V.

Lemma 8.1. Let V be an irreducible submodule of $C^\infty(G/K, C)$, then $V = (V \cap C^\infty(M)) \otimes_R C$ if and only if $\phi_V(G/K) \subset R$.

Proof. If $\phi_V(G/K) \subset R$, then $\phi_V \in C^\infty(M)$. Thus $G \cdot \phi_V \subset V \cap C^\infty(M)$. Let V_0 be the R-linear span of $G \cdot \phi_V$; then V_0 is a G-submodule of $C^\infty(M)$. Clearly $V = V_0 \otimes_R C$.

Suppose that $V \cap C^\infty(M) \neq (0)$. Then by Frobenius reciprocity (applied to R-modules), $V \cap C^\infty(M)$ has a K-fixed vector. Corollary 8.1 implies $\phi_V \in V \cap C^\infty(M)$. Now suppose $\phi_V \notin V \cap C^\infty(M)$; then by the above $V \cap C^\infty(M) = 0$.

We prove a result that tells us for what pairs (G, K), ϕ_V is always real.

Theorem 8.3. Let $\mathfrak{g} = \mathfrak{k} \oplus \mathfrak{p}$ as in the proof of Theorem 8.1. Let $\mathfrak{a} \subset \mathfrak{p}$ be a maximal abelian subsystem. Let $M^* = \{m \in K \mid \mathrm{Ad}(m)\mathfrak{a} \subset \mathfrak{a}\}$. Then ϕ_V is real valued for any irreducible subrepresentation of $C^\infty(G/K, C)$ if there is an $m \in M^*$ so that $\mathrm{Ad}(m)|_{\mathfrak{a}} = -I$.

Note. This condition is actually necessary and sufficient.*

Proof. We use the facts $\mathrm{Ad}(K)\mathfrak{a} = \mathfrak{p}$ and $\exp(\mathfrak{p})p_0 = G/K$ where $p_0 = eK$. These facts imply that if $A = \exp \mathfrak{a}$, then ϕ_V is completely determined by its values on Ap_0.

*This is a result of Professor Joseph Wolf.

Now by the proof of dim $V^K \neq (0)$, there is $v_0 \in V^K$ so that $\phi_V(gK) = \langle g \cdot v_0, v_0 \rangle$. Thus $\overline{\phi_V(gK)} = \phi_V(g^{-1}K)$ (the inner product is Hermitian). Suppose that there is $m \in M^*$ so that $\mathrm{Ad}(m)|\mathfrak{a} = -I$. If $a \in A$, then $\phi_V(ap_0) = \phi_V(map_0) = \phi_V(mam^{-1}p_0) = \phi_V(a^{-1}p_0)$. Now if $x \in \mathfrak{p}$, $x = \mathrm{Ad}(k)y$, $y \in \mathfrak{a}$ and $\phi_V(\exp\, xp_0) = \phi_V(k(\exp\, y)h_k^{-1}p_0) = \phi_V(\exp\, yp_0) = \phi_V(\exp\,(y)^{-1}p_0) = \phi_V(k\,\exp(-y)k^{-1}p_0) = \phi_V(\exp(-(x))p_0)$. Now this implies by the above that $\phi_V = \bar{\phi}_V$.

Corollary 8.2. If (G, K) is a symmetric pair corresponding to a rank 1 symmetric space, then all the zonal spherical functions of $C^\infty(G/K, C)$ are real valued.

Proof. If (G, K) is rank 1, then K acts transitively on the unit sphere of $T_{p_0}(G/K)$. Thus if $\mathfrak{a}$ is as in the proof of Theorem 8.3, dim $\mathfrak{a} = 1$. If $x \in \mathfrak{a}$ is a unit vector, then the orbit $\mathrm{Ad}(K)x$ must contain $-x$. Thus there is $k \in K$ so that $\mathrm{Ad}(k)x = -x$. Q.E.D.

Combining Lemma 8.1 with Corollary 8.2 we have

Corollary 8.3. Let (G, K) correspond to a rank 1 symmetric space. Let $V_\alpha^0 = \phi_\alpha(V) \cap C^\infty(G/K)$ where ϕ_α, V_α are as in Theorem 8.1. Then $C^\infty(G/K) = \sum_{\alpha \in I} V_\alpha^0$ and V_α^0 is not equivalent to V_β^0 for $\alpha \neq \beta$.

We give some concrete examples of this situation.

EXAMPLE 1. Let $G = SO(n+1)$, $K = SO(n)$. Then (G, K) is rank 1. Let V_α^0 be as in Corollary 8.3. Proposition 8.1 says that if $f_1, \ldots, f_{p+1}$ is an orthonormal basis of V (volume $G/K = p + 1$) and if $x : (S^n = G/K) \to S_1^p$ is defined by $x(gK) = (f_1(gK), \ldots, f_{p+1}(gK))$, then x is minimal. Applying Corollary 6.1, we find that f_i is a spherical harmonic of degree k on S^n for $i = 1, \ldots, p + 1$. Let $H^{k,n}$ be the space of all spherical harmonics of degree k on S^n. Then Corollary 8.3 says that $\sum_{k=0}^\infty H^{k,n}$ is dense in $C^\infty(S^n)$, and every class 1 representation of (G, K) is of the form $H^{k,n}$ for some k.

EXAMPLE 2. Let $G = SU(n + 1)$, $K = U(n)$. $G/K = CP^n$, complex projective space. (G, K) is again rank 1. We wish to give an explicit description of the V_α^0.

Let $S^{2n+1} \to CP^n$ be the Hopf fibration defined as follows:

We look at S^{2n+1} as the unit sphere in C^{n+1}. To each $p \in S^{2n+1}$ we assign the complex line $[p]$ through p and 0. Thus $\pi(p) = [p]$ where $q \in [p]$ if and only if $q \in e^{i\theta}p$ for some $\theta \in R$.

CP^n is isotropy irreducible and we may define a Riemannian structure on CP^n such that π is a Riemannian submersion. Indeed, let T^1 act on S^{2n+1} by left multiplication. Then T^1 acts transitively on each $\pi^{-1}(p)$.

Now let $f \in C^\infty(CP^n)$ be such that $\Delta_{CP^n} f = \lambda f$. Then $\Delta_{S^n}(f \circ \pi) = (\Delta_{CP^n} f) \circ \pi$. Thus $\Delta_{S^n}(f \circ \pi) = \lambda f \circ \pi$. This implies that $\lambda = -m(m + 2n)$ and that $f \circ \pi$ is a spherical harmonic of degree m by Example 1. Now $f \circ \pi = \sum a_{i_1 \ldots i_{n+1}, j_1, \ldots, j_{n+1}} z_1^{i_1} \cdots z_{n+1}^{i_{n+1}} \bar{z}_1^{j_1} \cdots \bar{z}_{n+1}^{j_{n+1}}$,

$$\sum i_k + \sum j_e = m,$$

and $Df \circ \pi = 0$. Now $(f \circ \pi)(e^{i\theta} p) = f \circ \pi(p)$. Thus $\sum i_k = \sum j_e$.

Let $\mathscr{H}^{m,n}$ be the collection of all harmonic polynomials on C^{n+1} of degree m in $z_1, \ldots, z_{n+1}$ and degree m in $\bar{z}_1, \ldots, \bar{z}_{n+1}$. Then $f \in \mathscr{H}^{m,n}$ is completely determined by its (well defined) restriction to CP^n and $C^\infty(CP^n) = \sum_{m=0}^{\infty} \mathscr{H}^{m,n}$. $\mathscr{H}^{m,n}$ is called the space of Hermitian harmonics in $n + 1$ variables. $\Delta_{CP^n}|_{\mathscr{H}^{m,n}} = 4m(m + n)I$.

We note that the Riemannian structure defined above on CP^n is that structure with sectional curvatures varying from 1 to 4. Thus if we are interested in the Riemannian structure with sectional curvatures varying from $\frac{1}{4}$ to 1 (i.e., holomorphic curvature 1) we must consider the operator $\frac{1}{4}\Delta_{CP^n}$ with eigenvalues $m(m + n)$.

We have proved

Proposition 8.2. Let $x : CP^n \to S^p \subset E^{p+1}$ be a minimal isometric immersion of a metric of constant holomorphic curvature K. Then

(a) there is a positive integer m so that $K = (2n)/(m(m + n))$;

(b) $x = (x_1, \ldots, x_{p+1})$ and x_i is a Hermitian harmonic of degree m.

9. The Classification Theorem

In this section, we give a description of minimal isometric immersions of compact symmetric spaces into spheres. We first consider $M = G/K$ where G is a compact connected subgroup, K is a closed subgroup of G, M is orientable, and the isotropy action of K is irreducible. Let $\langle , \rangle$ be some G-invariant Riemannian structure on M ($\langle , \rangle$ is unique up to scalar multiple and is thus naturally reductive). Let Δ be the Laplace–Beltrami operator of $(M, \langle , \rangle)$. Let, for each $\lambda \in R$, V_λ be the eigenspace with eigenvalue λ. Then $V_\lambda \neq 0$ only if $\lambda \leq 0$ (see Section 1). Also, $\dim V_\lambda < \infty$. Let $(f_1, f_2) = \int_M f_1 f_2 \omega$ where ω is the Riemannian volume element of $(M, \langle , \rangle)$ ($\omega = *1$). Let $p_\lambda + 1 = \dim V_\lambda$. For each λ such that $\lambda \neq 0$ and $V_\lambda \neq (0)$, let $\phi_1, \ldots, \phi_{p_\lambda + 1}$ be an orthonormal basis of V. Define

$$x_\lambda : M \to E^{p_\lambda + 1}$$

via

$$x_\lambda(q) = \frac{1}{(p_\lambda + 1)^{1/2}}(\phi_1(q), \ldots, \phi_{p_\lambda+1}(q)).$$

Then x_λ defines an isometric immersion of $(M, c\langle,\rangle)$ into $S_1^{p_\lambda}$, for some $c > 0$. (Indeed $\sum d\phi_i \otimes d\phi_i$ is an invariant nonzero bilinear form on $T(M)$; thus $\sum d\phi_i \otimes d\phi_i = c\langle,\rangle$, $c > 0$.) Now applying Takahashi's result, Corollary 5.2, we see that x_λ is a minimal immersion and $c = -\lambda/n$.

Theorem 9.1. Let $x : (M, c\langle,\rangle) \to S_1^q$ be a minimal isometric immersion so that $x(M)$ is not contained in a great sphere of S_1^q. Then there is an eigenvalue $\lambda < 0$ of the Laplace–Beltrami operator of $(M, \langle,\rangle)$ so that $c = -\lambda/n$ and a linear isometric injection $A : E^{q+1} \to E^{p_\lambda+1}$ and a linear map $B : E^{p_\lambda+1} \to E^{p_\lambda+1}$ so that $A \circ x = B \circ x_\lambda$.

Proof. The statement that $x(M)$ is not contained in a great sphere of S_1^q is just the statement that the height functions $\phi_1, \ldots, \phi_{q+1}$ of x $[x(p) = (\phi_1(p), \ldots, \phi_{q+1}(p))]$ are linearly independent. Now Corollary 5.2 implies that $(1/c)\Delta\phi_i = -n\phi_i$. Thus setting $\lambda = -cn$, we have the first assertion. Now since $\phi_1, \ldots, \phi_{q+1}$ are linearly independent, $q \le p_\lambda$. Let $\psi_1, \ldots, \psi_{q+1}$ be an orthonormalization of $\phi_1, \ldots, \phi_{q+1}$. Then $x = C\Phi$ where C is a nonsingular $(q+1) \times (q+1)$ matrix. Now extend $\psi_1, \ldots, \psi_{q+1}$ to an orthonormal basis, $\psi_1, \ldots, \psi_{p_\lambda+1}$ of V_λ. Then $(\psi_1, \ldots, \psi_{p_\lambda+1}) = (p_\lambda + 1)^{1/2} U \cdot x_\lambda$ where U is an orthogonal transformation of E. Let $A(y_1, \ldots, y_{p+1}) = (y_1, \ldots, y_{p+1}, 0, \ldots, 0) \in E^{p_\lambda+1}$ and let $\tilde{C}(z_1, \ldots, z_{p_\lambda+1}) = AC(z_1, \ldots, z_{q+1})$. Then we have $Ax = (p_\lambda + 1)^{1/2}\tilde{C}Ux$. Set $B = (p_\lambda + 1)^{1/2}\tilde{C}U$.

If (G, K) is a symmetric pair of compact type and if the isotropy representation splits into $k > 1$ irreducible parts, then there are many conformally inequivalent Riemannian structures that make G/K into a Riemannian symmetric space. Thus there are many eigenspace decompositions of $C^\infty(G/K, C)$ for the various Laplace–Beltrami operators of the many symmetric Riemannian structures. Thus the above simple classification for the isotropy irreducible case has no immediate generalization to the isotropy reducible case.

We give examples that give an idea of how we can use the classification theorem to control the isotropy reducible case. Let $M = M_1 \times M_2$, M_i an isotropy irreducible symmetric space of compact type or a 1-torus. Let $x_i : M_i \to S_1^{p_i} \subset E^{p_i+1}$ be a minimal isometric immersion of some multiple of the symmetric Riemannian structure of M_i.

EXAMPLE 1. Let $\theta \in R$ and set

$$x_\theta^+(p, q) = (\cos \theta x_1(p), \sin \theta x_2(q)),$$

then $x_\theta^+ : M_1 \times M_2 \to S_1^{p_1+p_2+1}$ is a minimal isometric immersion of a symmetric metric on $M_1 \times M_2$.

In particular, if $M_1 = S_1^{n_1}$ and $M_2 = S_1^{n_2}$ and if $x_i =$ identity on $S_1^{n_i}$, then x_θ^+ is a Clifford minimal immersion of $S^{n_1} \times S^{n_2}$ into $S_1^{n_1+n_2+1}$. Note that in this case it is an equivariant minimal hypersurface. In particular if $n_i = 1$, $i = 1, 2$, this gives the Clifford torus, minimally, equivariantly immersed as a hypersurface of S_1^3.

EXAMPLE 2. Let x_1, x_2 be as above and define

$$x^{\otimes}(p, q) = x_1(p) \otimes x_2(q) \in E^{p_1+1} \otimes E^{p_2+1} = E^{(p_1+1)(p_2+1)}.$$

Then $x^{\otimes} : M_1 \times M_2 \to S^{(p_1+1)(p_2+1)-1}$ is a minimal isometric immersion of a symmetric metric on $M_1 \times M_2$.

If again $M_i = S^1 \subset E^2$, $i = 1, 2$, and if x_i is the identity map then $x^{\otimes} : T^2 \to S_1^3 \subset E^4$ is a minimal equivariant immersion, giving the Clifford torus.

10. Rigidity Questions

Let M be a Riemannian manifold and let $x : M \to S_1^p \subset E^{p+1}$ be a minimal isometric immersion. Then x is said to be *rigid* if whenever $x_1 : M \to S_1^p$ is another minimal isometric immersion, there is an isometry U of S_1^p with S_1^p so that $U \circ x = x_1$. We note that in particular a rigid isometric immersion of a homogeneous space is equivariant.

It was expected that if the metric structure of M were severely limited, then any minimal isometric immersion of M into a sphere would be rigid. The constructions of Section 9 show that if M is a symmetric space but not isotropy irreducible, then nonrigidity for minimal isometric immersions of symmetric metrics on M is not hard to find.

If (G, K) is an isotropy irreducible symmetric pair of compact type but if each eigenspace for the Laplace–Beltrami operator of M is not irreducible as a representation of G, then using Corollary 5.2 it is again not hard to find examples of nonrigidity.

The condition that guarantees the irreducibility of each eigenspace for the Laplace–Beltrami operator is that (G, K) be rank 1. Indeed, we have seen this to be true if $G/K = S^n$ or CP^n. In [6], doCarmo and Wallach conjectured rigidity for rank 1 symmetric spaces. In [5], Chern et al.

indicated that it was highly likely for constant curvature spheres. For the remainder of this article, we will analyze these questions. We first introduce a weaker notion than rigidity which we will call linear rigidity.

Definition 10.1. Let M be a Riemannian manifold and let $x: M \to S_1^n \subset E^{n+1}$ be an isometric immersion of M into S_1^n. Then x is said to be *linearly rigid* if whenever $A : E^{n+1} \to E^{n+1}$ is a linear map so that

(1) $A(x(M)) \subset S_1^n$, and

(2) $A \circ x : M \to E^{n+1}$ is an isometric immersion,

then A is orthogonal.

Corollary 5.2 immediately implies that rigidity for minimal isometric immersions implies linear rigidity. We also note that for minimal isometric immersions, Corollary 5.2 says that (2) implies (1).

We indicate that in some cases linear rigidity is actually equivalent to rigidity for minimal isometric immersions.

Lemma 10.1. Let (G, K) be a pair of a compact Lie group G and a closed subgroup K of G so that $G/K = M$ is isotropy irreducible. Let Δ be the Laplace–Beltrami operator of some G-invariant Riemannian structure on M. Let λ be a nonzero eigenvalue for Δ and let $x_\lambda : M \to S_1^{p_\lambda} \subset E^{p_\lambda + 1}$ be the minimal isometric immersion defined as in Section 9. Then x_λ is rigid if and only if x is linearly rigid.

Proof. This lemma is just a restatement of Corollary 5.2.

Thus for isotropy irreducible symmetric spaces of compact type, rigidity and linear rigidity are essentially the same notions.

Let $M_n(R)$ be the space of all $n \times n$ matrices with real entries. Then $M_n(R)$ has a polar decomposition. That is, if $A \in M_n(R)$, A can be written uniquely in the form UP where U is orthogonal and P is symmetric and positive semidefinite, written $P \geq 0$. Using the polar decomposition of $M_n(R)$, we may reformulate the notion of linear rigidity.

Lemma 10.2. Let M be a Riemannian manifold. Let $x : M \to S_1^n \subset E^{n+1}$ be a minimal isometric immersion. Then x is linearly rigid if and only if whenever $A : E^{n+1} \to E^{n+1}$ is symmetric and positive semidefinite ($A \geq 0$) satisfying

(2′) $A \circ x : M \to E^{n+1}$

is an isometric immersion, then A is the identity.

Now we may identify the space of all symmetric mappings of E^{n+1} with $S^2(E^{n+1})$, the symmetric square of E^{n+1}, as follows: if $u, v \in E^{n+1}$, $uv \in S^2(E^{n+1})$ (the symmetric product of two vectors will be denoted uv), and if $t \in E^{n+1}$, we set $uv(t) = \frac{1}{2}\{\langle u, t\rangle v + \langle v, t\rangle u\}$ where $\langle , \rangle$ is the inner product on E^{n+1}. Under this identification, the inner product on $S^2(E^{n+1})$ is given by $(A, B) = \operatorname{tr} {}^tBA$ (where ${}^tB$ is the transpose of B). We note that if $A \in S^2(E^{n+1})$ and if $u, w \in E^{n+1}$, then $\langle Au, w\rangle = (A, uw)$.

Let $x : M \to S_1^n \subset E^{n+1}$ be an isometric immersion. We identify (as usual) $T_p(S_1^n)$ with $\{v \in E^{n+1} | \langle v, p\rangle = 0\}$. Let for each $p \in M$, $S^2(x_*(T_p(M)))$ be the symmetric square of $x_*(T_p(M))$ in $S^2(E^{n+1})$. Let $W_x = W$ be the subspace of $S^2(E^{n+1})$ spanned by $U_{p\in M} S^2(x_*(T_p(M)))$. Let W_1 be the orthogonal complement of W in $S^2(E^{n+1})$. Let $L = \{C \in W_1 | C + I \geq 0\}$.

Lemma 10.3. Let $x : M \to S_1^n$ be a minimal isometric immersion.

(a) If $A \in S^2(E^{n+1})$ and $A \geq 0$, then A satisfies (2′) of Lemma 10.2 if and only if $A^2 - I \in L$.

(b) Let W_0 be the space spanned in $S^2(E^{n+1})$ by the set $\{x(p)^2 \mid p \in M\}$. Then $W_0 \subset W$.

Proof. For each $p \in M$, $(A \circ x)_{*p} = A \circ x_{*p}$ by the above identifications. Let $X \in T_p(M)$. Then if A satisfies (2′), $\langle A \cdot x_{*p}(X), Ax_{*p}(X)\rangle = \langle x_{*p}(X), x_{*p}(X)\rangle$. Thus $(A^2, (x_{*p}(X))^2) = (I, (x_{*p}(X))^2)$. Thus $A^2 - I$ is orthogonal to $S^2(x_{*p}(T_p(M)))$. But p is arbitrary in M; thus $A^2 - I$ is orthogonal to W. Hence $A^2 - I \in L$. The sufficiency is proved by reversing the above steps.

The proof of (b) is an instructive exercise in the use of Corollary 5.2.

11. The Higher Fundamental Forms

Let $\overline{M}$ be a Riemannian manifold of constant curvature K. Let M be a Riemannian manifold. Let $x : M \to \overline{M}$ be an isometric immersion. We define the higher fundamental forms of x and the osculating spaces of x.

Let $p \in M$ and let $B_p : T_p(M) \times T_p(M) \to N_p(M)$ be the second fundamental form of x. We set $O_p^2(M)$ equal to the linear span of the image of B_p. We say that $p \in M$ is degree 2 regular if $O_p^2(M)$ is of maximal dimension. Let $R_2 \subset M$ be the space of all degree 2 regular elements of M. Then R_2

is open in M. Let $p \in R_2$. Let N_2 be normal projection in $N_p(M)$ relative to $N_p(M) = O_p^2 \oplus O_p^{2\perp}$ (we write $v \to v^{N_2} \in O_p^{\perp}$). Set $B_{2p} = B_p$ and define $B_{3_p}(X_1, X_2, X_3) = (\overline{\nabla}_{X_{1_p}} B_2(X_2, X_3))^{N_2}$ for $X_1, X_2, X_3 \in T_p(M)$ arbitrarily extended to vector fields on M. B_{3p} is well defined and defines a symmetric tensor field on R_2. Let O_p^3 be the linear span of the image of B_{3_p}. We call an element $p \in R_2$ degree 3 regular if dim O_p^3 is maximal. We define B_{i_p}, O_p^i for $i = 2, 3, \ldots$ by recursion as above on the space R_{i-1} of all degree $i-1$ regular elements of M. Clearly the above process must eventually stop since $\dim(T_p(M) + O_p^2 + O_p^3 + \cdots + O_p^m) \leq \dim T_p(\overline{M})$. Let m be the first integer ≥ 2 such that $B_m \not\equiv 0$ but $B_{m+1} \equiv 0$. Then we call m the degree of x and the set of all m-regular elements will be called the set of all completely regular points of M, denoted $M' = R_m$. We note that M' is open in M, $M' \neq \varnothing$. Let for each nonnegative integer k, $S^k(T_p(M))$ be the k-fold symmetric power of $T_p(M)$. The universal property of $S^k(T_p(M))$ says that for $p \in R_{k-1}$, B_k induces a linear map of $S^k(T_p(M)) \to O_k^p$. Let $S(T_p(M)) = \sum_{k=0}^{\infty} S^k(T_p(M))$; then $S(T_p(M))$ is an algebra under symmetric product. Let $S^+(T_p(M)) = \sum_{k=1}^{\infty} S^k(T_p(M))$. Then $S^+(T_p(M))$ is a subalgebra of $S(T_p(M))$. Let $Q_p : S(T_p(M)) \to T_p(M)^{\perp}$ for $p \in M'$ be defined by $Q = x_{*p} + B_{2_p} + \cdots + B_{m_p}$.

Let $r_p \in S^2(T_p(M))$ be defined as follows: let $e_1, \ldots, e_n$ be an orthonormal basis of $T_p(M)$, set $r_p = \sum e_i^2$. Then $p \to r_p$ defines a tensor field on M.

Lemma 11.1. If $x : M \to \overline{M}$ is minimal, then for each $p \in M'$, Kernel $Q_p \supset S(T_p(M))r_p$.

Proof. We need only prove that if $E_1, \ldots, E_n$ is an orthonormal moving frame in a neighborhood of $p \in M'$, then $B_{k+2}(E_{i_1} \cdots E_{i_k} r) = 0$ in a small neighborhood of p. This statement for $k = 0$ is just the statement of minimality. If it is true for $k-1$, then $B_{k+2}(E_{i_1} \cdots E_{i_k} r) = (\overline{\nabla}_{E_{i_1}}(B_{k+1}(E_{i_1} \cdots E_{i_k} r))^{N_{k+1}} = 0$. Thus the result is true by induction.

We give some examples of the notion of degree defined above.

EXAMPLE 1. Let

$$x_k : S^n_{\frac{n}{k(n+k-1)}} \to S_1^{p_k}$$

where $p_k + 1 = \dim H^{k,n}$ (see Section 8, Example 1 for notation) be the minimal isometric equivariant immersion. Then the degree of x_k is k.

EXAMPLE 2. Let

$$x_1 : CP^n_{\frac{2n}{n+1}} \to S^{n(n+2)-1}$$

be the equivariant minimal isometric immersion corresponding to the Hermitian harmonics of degree 1 (see Section 8, Example 2). Then x_1 has degree 2.

Proposition 11.1. Let M be a connected analytic Riemannian manifold and let $x : M \to S_1^n \subset E^{n+1}$ be an analytic isometric immersion so that $x(M)$ is not contained in a great sphere of S_1^n. If x is of degree ≤ 3, then x is linearly rigid.

Proof. Since x is analytic and not contained in a great sphere of S_1^n, we note that if $p \in M'$, then $Rp + Q(S(T_p(M))) = E^{n+1}$. This says that $Rp + x_{*p}(T_p(M)) + B_{2p}(S^2(T_p(M))) + B_{3p}(S^3(T_p(M))) = E^{n+1}$ for each $p \in M'$. In the notation of Section 10, we show that if $X \in T_p(M)$ and if $\sigma : (-\varepsilon, \varepsilon) \to M$ is the geodesic through p in M with tangent vector X, then p^2, $p\dot{\sigma}(0)$, $p\ddot{\sigma}(0)$, $p\dddot{\sigma}(0) \in W + W_0 = V$. Indeed, W_0 (see Lemma 10.3(b) for the definition) is the linear span in $S^2(E^{n+1})$ of all the q^2, $q \in x(M)$. Thus $\sigma(t)^2 \in V$ for all $t \in (-\varepsilon, \varepsilon)$. Hence $d/dt\, \sigma(t)^2 \in V$. Thus $2\dot{\sigma}(t)\sigma(t) \in V$, for $t \in (-\varepsilon, \varepsilon)$. This implies that $\dot{\sigma}(0)\cdot\sigma(0) \in V$. Now $d/dt\, \dot{\sigma}(t)\sigma(t) = \ddot{\sigma}(t)\sigma(t) + \dot{\sigma}(t)\dot{\sigma}(t)$. Now the definition of W implies $\dot{\sigma}(t)^2 \in V$. Thus $\ddot{\sigma}(t)\sigma(t) \in V$ for all $t \in (-\varepsilon, \varepsilon)$. In particular, $\ddot{\sigma}(0)p \in V$. Again, $d/dt\, \ddot{\sigma}(t)\sigma(t) = \dddot{\sigma}(t)\sigma(t) + \ddot{\sigma}(t)\dot{\sigma}(t)$. Also, $d/dt\dot{\sigma}(t)^2 = 2\ddot{\sigma}(t)\dot{\sigma}(t)$. Thus $\dddot{\sigma}(0)\sigma(0) = \dddot{\sigma}(0)p \in V$, as we asserted. Suppose $C \in V^\perp$ in $S^2(E^{n+1})$. Then $(C, p^2) = (C, p\dot{\sigma}(0)) = (C, p\ddot{\sigma}(0)) = (C, p\dddot{\sigma}(0)) = 0$. Thus $\langle Cp, p\rangle = \langle Cp, \dot{\sigma}(0)\rangle = \langle Cp, \ddot{\sigma}(0)\rangle = \langle Cp, \dddot{\sigma}(0)\rangle = 0$. But since x is of degree at most 3, this says $Cp = 0$. Now $p \in M'$ is arbitrary and M' is open in M. If $x(M')$ is contained in a proper subspace U of E^{n+1}, then $x(M) \subset U$ by analyticity. Thus by our assumption on x, we have $Cx(M) = 0$. Thus $C = 0$. This implies that $V = S^2(E^{n+1})$. Hence x is linearly rigid.

We note that if $(M, \langle,\rangle)$ is an analytic Riemannian manifold and $x: M \to S_1^n$ is a minimal isometric immersion, then x is analytic.

Corollary 11.1. Let $(M, \langle,\rangle)$ be an analytic Riemannian manifold and let $x : M \to S_1^n$ be a minimal isometric immersion not contained in a great sphere of S_1^n. If n-dim $M \leq 2$, then x is linearly rigid.

Note. If we replace minimal by analytic in the above statement, we have the same conclusion.

Corollary 11.2. Let $x : S_K^n \to S^p \subset E^{p+1}$ be a minimal isometric immersion not contained in a proper subspace of E^{p+1}. If $K = 1$, $(n/2(n+1))$, $(n/3(n+2))$, then $p = p_k$ ($k = 1, 2$, or 3) and there is an orthogonal transformation U of E^{p+1} so that $U \circ x = x_k$ where x_k is as in Example 1 above.

Corollary 11.3. Let

$$x : CP^n_{\frac{2n}{n+1}} \to S_1^p \subset E^{p+1}$$

be a minimal immersion so that $x(CP^n)$ is not contained in a subspace of E^{p+1}. Then $p = n(n+2) - 1$ and there is an orthogonal linear transformation U of E^{p+1} so that $x = U \circ x_1$ where x_1 is as in Example 2 above.

12. Rigidity and Nonrigidity for Spheres

We have seen in Corollary 11.2 and Corollary 6.1 that if U is an open connected subset of a constant curvature sphere S_K^n and if $x : U \to S_1^p$ is a minimal isometric immersion not contained in a great sphere, then $K = n/k(n+k-1)$ for some k, a positive integer. Furthermore, if $k = 1, 2, 3$, then x is rigid. We ask in this section, how far does this result extend? Our first result in this direction is

Theorem 12.1. Let U be as above and let $n = 2$. If $x : U \to S_1^p$ is a minimal isometric immersion such that x is not contained in a great hypersphere, then x is rigid.

The global version of this result is due to Calabi [2] and doCarmo and Wallach [6]. Due to Corollary 6.1, the global version implies the local version.

To prove Theorem 12.1 we actually prove a stronger result due to doCarmo and Wallach [6].

Proposition 12.1. Let $\{f_1, \ldots, f_p\}$ be a linearly independent set of spherical harmonics of degree k on S_1^2. Suppose that $\sum f_i^2$ as a function on S_1^2 is identically 1. Then $p = 2k + 1$ and $((2k+1)^{1/2})f_1, \ldots, ((2k+1)^{1/2})f_{2k+1}$ is an orthonormal basis of the space $H^{k,2}$ of spherical harmonics of degree k on S_1^2.

Proposition 12.1 combined with Corollary 6.1 immediately implies

Theorem 12.1. We prove Proposition 12.1 by analyzing the osculating spaces for equivariant minimal immersions. To this end, we begin our study of a more general situation. Suppose that G is a compact Lie group and that K is a closed subgroup of G. Let $G/K = M$. Let $x : M \to S^p \subset E^{p+1}$ be an equivariant immersion. Then it is clear that $M' = M$, that is, the higher order fundamental forms are all defined. Furthermore, if we let $x(eK) = v$, and if we denote the action of G on E^{p+1} by $(g, u) \to g \cdot u$, then $x(gK) = gv$, and extending the isotropy action of K to $S(T_{eK}(M))$, we have $B_j(k \cdot u) = k \cdot B_j(u)$ for $k \in K$, $u \in S^j(T_{eK}(M))$. Thus O^j_{eK} is K-invariant for $j = 1, 2, \ldots$. Now we assume that $x(M)$ is not contained in a great sphere of S^p_1. Then by analyticity, we have $E^{p+1} = Rv + O^1_{eK} + O^2_{eK} + \cdots + O^j_{eK}$, an orthogonal direct sum of K-invariant subspaces of E^{p+1} where m is the degree of x.

Now suppose that $G = SO(n+1)$ and $K = SO(n)$. Let $P_0 = eK \in SO(n+1)/SO(n) = S^n$. Then the isotropy action of $SO(n)$ on $T_{p_0}(S^n)$ is exactly the usual action of $SO(n)$ on E^n. Thus as a K-module, $S^j(T_{p_0}(S^n))$ is equivalent to the $SO(n)$ module of all homogeneous polynomials of degree j on E^n. Let r be as in Section 11. Then it is easy to check that as a K-module, $S^j(T_{p_0}(S^n)) = H^{j,n-1} \oplus S^{j-2}(T_p(S^n))r$. We have shown

Lemma 12.1. Let $x : S^n_K \to S^p_1 \subset E^{p+1}$ be a minimal equivariant, isometric immersion, not contained in a great sphere of S^p_1. If degree $(x) = m$, then as an $SO(n+1)$ module, E^{p+1} is equivalent to the space of spherical harmonics of degree m, $H^{m,n}$. Furthermore, as a $K = SO(n)$ module, $O^j_{p_0}$ is equivalent to the space of spherical harmonics of degree j on S^{n-1}.

The abstract statement $H^{m,n} = \sum_{j=0}^m H^{j,n-1}$ as an $SO(n-1)$ module is a special case of the branching theorem which we will describe later in this section. The critical point of Lemma 12.1 for our purposes is that $O^j_{p_0}$ is isomorphic with $H^{j,n-1}$.

Proposition 12.2. Let V be a class 1 representation of $(SO(n+1), SO(n)) = (G, K)$. Let $v \in V$ be a K-fixed unit vector. Let W_0 be the linear span of $G \cdot v^2$ in $S^2(V)$. Then W_0 contains all G-subrepresentations of $S^2(V)$ which are class 1 for (G, K).

Proof. Let $p_0 = eK$. Let $x : S^n \to S$ (the unit sphere of V) be defined by $x(gK) = g \cdot v$. Then x defines a minimal isometric immersion of $S^n_K \to S$. Set $V_i = O^i_{p_0}$, $i = 1, \ldots, m$ (m = degree of x = degree of the spherical harmonics that comprise V). Set $V_0 = Rv$. Then $V = V_0 + V_1 + \cdots + V_m$. By Lemma 12.1, $\mathrm{Hom}_K(V_i, V_j) = 0$ if $i \neq j$. Let now σ be an arbitrary geodesic through p_0. Then $((d^k/dt^k)x \circ \sigma)(0) \in V_0 + V_1 + \cdots + V_k$ by the definition of B_j, $j = 0, \ldots, k$.

Now suppose that $A : V \to V$ is linear, $A \geqq 0$, $A \in \mathrm{Hom}_K(V, V)$ ($kA = Ak$ for $k \in K$) and $A(G \cdot v) \subset S$. We show that A is the identity. Indeed, if $k \in K$, then $Akv = kAv$; hence $kAv = Av$. Thus Av is K-fixed which implies $Av = v$ since $A \geq 0$. By Schur's lemma, $AV_j \subset V_j$, $j = 0, 1, \ldots, m$ and $A|V_j = \lambda_j I$. Thus we need only show that $\lambda_j = 1$ for all j. We have shown that $\lambda_0 = 1$. Suppose $\lambda_0, \ldots, \lambda_{S-1}$ are all 1. We show $\lambda_S = 1$. Let σ be an arbitrary nontrivial geodesic through p_0 in S^n. Let $\gamma(t) = x(\sigma(t))$. Set $\gamma^{(j)}(t) = (d^j\gamma/dt^j)(t)$. We note that by the definition of B_S, $\gamma^{(S)}(0)$ projects nontrivially on V_S. Thus it is enough to prove that $\langle A\gamma^{(S)}(0), A\gamma^{(S)}(0)\rangle = \langle \gamma^{(S)}(0), \gamma^{(S)}(0)\rangle$.

Now

$$0 = \frac{d^{2S}}{dt^{2S}}\langle A\gamma, A\gamma\rangle = \sum_{j=0}^{2S} \binom{2S}{j} \langle A\gamma^{(j)}, A\gamma^{(2S-j)}\rangle.$$

Thus

$$\langle A\gamma^{(S)}(0), A\gamma^{(S)}(0)\rangle = \sum_{j=0}^{S-1} a_j^S \langle A\gamma^{(j)}(0), A\gamma^{(2S-j)}(0)\rangle \tag{12.1}$$

with

$$a_j^S = -2\binom{2S}{j}\binom{2S}{S}^{-1}$$

which is independent of A. In particular, (12.1) is true for $A = I$. Now if $u > S - 1$, then $\gamma^{(u)}(0) = w_0 + w_1 + \cdots + w_u$, $w_j \in V_j$, $j = 0, 1, \ldots, u$. If $i > S - 1$, $j \leq S - 1$, then $\langle \gamma^{(j)}(0), Aw_i\rangle = 0$ by Schur's lemma. Thus, for $j \leq S - 1$, $u > S - 1$, we have

$$\begin{aligned}\langle A\gamma^{(j)}(0), A\gamma^{(u)}(0)\rangle &= \langle \gamma^{(j)}(0), A\gamma^{(u)}(0)\rangle \\ &= \sum_{i=1}^{S-1} \langle \gamma^{(j)}(0), Aw_i\rangle = \sum_{i=1} \langle \gamma^{(j)}(0), w_i\rangle = \langle \gamma^{(j)}(0), \gamma^{(u)}(0)\rangle.\end{aligned}$$

Thus,

$$\begin{aligned}\langle A\gamma^{(S)}(0), A\gamma^{(S)}(0)\rangle &= \sum_{j=1}^{S-1} a_j^S \langle A\gamma^{(j)}(0), \gamma^{(2S-j)}(0)\rangle \\ &= \sum_{j=0}^{S-1} a_j^S \langle \gamma^{(j)}(0), \gamma^{(2S-j)}(0)\rangle = \langle \gamma^{(S)}(0), \gamma^{(S)}(0)\rangle.\end{aligned}$$

Thus the induction is complete and $A = I$.

We now complete the proof of Proposition 12.2. Suppose $U \subset S^2(V)$ is a G-submodule which is Class 1 for (G, K) and that U is not contained in W_0. Then by orthogonal projection, we may assume that $U \perp W_0$. Let $C \in U$ be K-fixed. Then $C \in \mathrm{Hom}_K(V, V)$. Let $t > 0$ be so small that $1 + tC \geq 0$. Then $(1 + tC, g \cdot v^2) = 1$ for all $g \in G$. Thus if $B = (1 + tC)^{1/2}$, then $\langle Bgv, Bgv \rangle = 1$ for all $g \in G$. Thus by the above, $B = I$. Hence $1 + tC = I$, thus $C = 0$. Hence $U = (0)$. Q.E.D.

Proposition 12.1 follows directly from Proposition 12.2. Indeed, *every* representation of $SO(3)$ is class 1 for $SO(2)$. Thus $W_0 = S^2(V)$, for any class 1 representation of $(SO(3), SO(2))$. Let $f_1, \ldots, f_p$ be as in Proposition 12.1. Let $\phi_1, \ldots, \phi_{2k+1}$ be an orthonormal basis of $H^{k,2}$. Then $\sum \phi_i^2 = 2k + 1$. Thus if $x(z) = (1/(2k + 1)^{1/2})\,(\phi_1(z), \ldots, \phi_{2k+1}(z))$, $x : S^2 \to S^{2k}$, is an equivariant minimal immersion of $S^2_{2/(k(k+1))}$ into S^{2k}. $(f_1, f_2, \ldots, f_p, 0, \ldots, 0) = Ax$ for some $A \in M_{2k+1}(R)$. Let $w \to g \cdot w$ denote the action of $SO(3)$ on E^{2k+1} induced by x, and $v = x(eK)$, then $\langle ({}^tAA - I)g \cdot v, g \cdot v \rangle = 0$ for $g \in G$ $SO(3)$. Thus $({}^tAA - I, (gv)^2) = 0$. But then ${}^tAA - I$ is orthogonal to $W_0 = S^2(V)$. Hence ${}^tAA = I$. This proves Proposition 12.1.

Our next result shows that Theorem 12.1 and Corollary 11.2 are the *only* rigidity statements true for minimal isometric immersions of constant curvature spheres into constant curvature spheres.

Let us say that two minimal immersions $x : S^n_K \to S^p_1$ and $y : S^n_K \to S^p_1$ are equivalent if there is an isometry A of S^p so that $A \circ x = y$.

Theorem 12.2. (doCarmo, Wallach [*8*]). For each $n \geq 3$ and $K > 0$, there is a vector space $W_{n,K}$ and a compact convex body $L_{n,K} \subset W_{n,K}$ that parametrizes (smoothly) the set of all inequivalent minimal isometric immersions $x : S^n_K \to S^p$ that are not contained in a great sphere. If $K \neq 1$, $(n/2(n + 1))$, $(n/3(n + 2))$ and if $L_{n,K} \neq \varnothing$, then $\dim L_{n,K} \geq 18$. The interior points of $L_{n,K}$ correspond to the minimal immersions where p is maximal

The meaning of the smoothness is that the map $L_{n,K} \times S^n_K \to S^{p_K}_1(p_K$ is the maximum of the above defined p's) given by $(x, z) \to x(z)$ is smooth. Clearly, by Corollary 6.1, it is implied that $L_{n,K} \neq \varnothing$ only if $K = (n/k(n + k - 1))$ for some integer $k \geq 1$. The existence of $L_n = L_{n,(n/k(n+k-1))}$ with the required properties is a direct consequence of Lemma 10.3(a). We recall the construction leading to Lemma 10.3(a). Let $x_k = x_{n/k(n+k-1))} : S_{(n/k(n+k-1))} \to S^{p_k} \subset E^{p_k+1}$ be the equivariant minimal immersion defined as in Section 9. That is, let $\phi_1, \ldots, \phi_{p_k+1}$ be an orthonormal basis of $H^{k,n}$, the space of spherical harmonics of degree k on S^n. Set $x_k = (1/(p_k + 1)^{1/2})(\phi_1, \ldots, \phi_{p_k+1})$. The action on E^{p_k+1} of $SO(n+1)$ is defined as follows: if $g \in SO(n+1)$, $g \cdot \phi_i = \sum g_{ji}\phi_j$. Set $\rho(g) = (g_{ij})$. Then $x(g) = \rho(g)x(z)$ for $z \in S^n$, $g \in SO(n+1)$. The space W_n is the subspace of $S^2(E^{p_k+1})$ orthogonal to the subset $\cup_{p \in S^n}(x_{k_*p}(T(S^n)))$. $L_n^k = \{C \in W_n^k | C + I \geq 0\}$ (≥ 0 means symmetric positive definite). We note that if $V_1 = x_{k*eK}(T_{eK}(S^n))$, then $\cup_{p \in S^n} S^2(x_{k_*p}(T(S^n))) = G \cdot S^2(V_1)$. Thus W_n^k is a G-submodule of $S^2(E^{p_k+1})$. If $C \in L_n^k$, then the corresponding minimal isometric immersion of $y_C : S^n_{(n/k(n+k-1))} \to S^{p_k}$ is given by $y_C = (C + I)^{1/2} \circ x_k$. The correspondence $C \to y_C$ clearly satisfies the conditions of Theorem 12.2 for the parametrization. We therefore need only prove dim $W_n^k \geq 18$ for $n \geq 3$, $k \geq 4$. Let V_j be the $K = SO(n)$ module of spherical harmonics of degree j. We need a preliminary lemma. Let V^k be E^{p_k+1} with the action of $SO(n+1)$ defined above.

Lemma 12.2. Let U_n^k be the sum of all G-submodules $Z \subset S^2(V^k)$ such that Z contains no K-submodules isomorphic with V_0 or V_2. Then $U_n^k \subset W_n^k$.

Proof. Let $X = \{u \in S^2(V^k) | (u, S^2(V_1)) = 0\}$. Then as a K-module $S^2(V^k) = S^2(V_1) \oplus X$. Let $P : S^2(V^k) \to S^2(V_1)$ be the corresponding projection. Let $\Gamma(S^2(V_1))$ be the orthogonal $SO(n+1) = G$-module induced by $S^2(V_1)$ (see Section 8 for the definition). If $u \in S^2(V^k)$, define $f_u : G \to S^2(V_1)$, via $f_u(g) = P(gu)$ for $g \in G$. Then $f_u \in \Gamma(S^2(V_1))$. Define $\psi : S^2(V^k) \to \Gamma(S^2(V_1))$ by $\psi(u) = f_u$. Then $\psi(g_0 u)(g) = f_{g_0 u}(g) = P(gg_0 u)$ $= f_u(gg_0) = (g_0 \cdot f_u)(g)$ for g_0, $g \in G$, $u \in S^2(V^k)$. Thus ψ is a G-module homomorphism. Ker $\psi = W_n^k$. Indeed, $u \in S^2(V^k)$ and if $\psi(u) = 0$, then $f_u(g) = 0$ for all $g \in G$. Thus $P(gu) = 0$ for all $g \in G$. This says that $(gu, S^2(V_1)) = 0$ for all $g \in G$. Thus by definition of W_n^k, $u \in W_n^k$. Let T be the linear span of $G \cdot S^2(V_1)$ in $S^2(V^k)$ ($T = W_n^{k\perp}$ in $S^2(V^k)$), then $\psi : T \to \Gamma(S^2(V_1))$ is a K-module injection. This implies that dim $\mathrm{Hom}_G(U_n^k, T) \leq$ dim $\mathrm{Hom}_G(U_n^k, \Gamma(S^2(V_1))) =$ dim $\mathrm{Hom}_K(U_n^k, S^2(V_1))$ by Frobenius reciprocity. Now as a K-module, $S^2(V_1) = V_0 + V_2$. Thus by hypothesis on U_n^k, dim $\mathrm{Hom}_K(U_n^k, S^2(V_1)) = 0$. Hence $\mathrm{Hom}_G(U_n^k, T) = 0$. This implies $U_n^k \perp T$. Hence $U_n^k \subset W_n^k$.

To complete the proof of Theorem 12.2, we need some results in the representation theory of $SO(n+1)$ which can be found for example in Boerner [*1*]. We first give the classification of representations of $SO(n+1)$ $= G$. Let $T \subset G$ be the subgroup consisting of matrices of the form

$$\begin{bmatrix} A_1 & & & & & \\ & A_2 & & & 0 & \\ & & \cdot & & & \\ & & & \cdot & & \\ & & & & \cdot & \\ & & & & A_p & \\ & 0 & & & & 1 \end{bmatrix} \quad \text{if} \quad n = 2p$$

or

$$\begin{bmatrix} A_1 & & & & & \\ & A_2 & & & 0 & \\ & & \cdot & & & \\ & & & \cdot & & \\ & & & & \cdot & \\ 0 & & & & & A_p \end{bmatrix} \quad \text{if} \quad n = 2p - 1$$

where

$$A_i = \begin{bmatrix} \cos t_i & \sin t_i \\ -\sin t_i & \cos t_i \end{bmatrix}.$$

Let E_{ij} be the $(n + 1) \times (n + 1)$ matrix with 1 in the i, j position and 0 elsewhere. Set $h_i = E_{2i-1,2i} - E_{2i,2i-1}$, $i = 1, \ldots, p$. Then if $\mathfrak{h}$ is the Lie algebra of T, $h_1, \ldots, h_p$ is a basis of $\mathfrak{h}$. Let $\mathfrak{h}^*$ be the real dual of $\mathfrak{h}$, and let $\phi_1, \ldots, \phi_p$ be the dual basis for $h_1, \ldots, h_p$. Order the elements of $\mathfrak{h}^*$ lexicographically relative to $\phi_1, \ldots, \phi_p$.

Let W be a continuous complex G-module. Then a weight for W is an element $\phi \in \mathfrak{h}^*$ such that $0 \neq W_\phi = \{w \in W \mid h \cdot w = i\phi(h)w \text{ for all } h \in \mathfrak{h}\}$.

Suppose now that W is irreducible. Then $W = \sum W_\phi$, ϕ a weight for W. Let ϕ_W be the highest (largest) weight of W relative to the above order. Then $\phi_W = 2\pi \sum m_i \phi_i$ where $m_i = p_i/2$, p_i an integer, and

$(*)_{2p}$ $\quad m_1 \geq m_2 \geq \cdots \geq m_p \geq 0,$

$(*)_{2p-1}$ $\quad m_1 \geq m_2 \geq \cdots \geq m_{p-1} \geq |m_p|.$

Theorem 12.3. (E. Cartan, c.f. [*1*]). Let $\phi = 2\pi \sum m_i \phi_i$ subject to $(*)_{2p}$ if $n = 2p$ or $(*)_{2p-1}$ if $n = 2p - 1$. Then ϕ is the highest weight for an irreducible complex G-module. If W_1, W_2 are two irreducible complex G-modules, then W_1 is equivalent to W_2 if and only if $\phi_{W_1} = \phi_{W_2}$.

Let $(m) = (m_1, \ldots, m_p)$, m_i half-integers subject to $(*)_{2p}$ or $(*)_{2p-1}$ depending on whether $n = 2p$ or $2p - 1$. Let ${}_nV^{(m)}$ be the irreducible $SO(n + 1)$ module with highest weight $2\pi \sum m_i \phi_i$.

Theorem 12.4. (The Branching Theorem, c.f. [*1*]). As an $SO(n)$, $SO(n+1)$ module, ${}_nV^{(m)} = \sum_{(m')}({}_{n-1}V^{(m')})$ summed over all indices (m') such that

$$m_1 \geq m_1' \geq \cdots \geq m_p \geq |m_p'| \quad \text{if} \quad n = 2p,$$
$$m_1 \geq m_1' \geq \cdots \geq m_{p-1}' \geq |m_p| \quad \text{if} \quad n = 2p - 1.$$

m_i' is half integral only if m_i is half integral.

Now let ${}_nV^k$ be the complex $SO(n+1)$ module of spherical harmonics of degree k. Then $\phi_{{}_nV^k} = 2\pi k \phi_1$.

Lemma 12.3. $\dim_C (U_k^n \otimes_R C) = \dim_C \sum_{j=2}^{[k/2]} {}_nV^{(2k-2j,2j,0,\ldots,0)} + \dim_C U^{k-1}$, where $[k/2]$ is the largest integer less than or equal to $k/2$.

Proof. In the Appendix to doCarmo and Wallach [*8*], it is proved that as an $SO(n+1)$ module,

$$S^2({}_nV^k) \otimes_R C = \sum_{j=0}^{[k/2]} V^{(2k-2j,2j,0,\ldots,0)} + S^2({}_nV^{k-1}).$$

Now ${}_nV^{(2k-2j,2j,0,\ldots,0)}$ contains ${}_{n-1}V^{(0,0,\ldots,0)}$ or ${}_{n+1}V^{(2,0,\ldots,0)}$ only if $j = 0, 1$, by Theorem 12.4.

We can now prove Theorem 12.2. Indeed, let $n \geqq 3$, $k \geq 4$. Then $\dim L_n^k = \dim W_n^k \geq \dim U_n^k \geq \dim U_n^4 \geq \dim U_3^4 = \dim_3 V^{(4,4)} = 9 + 9 = 18$.

The author would like to thank the members of the audience of the short course at Washington University, St. Louis (especially Professors Coifman and Jensen). Their comments were invaluable aids in the preparation of this exposition.

References

[*1*] H. Boerner, "Representations of Groups," North Holland Publishing Co., Amsterdam, 1965.

[*2*] E. Calabi, Minimal immersions of surfaces in Euclidean spheres, *J. Diff. Geom.*, **1** (1967), 111–125.

[*3*] E. Cartan, Sur la détermination d'un système orthogonal complet dans un espace de Riemann symétrique clos, *Rend. Circ. Mat. Palermo*, **53** (1929), 217–252.

[*4*] S. S. Chern, Minimal submanifolds in a Riemannian manifold, University of Kansas, Dept. of Math., Technical Report 19 (New Series), 1968.

[*5*] ———, M. doCarmo, and S. Kobayashi, Minimal submanifolds of a sphere with second fundamental form of constant length. To appear in the Stone Jubilee Volume.

[*6*] M. doCarmo and N. Wallach, Representations of compact groups and minimal immersions in spheres, *J. Diff. Geom.* **4** (1970), 91–104.

[*7*] ———, Minimal immersions of spheres into spheres, *Proc. Nat. Acad. Sci. U.S.A.* **63** (1969), 640–642.

[*8*] ———, Minimal immersions of spheres into spheres, *Ann. Math.*, **95** (1971), 43–62.

[*9*] W. Y. Hsiang, On the compact, homogeneous minimal submanifolds, *Proc. Nat. Acad. Sci. U.S.A.*, **56** (1966), 5–6.

[*10*] ———, H. B. Lawson, Minimal submanifolds of low cohomogeneity. To appear.

[*11*] H. B. Lawson, Local rigidity theorems for minimal hypersurfaces, *Ann. of Math.*, (2) **89** (1969), 187–197.

[*12*] G. D. Mostow, Equivariant embeddings of Euclidean space, *Ann. of Math.*, **65** (1957), 432–446.

[*13*] B. O'Neil, The fundamental equations of a submersion, *Michigan Math. J.*, **13** (1966), 459–469.

[*14*] R. S. Palais, Imbeddings of compact differentiable transformation groups in orthogonal representations, *J. Math. Mech.*, **6** (1957), 673–678.

[*15*] J. Simons, Minimal varieties in Riemannian manifolds, *Ann. of Math.*, **88** (1968), 62–105.

[*16*] T. Takahashi, Minimal immersions of Riemannian manifolds, *J. Math. Soc. Japan*, **18** (1966), 380–385.

[*17*] N. Wallach, Extension of locally defined minimal immersions of spheres into spheres, *Arch. Math.* **21** (1970), 210–213.

SPHERICAL FUNCTIONS ON SEMISIMPLE LIE GROUPS

RAMESH GANGOLLI

University of Washington
Seattle, Washington 98105

Preface

These notes grew out of lectures given by the author at Washington University, St. Louis, in the spring of 1970.

The purpose of the lectures was to give a synthesized account showing the motivations, problems, and results of the theory of spherical functions, as developed principally by Harish-Chandra. Thus the aim of these lectures is, frankly, to make a survey rather than to present current research. Nevertheless, I hope that I have reported on most of the recent developments in this theory, which is now taking a recognizable shape.

In an exposition of this type, it is hard to strike a balance between giving too much detail and giving too little. The general tenor of these notes is that of a rapid introduction. Since my audience consisted of a mixed bag of people of different primary interests, I have tried to sketch the proofs of the elementary but basic theorems of this subject but have not attempted to give the proofs of the deep results of Harish-Chandra. Rather, I have tried to give an account of the principal ideas and motivations, a sketch of the methods where possible, and accurate definitions and statements of results, cheerfully omitting the proofs when giving them would have obscured the overall pattern. I hope I have been careful in giving references to the proofs.

It is a pleasure to thank the various members of my audience who patiently sat through the lectures. I have benefited from many of their comments and especially from those of R. Coifman.

Part I: Dual Object for Harmonic Analysis of Spherical Functions

1. Introduction

In classical Fourier analysis on the d-dimensional torus T^d, one has the formulas

$$\hat{f}(n) = 1/(2\pi)^d \int_{T^d} f(x)e^{-in\cdot x}\,dx, \qquad f \in L_2(T^d), \quad n \in Z^d \tag{1.1}$$

$$f(x) \sim \sum \hat{f}(n)e^{in\cdot x} \tag{1.2}$$

$$\frac{1}{(2\pi)}\int_{T^d} |f(x)|^2\,dx = \sum |\hat{f}(n)|^2. \tag{1.3}$$

Similarly on R^d, we have

$$\hat{f}(\lambda) = (2\pi)^{-d/2} \int_{R^d} f(x) e^{-i\lambda \cdot x}, \qquad f \in L_2(R^d), \quad \lambda \in R^d \tag{1.4}$$

$$f(x) \sim \int_{R^d} \hat{f}(\lambda) e^{i\lambda \cdot x} \, d\lambda \tag{1.5}$$

$$\int_{R^d} |f(x)|^2 \, dx = \int_{R^d} |\hat{f}(\lambda)|^2 \, d\lambda. \tag{1.6}$$

The twiddles indicate that one has equality in some suitable sense. For example, if f were L_2, then one has equality in the mean while if f were nice enough, say C^∞ with compact support, then the equality is literal. The inversion formulas (1.2) and (1.5) may be regarded as an effort to expand an "arbitrary" function in terms of the special functions $\{e^{in \cdot x}, n \in Z^d\}$ and $\{e^{i\lambda \cdot x}, \lambda \in R^d\}$, respectively.

Next let G be a compact Lie group, and let $L_p^{\#}(G)$ be the subspace of $L_p(G)$ consisting of central functions, i.e., those f for which $f(yxy^{-1}) = f(x)$, $x, y \in G$. Let $\hat{G}$ be the set of equivalence classes of irreducible unitary representations of G, and for $\delta \in \hat{G}$, let χ_δ be the character of δ, and $d(\delta)$ its degree. Then one has the formulas

$$\hat{f}(\delta) = \frac{1}{d(\delta)} \int_G f(x) \overline{\chi_\delta(x)} dx, \qquad f \in L_2^{\#}(G), \quad \delta \in \hat{G} \tag{1.7}$$

$$f(x) \sim \sum_{\delta \varepsilon \hat{G}} \hat{f}(\delta) \chi_\delta(x) \, d(\delta) \tag{1.8}$$

$$\int_G |f(x)|^2 \, dx = \sum_{\delta \varepsilon \hat{G}} |\hat{f}(\delta)|^2 \, d(\delta)^2, \qquad f \in L_2^{\#}(G). \tag{1.9}$$

Here (1.8) may be regarded as an expansion of f in terms of the functions $\{\chi_\delta, \delta \in \hat{G}\}$.

We may abstract certain features of these three situations. In each case we are given a locally compact topological group (e.g., T^d, R^d, G) and various function spaces naturally associated with the group (e.g., $L_2(T^d)$, $L_2{}^{\#}(G)$). One also has a "dual object" (e.g., Z^d, R^d, $\hat{G}$) and members of this dual object are in natural correspondence with certain special functions on the group (e.g., $n \leftrightarrow e^{in \cdot x}$, $\delta \leftrightarrow \chi_\delta(x)$). Very roughly speaking, harmonic analysis consists of understanding the relation between the group and its "dual object." Slightly more precisely, the mappings $f \to \hat{f}$ illustrated above set up a correspondence between various types of function spaces on the group and other function spaces on the dual object, and a large part of the game consists of making inferences about the nature of f from data on its Fourier transform $\hat{f}$, and vice versa. In short, one wants to study the map $f \to \hat{f}$ as fully as possible. There are also structural questions one might like to ask, such as "to what extent does the dual object characterize the group"?

Now imagine that we are given a locally compact group G and a certain space $\mathscr{F}$ of functions on G. Loosely speaking, when we talk of harmonic analysis of functions in $\mathscr{F}$, we mean the accomplishment of the following steps.

(a) The construction of a suitable dual object for the pair $(G, \mathscr{F})$.

(b) The construction of special functions (such as $e^{in \cdot x}$, $\chi_\delta(x)$ in the examples above) corresponding to each member of this dual object. Presumably these functions would have some natural relation with the space $\mathscr{F}$.

(c) Definition of the Fourier transform $f \to \hat{f}$, setting up a correspondence between members of $\mathscr{F}$ and functions on the dual object.

(d) Study of the map $f \to \hat{f}$. One may ask, what is the natural domain of this map? Is it defined on all of $\mathscr{F}$? If so, can one extend it to bigger spaces of functions? What can we say about f given $\hat{f}$, and vice versa?

(e) To what extent does the dual object determine the pair $(G, \mathscr{F})$?

If $(G, \mathscr{F})$ is arbitrary as above, almost nothing systematic is known about this sequence of questions in this generality. We shall be dealing with a very special case of this situation, which we now specify.

We shall start with a connected noncompact semisimple Lie group G with finite center, and let K be a maximal compact subgroup of G. Then G/K is a symmetric space of noncompact type. We shall denote by $C_c(G)$, $C_c^\infty(G)$, $C^\infty(G)$, $L_p(G)$ the customary function spaces on G. We shall denote by $C_c(K\backslash G/K)$ the subspace of $C_c(G)$ consisting of those f such that $f(k_1xk_2) = f(x)$, $x \in G$, k_1, $k_2 \in K$. A function f having this last property, viz. $f(k_1xk_2) = f(x)$ is said to be *spherical* or *bi-invariant under* K. We shall denote by $C_c^\infty(K\backslash G/K)$, $C^\infty(K\backslash G/K)$, $L_p(K\backslash G/K)$, the subspace of $C^\infty(G)$, $C_c^\infty(G)$, $L_p(G)$ consisting of spherical functions. Then my aim in these lectures is to study the harmonic analysis of functions in the spaces $C_c(K\backslash G/K)$, $C^\infty(K\backslash G/K)$, $L_p(K\backslash G/K)$, etc. In keeping with what I have said above, I shall discuss each of the steps (a)–(e) for the pair $(G, C_c(K\backslash G/K))$, and also discuss analogues of (1.1)–(1.9) for this situation.

2. Elementary Spherical Functions

In this section we shall discuss the construction of a dual object suitable for the harmonic analysis of $C_c(K\backslash G/K)$. By analogy with the examples mentioned above, we shall seek functions analogous to $e^{i\lambda\cdot x}$, $e^{in\cdot x}$, etc. It will be helpful to recall some features of the examples. Thus if we put $\varphi_\lambda(x) = e^{i\lambda\cdot x}$, $\varphi_n(x) = e^{in\cdot x}$, we have
(2.1) Each φ_λ is an eigenfunction of every differential operator $P(D) = P(D_1, \ldots, D_d)$ with constant coefficients on R^d. Indeed $P(D)\ \varphi_\lambda = P(i\lambda)\varphi_\lambda$ and the map $P(D) \to P(i\lambda)$ is a homomorphism of the algebra of such differential operators. Similarly for T^d.
(2.2) Each φ_λ is an eigenfunction of a large class of convolution operators. Indeed $f * \varphi_\lambda = \hat{f}(\lambda)\varphi_\lambda$ for $f \in C_c(R^d)$. Similarly $f * \varphi_n = \hat{f}(n)\varphi_n$, $f \in C(T^d)$.
(2.3) φ_λ determines a homomorphism of the algebra $C_c(R^d)$ via the mapping $f \to (f * \varphi_\lambda)(0) = \hat{f}(\lambda)$. Similarly for T^d.
(2.4) $\varphi_\lambda(x + y) = \varphi_\lambda(x)\ \varphi_\lambda(y)$; $\varphi_n(x + y) = \varphi_n(x)\varphi_n(y)$.

In the case of a compact Lie group G and the algebra $C^\#(G)$, if we put $\varphi_\delta = \chi_\delta/d(\delta)$, we have
(2.1)′ φ_δ is an eigenfunction of every *central* differential operator D on G. Thus $D\varphi_\delta = \hat{D}(\delta)\varphi_\delta$ with $\hat{D}(\delta) \in \mathbf{C}$. The map $D \to \hat{D}(\delta)$ is a homomorphism of the algebra of *central* differential operators on G.
(2.2)′ φ_δ is an eigenfunction of a large class of convolution operators. $f * \varphi_\delta = \hat{f}(\delta)\varphi_\delta$, $\hat{f} \in C^\#(G)$.
(2.3)′ Each φ_δ determines a homomorphism of the algebra $C^\#(G)$ via the map $f \to \hat{f}(\delta) = (f * \varphi_\delta)(e)$.

In the case of φ_δ, a simple algebraic property like (2.4) is, of course, missing, but (2.4) does have a sort of analogue. Namely if we view (2.4) as a functional equation, we have an analogous functional equation

$$\int_G \varphi_\delta(uyu^{-1}x)\, du = \varphi_\delta(y)\varphi_\delta(x) \tag{2.4$'$}$$

which reduces to (2.4) if G is commutative. These properties should be borne in mind in the discussion that follows.

In the case of interest to us, the fundamental point of departure is to notice

Proposition 2.1 (Gelfand [5]): The convolution algebra $C_c(K\backslash G/K)$ is commutative.

This comes about due to a structural fact about the pair G, K. Namely if $\mathfrak{g}_0$, $\mathfrak{k}_0$ are the Lie algebras of G, K, respectively, and if $\mathfrak{g}_0 = \mathfrak{k}_0 + \mathfrak{p}_0$ is the Cartan decomposition of $\mathfrak{g}_0$, and if we put $P = \exp \mathfrak{p}_0$, then $G = KP$ uniquely, i.e., each $x \in G$ can be factored in the form $x = kp$, $k \in K$, $p \in P$, uniquely. Further the mapping $\sigma(kp) = kp^{-1}$ is an involutive automorphism of G. Now let α be the antiautomorphism of G, $\alpha: x \to x^{-1}$. Then both α and σ act on $C_c(K\backslash G/K)$. $f^\alpha(x) = f(x^{-1})$, $f^\sigma(x) = f(\sigma(x))$. Now if $f \in C_c(K\backslash G/K)$, we have $f^\alpha(x) = f(x^{-1}) = f(p^{-1}) = f(\sigma(x)) = f^\sigma(x)$. Thus $f^\alpha = f^\sigma$, but since α is an *anti* automorphism we have $(f * g)^\alpha = g^\alpha * f^\alpha = g^\sigma * f^\sigma = (g * f)^\sigma = (g * f)^\alpha$. Hence $f * g = g * f$ for f, $g \in C_c(K\backslash G/K)$. Obviously similar statements hold about other algebras such as $C^\infty(K\backslash G/K)$, $L_1(K\backslash G/K)$, etc.

We topologize the algebra $C_c(K\backslash G/K)$ as follows. Say $f_n \to f$ if the supports of all the f_n are contained in a fixed compact set U and the convergence is uniform on U. With this topology, $C_c(K\backslash G/K)$ is a topological algebra.

The functions we are seeking will be constructed via continuous homomorphisms of this algebra. Let φ be a continuous linear functional on this algebra. Then we may identify φ with a unique complex Borel measure (also called φ) such that (i) φ is finite on compact sets, (ii) φ is spherical, i.e., $\varphi(k_1 E k_2) = \varphi(E)$ for any Borel E and k_1, $k_2 \in K$. Thus $\varphi(f) = \int_G f(x)\, d\varphi(x)$.

Definition 2.2 A complex Borel measure φ on G is said to be *elementary spherical* if (i) φ is spherical, i.e., $\varphi(k_1 E k_2) = \varphi(E)$, E Borel, k_1, $k_2 \in K$. (ii) φ defines a continuous nontrivial homomorphism of the algebra $C_c(K\backslash G/K)$ via the mapping $\hat{\varphi}: f \to \int_G f(x^{-1})\, d\varphi(x)$.

Let φ be an elementary spherical measure. Then there is $f_0 \in C_c(K\backslash G/K)$ such that $\hat{\varphi}(f_0) = 1$. Consider the function $f_0 * \check{\varphi}$. If μ is the measure $(f_0 * \varphi)(x)\, dx$, then we have

$$\begin{aligned}
\mu(\tilde{f}) &= \int_G f(x^{-1})\,(f_0 * \varphi)\,(x)\, dx \\
&= (f * f_0 * \varphi)\,(e) \\
&= \varphi((f * f_0)^{\sim}) \\
&= \hat{\varphi}(f * f_0) \\
&= \hat{\varphi}(f)\hat{\varphi}(f_0) \qquad \text{because} \quad \varphi \text{ is elementary}
\end{aligned}$$

$$= \hat{\varphi}(f)$$
$$= \varphi(\tilde{f}) \qquad \text{for} \quad f \in C_c(K\backslash G/K). \tag{2.5}$$

Thus the measure μ and the measure φ agree on $C_c(K\backslash G/K)$. Since both μ and φ are spherical, it follows that $\mu = \varphi$. But μ has a continuous-density, viz. $f_0 * \varphi$. Thus we have

Proposition 2.3 (see, e.g., Tamagawa [*21*]): Let φ be an elementary spherical measure. Then φ is absolutely continuous with respect to the Haar measure dx of G. Moreover the density of φ with respect to dx is a uniquely determined continuous spherical function.

In view of this proposition, we can identify an elementary spherical measure φ with a unique continuous spherical function which we shall again call $\varphi(x)$. Thus $d\varphi(x) = \varphi(x)\,dx$. $\varphi \in C(K\backslash G/K)$.

Definition 2.4: Let $\varphi \in C(K\backslash G/K)$. We say that φ is an elementary spherical function if $\varphi(x)\,dx$ is an elementary spherical measure.

We shall see shortly that the elementary spherical functions are analogous to the special functions $e^{i\lambda\cdot x}$, etc., considered above. Indeed we have

Proposition 2.5: Let $\varphi \in C(K\backslash G/K)$. Then φ is an elementary spherical function if and only if φ satisfies any one of the following equivalent conditions:

(i) $\varphi(e) = 1$, and for each $f \in C_c(K\backslash G/K)$, we have $f * \varphi = \hat{\varphi}(f)\varphi$, where $\hat{\varphi}(f)$ is the complex number $\int_G f(x^{-1})\varphi(x)\,dx$, i.e., $\hat{\varphi}(f) = (f * \varphi)(e) = \varphi(\tilde{f})$.

(ii) $\int_K \varphi(xky)\,dk = \varphi(x)\varphi(y)$, $x, y \in G$, and $\varphi \not\equiv 0$.

Note that (i) says that φ is an eigenfunction of a large class of convolution operators. The property (ii) is analogous to the functional equation (2.4)′.

We will sketch a proof of the above proposition. Indeed suppose that φ is elementary spherical, and let $f \in C_c(K\backslash G/K)$. Then for any $g \in C_c(K\backslash G/K)$, $(f * \varphi)(\tilde{g}) = (g * f * \varphi)(e) = \hat{\varphi}(f * g) = \hat{\varphi}(f)\hat{\varphi}(g) = \varphi(\tilde{f})\varphi(\tilde{g})$. This proves that $f * \varphi = \hat{\varphi}(f)\varphi$. By evaluating both sides at e we conclude that $\varphi(e) = 1$. Thus φ elementary spherical $\Rightarrow$(i). Conversely if φ satisfies condition (i), then the map $f \to (f * \varphi)(e) = \hat{\varphi}(f)$ is a continuous homomorphism of $C_c(K\backslash G/K)$, because $(f * g * \varphi)(e) = \hat{\varphi}(g)(f * \varphi)(e) = \hat{\varphi}(g)\hat{\varphi}(f)$ since $\varphi(e) = 1$. It is obvious that $\hat{\varphi}$ is a nontrivial homomorphism of $C_c(K\backslash G/K)$.

Similarly let φ be elementary spherical. Then we have

$$\hat{\varphi}(f * g) = \int_G \int_G f(x^{-1})g(y^{-1})\varphi(xy)d\,x\,dy$$
$$= \int_G \int_G f(x^{-1})g(y^{-1})\left(\int_K \varphi(xky)\,dk\right)dx\,dy.$$

On the other hand, $\varphi(f * g) = \hat{\varphi}(f)\hat{\varphi}(g) = \int_G\int_G f(x^{-1})g(y^{-1})\varphi(x)\varphi(y)\,dx\,dy$. This being true for all $f, g \in C_c(K\backslash G/K)$, it follows by approximation on $C_c(K \times K\backslash G \times G/K \times K)$ that $\int_K \varphi(xky)\,dk = \varphi(x)\varphi(y)$. Moreover, by evaluating this last equation at e, we conclude that $\varphi(e) = \varphi(e)^2$, so $\varphi(e) = 0$ or 1. But if $\varphi(e) = 0$, then $\varphi(y) = \int_K \varphi(ky)\,dk = \int_K \varphi(eky)dk = \varphi(e)\varphi(y) = 0$. So $\varphi \equiv 0$; then φ would be trivial. This shows that φ elementary spherical $\Rightarrow$ (ii). The converse is left to the reader.

Note that if φ is elementary spherical, then the mapping $f \to (f * \varphi)(e)$ is the nontrivial homomorphism of $C_c(K\backslash G/K)$ corresponding to φ. Thus the above proposition shows that the elementary spherical functions enjoy properties analogous to (2.2)–(2.4).

Notice also that we have not yet talked of any property that is analogous to (2.1), (2.1)′. We will now consider this aspect.

We may regard φ as a function on the symmetric space G/K. Now G acts transitively on G/K and the algebra $\mathcal{D}(G/K)$ of differential operators on G/K that commute with this action of G is evidently analogous to the algebra of differential operators with constant coefficients on R^d. Thus one could hope that if $D \in \mathcal{D}(G/K)$ and if φ were a C^∞ function on G/K, then we would have $D\varphi = \lambda_D\varphi$ with $\lambda_D \in \mathbf{C}$. We shall see presently that this is the case. Right now we would like to discuss a few facts about the algebra $\mathcal{D}(G/K)$. For various purposes it will be convenient to relate elements of $\mathcal{D}(G/K)$ to certain left invariant differential operators on G.

Let $\mathcal{G}$ be the universal enveloping algebra of the complexification $\mathfrak{g}$ of $\mathfrak{g}_0$. Then each $D \in \mathcal{G}$ can be thought of as a left-invariant differential operator on $C^\infty(G)$. Let $C_K^\infty(G)$ be the subspace of $C^\infty(G)$ consisting of those f for which $f(xk) = f(x)$, $x \in G$, $k \in K$. If π is the natural projection from $G \to G/K$, then the mapping $f \to f' = f \circ \pi$ is an isomorphism of $C^\infty(G/K)$ (the space of C^∞ functions on the manifold G/K) with $C_K^\infty(G)$. Now let $\mathcal{Q}$ be the centralizer of $\mathfrak{k}_0$ in $\mathcal{G}$. Then evidently $\mathcal{Q}$ consists of those differential operators on $C^\infty(G)$ that commute with left translations by elements of G and with right translations by elements of K. Clearly $C_K^\infty(G)$ is stable under each $D \in \mathscr{Q}$ and since $C_K^\infty(G) \cong C^\infty(G/K)$, the restriction D_0 of D to $C_K^\infty(G)$ induces a differential operator D' on $C^\infty(G/K)$. Obviously D' is left invariant under the action of G on $C^\infty(G/K)$; hence $D' \in \mathscr{D}(G/K)$, and D' depends only on the restriction D_0 of D to $C_K^\infty(G)$.

Proposition 2.6: The mapping $D_0 \to D'$ of the algebra $\mathscr{Q}_0 = \{D_0, D \in \mathscr{Q}\}$ to $\mathscr{D}(G/K)$ is an isomorphism onto.

We will not prove this proposition here. It involves standard differential geometry and can be found, e.g., in Ref. [*11a*, p. 390].

In view of this proposition we shall identify $C_K^\infty(G)$ with $C^\infty(G/K)$ and $\mathscr{Q}_0$ with $\mathscr{D}(G/K)$. Further, if $D \in \mathscr{Q}$ and $\varphi \in C^\infty(G/K)$, then $D_0\varphi = D\varphi$, so sometimes we may even slur over the distinction between an element D of $\mathscr{Q}$ and its restriction D_0 to $C^\infty(G/K)$. These imprecisions do not generally cause any confusion.

Proposition 2.7 (Gelfand [*5*], Selberg [*19*]): Let φ be an elementary spherical function. Then $\varphi \in C^\infty(K\backslash G/K)$. Moreover if $D \in \mathscr{Q}$, then $D\varphi = \lambda_D\varphi$ where $\lambda_D \in \mathbf{C}$. The mapping $D \to \lambda_D$ is a homomorphism of $\mathscr{Q}$.

Indeed, if φ is as hypothesized, there exists $f_0 \in C_c^\infty(K\backslash G/K)$ such that $\hat{\varphi}(f_0) = 1$. Then $f_0 * \varphi = \hat{\varphi}(f_0)\varphi = \varphi$ by Proposition 2.5. But $f_0 * \varphi \in C^\infty(K\backslash G/K)$; hence $\varphi \in C^\infty(K\backslash G/K)$. Now let $D \in \mathscr{Q}$. Then D commutes with right translations by $k \in K$. Hence $Df_0 \in C^\infty(K\backslash G/K)$. Now $D\varphi = D(f_0 * \varphi) = D(\varphi * f_0) = \varphi * Df_0 = Df_0 * \varphi = \hat{\varphi}(Df_0)\varphi$ by Proposition 2.5 again. Hence the second assertion follows upon setting $\lambda_D = \hat{\varphi}(Df_0)$. It is obvious that λ_D is independent of the choice of f_0, and that $D \to \lambda_D$ is a homomorphism of $\mathscr{Q}$.

Let us note that if $D' \in \mathscr{D}(G/K)$, then there is a $D \in \mathscr{Q}$ whose restriction to $C^\infty(G/K)$ is D'. Hence $D'\varphi = D\varphi = \lambda_D\varphi$.

Actually it turns out that the converse of the above proposition is also true. Precisely, we have

Proposition 2.8: Suppose $\varphi\ (\neq 0) \in C^\infty(K\backslash G/K)$ and that for each $D \in \mathscr{Q}$, $D\varphi = \lambda_D\varphi$ with $\lambda_D \in \mathbf{C}$. Then φ is proportional to an elementary spherical function.

We will sketch a proof of this. The first observation is that φ is analytic. This comes about because there is an elliptic operator in $\mathcal{Q}$. Indeed if ω is the Casimir operator of $\mathfrak{g}_0$ and if θ is the Casimir operator of $\mathfrak{k}_0$, the element $\omega - 2\theta$ is elliptic. Indeed $\omega - 2\theta = X_1^2 + X_2^2 + \cdots + X_n^2 + Y_1^2 + \cdots + Y_m^2$ where the X's and the Y's are orthonormal bases of $\mathfrak{k}_0$ and $\mathfrak{p}_0$ with respect to $-B$ and B, respectively. (Here B is the Killing form of $\mathfrak{g}_0$.) Since φ is an eigenfunction of every operator in $\mathcal{Q}$, it follows from the theorem of S. Bernstein and F. John [*13*, p. 57] that φ is analytic. Now let $D \in \mathcal{G}$. Then the operators $\{\mathrm{Ad}\,(k)D, k \in K\}$ lie in a finite dimensional subspace of $\mathcal{G}$ because K is compact. Hence if we put $D_0 = \int_K \mathrm{Ad}\,(k)D\,dk$, we see that the map $D \to D_0$ is a projection, mapping $\mathcal{G}$ onto $\mathcal{Q}$. If f is any function in $C^\infty(G/K)$, one checks quickly that $(Df)(e) = (D_0 f)(e)$ for any $D \in \mathcal{G}$.

We now claim that $\varphi(e) \neq 0$. If it were, then $(D_0\varphi)(e) = \lambda_{D_0}\varphi(e) = 0$ for each $D_0 \in \mathcal{Q}$. But this would mean that for $D \in \mathcal{G}$, $(D\varphi)(e) = (D_0\varphi)(e) = 0$. Since φ is analytic, this would imply that $\varphi \equiv 0$ by Taylor's theorem. Hence $\varphi(e) \neq 0$. We can assume therefore that $\varphi(e) = 1$, so far as the last part of the proposition is concerned. We shall show that φ satisfies the functional equation $\int_K \varphi(xky)\,dk = \varphi(x)\varphi(y)$. Fix x and let $f(y)$ be the function $f(y) = \int_K \varphi(xky)\,dk$. It is obvious that $f \in C^\infty(K\backslash G/K)$. Moreover, since K is compact, and φ analytic, f is likewise analytic. Now consider the function $F(y) = f(y) - \varphi(x)\varphi(y)$. We have $F(e) = f(e) - \varphi(x)\varphi(e) = \int_K \varphi(xk)\,dk - \varphi(x) = 0$ because $\varphi \in C^\infty(K\backslash G/K)$. Moreover, if $D \in \mathcal{G}$ and D_0 is as above, we have $(D_0 f)(y) = \int_K (D_0\varphi)(xky)\,dk = \lambda_{D_0} \int \varphi(xky)\,dk = \lambda_{D_0} f(y)$.

Hence

$$\begin{aligned}(D_0F)(e) &= (D_0f)(e) - \varphi(x)(D_0\varphi)(e)\\ &= \lambda_{D_0} f(e) - \varphi(x)\lambda_{D_0}\varphi(e)\\ &= \lambda_{D_0}(\varphi(x)\varphi(e) - \varphi(x)\varphi(e))\\ &= 0 \end{aligned} \tag{2.6}$$

this being true for all $D_0 \in \mathcal{Q}$. Now since $F \in C^\infty(G/K)$, we have for any $D \in \mathcal{G}$, $(DF)(e) = (D_0F)(e)$. Hence $(DF)(e) = 0$ for all $D \in \mathcal{G}$. Since F is analytic this means that $F \equiv 0$. This proves that φ satisfies the functional equation, so φ is an elementary spherical function.

The argument in the above proof shows incidentally that an elementary spherical function is determined by its system of eigenvalues. More precisely,

Corollary 2.9: Let φ_1, φ_2 be elementary spherical functions and suppose that for each $D_0 \in \mathscr{Q}$ $D_0\varphi_1 = \lambda_{D_0}\varphi_1$ and $D_0\varphi_2 = \lambda_{D_0}\varphi_2$. Then $\varphi_1 = \varphi_2$.

Indeed if we put $\varphi = \varphi_1 - \varphi_2$, φ is analytic, $D_0\varphi = \lambda_{D_0}\varphi$, and $\varphi(e) = 0$. Hence $(D_0\varphi)(e) = 0$ for all $D_0 \in \mathscr{Q}$. But for any $D \in \mathscr{G}$, $(D\varphi)(e) = (D_0\varphi)(e) = 0$, $D_0 \in \mathscr{Q}$. So $\varphi \equiv 0$ by analyticity.

The above discussion brings out the analogy between the elementary spherical functions and the functions $e^{i\lambda \cdot x}$, etc., mentioned in Section 1. One may hope then that the set of elementary spherical functions would be a suitable dual object for the harmonic analysis of $C_c(K\backslash G/K)$. However, one would wish to parametrize this set by means of more intrinsic data about (G, K) and by means of the representation theory of G. The main aim of the rest of this section is to discuss such a parametrization, due to Harish-Chandra. Incidentally it will also set the stage for definition of the Fourier transform for spherical functions in Section 4.

Let $\mathfrak{g}_0 = \mathfrak{k}_0 + \mathfrak{p}_0$ be the Cartan decomposition, and let $\mathfrak{h}_{\mathfrak{p}_0}$ be a maximal Abelian subspace of $\mathfrak{p}_0$. Extend $\mathfrak{h}_{\mathfrak{p}_0}$ to a maximal Abelian subalgebra $\mathfrak{h}_0$ of $\mathfrak{g}_0$. Then $\mathfrak{h}_0 = \mathfrak{h}_{\mathfrak{k}_0} + \mathfrak{h}_{\mathfrak{p}_0}$ where $\mathfrak{h}_{\mathfrak{k}_0} = \mathfrak{h}_0 \cap \mathfrak{k}_0$. Complexify $\mathfrak{g}_0$, $\mathfrak{k}_0$, etc., to $\mathfrak{g}$, $\mathfrak{k}$, $\mathfrak{h}$, $\mathfrak{h}_{\mathfrak{p}}$, $\mathfrak{h}_{\mathfrak{k}}$, etc. Then $\mathfrak{h}$ is a Cartan subalgebra of $\mathfrak{g}$. Let Δ be the set of nonzero roots of $\mathfrak{g}$ $w{\cdot}r{\cdot}t{\cdot}\mathfrak{h}$; then every root in Δ is real on $\mathfrak{h}_{\mathfrak{p}_0} + i\mathfrak{h}_{\mathfrak{k}_0} = \mathfrak{h}^*$. Fix compatible lexicographic orderings on $\mathfrak{h}^*$ and $\mathfrak{h}_{\mathfrak{p}_0}$ and order their duals in the usual way by means of these orderings. Let Δ_+ be the set of positive roots in Δ under this ordering. Then $\Delta_+ = P_+ \cup P_-$ where P_+ is the set of those roots in Δ_+ which do not vanish identically on $\mathfrak{h}_{\mathfrak{p}_0}$, and P_- is the set of those which do. Let $\mathfrak{n} = \sum_{\alpha \in P_+} \mathfrak{g}^\alpha$ where $\mathfrak{g}^\alpha$ is the root-space of α, and let $\mathfrak{n}_0 = \mathfrak{n} \cap \mathfrak{g}_0$. Then $\mathfrak{n}_0$ is a nilpotent subalgebra of $\mathfrak{g}_0$. Moreover we have the Iwasawa decomposition theorem, viz.: $\mathfrak{g}_0 = \mathfrak{k}_0 + \mathfrak{h}_{\mathfrak{p}_0} + \mathfrak{n}_0$ is a direct sum, $\mathfrak{h}_{\mathfrak{p}_0} + \mathfrak{n}_0$ is a solvable Lie algebra, and $\mathfrak{n}_0$ is an ideal in it. If $A = \exp \mathfrak{h}_{\mathfrak{p}_0}$ and $N = \exp \mathfrak{n}_0$, then both A, N are simply connected; indeed A is a vector group and $G = KAN$, uniquely, i.e., the mapping $(k, a, n) \to kan$ of $K \times A \times N$ into G is a diffeomorphism onto. Now for any $x \in G$, let $H(x)$ be the unique element of $\mathfrak{h}_{\mathfrak{p}_0}$ such that $x = k \cdot \exp H(x) \cdot n$, $k \in K$, $n \in N$. Let Λ_0 be the dual of $\mathfrak{h}_{\mathfrak{p}_0}$, i.e., the space of real-valued linear functionals on $\mathfrak{h}_{\mathfrak{p}_0}$ and let be its complexification. We may view Λ as the space of complex-valued R-linear functions on $\mathfrak{h}_{\mathfrak{p}_0}$.

Let $S(\mathfrak{h}_{\mathfrak{p}})$ be the symmetric algebra over $\mathfrak{h}_{\mathfrak{p}}$. If $p \in S(\mathfrak{h}_{\mathfrak{p}})$, then we can regard p as a polynomial function on $\mathfrak{h}_{\mathfrak{p}}$. Moreover the Killing form B can be used to identify Λ and $\mathfrak{h}_{\mathfrak{p}}$. Hence one may speak of the value of p at any $\nu \in \Lambda$.

This being said we turn to the problem to parametrizing the elementary spherical functions. The starting point is the observation that given a $\nu \in \Lambda$, $\nu: \mathfrak{h}_{\mathfrak{p}0} \to \mathbf{C}$, one can construct from this data a function ψ_ν which is (i) spherical, (ii) an eigenfunction of each operator $D \in \mathcal{Q}$. First, a lemma.

Lemma 2.10: $\mathcal{G} = (\mathfrak{k}_0\mathcal{G} + \mathcal{G}\mathfrak{n}_0) \oplus \mathcal{H}_{\mathfrak{p}0}$, where $\oplus$ stands for direct sum, and $\mathcal{H}_{\mathfrak{p}0}$ is the enveloping algebra of $\mathfrak{h}_{\mathfrak{p}0}$.

The proof comes, essentially, from the Poincaré–Birkhoff–Witt theorem [*11a*, p. 427].

Thus, given $D \in \mathcal{G}$, there is a unique $D_\mathfrak{h} \in \mathcal{H}_{\mathfrak{p}0}$ such that $D - D_\mathfrak{h} \in \mathfrak{k}_0\mathcal{G} + \mathcal{G}\mathfrak{n}_0$.

Now let $\nu \in \Lambda$, and let χ_ν be the unique homomorphism of $S(\mathfrak{h}_{\mathfrak{p}0})$ which extends ν and sends $1 \to 1$. Consider the function $F_\nu(x) = \exp \nu(H(x))$ on G. Since $H(xn) = H(x)$ for all $n \in N$, it is obvious that $DF_\nu = 0$ if $D \in \mathcal{G}\mathfrak{n}_0$. Also, if $D \in \mathcal{H}_\mathfrak{p}$, we have $DF_\nu = \chi_\nu(D)F_\nu$, because $H(x \exp H) = H(x) + H$, $H \in \mathfrak{h}_{\mathfrak{p}0}$. Thus, if $D \in \mathcal{Q}$ and if $D \in D_\mathfrak{h} + D_1 + D_2$ where $D_\mathfrak{h} \in \mathcal{H}_{\mathfrak{p}0}$, $D_1 \in \mathfrak{k}_0$, $D_2 \in \mathcal{G}\mathfrak{n}_0$, then $DF_\nu = \chi_\nu(D)F_\nu + D_1F_\nu$. Now since $D_1 \in \mathfrak{k}_0\mathcal{G}$, it follows that $\int_K(D_1F_\nu)(xk)\,dk$ is zero. Hence we get

$$\int_K (DF_\nu)(xk)\,dk = \chi_\nu(D_\mathfrak{h}) \int_K F_\nu(xk)\,dk \tag{2.7}$$

but since $D \in \mathcal{Q}$, D commutes with right translations by k; hence, if we put $\psi_\nu(x) = \int_K F_\nu(xk)\,dk$, we see that $(D\psi_\nu)(x) = D(\int_K F_\nu(xk)\,dk) = \int_K DF_\nu(xk)\,dk = \chi_\nu(D_\mathfrak{h})\,\psi_\nu(x)$. Thus we have

Proposition 2.11: For any $\nu \in \Lambda$, the function $\psi_\nu(x) = \int_K \exp \nu(H(xk))dk$ is an elementary spherical function on G.

See Refs. [*10d*] and [*11a*, p. 428].

However, two distinct $\nu \in \Lambda$ may give rise to the same ψ_ν. Indeed the situation is as follows. Let W be the group of linear transformations of $\mathfrak{h}_{\mathfrak{p}0}$ induced by those elements of $\mathrm{Ad}(K)$ that stabilize $\mathfrak{h}_{\mathfrak{p}0}$. It is known that W is a finite group [*11a*] and is known as the Weyl group of the pair (G, K). It acts on $\mathfrak{h}_\mathfrak{p}$, Λ_0, Λ, etc., in a natural way. Let $\rho = \frac{1}{2}\sum_{\alpha \varepsilon P_+} \alpha$. Then,

Proposition 2.12: Let $\nu, \nu' \in \Lambda$ be such that $(\nu + \rho) = s(\nu' + \rho)$ for some $s \in W$. Then $\psi_\nu = \psi_{\nu'}$.

We will sketch a proof. The Haar measures dx of G, and da, dn of A, N, respectively, can be so normalized that $dx = e^{2\rho(\log a)}\, dk\, da\, dn$, $x = kan$, i.e., for $f \in C_c(G)$, we have

$$\int_G f(x)\,dx = \int_K \int_A \int_N f(kan) e^{2\rho(\log a)}\, dk\, da\, dn.$$

Since ψ_ν, $\psi_{\nu'}$ are spherical, the proposition will follow if we can show that for $f \in C_c(K\backslash G/K)$, we have $\int_G \psi_\nu(x) f(x)\,dx = \int_G \psi_{\nu'}(x) f(x)\,dx$. Now, by the definition of ψ_ν, one gets

$$\begin{aligned}
\int_G \psi_\nu(x) f(x)\,dx &= \int_G \int_K e^{\nu(H(xk))}\, dk\, f(x)\,dx \\
&= \int_G e^{\nu(H(xk))} f(x)\,dx \\
&= \int_K \int_A \int_N e^{\nu(H)(kan))} f(kan)\, e^{2\rho(\log a)}\, dk\, da\, dn \\
&= \int_A \int_N e^{\nu(\log a) + 2\rho(\log a)} f(an)\, da\, dn \\
&= \int_A e^{(\nu+\rho)(\log a)} \left(e^{\rho(\log a)} \int_N f(an)\, dn \right) da \\
&= \int_A e^{(\nu+\rho)(\log a)} F_f(a)\, da
\end{aligned} \tag{2.8}$$

where

$$F_f(a) = e^{\rho(\log a)} \int_N f(an)\, dn. \tag{2.9}$$

In the above calculation, we have used the fact that $H(kan) = \log a$ and that f is bi-invariant under K. Similarly,

$$\int_G \psi_{\nu'}(x) f(x)\,dx = \int_A e^{(\nu'+\rho)(\log a)} F_f(a)\, da.$$

However it is known that $F_f(a^s) = F_f(a)$, all $s \in W$ (here a^s is the image of a under s). It follows that if $\nu + \rho = s(\nu' + \rho)$ for some $s \in W$, then $\int_G \psi_\nu f = \int_G \psi_{\nu'} f$, and hence $\psi_\nu = \psi_{\nu'}$.

Corollary 2.13: Let $\nu \in \Lambda$, and let φ_ν be defined by $\varphi_\nu(x) =$

$\int_K \exp(i\nu - \rho)(H(xk))\, dk$; then φ_ν is an elementary spherical function. Moreover, if $s \in W$, then $\varphi_\nu = \varphi_{s\nu}$.

The proof follows immediately from the above proposition.

Having available a fairly large class of elementary spherical functions, one might ask if these are all the elementary spherical functions there can possibly be. Before we state a result, consider the following: Let $D \in \mathcal{Q}$ and $\nu \in \Lambda$. Let φ_ν be the elementary spherical function constructed above. Then $D\varphi_\nu = \chi_{i\nu-\rho}(D_{\mathfrak{h}})\varphi_\nu$. Consider the homomorphism $D \to \chi_{i\nu-\rho}(D_{\mathfrak{h}})$. It is a composition of the following three maps: $D \to D_{\mathfrak{h}}$; the automorphism σ of $\mathcal{H}_{\mathfrak{p}0}$ which sends $H \to H - \rho(H)$ for each $H \in \mathfrak{h}_{\mathfrak{p}0}$; and "evaluation at $i\nu$." Thus if we define a map γ by $\gamma(D) = \sigma(D_{\mathfrak{h}})$, then $D\varphi_\nu = \gamma(D)(i\nu)\varphi_\nu$. The map γ maps $\mathcal{Q}$ into $S(\mathfrak{h}_{\mathfrak{p}0})$. But, since $\varphi_{s\nu} = \varphi_\nu$ for each $s \in W$, it is clear that $\gamma(D)(is_\nu) = \gamma(D)(i_\nu)$ for all $s \in W$. Thus $\gamma(D)$ is in the ring $I(\mathfrak{h}_{\mathfrak{p}0})$ of W-invariant elements of $S(\mathfrak{h}_{\mathfrak{p}0})$. In other words γ maps $\mathcal{Q}$ into $I(\mathfrak{h}_{\mathfrak{p}0})$. One now has the following theorem [*11a*, p. 428 ff.].

Theorem 2.14 (Harish-Chandra, Chevalley): The mapping γ is a homomorphism of $\mathcal{Q}$ onto $I(\mathfrak{h}_{\mathfrak{p}0})$, and the kernel of γ is $\mathcal{Q} \cap \mathcal{G}\mathfrak{k}_0$. Moreover the algebra $\mathcal{Q}/\mathcal{Q} \cap \mathcal{G}\mathfrak{k}_0$ is naturally isomorphic to $\mathcal{D}(G/K)$. Thus $\mathcal{D}(G/K) \cong I(\mathfrak{h}_{\mathfrak{p}0})$. In particular, $\mathcal{D}(G/K)$ is commutative.

We refer to Refs. [*10d*] and [*11a*, p. 429] for the proofs of this theorem.

Using this theorem, we can show that the construction we made above does give *all* the elementary spherical functions. Let γ^* be the map γ lifted to $\mathcal{Q}/\mathcal{Q} \cap \mathcal{G}\mathfrak{k}_0$, so that γ^* is an isomorphism of this algebra with $I(\mathfrak{h}_{\mathfrak{p}0})$. Now let φ be an elementary spherical function; then $D\varphi = \lambda_D\varphi$ and $D \to \lambda_D$ is a homomorphism of $\mathcal{Q}$ which is trivial on $\mathcal{Q} \cap \mathcal{G}\mathfrak{k}_0$. Let $a \in I(\mathfrak{h}_{\mathfrak{p}0})$, and define a homomorphism χ of $I(\mathfrak{h}_{\mathfrak{p}0})$ by $\chi(a) = \lambda_D$ where D is any element of $\mathcal{Q}$ such that $\gamma(D) = a$. Then χ is well defined. Now $S(\mathfrak{h}_{\mathfrak{p}0})$ is integral over $I(\mathfrak{h}_{\mathfrak{p}0})$, so we can extend χ to a homomorphism of $S(\mathfrak{h}_{\mathfrak{p}0})$; hence there exists $\mu \in \Lambda$ such that $\chi(a) = a(\mu)$ (evaluation at μ). Now consider the elementary spherical function $f_\mu = \int_K e^{(\mu-\rho)(H(xk))}\, dk$. Then by the matter above, $Df_\mu = \gamma(D)(\mu) f_\mu$ for $D \in \mathcal{Q}$ but $\gamma(D)(\mu) = \chi(\gamma(D)) = \lambda_D$. Thus $Df_\mu = \lambda_D f_\mu$. Hence the elementary spherical functions f_μ and φ have the same eigenvalues. It follows that $\varphi = f_\mu$. Thus the above construction exhausts all the elementary spherical functions. Moreover, the elementary functions φ_ν and $\varphi_{\nu'}$ are the same only if $s\nu = \nu'$ for some $s \in W$. Summarizing, we have

Theorem 2.15 (Harish-Chandra [*10d*]): For any $\nu \in \Lambda$, the function

$$\varphi_\nu(x) = \int_K \exp(i\nu - \rho)(H(xk))\, dk \tag{2.10}$$

is an elementary spherical function. Every elementary spherical function is of this form. Moreover, $\varphi_\nu = \varphi_{\nu'}$ if and only if $\nu = s\nu'$ for some $s \in W$.

This means that the elementary spherical functions are parametrized by the space Λ/W of orbits of W on Λ. This space is a natural dual object for $C_c(K\backslash G/K)$.

Obviously, given $\nu \in \Lambda$, the corresponding φ_ν defines a continuous homomorphism of $C_c(K\backslash G/K)$ as follows:

$$f \to \int_G f(x^{-1})\varphi_\nu(x)\, dx. \tag{2.11}$$

If we write $\hat{f}(\nu)$ for the right side, then the function $\hat{f}$ is the so-called Fourier-transform of f. Obviously, as a function on Λ, $\hat{f}$ is invariant under W.

3. Representation Theory and Elementary Spherical Functions

In this section, we shall summarize some facts about the connections between representations of G on Banach or Hilbert spaces and elementary spherical functions.

Definition 3.1: Let B be a Banach space. By a representation π of G on B, we mean a mapping π which sends each $x \in G$ to a bounded linear operator $\pi(x)$ on B such that (i) $\pi(xy) = \pi(x)\pi(y)$; (ii) $\pi(e) = I$; (iii) for each $b \in B$, the function $x \to \pi(x)b$ is continuous. If B is a Hilbert space and each $\pi(x)$ is a unitary operator, then π is said to be a unitary representation.

Definition 3.2: A representation π is said to be topologically irreducible if there is no nontrivial closed subspace of B stable under $\{\pi(x),\ x \in G\}$.

It is said to be topologically completely irreducible if the linear span of the set $\{\pi(x), x \in G\}$ is dense in the set of all bounded linear operators, in the strong operator topology.

Historically, unitary representations of G were studied before Banach representations. Though nowadays it would be slightly more economical to reverse this order for discussion, it will serve our eccentric purpose to preserve it.

Let π be an irreducible unitary representation of G on a Hilbert space H. Consider the restriction of π to K. This is a unitary representation of K on H. Since K is compact, this representation splits into a direct sum of irreducible representations of K (in the Hilbert space sense). Now suppose that the trivial representation $\mathbf{1}: K \to 1$ occurs in this reduction. This merely means that there is a unit vector v in H such that $\pi(k)\, v = v, k \in K$. Then the function $\varphi(x) = \langle v, \pi(x)v\rangle$ is easily seen to be a positive definite spherical function on G. This suggests that there is a connection between positive definite spherical functions on G and representations π with the above property.

Definition 3.3: Let π be an irreducible unitary representation of G on a Hilbert space H. We say that π is of class $\mathbf{1}$ with respect to K if the trivial representation $\mathbf{1}: K \to 1$ occurs in the reduction of the restriction of π to K.

Let π be an irreducible unitary representation of G on H, and suppose π is of class $\mathbf{1}$. Then there is a $v \in H$ such that $\pi(k)v = v, k \in K$. A remarkable fact in our situation is that such a v is essentially unique.

Proposition 3.4: Let π be an irreducible unitary representation of G, on H and let H_1 be the subspace of H defined by $H_1 = \{v \in H | \pi(k)v = v$ $k \in K\}$. Then $\dim H_1 \leqq 1$.

We will sketch a proof. Extend π to a representation of $C_c(G)$ in the usual way, i.e., $f \to \pi_f = \int_G f(x)\pi(x)\, dx$. Then a closed subspace of H is stable under $\{\pi(x), x \in G\}$ if and only if it is stable under $\{\pi_f, f \in C_c(G)\}$. Now, it is easy to see that if $f \in C_c(K\backslash G/K)$, then π_f maps H_1 into itself. Hence, since $C_c(K\backslash G/K)$ is commutative, the family $\{\pi_f, f \in C_c(K\backslash G/K\}$ is a commutative algebra of normal operators on H_1. Now suppose dim

$H_1 > 1$. Then one can decompose $H_1 = U + V$ where U, V are nontrivial subspaces stable under $\{\pi_f, f \in C_c(K\backslash G/K)\}$. But then if $u \in U$ is a nonzero vector, the subspace $\{\pi_f u, f \in C_c(G)\}$ is stable under each π_g, $g \in C_c(G)$ and is orthogonal to V, because $\langle \pi_f u, v\rangle = \int_G \langle \pi(x)u, v\rangle f(x)\, dx = \int_G \langle \pi(x)u, v\rangle f^{\#}(x)\, dx$ where $f^{\#}(x) = \int_K\int_K f(kxk')\, dk\, dk'$. (Note here that u, v are elements of H_1.) But this last expression is equal to $\langle \pi_{f_+}u, v\rangle$ which equals zero because $\pi_{f_+}u \in U$, $v \in V$. It follows that the closure of $\{\pi_f u, f \in C_c(G)\}$ is a nontrivial π-stable subspace of H. This is a contradiction.

This pretty result is due in this form to Gelfand and Naimark [6]. For the case of symmetric pairs (G, K) of compact type, this result was known to E. Cartan; it is of great importance to the contents of this section. The above proposition shows that if π above is of class **1**, then the space H_1 is 1-dimensional. Let $v \in H_1$ be a unit vector, and consider the function $\varphi_\pi(x) = \langle \pi(x)v, v\rangle$. It is obvious that φ_π is positive definite and spherical. Moreover, it does not depend on the choice of the unit vector v in H_1, since H_1 is one-dimensional. So φ_π is unambiguously defined by π.

Proposition 3.5: Let π be an irreducible unitary representation of G on H, of class **1**, and let φ_π be the function $\varphi_\pi(x) = \langle \pi(x)v, v\rangle$ where $v \in H_1$ is the essentially unique unit vector in H, fixed by $\pi(k)$, $k \in K$. Then φ_π is an elementary spherical function.

Indeed $\varphi(e) = 1$ is obvious. We have

$$\int_K \varphi_\pi(xky)\, dk = \int_K \langle \pi(xky)v, v\rangle\, dk$$

$$= \left\langle \int_K \pi(ky)\, v\, dk,\ \pi(x^{-1})v \right\rangle \tag{3.1}$$

but $\int_K \pi(ky)v\, dk$ is obviously in H_1 by the very definition of H_1. Hence it equals $C_y v$ where C_y is a complex number, because $H_1 = \mathbf{C}v$. So

$$\int_K \varphi_\pi(xky)\, dk = C_y \langle v, \pi(x^{-1})v\rangle$$

$$= C_y \langle \pi(x)v, v\rangle$$

$$= C_y \varphi_\pi(x). \tag{3.2}$$

Setting $x = e$ in this, we find $C_y = \varphi_\pi(y)$. Hence φ_π satisfies the functional equation and is elementary.

We may call φ_π the elementary spherical function determined by π.

Conversely, if we are given an elementary positive definite spherical function φ, we can construct an irreducible representation π of class **1** such that φ is the elementary spherical function determined by π. Indeed, given any continuous positive definite function φ on G such that $\varphi(e) = 1$, the classical construction of Gelfand-Raikov [*7*, *11a*, p. 413] shows that there exists a Hilbert space H, a unitary representation π of G on H, and a unit vector $v \in H$ such that $\varphi(x) = \langle \pi(x)v, v\rangle$. Let us briefly recall this construction. Let V_φ be the linear span of all left translates of φ; on V_φ one defines an inner product as follows: If $f = \sum_i \alpha_i(x_i\varphi)$ and $g = \sum_j \beta_j (y_j\varphi)$ where $x_i \in G$, $y_j \in G$ (here $x_i\varphi$ denotes the left translate of φ), then define $\langle f, g\rangle = \sum_{i,j} \alpha_i \bar{\beta}_j \varphi(x_i^{-1}x_j)$. Then G acts on V_φ by left translation, and the subspace $N = \{f, \langle f, f\rangle = 0\}$ is stable under this action. Thus this action induces a representation of G on the factor space V_φ/N, which preserves the inner product inherited by this space from V_φ. The completion of this pre-Hilbert space gives a Hilbert space H, on which this action of G gives a unitary representation π. If v is the vector in H that comes from $\varphi \in V_\varphi$, then unscrambling these definitions gives $\varphi(x) = \langle \pi(x)v, v\rangle$.

Now, if φ is elementary spherical, it is obvious that $\pi(k)v = v$, $k \in K$. Thus π is of class **1**. Moreover, π is irreducible. Indeed, let E_1 be the operator $\int_K \pi(k)\, dk$; then using the definition of the inner product in H, we find that $\langle E_1\pi(x)v, w\rangle = \varphi(x)\langle v, w\rangle$ for any $w \in H$. Hence $E_1\pi(x)v = \varphi(x)v$. Since H is spanned by $\pi(x)v$, it follows that $E_1 H = \mathbf{C}v$. Now let H_0 be sum of all the π-stable closed subspaces of H that are annihilated by E_1 and $H_0^\perp$ be its orthocomplement. Then $H_0^\perp$ is π-stable. If U is any closed π-stable subspace of $H_0^\perp$, then $E_1 U \neq \{0\}$. But since $E_1 H = \mathbf{C}v$, this means that $v \in E_1 U$. Since $E_1 U$ is π-stable, $\pi(x)v \in E_1 U$, and so $H \subset E_1 U$. Hence $U = H$. This shows that $H_0^\perp = H$, $H_0 = \{0\}$, and π is irreducible. Putting this together with Proposition 3.5, we have

Theorem 3.6 (Gelfand and Naimark [*6*]): Let π be an irreducible unitary representation of class **1** of G on H; then the function $\varphi_\pi(x) = \langle \pi(x)v, v\rangle$ (where v is such that $\pi(k)v = v$, $\|v\| = 1$) is an elementary positive definite spherical function. Conversely, given an elementary positive definite spherical function φ on G, there exists an irreducible unitary representation π of G on a Hilbert space H such that $\varphi = \varphi_\pi$. π is unique to within unitary equivalence.

In view of this connection between elementary positive definite spherical functions, one naturally asks the following question:

Which of the elements $\lambda \in \Lambda$ have the property that the corresponding elementary spherical function φ_λ is positive definite?

Suppose $\lambda \in \Lambda$ is such that φ_λ is positive definite; then $\varphi_\lambda(x) = \varphi_\lambda(x^{-1})$. Now the formula $\varphi_\lambda(x) = \int_K \exp(i\lambda - \rho)(H(xk))\, dk$ shows that $\overline{\varphi_\lambda(x)} = \varphi_{\bar{\lambda}}(x)$. (Here, we recall that $\Lambda = \Lambda_0 + i\Lambda_0$, so that $\lambda = \lambda_R + i\lambda_I$, λ_R, $\lambda_I \in \Lambda_0$, so $\bar{\lambda}$ has the natural meaning $\bar{\lambda} = \lambda_R - i\lambda_I$.) Thus if φ_λ is positive definite, we must have $\varphi_{-\bar{\lambda}}(x) = \varphi_\lambda(x^{-1})$. On the other hand, a result of Harish-Chandra [*10d*, p. 294] shows that for any $\lambda \in \Lambda$, $\varphi_\lambda(x) = \varphi_{-\lambda}(x^{-1})$. Hence we see that $\varphi_{-\bar{\lambda}}(x) = \varphi_{-\lambda}(x)$. Thus we must have $\lambda = s\bar{\lambda}$ for some $s \in W$.

A particular instance when this condition is satisfied is when $\lambda = \bar{\lambda}$, i.e., $\lambda \in \Lambda_0$. In this case, the elementary spherical functions were known to be positive definite since the publication of Ref. [*10a*].

This was done as follows: Let M be the centralizer of A in K, and let B be the subgroup MAN. Suppose $\lambda \in \Lambda$. Then defines a quasicharacter χ_λ (i.e., a not-necessarily-unitary homomorphism into **C**) of B, as follows: $\chi_\lambda\,(m \exp H\, n) = \exp i\lambda(H)$. Now let H_λ be the space of measurable functions on G satisfying

(i) $f(bx) = \chi_{\lambda - i\rho}(b) f(x)$ for almost all x,

(ii) $\|f\|^2 = \int_K |f(k)|^2\, dk < \infty$.

Then H_λ is a Hilbert space with this norm. The right action of G on H_λ gives a representation π_λ of G on H_λ. Actually π_λ is just the representation induced from B by means of χ_λ. π_λ is not unitary in general. However, if λ is real, then the character is unitary and π is also unitary. Harish-Chandra showed that π_λ is of class **1** in that case and its elementary spherical function is φ_λ [*10a*, p. 242]. *Hence φ_λ is positive definite if $\lambda \in \Lambda_0$.*

However, it was known for a long time that $\{\varphi_\lambda;\ \lambda \in \Lambda_0\}$ does not exhaust the set of positive definite elementary spherical functions. Indeed $\varphi_{-i\rho} \equiv 1$ and $-i\rho \notin \Lambda_0$. In fact for $G = SL(2, R)$, Bargmann constructed a new inner product on each $\{H_\lambda,\ 0 \leqq -i\lambda \leqq \frac{1}{2}\}$ and showed that with respect to this new inner product, π_λ was unitary. See also Takahashi [*20*] who considered the case $G = SO(p, 1)$.

Recently Kostant [*14b*] showed that the condition "$\lambda = s\bar{\lambda}$ for some $s \in W$," is sufficient for the existence of a new inner product on H_λ with respect to which the representation π_λ is unitary. Kostant's methods are heavily algebraic, relying on a careful study of the representations of $\mathcal{G}$ to which the π_λ give rise. Unlike in the case of $SL(2, R)$ and $SO(p, 1)$, the new inner product is only shown to exist abstractly, not exhibited explicitly

[*14b*]. More recently, Lipsman [*16*] has been able to show, among other things, how to construct this inner product explicitly for a large class of $\lambda \notin \Lambda_0$ which satisfy the condition $\lambda = s\bar{\lambda}$ for some $s \in W$.

A question which arises naturally from the above discussion is the following. What is the connection between elementary spherical functions (not necessarily positive definite) and representations of G on a Banach space? In particular, is it analogous to the above connection between *positive definite* elementary spherical functions and unitary representations of class **1**? The answer is, substantially, yes. We shall summarize some of the results.

Let π be an irreducible representation of G on a Banach space B. Let $\hat{K}$ equal the set of equivalence classes of irreducible representations of K.

Definition 3.7: A vector $v \in B$ is said to be K-finite under π if the span of $\{\pi(k)v, k \in K\}$ is finite dimensional.

Definition 3.8: Let $\delta \in \hat{K}$. A vector $v \in B$ is said to transform according to δ under π if (i) v is K-finite and (ii) the reduction of the restriction of $\pi(K)$ to the finite dimensional space spanned by $\{\pi(k)v, k \in K\}$ contains irreducible components all of class δ. We denote by B_δ the subspace of B consisting of vectors that transform under π according to δ.

Because of the Peter-Weyl theorem, one has

Proposition 3.9: Let π be a representation of G on a Banach space B. Let $B_\omega = \sum_{\delta \in \hat{K}} B_\delta$. Then B_ω is dense in B.

Definition 3.10: We say that π is of class δ if $B_\delta \neq \{0\}$.

Now suppose that π is of class **1**. Then there is a $v \in B$ such that $\pi(k)v = v$, $k \in K$. Evidently, if we want to associate to π an elementary spherical function in a canonical way, we should like to have some uniqueness for this v. In other words, we ask: Under what conditions is B_1 one dimensional? As we saw in Proposition 3.4, if B were a Hilbert space and π a unitary representation, irreducibility of π would be enough. However, if π is a Banach representation, this may not be enough. But, if π satisfies an additional condition, it turns out that B_1 will be one dimensional and much more will be true. We shall formulate this condition and then state a theorem.

Definition 3.11: Given π as above, a subspace $U \subset B$ is said to be differentiable under π if for each $X \in \mathfrak{g}_0$ and each $u \in U$, the limit $\lim_{t \to 0} (\pi(\exp tX) u - u)/t$ exists and belongs to U.

It follows that the sum of two subspaces differentiable under π is again differentiable under π. Hence there is a unique largest subspace of B differentiable under π. We call it B_∞. Then for $X \in \mathfrak{g}_0$, if we define the endomorphism $\pi_\infty(X)$ of B_∞ by $\pi_\infty(X)u = \lim_{t \to 0} (\pi(\exp tX)u - u)/t$, the map $\pi_\infty: X \to \pi_\infty(X)$ is a representation of $\mathfrak{g}_0$, and hence, of $\mathscr{G}$ on B_∞.

Definition 3.12: We say π is *quasisimple* if for each $z \in \mathscr{Z}$ (the center of $\mathscr{G}$), the endomorphism $\pi_\infty(z)$ is a scalar multiple of the identity. Thus $\pi_\infty(z) = \chi(z)I$, $\chi(z) \in \mathbf{C}$. The map $\chi: z \to \chi(z)$ is clearly a homomorphism of $\mathscr{Z}$ into $\mathbf{C}$. It is called the *infinitesimal character* of π.

We remark that if π is irreducible and unitary, then it is known that π is quasisimple. See, e.g., Segal [*18a*].

Theorem 3.13 (Harish-Chandra [*10b*]): Let π be a topologically irreducible quasisimple representation of G on a Banach space B. Then

(i) For each $\delta \in \hat{K}$, $\dim B_\delta < \infty$.

(ii) There exists an integer N such that $\dim B_\delta \leqq N(d(\delta)^2)$, $\delta \in \hat{K}$.

(iii) Let $\hat{K}_F$ be the subset of $\hat{K}$ consisting of those $\delta \in \hat{K}$ which occur in the reduction of the restriction to K of some finite-dimensional representation σ of G. Then for $\delta \in \hat{K}_F$, $\dim B_\delta \leqq d(\delta)^2$.

The proof of this theorem is very deep. It uses a formidable amount of machinery, including the theory of analytic vectors and a detailed knowledge of the structure of the enveloping algebra $\mathscr{G}$. We shall not go into the proof.

We note, however, that the class of the trivial representation of K is certainly contained in $\hat{K}_F$. Hence from (iii) of the theorem, we get

Corollary 3.14: If π is irreducible and quasisimple, then $\dim B_1 \leqq 1$.

In particular, if π is of class **1**, then dim $B_{\mathbf{1}} = 1$.

Now suppose π is an irreducible quasisimple representation of G, of class **1**. Let $E_{\mathbf{1}}$ be the endomorphism $\int_K \pi(k)\, dk$. Then E_1 is a projection whose range is $B_{\mathbf{1}}$. For any $x \in G$, the endomorphism $E_{\mathbf{1}}\, \pi(x)\, E_{\mathbf{1}}$ maps $B_{\mathbf{1}}$ into itself. So if v is a nonzero vector element of $B_{\mathbf{1}}$, $E_{\mathbf{1}}\, \pi(x)\, E_{\mathbf{1}} v$ is a multiple of v, since dim $B_{\mathbf{1}} = 1$. We define the function $\varphi_\pi(x)$ by

$$E_{\mathbf{1}}\, \pi(x)\, E_{\mathbf{1}} v = \varphi_\pi(x) v.$$

Clearly $\varphi_\pi(x)$ is the trace of the restriction to $B_{\mathbf{1}}$ of the endomorphism $E_{\mathbf{1}}\, \pi(x)\, E_{\mathbf{1}}$, so it does not depend on the choice of v.

Proposition 3.15: $\varphi_\pi(x)$ is an elementary spherical function. Indeed $\varphi_\pi(e) = 1$ is obvious. Moreover,

$$\begin{aligned}
\left(\int_K \varphi_\pi(xky) dk \right) v &= \int_K \varphi_\pi(xky) v \, dk \\
&= \int_K E_{\mathbf{1}}\, \pi(xky)\, E_{\mathbf{1}}\, v \, dk \\
&= E_{\mathbf{1}}\, \pi(x) \left(\int_K \pi(k)\, dk \right) \pi(y)\, E_{\mathbf{1}}\, v \\
&= E_{\mathbf{1}}\, \pi(x)\, E_{\mathbf{1}}\, \pi(y)\, E_{\mathbf{1}} v \\
&= (E_{\mathbf{1}} \pi(x))\, \varphi_\pi(y)\, v \\
&= \varphi_\pi(y)\, E_{\mathbf{1}} \pi(x)\, v = \varphi_\pi(y)\, \varphi_\pi(x)\, v
\end{aligned} \tag{3.3}$$

which verifies the functional equation.

We may say that π determines the elementary spherical function φ_π. The question that arises is: To what extent does φ_π determine π? In the case of unitary representations, the answer is as satisfactory as one could wish. In the present case, the known result is weaker. To state it, we need a definition.

Definition 3.16: Let π_i be representations of A on Banach spaces B_i, $i = 1, 2$. We say that π_1 is infinitesimally equivalent to π_2 if there exist subspaces W_i of $B_{i\infty}$, $i = 1, 2$, and a one-to-one linear map A of W_1 onto W_2 such that

(i) W_i is dense in B_i, $i = 1, 2$.

(ii) W_i is stable under $\pi_{i\infty}(a)$, $a \in \mathscr{G}$; $i = 1, 2$.

(iii) For each $a \in \mathscr{G}$, we have $A \circ \pi_{1\infty}(a) = \pi_{2\infty}(a) \circ A$ as linear maps from W_1 to W_2.

It is known that if the π_i are unitary, then infinitesimal equivalence implies unitary equivalence.

The theorem is as follows:

Theorem 3.17 (Harish-Chandra [*10b*]): Let π_1, π_2 be topologically irreducible quasisimple representations of G on Banach spaces B_1, B_2, respectively. Suppose both are of class **1**. Then they are infinitesimally equivalent if and only if $\varphi_{\pi_1} \equiv \varphi_{\pi_2}$.

Actually somewhat more is true. As we saw above, if π is topologically irreducible and quasisimple, then dim $B_\delta < \infty$. This means that if E_δ is the projection onto B_δ defined by $E_\delta = d(\delta)\int_K \pi(k)\bar{\chi}_\delta(k)\, dk$ where χ_δ is the character and $d(\delta)$ is the degree of δ, then the endomorphism $E_\delta \pi(x)\, E_\delta$ leaves B_δ invariant. Hence, since dim $B_\delta < \infty$, the restriction of this endomorphism to B_δ has a trace which we denote by $\operatorname{tr}(E_\delta\, \pi(x)\, E_\delta)$. Now let $\varphi_\delta(x) = \operatorname{tr}(E_\delta \pi(x)\, E_\delta)$. Then we say that φ_δ is the spherical function of type δ associated to π. The result above is a special case of a more general theorem which states that if $\varphi_{\pi_1}^{\delta_1} \equiv \varphi_{\pi_2}^{\delta_2}$ for some classes δ_1, δ_2 occurring in π_1, π_2 respectively, then $\delta_1 = \delta_2$ and π_1 is infinitesimally equivalent to π_2, and conversely. This theorem is proved in [*10b*, Theorem 1].

Finally there remains the question of constructing, given an elementary spherical function φ_λ, $\lambda \in \Lambda$, an irreducible quasisimple representation whose elementary spherical function is φ_λ. The representation π_λ that we described above does indeed have this property thanks to the result of Kostant which proves that it is irreducible [*14a*]. See also Wallach [*23a*]. The fact that π_λ contains φ_λ as one of its elementary spherical functions was known earlier [*10b*].

To summarize, one may say that elementary spherical functions correspond to irreducible quasisimple Banach representations of G, while the positive definite elementary spherical functions correspond to irreducible unitary representations of G. Moreover, φ_λ is positive definite if and only if $\lambda = s\bar{\lambda}$ for some $s \in W$.

A deeper study of the functions φ_λ^δ was made by Rader [*17*] in a recent thesis.

Note added in proof: I gather that Rader has shown that the conclusion of Theorem 3.13(iii) holds for all $\delta \in \hat{K}$, not merely for $\delta \in \hat{K}_F$.

Part II: Harmonic Analysis of Spherical Functions

1. Introduction

Let $f \in C_c(K \backslash G/K)$, $\lambda \in \Lambda$, and let φ_λ be the elementary spherical function corresponding to λ. Then the discussion of the last chapter suggests that we define the Fourier transform $\hat{f}$ of f to be the function on Λ;

$$\hat{f}(\lambda) = \int_G f(x)\, \varphi_\lambda(x^{-1})\, dx. \tag{1.1}$$

Our object in Part II is to study the map $f \to \hat{f}$ not only on the function space $C_c(K \backslash G/K)$ but also on larger spaces of spherical functions.

By the very definition of φ_λ by means of the integral formula

$$\varphi_\lambda(x) = \int_K \exp(i\lambda - \rho)\,(H(xk))\, dk, \qquad \lambda \in \Lambda \tag{1.2}$$

one sees that for fixed x, $\varphi_\lambda(x)$ is an entire function of $\lambda \in \Lambda$, if Λ is regarded as a complex l-space with the obvious analytic structure. (Here $l = \dim_R \mathfrak{h}_{\mathfrak{p}_0}$.) It follows that for $f \in C_c(K \backslash G/K)$, $\hat{f}$ is an entire holomorphic function of $\lambda \in \Lambda$.

Now the space $C_c(K \backslash G/K)$ can be completed in an enormous variety of norms to yield various Banach spaces of spherical functions on G, and it is pertinent to ask if the mapping $f \to \hat{f}$ can be extended continuously to any of these completions. Of course, one cannot expect that when so extended, the function $\hat{f}$ will be defined for all $\lambda \in \Lambda$. Indeed $\hat{f}$ will be defined only for those λ for which the functional $f \to \hat{f}(\lambda)$ admits a continuous extension to the completion. This will in turn depend upon the properties of the corresponding φ_λ.

The most classical completions of $C_c(K \backslash G/K)$ in the above manner are the spaces $L_p(K \backslash G/K)$. Here I shall discuss only the spaces $L_1(K \backslash G/K)$ and $L_2(K \backslash G/K)$. For $L_1(K \backslash G/K)$, only the beginnings of a theory of the map $f \to \hat{f}$ exist, and I shall summarize very quickly in Section 2 what is known at the present time. In Sections 3–6 I shall summarize the theory of the map $f \to \hat{f}$ for $L_2(K \backslash G/K)$, for which the theory is now reasonably complete. In particular I shall give an account of Harish-Chandra's Plancherel formula for $L_2(K \backslash G/K)$. The route that I have chosen leads to this Plancherel formula via a theorem that characterizes the Fourier transform $\hat{f}(\lambda)$ of a function $f \in C_c^\infty(K \backslash G/K)$. This theorem is analogous to the classical Paley–

Wiener theorem. Finally in Section 7 I shall return to the space $L_1(K\backslash G/K)$ and discuss some unanswered questions.

2. The Algebra L_1 $(K\backslash G/K)$

The fact that $L_1(K\backslash G/K)$ is an algebra needs no comment. Moreover, by what we saw in Part I, it is a commutative Banach algebra. It follows that Gelfand's theory of commutative Banach algebras applies to it. The first question is: What is the maximal ideal space of $L_1(K\backslash G/K)$?

Let $\hat{\varphi}$ be a continuous homomorphism of $L_1(K\backslash G/K)$ onto $\mathbf{C}$. Then in particular $f \to \hat{\varphi}(f)$ is a continuous linear functional on $L_1(K\backslash G/K)$. We extend this to a continuous linear functional $\hat{\varphi}'$ on $L_1(G)$ as follows. $\hat{\varphi}'(f) = \hat{\varphi}(f^{\#})$ where $f^{\#}(x) = \int_K\int_K f(kxk')\,dk\,dk'$. Then it follows that there exists a $\hat{\varphi}' \in L_\infty(G)$ such that $\hat{\varphi}'(f) = \int_G f(x)\hat{\varphi}'(x^{-1})\,dx$ for all $f\in L_1(G)$. Let $\varphi(x) = \int_K\int_K \varphi'(k_1xk_2)\,dk_1$. Then $\varphi \in L_\infty(K\backslash G/K)$. Moreover, if $f\in L_1(K\backslash G/K)$, we have $\hat{\varphi}(f) = \hat{\varphi}'(f) = \int_G f(x)\hat{\varphi}'(x^{-1})\,dx = \int_G f(x)\varphi(x^{-1})\,dx$. Thus, given a continuous homomorphism $\hat{\varphi}$ of $L_1(K\backslash G/K)$ to $\mathbf{C}$, there is a bounded spherical function φ such that $\hat{\varphi}(f) = \int_G f(x)\varphi(x^{-1})dx$. Now the homomorphism $\hat{\varphi}$ is continuous on $L_1(K\backslash G/K)$. If we restrict it to $C_c(K\backslash G/K)$, then it is obvious that it gives us a continuous homomorphism of $C_c(K\backslash G/K)$ (in its topology as defined in Part I). Hence by the results above we can conclude that φ is an elementary spherical function. Conversely a bounded elementary spherical function evidently determines a homomorphism of $C_c(K\backslash G/K)$ which is continuous in the L_1-norm. It follows that it defines a continuous homomorphism of $L_1(K\backslash G/K)$ onto $\mathbf{C}$. We have

Proposition 2.1: The maximal ideal space $\mathcal{M}$ of $L_1(K\backslash G/K)$ is in one-to-one correspondence with the set $\mathcal{B}$ of all bounded elementary spherical functions on G, as follows: If φ is such a function, then the mapping $f \to \hat{\varphi}(f) = \int_G f(x^{-1})\varphi(x)\,dx$ is a continuous homomorphism of $L_1(K\backslash G/K)$ whose kernel M is the maximal ideal corresponding to φ. Conversely a given maximal ideal M in $L_1(K\backslash G/K)$ determines a homomorphism $\hat{\varphi}_M$ in the usual way, and there is a unique bounded elementary spherical function φ_M such that $\varphi_M(f) = \int_G f(x^{-1})\varphi_M(x)\,dx$, $f\in L_1(K\backslash G/K)$. The maps $\varphi \to M_\varphi$ and $M \to \varphi_M$ are inverse to each other and define a one-to-one correspondence of $\mathcal{B}$ onto $\mathcal{M}$.

In view of the fact that the set of all elementary spherical functions is parametrized by Λ/W, it is natural to ask: For which $\lambda \in \Lambda$ is it true that φ_λ is *bounded*? If we could identify this subset of Λ, we would then get a natural parametrization of the maximal ideal space $\mathscr{M}$ by a subset of the complex l-space Λ.

Let $\lambda \in \Lambda$; then since $\Lambda = \Lambda_0 + i\Lambda_0$, we can write $\lambda = \xi + i\eta$, ξ, $\eta \in \Lambda_0$. Thus ξ, η are real valued on $\mathfrak{h}_{\mathfrak{p}_0}$. Moreover, since ρ is real valued on $\mathfrak{h}_{\mathfrak{p}_0}$, it follows from the formula (1.2) that $\varphi_{i\eta}(x) > 0$ and

$$|\varphi_{\xi+i\eta}(x)| \leqq \varphi_{i\eta}(x), \qquad x \in G. \tag{2.1}$$

Now, if $\eta = -\rho$, Eq. (1.2) shows that $\varphi_{-i\rho} \equiv 1$. Hence $\varphi_{\xi+i\rho}$ is bounded for any $\xi \in \Lambda_0$. Since $\varphi_\lambda = \varphi_{s\lambda}$ for any $s \in W$, it follows that $\varphi_{\xi+is\rho}$ is bounded for any $\xi \in \Lambda_0$ and any $s \in W$. Now let C_ρ be the closed convex hull of the points $\{s\rho, s \in W\}$ in Λ_0. Then, because the function $\lambda \to \varphi_\lambda(x)$ is entire holomorphic for each fixed x, a simple argument involving the maximum principal for holomorphic functions shows that the function $\dot{\varphi}_{\xi+i\eta}$ is bounded for any $\xi \in \Lambda_0$ and $\eta \in C_\rho$. This argument, due to E. Stein, shows that for each $\lambda \in \Lambda_0 + iC_\rho$, φ_λ is bounded. Recently, Helgason and Johnson showed that the converse is also true.

Proposition 2.2 (Helgason and Johnson [*12*]): Let C_ρ be the closed convex hull of the points $\{s\rho, s \in W\}$ in Λ_0. Then φ_λ is bounded if and only if $\lambda \in \Lambda_0 + iC_\rho$. Thus the maximal ideal space $\mathscr{M}$ of $L_1(K\backslash G/K)$ is parametrized by the orbits of W on this domain $\Lambda_0 + iC_\rho$. Moreover, for any λ in the set $\Lambda_0 + iC_\rho$, $|\varphi_\lambda| \leqq 1$.

We shall denote by Λ_B the set $\Lambda_0 + iC_\rho$. Thus, φ_λ is bounded if and only if $\lambda \in \Lambda_B$, and Λ_B/W parametrizes the space $\mathscr{M}$.

The Fourier transform $\hat{f}$ can now be defined for each $f \in L_1(K\backslash G/K)$ as a W-invariant continuous function on Λ_B.

An important fact about the algebra $L_1(K\backslash G/K)$ is contained in the following:

Proposition 2.3: The algebra $L_1(K\backslash G/K)$ is semisimple, i.e., the Fourier transform $f \to \hat{f}$ is injective on $L_1(K\backslash G/K)$.

The proof is sketched in Ref. [*11a*, p. 453]. The above proposition shows that if $\hat{f}(\lambda) = 0$ for all $\lambda \in \Lambda_B$ and if $f \in L_1(K\backslash G/K)$, then $f = 0$. However, somewhat more is true. The chief reason is that for $f \in$

$L_1(K\backslash G/K)$, the function $\hat{f}(\lambda)$ is not merely continuous on Λ_B but holomorphic in the interior of Λ_B. Thus

Proposition 2.4: For $f \in L_1(K\backslash G/K)$, $\hat{f}$ is holomorphic in the interior of Λ_B and continuous on Λ_B.

Indeed, we can identify Λ_B with a subdomain of $\mathbf{C}^l$ in the obvious way. Let $\lambda_1, \ldots, \lambda_l$ be the coordinates of λ, and let D be the polydisk $\{(\mu_1, \ldots, \mu_n) | \mu_i \in \mathbf{C}, |\mu_i - \lambda_i| \geqq r_i\}$ with r_i so small that D is contained in the interior of Λ_B. For each $x \in G$, $\mu \to \varphi_\mu(x)$ is an entire function of μ. Hence one has

$$\varphi_\lambda(x) = \frac{1}{(2\pi i)^l} \iint \cdots \int \frac{\varphi_\mu(x)}{\prod_{i=1}^{l} (\mu_i - \lambda_i)} \, d\mu_1 \ldots d\mu_l$$

by the Cauchy integral formula, where the integrals are a complex contour integral on the sets $\{\mu_i, |\mu_i - \lambda_i| = r_i\}$. If we multiply this by $f(x)$, integrate over G, and use Fubini, not forgetting that D is in Λ_B and so all the functions φ_μ have a uniform bound, we arrive at a Cauchy integral formula for $\hat{f}$, which shows that $\hat{f}$ is holomorphic at λ.

Thus the Fourier transform represents $L_1(K\backslash G/K)$ as an algebra of W-invariant holomorphic functions on the interior of Λ_B that have continuous extensions to Λ_B. We note in particular that there are no functions in $L_1(K\backslash G/K)$ whose Fourier transforms have compact support. Thus $L_1(K\backslash G/K)$ cannot be a completely regular Banach algebra.

It is therefore to be expected that the ideal theory of this algebra is likely to be deep. Very little is known about $L_1(K\backslash G/K)$ and its Fourier transforms. We shall come back to this in Section 7. Right now, we notice the following strengthening of Proposition 2.3.

Proposition 2.5: Suppose $f \in L_1(K\backslash G/K)$ and $\hat{f}(\lambda) = 0$ for $\lambda \in \Lambda_0$; then $f = 0$.

Indeed Λ_0 is in the interior of Λ_B. So since $\hat{f}$ is holomorphic there, it follows that $\hat{f} = 0$ on Λ_B. Hence $f = 0$ by Proposition 2.3. (Note that Λ_B is connected!)

The above more or less exhausts the information that is available at the present time about the algebra $L_1(K\backslash G/K)$ in general. In the case of

$G = SL(2, R)$, Ehrenpreis and Mautner [*2*] have obtained more far-reaching results. We shall formulate analogues of some of these in Section 7.

3. The Abstract Plancherel Formula for Spherical Functions

We want to consider the following question: Can the map $f \to \hat{f}$ be extended from $C_c(K\backslash G/K)$ to $L_2(K\backslash G/K)$? If so, what is the appropriate dual object on which $\hat{f}$ will live? Moreover, can one get an analogue of the Plancherel formulas (1.3), (1.6), etc., in our situation?

In the classical situations, the functions $e^{i\lambda\cdot x}$, etc., are positive definite and the dual object appropriate for $L_2(R)$, etc., is the set of all these positive definite functions. We shall soon see that an analogous result holds in the case of $L_2(K\backslash G/K)$.

The most economical way for our purposes seems to be to get this result as a consequence of the abstract Plancherel theorem for G.

Let $\hat{G}$ be the set of equivalence classes of irreducible unitary representations of G. For any class $\alpha \in \hat{G}$ and a representation $\pi \in \alpha$, it is known that the operator $\pi(f) = \int_G \pi(x) f(x)\, dx$ is of trace class for each $f \in C_c^\infty(G)$. Moreover, tr $\pi(f)$ depends only on the class α and not on the choice of $\pi \in \alpha$, and if we put $T_\alpha(f) = \text{tr}\ \pi(f)$, the map $T_\alpha : f \to T_\alpha(f)$ is a distribution on $C_c^\infty(G)$. These results are proved in Harish-Chandra [*10c*].

According to the abstract Plancherel theorem for G, as proved by Segal [*18b*], there exists a canonical structure of a measure space on $\hat{G}$ and a unique measure μ on $\hat{G}$ such that for each $f \in C_c^\infty(G)$ the function $\alpha \to T_\alpha(f)$ is measurable, and

$$\int_{\hat{G}} |f(x)|^2\, dx = \int_{\hat{G}} T_\alpha(f * f^*)\, d\mu(x) \tag{3.1}$$

where $f^*(x) = \bar{f}(x^{-1})$.

We shall now complete $T_\alpha(f * f^*)$ for an $f \in C_c^\infty(K\backslash G/K)$. Let π be a representation in the class α, and suppose H_π is the Hilbert space on which π acts. Since π is unitary and irreducible, π is quasisimple, cf. Part I, Section 3. Hence $H_\pi = \sum_{\delta \in \hat{K}} H_\pi(\delta)$ where (i) $H_\pi(\delta)$ is finite dimensional for each $\delta \in \hat{K}$, (ii) $H_\pi(\delta)$ is the subspace of H_π consisting of vectors that transform under π according to δ and the sum is an orthogonal direct sum. Moreover, we know that dim $H_\pi(\mathbf{1}) \leqq 1$, cr. Part I, Section 3. Let v_0, $v_1, \ldots$ be an orthonormal basis of H_π that is consistent with the above decomposition.

Proposition 3.1: Suppose π is not of class 1. Then $T_\alpha(f) = \text{tr}\ \pi(f) = 0$ for any $f \in C_c^\infty(K\backslash G/K)$.

Indeed, we have tr $\pi(f) = \sum_{i=0}^{\infty} \langle \pi(f)v_i, v_i \rangle$. Let E_δ be the operator $d(\delta) \int_K \bar{\chi}_\delta(k)\, \pi(k)\, dk$ for any $\delta \in \hat{K}$. Then E_δ is a self-adjoint projection whose range is $H_\pi(\delta)$. Since **1** does not occur in π, we have that $H_\pi(\mathbf{1}) = \{0\}$ so that the projection $E_{\mathbf{1}} = \int_K \pi(k)\, dk$ must equal zero. However, we have $\pi(f) = \int_G f(x)\, \pi(x)\, dx = \int_G f(x) \int_K \pi(xk)\, dx$, because $f \in C^\infty(K\backslash G/K)$. This last expression is equal to $\pi(f)E_{\mathbf{1}}$ and so it equals zero. Thus $\pi(f) = 0$, so $\langle \pi(f)v_i, v_i \rangle = 0$ all i. Hence $T_\alpha(f) = 0$.

We see thus that in the integral $\int_{\hat{G}} T_\alpha(f * f^*)\, d\mu(\alpha)$, the integrand T_α $(f^* f^*)$ is zero for all α that are not of class **1**, provided $f \in C_c^\infty(K\backslash G/K)$. Let $\hat{G}_1$ be the subset of $\hat{G}$ consisting of those $\alpha \in \hat{G}$ which are of class **1**. Then $\hat{G}_1$ is known to be measurable, and we obtain

$$\int_G |f(x)|^2\, dx = \int_{\hat{G}\alpha} T_\alpha(f * f^*)\, d\mu(\alpha). \tag{3.2}$$

The set $\hat{G}_1$ of equivalence classes of irreducible unitary representations of G is, as we have seen in Part I, in one-to-one correspondence with the set of elementary positive definite spherical functions. This latter set can be identified with the space Λ_P/W where Λ_P is the subset of Λ consisting of λ such that φ_λ is positive definite. Choose the following identification of Λ_P/W with $\hat{G}_1$. For any $\lambda \in \Lambda_P$, let λ^* denote its image under the projection $\Lambda_P \leftrightarrow \Lambda_P/W$, and let $\alpha(\lambda^*)$ denote the unique class in $\hat{G}_1$ whose elementary positive definite spherical function is $\bar{\varphi}_\lambda$. (Note that $\bar{\varphi}_\lambda$ is positive definite and depends only on λ^*.) Then $\lambda^* \to \alpha(\lambda^*)$ is an identification of Λ_P/W with $\hat{G}_1$.

The results of Ref. [*10g*] show that this identification is natural, i.e., the measure space structure that Λ_P/W acquires from $\hat{G}_1$ via this identification is the same as the measure space structure of Λ_P/W as the quotient under W of the measurable subset Λ_P of Λ. Thus the measure μ induces a measure on Λ_P/W which we continue to call μ, and we may write (3.2) as

$$\int_G |f(x)|^2\, dx = \int_{\Lambda_P/W} T_{\alpha(\lambda^*)}\, (f * f^*)\, d\mu(\lambda^*). \tag{3.3}$$

We will now compute $T_{\alpha(\lambda^*)}\,(f * f^*)$ in terms of λ^*. Let π be a representation of class $\alpha(\lambda^*)$. Then $H_\pi = \sum_{\delta \in K} H_\pi(\delta)$ and $\dim H_\pi(\mathbf{1}) = 1$. Let v_0, $v_1, \ldots$ be an orthonormal basis compatible with the above direct sum, and let $v_0 \in H_\pi(\mathbf{1})$, so that $H_\pi(\mathbf{1}) = \mathbf{C}v_0$. Then

$$
\begin{aligned}
T_{\alpha(\lambda^*)}(f*f^*) &= \operatorname{tr} \pi(f)\, \pi(f^*) \\
&= \sum_{i=0}^{\infty} \langle \pi(f)v_i, \pi(f)v\rangle_i \\
&= \sum_{i=0}^{\infty} |\pi(f)v_i|^2. \qquad (3.4)
\end{aligned}
$$

Now fix $i > 0$; then $v_i \in H_\pi(\delta)$ for some $\delta \neq \mathbf{1}$. Thus if E_δ and $E_{\mathbf{1}}$ are projections onto $H_\pi(\delta)$ and $H_\pi(\mathbf{1})$, it is obvious that $E_\delta E_{\mathbf{1}} = E_{\mathbf{1}} E_\delta = 0$. However, since $f \in C_c^\infty(K\backslash G/K)$, we have $\pi(f) = \pi(f)E_1$ as we saw in Proposition 3.1 above, and since $v_i \in H_\pi(\delta)$, we have $v_i = E_\delta v_i$. Thus $\pi(f)v_i = \pi(f)E_{\mathbf{1}}E_\delta v_i = 0$, because $E_{\mathbf{1}}E_\delta = 0$. Thus $|\pi(f)v_i|^2 = 0$ for all $i > 0$. Hence $T_{\alpha(\lambda^*)}(f*f^*) = |\pi(f)v_i|^2$. Now consider $\pi(f)v_0$. Since $f \in C_c^\infty(K\backslash G/K)$, we have $E_{\mathbf{1}}\pi(f) = \pi(f)$. So $E_{\mathbf{1}}\pi(f)v_0 = \pi(f)v_0$. Thus $\pi(f)v_0 \in H_\pi(\mathbf{1})$. Hence $\pi(f)v_0 = cv_0$, $c \in \mathbf{C}$. To see what c must be, we have

$$
\begin{aligned}
c &= c\langle v_0, v_0\rangle \\
&= \langle cv_0, v_0\rangle \\
&= \langle \pi(f)v_0, v_0\rangle \\
&= \int_G f(x)\,\langle \pi(x)v_0, v_0\rangle\, dx \qquad (3.5)
\end{aligned}
$$

Now $\langle \pi(x)v_0, v_0\rangle$ is the unique positive definite elementary spherical function associated to π. Since $\pi \in \alpha(\lambda^*)$, it follows from the definition of $\alpha(\lambda^*)$ that this spherical function is $\bar{\varphi}_\lambda(x)$. Thus

$$
c = \int_G f(x)\, \overline{\varphi_\lambda(x)}\, dx = \hat{f}(\lambda) \qquad (3.6)
$$

where we used $\overline{\varphi_\lambda(x)} = \varphi_\lambda(x^{-1})$ since φ_λ is positive definite. Thus $\pi(f)v_0 = \hat{f}(\lambda)v_0$, and so $|\pi(f)v_0|^2 = |\hat{f}(\lambda)|^2$. Notice that $\hat{f}(\lambda)$, being W-invariant, depends only on λ^*. Thus

$$
\int_G |f(x)|^2\, dx = \int_{\Lambda_P/W} |\hat{f}(\lambda^*)|^2\, d\mu(\lambda^*) \qquad (3.7)
$$

or, if we pull back the measure μ to Λ_P (and still call it μ!),we get

$$
\int_G |f(x)|^2\, dx = \frac{1}{w}\int_{\Lambda_P} |\hat{f}(\lambda)|^2\, d\mu(\lambda) \qquad (3.8)
$$

where w is an order of W, and $f \in C_c^\infty(K\backslash G/K)$.

This is the abstract Plancherel formula for spherical functions.

It shows that the map $f \to \hat{f}$ sends a function $f \in C_c^\infty(K\backslash G/K)$ to a function $\hat{f} \in L_2(\Lambda_P, (1/w)\, d\mu)$ and the mapping preserves the L_2-norm. Thus one can extend it in the usual way to an isometry of $L_2(K\backslash G/K)$ with some closed subspace of $L_2(\Lambda_P, (1/w)\, d\mu)$.

A number of questions arise: First, the above shows only that the measure μ exists and is unique. There remains the problem of computing this measure explicitly in terms of the structure of F. Second, the range of the map $f \to \hat{f}$ when extended to $L_2(K\backslash G/K)$ is not specified above. One should determine if it is all of $L_2(\Lambda_P, (1/w)\, d\mu)$. Lastly, one should, in analogy with classical situations, expect an inversion formula

$$f(x) = \frac{1}{w} \int_{\Lambda_P} \hat{f}(\lambda)\varphi_\lambda(x)\, d\mu(\lambda) \tag{3.9}$$

to hold in a suitable sense. For example, if f is in L_2 it might hold in the mean, while if f is in $C_c^\infty(K\backslash G/K)$ it should hold literally.

That all these questions have satisfactory answers is the content of Harish-Chandra's work in this subject. Indeed, Harish-Chandra was able to infer, by analogy with the theory of eigenfunction expansions, what the measure μ might be. We shall deal with this analogy in Section 4. Moreover, having inferred this, Harish-Chandra was able to establish the inversion formula (3.9), from which the Parseval relation follows. Also, as we shall see below, these considerations will automatically lead to the fact that $f \to \hat{f}$ is onto $L_2(\Lambda_P, (1/w)\, d\mu)$. We shall describe these matters in Sections 5 and 6.

Let me remark that it is possible to obtain the abstract Plancherel formula for $L_2(K\backslash G/K)$ without using the abstract Plancherel formula for G. For example, one may take the approach of [*10g*] or [*11f*]. However, in order to see quickly what the real problems are, it seemed to me more succinct to take the above approach.

4. The c-Function of Harish-Chandra

In order to understand the motivation behind Harish-Chandra's definition of the c-function alluded to above, let us consider first the special case $G = SL(2, R)$, $K = SO(2)$.

In this special case G/K is just the hyperbolic plane of Poincaré–Lobachevsky. Indeed, G acts transitively on the upper half-plane $H = \{z \in \mathbf{C};\ \operatorname{Im} z > 0\}$ and the isotropy subgroup of the point $i \in H$ is isomorphic to

$SO(2)$. Identifying H with G/K, one can check that the only Riemannian metric on H which is invariant under G is, up to constant multiples, the hyperbolic metric of Poincaré. A function f on G such that $f(xk) = f(x)$, $x \in G$, $k \in K$ may be viewed as a function on H. Moreover, if f is spherical on G, then it may be viewed as a function on H that is constant on K-orbits in H. But the orbits of K in H are spheres (in the hyperbolic metric) with center i. Hence f is spherical if and only if, when viewed as a function on H, f is constant on each sphere centered at i. (This incidentally is the original reason for the terminology "spherical function.") This means that f is a radial function on H, i.e., the value of f at a point $z \in H$ depends only on the hyperbolic distance of z from i. Thus we can view f as a function on the half line $[0, \infty)$ if we wish.

In terms of the structure of G, here is what we have: Given $x \in G$, we can write $x = kp$, with k orthogonal, p positive definite symmetric. Hence $p = k_2^{-1} d\, k_2$ with k_2 orthogonal and d diagonal. Hence $x = k_1\, d\, k_2$ with $k_1 = k\, k^{-1}$. Obviously k_1, k_2 are orthogonal. If f is a spherical function, then $f(x) = f(d)$. Note that d is not uniquely determined by x. Indeed, if λ^2, λ^{-2} are the eigenvalues of ${}^t xx$, then d could be either $\begin{pmatrix} \lambda & 0 \\ 0 & \lambda^{-1} \end{pmatrix}$ or $\begin{pmatrix} \lambda^{-1} & 0 \\ 0 & \lambda \end{pmatrix}$. This is the full extent of the ambiguity, however, and if we call the first of these d, then we have $f(x) = f(d) = f(d^{-1})$. Clearly f is determined completely on G by its restriction to the subgroup A of diagonal matrices, and this restriction of f is invariant under the map $d \to d^{-1}$ of A. Note that the Weyl group W is just the 2-element group in this case, and the map $d \to d^{-1}$ is induced by the action of the nontrivial element of W on A. Thus we see that a spherical f gives rise to a W-invariant function on A via the restriction map. Conversely, given a W-invariant function on A, one gets a spherical function on G in the obvious way. We may think of f as a function on A/W. In the present case, we can parametrize A as follows:

$$A = \left\{ \begin{pmatrix} \exp r/2 & 0 \\ 0 & \exp -r/2 \end{pmatrix} \middle| r \in \mathbf{R} \right\}.$$

Clearly $A \cong \mathbf{R}$ and the map $d \to d^{-1}$ of A corresponds to $r \to -r$. Thus f can be thought of as a function on $\mathbf{R}/W = [0, \infty]$. If f is continuous, it will be determined by its values on $A^+ = \left\{ \begin{pmatrix} \exp r/2 & 0 \\ 0 & \exp -r/2 \end{pmatrix} ; r > 0 \right\}$.

All the above has an appropriate generalization for general semisimple G. Indeed, in general, given $x \in G$, we can write $x = k_1 a k_2$ with k_1,

$k_2 \in K$, $a \in A$. a is not unique; however, if a' is another element in A such that $x = k_1\, a'\, k_2$, then $a' = a^s$ for some $s \in W$. Thus a spherical function f on G gives rise, via restriction, to a W-invariant function on A. Moreover, under the exponential map, A is isomorphic with $\mathfrak{h}_{\mathfrak{p}_0}$, so we may think of f as living on $A/W \cong \mathfrak{h}_{\mathfrak{p}_0}/W$. Now let $\mathfrak{h}_{\mathfrak{p}_0}^+ = \{H \in \mathfrak{h}_{\mathfrak{p}_0} | \alpha(H) > 0$ for $\alpha \in \mathfrak{p}_+\}$ and let $A_+ = \exp \mathfrak{h}_{\mathfrak{p}_0}^+$. Then A_+ is open, connected, and the image of A_+ under the projection map $A \to A/W$ is dense in A/W. Thus if φ is continuous and spherical, it will be determined by its restriction to A_+.

Returning to the special case $G = SL(2, R)$, consider an elementary spherical function φ_λ on G. It is an eigenfunction of each $D \in \mathscr{D}(G/K)$. However, in this case, $\mathscr{D}(G/K)$ is generated by 1 and ω, where ω is the Laplace–Beltrami operator of G/K. Thus φ_λ is an eigenfunction of ω. We saw above that, via restriction, φ_λ gives rise to a function on $(0, \infty)$. Call this function $f_\lambda(r)$, $r \in (0, \infty)$. (The parametrization above was chosen so that r is the hyperbolic distance of xK from K.) On the other hand, the Laplace–Beltrami operator ω can be written in geodesic polar coordinates (r, θ) as $\omega = \omega_r + f(r)\omega_\theta$ where ω_r is the radial part of ω, ω_θ the angular part, and $f(r)$ depends on the structure of G. It is obvious that f_λ is an eigenfunction of the operator ω_r, because ω_θ will annihilate f_λ. Thus $\omega f_\lambda = \lambda(\omega_r) f_\lambda$. If we compute ω_r, using these geodesic coordinates, we find that ω_r is just the classical Legendre operator, viz.,

$$1/sh\, r\, d/dr\, (sh\, r\, d/dr) \qquad \text{and} \qquad \lambda_\omega = -(\lambda^2 + 1/4).$$

(Note here that $\lambda \in \Lambda$, and in this case $\dim_{\mathbf{C}} \Lambda = 1$, so that we can view λ as a complex number.) Thus f_λ is a solution of

$$1/sh\, r\, d/dr(sh\, r\, d/dr) f_\lambda = -(\lambda^2 + 1/4) f_\lambda \tag{4.1}$$

and f_λ is continuous at $r = 0$. In other words, f_λ is just the classical Legendre function $P_{i\lambda + 1/2}(chr)$.

Following the decomposition $G = KAK$, one has the formula $dx = sh\, r\, dk_1\, dr\, dk_2$ for the Haar measure dx of G, and thus the space $L_2(K\backslash G/K)$ can be viewed as the space of functions on $(0, \infty)$ which are square-integrable with respect to the measure $sh\, r\, dr$ on $(0, \infty)$. Thus the problem of expanding a general element $f \in L_2(K\backslash G/K)$ in terms of the φ_λ is in this case the same as the problem of expanding a general element $f \in L_2((0, \infty), sh\, r\, dr)$ in terms of the functions $f_\lambda = P_{i\lambda + 1/2}(\cosh r)$. Let us note that the measure $sh\, r\, dr$ is precisely the canonical measure associated with the Sturm–Liouville operator $1/sh\, r\, d/dr\, (sh\, r\, d/dr)$.

Stated this way, the problem is seen to be one in the theory of eigenfunction expansions associated with classical Sturm–Liouville operators. The theory of such expansions is classical and well-known since Weyl–Kodaira–Titchmarsh. The present situation is a special case of that theory. Let us consider some of the features in the present case.

As to the Legendre operator, it can be written as $(d^2/dr^2) + \coth r(d/dr)$. Thus, the differential equation satisfied by f_λ is

$$\frac{d^2u}{dr^2} + \coth r \frac{du}{dr} = -(\lambda^2 + \tfrac{1}{4})u. \tag{4.2}$$

On noticing that coth r has an expansion as a power series in e^{-r}, viz.: $\coth r = (1 + e^{-2r})(1 - e^{-2r})^{-1} = 1 + 2\sum_{k \geqq 1} e^{-2kr}$, one can seek to construct a solution Φ_λ of (4.2) of the form $\exp(i\lambda - 1/2)r \sum_{n \geqq 0} a_n e^{-nr}$. Here the term $\exp(i\lambda - 1/2)r$ is motivated by the eigenvalue $-(\lambda^2 + 1/4)$ in (4.2) One can determine the coefficients a_n in this expansion recursively, and thus determine a solution Φ_λ of this form. Moreover since (4.2) is invariant under $\lambda \to -\lambda$, it turns out that $\Phi_{-\lambda}$ is also a solution of (4.2), and that Φ_λ and $\Phi_{-\lambda}$ are linearly independent solutions of (4.2) on $(0, \infty)$. Thus f_λ must be a linear combination of Φ_λ and $\Phi_{-\lambda}$, and so

$$f_\lambda = c_+(\lambda)\Phi_\lambda + c_-(\lambda)\Phi_{-\lambda}. \tag{4.3}$$

But, since $f_\lambda = P_{i\lambda+1/2}$, we know that $f_\lambda = f_{-\lambda}$ so we can conclude that $c_-(\lambda) = c_+(-\lambda) = \overline{c_+(\lambda)}$. Thus

$$f_\lambda = c_+(\lambda)\Phi_\lambda + \overline{c_+(\lambda)}\Phi_{-\lambda}. \tag{4.4}$$

Actually $c_+(\lambda)$ can be computed explicitly. See, e.g., Ref. [3].

Now, if $f \in L_2((0, \infty), sh\, r\, dr)$, and we put

$$\hat{f}(\lambda) = \int_0^\infty f(r)P_{-i\lambda+1/2}(ch\, r)sh\, r\, dr \tag{4.5}$$

then this is just the classical Mehler–Legendre transform, which can be "inverted" by

$$f(r) = 1/2 \int_0^\infty \hat{f}(\lambda)P_{i\lambda+1/2}(ch\, r)\, \lambda \tanh(\pi\lambda)\, d\lambda \tag{4.6}$$

and the measure $1/2\lambda \tanh(\pi\lambda)\, d\lambda$ is in this case exactly equal to $|c_+(\lambda)|^{-2} d\lambda$.

Thus in this case we may say that the measure that inverts the transform

(4.5) is the measure $|c_+(\lambda)|^{-2}\, d\lambda$, and $c_+(\lambda)$ is the leading coefficient in the asymptotic expansion of $f_\lambda = P_{i\lambda+1/2}$.

If we now consider general G, and for $\lambda \in \Lambda$, consider the elementary spherical function φ_λ, we see that under the restriction map we get a function on A_+. Moreover, we can also use this restriction map to transform spherical measures, or spherical distributions on G, to W-invariant measures or distributions on A_+. The point is that if D is a differential operator in $\mathcal{Q}$, we can regard D as a spherical distribution, and then the restriction map can be used to pull D down to a W-invariant differential operator $\delta'(D)$ on A_+. $\delta'(D)$ may be thought of as the radial part of D. $\delta'(D)$ does not have constant coefficients on A_+. Indeed the coefficients are singular on the boundaries on A_+. However, if we denote the restriction of φ_λ to A_+ by φ_λ again, then on A_+, φ_λ satisfies the following system of differential equations with variable coefficients.

$$\delta'(D)\varphi_\lambda = \lambda_D \varphi_\lambda, \qquad D \in \mathcal{D}(G/K). \tag{4.7}$$

These are analogous to (4.2). Note here that $\lambda_D = \gamma(D)(i\lambda)$ where γ is the isomorphism of $\mathcal{D}(G/K)$ with $I(\mathfrak{h}_{\mathfrak{p}_0})$ described in Part I.

This leads to a consideration of the system of differential equations,

$$\delta'(D)u = \gamma(D)(i\lambda)u, \qquad D \in \mathcal{D}(G/K) \tag{4.8}$$

for a fixed $\lambda \in \Lambda$. Here, by using detailed knowledge of the structure of $\mathcal{G}$ and $\mathcal{D}(G/K)$, one has the following result:

Theorem 4.1 (Harish-Chandra [*10d*]): The system (4.8) has at most w linearly independent solutions on A_+, where w is an order of W.

Thus if one can somehow construct w solutions, one would then have a basis for the space of solutions of (4.8) in terms of which φ_λ could be expressed. Moreover, suppose one has somehow constructed a solution Φ_λ of (4.8); then since $\gamma(D)(i\lambda) = \gamma(D)(is\lambda)$, $s \in W$, one expects that $\Phi_{s\lambda}$ might also be a solution of (4.8).

How can one construct the solutions aspired to above? Here Harish-Chandra was able to use the fact that if ω is the Casimir operator of $\mathcal{G}$ so that ω is automatically in $\mathcal{Q}$, then $\delta'(\omega)$ can be explicitly computed. In fact the result is

$$\delta'(\omega) = \sum_{i=1}^{l} H^2 + \sum_{\alpha \in P_+} \coth \alpha(H) \cdot H_{\bar{\alpha}}. \tag{4.9}$$

Here $H_1, \ldots, H_l$ is an orthonormal basis for $\mathfrak{h}_{\mathfrak{p}0}$, $\bar{\alpha}$ is the restriction of α to $\mathfrak{h}_{\mathfrak{p}0}$, and $H_{\bar{\alpha}}$ is the unique element of $\mathfrak{h}_{\mathfrak{p}0}$ such that $B(H_{\bar{\alpha}}, H) = \alpha(H)$ for $H \in \mathfrak{h}_{\mathfrak{p}0}$. Here, the elements of $S(\mathfrak{h}_{\mathfrak{p}0})$ on the right side are to be regarded as differential operators on A_+ in the usual way. One can also compute $\gamma(\omega)(i\lambda)$ and find that $\gamma(\omega)(i\lambda) = -(\langle\lambda, \lambda\rangle + \langle\rho, \rho\rangle)$.

Thus, among the equations of the system (4.8), we find the equation

$$\left(\sum_{i=1}^{l} H^2 + \sum_{\alpha\in P_+} \coth^2 \alpha(H)\ H_{\bar{\alpha}})u\right) = -(\langle\lambda, \lambda\rangle + \langle\rho, \rho\rangle)u. \qquad (4.10)$$

This is analogous to the Legendre equation. Now let $\sum$ be the set of restrictions of elements of P_+ to $\mathfrak{h}_{\mathfrak{p}0}$, and let $\alpha_1, \ldots, \alpha_l$ be the simple elements in $\sum$. Then for any $\alpha \in P_+$, $\bar{\alpha} = \sum_{i=1}^{l} n_i\, \alpha_i$ where n_i are nonnegative integers. Thus $\alpha(H) = \sum_{i=1}^{l} n_i\alpha_i(H)$, and $H_{\bar{\alpha}} = \sum_{i=1}^{l} n_i\, H_{\alpha_i}$ where H_{α_i} is defined by $B(H_{\alpha_i}, H) = \alpha_i(H)$ for $H \in \mathfrak{h}_{\mathfrak{p}0}$. Now note that each of the coefficients $\coth \alpha(H)$ in (4.10) can be expanded in a series of the form $\sum_{\mu\in L} b_\mu e^{\mu(H)}$ where μ ranges over the semilattice $L = \{\mu | \mu = \sum_{i=1} m_i\alpha_i$, $m_i \geqq 0, m_i \in Z\}$. Thus one may hope to construct a solution Φ_λ of (4.10) of the form $\exp(i\lambda - \rho)(H) \sum_{\mu\in L} \Gamma_\mu(\lambda) \exp -\mu(H)$, in analogy with the case $G = SL(2, R)$. Actually, if one simply substitutes this expression for u in (4.10), and equates coefficients, then one gets a recursion relation for the coefficients $\Gamma_\mu(\lambda)$ of the following form:

$$\{\langle\mu, \mu\rangle - 2\langle\mu, \lambda\rangle\}\, \Gamma_\mu(\lambda)$$
$$= 2 \sum_{\alpha\in P_+} \sum_{k\geqq 1} \Gamma_{\mu-2k\bar{\alpha}}(\lambda)\, \{\langle\mu + \rho - 2k\bar{\alpha}, \bar{\alpha}\rangle - \langle\bar{\alpha}, \lambda\rangle\}. \qquad (4.11)$$

Here, $\bar{\alpha}$ denotes the restriction of α to $\mathfrak{h}_{\mathfrak{p}0}$ and k runs over those integers equal to or greater than 1 such that $\mu\text{-}2k\bar{\alpha} \in L - \{0\}$. This relation is a recurrence in the following sense: Given $\mu = \sum m_i\alpha_i \in L$, let $m(\mu) = \sum_{i=1}^{l} m_i$ be the *level* of μ. Let μ be of level r. Then all the terms on the right side of (4.11) are of the form $\Gamma_\nu(\lambda)$ with $m(\nu) < r$. Thus (4.11) determines $\Gamma_\mu(\lambda)$ if the $\Gamma_\nu(\lambda)$ are known for all ν of level less than r. Putting $\Gamma_0 \equiv 1$, we see that the $\Gamma_\mu(\lambda)$ are rational functions of λ. They are not everywhere defined. However, it is seen easily that the set Λ' where they are all defined is an open dense connected set in Λ, being the complement of the union of a countable number of hyperplanes.

In Ref. [*10d*], Harish-Chandra was able to estimate the functions $\Gamma_\mu(\lambda)$ for $\lambda \in \Lambda'$, and using the estimate, he showed how one can construct a solution of (4.10). We can state

Proposition 4.2: Let U be a compact subset of Λ'. Then there is a constant $\mathbf{C}$ such that $|\Gamma_\mu(\lambda)| \leq C^{m(\mu)}$ for $\lambda \in U$. Hence, we can find a real number $M > 0$ such that if $H \in \mathfrak{h}_{\mathfrak{p}0}(M) \overset{\text{def.}}{=} \{H | H \in \mathfrak{h}_{\mathfrak{p}0}, \alpha_i(H) > M, i = 1, \ldots l\}$, then the series $\sum_{\mu \in L} \Gamma_\mu(\lambda) \exp(i\lambda - \rho - \mu)(H)$ converges absolutely, uniformly for $(\lambda, H) \in U \times \mathfrak{h}_{\mathfrak{p}0}(M)$; if $\Phi_\lambda(h)$ is the sum of that series at $H = \log h$, then Φ_λ is a solution of (4.10).

Moreover, one can arrange by shrinking if necessary that Λ' be W-invariant, so that if $\lambda \in \Lambda'$, then $s\lambda \in \Lambda'$ for each $s \in W$. The shrunk set will still be open dense, etc. Thus along with Φ_λ we can define functions $\Phi_{s\lambda}$, $s \in W$.

The functions $\Phi_{s\lambda}$ satisfy (4.10). The remarkable fact is that one can show that they satisfy the entire system (4.8) on $A(M)$ (which equals $\exp \mathfrak{h}_{\mathfrak{p}0}(M)$) and are linearly independent.

Theorem 4.3: The functions $\Phi_{s\lambda}$, $s \in W$ satisfy the Eqs. (4.8) and are linearly independent functions on $A(M)$.

It follows from Theorem 4.1 that these functions form a basis of solutions of (4.8) on $A(M)$. It follows that for $\lambda \in \Lambda'$,

$$\varphi_\lambda(h) = \sum_{s \in W} c_s(\lambda)\, \Phi_{s\lambda}(h) \tag{4.12}$$

for suitable complex numbers $c_s(\lambda)$. Actually it can be shown that if $c(\lambda) = c_1(\lambda)$ where 1 is the identity element of W, then $c_s(\lambda) = c(s\lambda)$. Thus we have

$$\varphi_\lambda(h) = \sum_{s \in W} c(s\lambda)\, \Phi_{s\lambda}(h), \qquad h \in A(M). \tag{4.13}$$

The details of all this are in [*10d*].

In analogy with the case of the Sturm–Liouville operator, Harish-Chandra proposed to show that the Plancherel measure $d\mu(\lambda)$ of Section 3 equals $|c(\lambda)|^{-2}\, d\lambda$ and lives on Λ_0.

If one reflects about the above definitions of $c(\lambda)$, one sees that it is quite unsuitable for computation, since it is abstractly defined in terms of the basis $\{\Phi_{s\lambda}, s \in W\}$. To circumvent this, Harish-Chandra established an integral formula for $c(\lambda)$, which we will describe briefly. Let $\mathfrak{n} =$

$\sum_{\alpha\in\rho_+}\mathfrak{g}^{-\alpha}$ $\mathfrak{n}_0 = \mathfrak{n}\cap\mathfrak{g}_0$, and let $\overline{N} = \exp \mathfrak{n}_0$. Then one can normalize the Haar measure $d\bar{n}$ on $\overline{N}$ by the condition $\int_{\mathfrak{g}} \exp -2\rho(H(\bar{n}))\, d\bar{n} = 1$. (This integral converges.) With this normalization, one has

$$c(\lambda) = \int_{\mathfrak{g}} \exp -(i\lambda + \rho)(H(\bar{n}))\, d\bar{n} \tag{4.14}$$

whenever the integral converges absolutely.

Using this integral formula, one can compute $c(\lambda)$ explicitly. We shall not describe the method of computation, which is due to Gindikin and Karpelevic [*8*]. But the result is as follows:

$$c(\lambda) = I(i\lambda)/I(\rho_*) \tag{4.15}$$

where

$$I(\nu) = \prod_{\alpha\in\Sigma} \beta\left(\frac{m_\alpha}{2}, \frac{m_{\alpha/2}}{4} + \frac{\langle \nu, \alpha\rangle}{\langle \alpha, \alpha\rangle}\right), \qquad \nu \in \Lambda. \tag{4.16}$$

We remind the reader that m_α is the number of elements in P_+ whose restriction to $\mathfrak{h}_{\mathfrak{p}_0}$ equals α, and ρ_* is the restriction of $\rho = \frac{1}{2}\sum_{\alpha\in P}\alpha$ to $\mathfrak{h}_{\mathfrak{p}_0}$.

Particular cases of this formula had been proved earlier by Bhanu-Murthy [*1a, 1b*], whose method was generalized by Gindikin and Karpelevic [*8*].

It follows from this formula that $c(\lambda)^{-1}$ is a tempered distribution on Λ_0. We shall have to use this fact below.

5. The Schwarz Space

The analogy with the Sturm–Liouville operators which was touched upon in Section 4 suggests that the Fourier transform $f \to \hat{f}$ ought to map $L_2(K\backslash G/K)$ unitarily onto the space $I_2(\Lambda_0, w^{-1}|c(\lambda)^{-2}|\, d\lambda)$ of W-invariant measurable functions on Λ_0 which are square integrable with respect to the measure $w^{-1}|c(\lambda)|^{-2}\, d\lambda$. Here $d\lambda$ is a suitably normalized Euclidean measure on Λ_0. (Recall that the Cartan form gives us a Euclidean structure on Λ_0.) However, a priori, the map $f \to \hat{f}$ is not defined for every $f \in L_2(K\backslash G/K)$. So one has to use a suitable dense subspace of $L_2(K\backslash G/K)$ on which the map is defined, show that the map sends this subspace unitarily onto a dense subspace of $I_2(\Lambda_0, w^{-1}|c(\lambda)|^{-2}\, d\lambda)$, and extend the map $f \to \hat{f}$ to all of $L_2(K\backslash G/K)$ in the usual way.

Moreover, if such a dense subspace of $L_2(K\backslash G/K)$ can be found, one would also like to have, at least for f in this subspace, the following pair of inverse formulas:

$$\hat{f}(\lambda) = \int_G f(x)\, \varphi_\lambda(x^{-1})\, dx, \qquad \lambda \in \Lambda_0 \tag{5.1}$$

$$f(x) = w^{-1} \int_{\Lambda_0} \hat{f}(\lambda)\varphi_\lambda(x)|c(\lambda)|^{-2}\, d\lambda, \qquad x \in G. \tag{5.2}$$

The analogy of these formulas with the classical formulas for the Fourier transform is obvious.

Harish-Chandra was able to define such a subspace of $L_2(K\backslash G/K)$ in analogy with the Schwarz space of rapidly decreasing C^∞ functions on $\mathbf{R}^n$. (In Ref. [*10e*] he called this space $I(G)$.) We shall devote this section to a discussion of this space. It will be called the Schwarz space of spherical functions and will be denoted by $\mathscr{C}(K\backslash G/K)$. We shall first try to motivate its definition.

Let $\mathscr{S}(\Lambda_0)$ be the space of C^∞ functions on Λ_0 which, together with their derivatives, decrease rapidly on Λ_0. Thus, a function f in $C^\infty(\Lambda_0)$ is in the space $\mathscr{S}(\Lambda_0)$ if and only if, given a differential operator D with constant coefficients on Λ_0 and an integer $k \geqq 0$, we have

$$\operatorname*{Sup}_{\lambda \varepsilon \Lambda_0} ((1 + \|\lambda\|)^k |(Df)(\lambda)|) < \infty \tag{5.3}$$

where $\|\cdot\|$ stands for the Euclidean norm on Λ_0. The functional $f \to \tau_{D,k}(f) = \sup_{\lambda \in \Lambda_0} (1 + \|\lambda\|)^k |(Df)(\lambda)|$ is a seminorm on $\mathscr{S}(\Lambda_0)$ and the family of all these seminorms for various D, k defines a locally convex topology on $\mathscr{S}(\Lambda_0)$. It is obvious that $C_c^\infty(\Lambda_0) \subset \mathscr{S}(\Lambda_0)$.

Now let $I_c^\infty(\Lambda_0)$ denote the space of W-invariant functions in $C_c^\infty(\Lambda_0)$, and put $I_c^\infty(\Lambda_0) = C_c^\infty(\Lambda_0) \cap I_c^\infty(\Lambda_0)$, $\mathscr{I}(\Lambda_0) = \mathscr{S}(\Lambda_0) \cap I_c^\infty(\Lambda_0)$. Then we have $I_C^\infty(\Lambda_0) \subset \mathscr{I}(\Lambda_0)$. Moreover, since $c(\lambda)^{-1}$ is a tempered distribution on Λ_0, it follows that $\mathscr{I}(\Lambda_0)$ is contained in $I_2(\Lambda_0, w^{-1}|c(\lambda)|^{-2}\, d\lambda)$ and indeed is dense in the latter space.

The definition of $\mathscr{C}(K\backslash G/K)$, which will be described below, is motivated by the hope that the Fourier transform $f \to \hat{f}$ will map it bijectively onto $\mathscr{I}(\Lambda_0)$, and the inverse formulas (5.1) and (5.2) will hold for f in $\mathscr{C}(K\backslash G/K)$.

Suppose for the moment that $a \in \mathscr{I}(\Lambda_0)$ and let us define the function φ_a on G by the formula

$$\varphi_a(x) = w^{-1} \int_{\Lambda_0} a(\lambda)\varphi_\lambda(x)|c(\lambda)|^{-2}\, d\lambda. \tag{5.4}$$

What kind of function is φ_a? Obviously since $a \in \mathscr{I}(\Lambda_0)$ and $|c(\lambda)|^{-2}$ is tempered, the integral converges and φ_a is spherical. Moreover, the differentiability and growth properties of φ_a will evidently depend on the corresponding properties of φ_λ. In this connection one has the following result [*10d*, p. 294]:

Proposition 5.1: Let $\lambda \in \Lambda_0$, and $D \in \mathscr{G}$. Then there exists a constant M and an integer d such that

$$|(D\varphi_\lambda)(x)| \leqq M(1 + \|\lambda\|)^d \varphi_0(x). \tag{5.5}$$

Thus we would expect that $\sup_{x \in G} \varphi_0(x)^{-1}|(D\varphi_a)(x)| < \infty$. Moreover, in analogy with $\mathbf{R}^n$, we would also expect that this last function, namely, $\varphi_0(x)^{-1}|D\varphi_a(x)|$ would decrease rapidly. The question is: How should one formulate the notion of "decay faster than a polynomial" on the group G? This is done as follows. Set $x \in G$. Then $x = k \exp X$ with $k \in K$ and $X \in \mathfrak{p}_0$. There is a unique K-invariant inner product induced by the Cartan-Rilling form on $\mathfrak{p}_0$. Let $\|\cdot\|$ denote the norm induced by this inner product. Let $\sigma(x) = \|X\|$. Then the function σ behaves like a sort of norm on G and the condition "decay faster than any polynomial" can be formulated to mean decay faster than any power of $1 + \sigma(x)$ on the group G. Thus, after Harish-Chandra,

Definition 5.2: The Schwarz space $\mathscr{C}(K\backslash G/K)$ consists of those functions $f \in C^\infty(G)$ such that

(i) f is spherical; so that $f \in C^\infty(K\backslash G/K)$.

(ii) Given any integer $r \geqq 0$ and any $D \in \mathscr{G}$ we have

$$\tau_{D,r}(f) = \sup_{x \in G} ((1 + \sigma(x))^r \varphi_0(x)^{-1}|(Df)(x)|) < \infty.$$

Notice that in Ref. [*10e*] this space is called $I(G)$. Actually a slightly different function is used in [*10e*] instead of the function $\varphi_0(x)^{-1}$. However this makes no difference. The seminorms $\tau_{D,r}$ for varying D, r define a locally convex topology on the space $\mathscr{C}(K\backslash G/K)$.

Having defined $\mathscr{C}(K\backslash G/K)$, one now has the difficult task of establishing (5.1) and (5.2) for f in this space. In this connection one has

Theorem 5.3: Let $a \in \mathscr{I}(\Lambda_0)$ and let φ_a be as in (5.4). Then $\varphi_a \in \mathscr{C}(K\backslash G/K)$, and the map $a \to \varphi_a$ is a continuous map of $\mathscr{I}(\Lambda_0)$ into $\mathscr{C}(K\backslash G/K)$.

The proof of this theorem is very deep, and we will not go into it. Let it suffice to say that it depends crucially on a detailed study of the function $\varphi_\lambda(h)$ and its derivatives for large h. Indeed the above theorem is one of the main results of [*10e*].

On the other hand, suppose $f \in \mathscr{C}(K\backslash G/K)$; then f need not be in $L_1(K\backslash G/K)$; therefore, for a fixed $\lambda \in \Lambda_0$, it is not obvious that the integral $\int_G f(x)\overline{\varphi_\lambda(x)}\, dx$ makes sense. In this regard, one has

Proposition 5.4: Suppose $f \in \mathscr{C}(K\backslash G/K)$. Then for any $\lambda \in \Lambda_0$, the integral $\int_G f(x)\overline{\varphi_\lambda(x)}\, dx$ converges absolutely. Indeed, if $f \in \mathscr{C}(K\backslash G/K)$, then for any $a \in A$, the integral

$$F_f(a) = \exp \rho(\log a) \int_N f(an)\, dn \tag{5.6}$$

converges absolutely and we have

$$\hat{f}(\lambda) = F_f^*(\lambda) \tag{5.7}$$

where F_f^* is the Euclidean Fourier transform on the vector group A. Finally, the map $f \to \hat{f}(\lambda)$ defines a continuous functional on $\mathscr{C}(K\backslash G/K)$.

Note: Here $F_f^*(\lambda) = \int_A \exp -i\lambda(\log a) \cdot F_f(a)\, da$.

Now suppose $a \in \mathscr{I}(\Lambda_0)$. Then according to Theorem 5.3 φ_a is in $\mathscr{C}(K\backslash G/K)$. Thus $\hat{\varphi}_a(\lambda) = F^*_{\varphi_a}(\lambda)$ is well defined. If the inversion formulas (5.1) and (5.2) are to hold, then we expect to have $\hat{\varphi}_a = a$. This last fact was established by Harish-Chandra [*10e*] after a difficult analysis.

Theorem 5.5: Let $a \in \mathscr{I}(\Lambda_0)$ and $\lambda \in \Lambda_0$. Then the map $T_\lambda\colon a \to \hat{\varphi}_a(\lambda)$ is a distribution on $\mathscr{I}(\Lambda_0)$. Indeed $T_\lambda(a) = a(\lambda)$. Thus $\hat{\varphi}_a = a$.

The proof of this theorem is quite as deep as that of Theorem 5.3. It is the second main result of [*10e*]. We will not go into the details here. However, we refer to Helgason [*11b*, *11e*] for a further and more complete discussion of these matters.

Granted the above theorems, in order to establish (5.1) and (5.2), it would be enough to show that each $f \in \mathscr{C}(K\backslash G/K)$ can be written as φ_a for a suitable $a \in \mathscr{I}(\Lambda_0)$.

In [*10e*, p. 612] Harish-Chandra states this as a conjecture. Moreover, later, in [*10f*, p. 48], he proved this conjecture to be true. However, his proof of this conjecture involves the full machinery of representation theory of G, especially the discrete series. In the next section I shall outline a proof which is somewhat more direct.

6. The Inversion Formula and the Paley–Wiener Theorem

In order to have an inversion formula such as (5.2) for $f \in \mathscr{C}(K\backslash G/K)$, it is sufficient to have it for a subspace of $\mathscr{C}(K\backslash G/K)$ which is dense in $L_2(K\backslash G/K)$. The most obvious such subspace is $C_c^\infty(K\backslash G/K)$. In this section, we will describe how one can get such an inversion formula for $f \in C_c^\infty(K\backslash G/K)$. The proof proceeds via attempting to characterize the Fourier transform $\hat{f}$ of such f. Hence the method incidentally yields an analogue of the classical theorem of Paley–Wiener which characterizes the Fourier transforms of functions $C_c^\infty(R^n)$.

The idea of the method goes back to a paper of Helgason [*11c*]. In that paper Helgason was able to settle the above problem in the following two cases: (i) G a complex group, or (ii) $G|K$ has rank 1. By improving upon certain estimates in Helgason's paper, I was recently able to show that one can use the same method even if rank $(G|K) > 1$.

Suppose $f \in C_c^\infty(R^n)$. Then its Fourier–Laplace transform $\hat{f}(\lambda) = \int_{R^n} f(x)\, e^{-i\lambda x}\, dx$ is defined not merely for $\lambda \in R^n$, but for all $\lambda = \xi + i\eta$ with $\xi, \eta \in R^n$. The classical theorem of Paley and Wiener states that $f \in C_c^\infty(R^n)$ and support $(f) \subset \{x \in R^n, |x| \leqq B\}$ if and only if its Fourier–Laplace transform $\hat{f}(\xi + i\eta)$ satisfies the two conditions (i) $\hat{f}$ is an entire holomorphic function, and (ii) given any integer $N > 0$, there is a constant C_N so that

$$|\hat{f}(\xi + i\eta)| \leqq C_N\, (1 + (|\xi|^2 + |\eta|^2)^{1/2})^{-N} \exp B|\eta|.$$

We shall formulate an analogue of this theorem and indicate a proof of it. We shall also see how this leads to an inversion formula and the Plancherel theorem for $L_2(K\backslash G/K)$.

Suppose $f \in C_c^\infty(K\backslash G/K)$. Then the restriction of f to A also has compact support. Let us say that f has support in the ball of radius R if the restriction of f to A has support in the ball of radius R in A (the obvious metric on A). Suppose that we take such a function f. It is not hard to see then that the function

$$F_f(a) = \exp \rho(\log a) \int_N f(an)\, dn, \qquad a \in A \tag{6.1}$$

also has support in the ball of radius R in A. See, e.g., Ref. [*11c*].

It follows from the easy half of the classical Paley–Wiener theorem that $F_f^*(\lambda)$, the Euclidean Fourier transform of F_f (recall that A is a vector group) is defined not merely for $\lambda \in \Lambda_0$ but for all $\lambda = \xi + i\eta$ with ξ, $\eta \in \Lambda_0$, i.e., for all $\lambda \in \Lambda$, and moreover that it is a W-invariant entire holomorphic function of $\lambda \in \Lambda$. Finally one sees by the same theorem that given any integer $N > 0$, there is $C_N > 0$ such that

$$|F_f^*\,(\xi + i\eta)| \leqq C_N\,(1 + (|\xi|^2 + |\eta|^2)^{1/2})^{-N} \exp R|\eta|.$$

However, as we have seen before, $F_f^*(\lambda)$ is exactly the same as $\hat{f}(\lambda)$. It follows that if $f \in C_c^\infty(K\backslash G/K)$ and if f is supported in the ball of radius R, then the spherical Fourier transform $\hat{f}$ is an entire holomorphic function of $\lambda \in \Lambda$ and given $N > 0$ we can find $C_N > 0$ such that

$$|\hat{f}(\xi + i\eta)| \leqq C_N(1 + (|\xi|^2 + |\eta|^2)^{1/2})^{-N} \exp R|\eta|, \qquad \xi, \eta \in \Lambda_0. \quad (6.2)$$

The thrust of the Paley–Wiener theorem in this context is to get the converse of this result. Namely, suppose we are given a W-invariant entire holomorphic function $b(\xi + i\eta)$, ξ, $\eta \in \Lambda_0$, satisfying

$$|b(\xi + i\eta)| \leqq C_N(1 + (|\xi|^2 + |\eta|^2)^{1/2})^{-N} \exp R|\eta|, \qquad \xi, \eta \in \Lambda_0 \quad (6.3)$$

for each $N > 0$; can we then assert that b is the spherical Fourier transform of a unique function $f \in C_c^\infty(K\backslash G/K)$ whose support is in the ball of radius R? We shall see below that the answer is yes.

To see this, suppose we are given a function b on Λ such as above. It is clear that the restriction of b to Λ_0 is in $\mathscr{I}(\Lambda_0)$. Hence we can form the function

$$\varphi_b(x) = w^{-1} \int_{\Lambda_0} b(\lambda)\varphi_\lambda(x)|c(\lambda)|^{-2}\, d\lambda. \quad (6.4)$$

It follows from the results quoted in the last section that $\varphi_b \in \mathscr{C}(K\backslash G/K)$ and that $\hat{\varphi}_b = b$. One now proves a crucial lemma.

Lemma 6.1: Under the above conditions on b, φ_b has support contained in the ball of radius R. Hence $\varphi_b \in C_c^\infty(K\backslash G/K)$.

We will indicate very briefly the main ideas behind the proof. Details can be found in Refs. [*11c*] and [*4*].

Let h be a regular element of A. Recall that for sufficiently far away from the walls of A^+, i.e., for $h \in A^+(M)$,

$$\varphi_\lambda(h) = \sum_{s \in W} c(s\lambda)\Phi_{s\lambda}(h), \qquad \lambda \in \Lambda_0'. \tag{6.5}$$

Actually though, one can show that the series defining $\Phi_\lambda(h)$ converges to an analytic function of h not merely for $h \in A^+(M)$ but for all $h \in A^+$. Thus, since both sides of (6.5) are analytic functions on the connected set A^+, they must agree on A^+ (since they agree on $A^+(M)$). On the other hand, one has also $|c(\lambda)|^{-2} = c(\lambda)^{-1}c(-\lambda)^{-1} = c(s\lambda)c(-s\lambda)$ by a result in [*10h*]. Hence, using the W-invariance of b, one can show that

$$\varphi_b(h) = \int_{\Lambda_0} b(\lambda)\, c(-\lambda)^{-1}\, \Phi_\lambda(h)\, d\lambda. \tag{6.6}$$

Now, $\Phi_\lambda(h) = \sum_{\mu \in L} \Gamma_\mu(\lambda) \exp(i\lambda - \rho - \mu)(\log h)$; thus

$$\varphi_b(h) = \int_{\Lambda_0} b(\lambda)\, c(-\lambda)^{-1} \sum_{\mu \in L} \Gamma_\mu(\lambda) \exp(i\lambda - \rho - \mu)(\log h)\, d\lambda. \tag{6.7}$$

If the intergretation and the summation in this equation can be interchanged, then we would have

$$\varphi_b(h) = \sum_{\mu \in L} \exp - (\rho + \mu)(\log h) \int_{\Lambda_0} \Gamma_\mu(\lambda) b(\lambda) c(-\lambda)^{-1} \exp i\lambda(\log h)\, d\lambda \tag{6.8}$$

or, if we write

$$B_\mu(h) = \int_{\Lambda_0} \Gamma_\mu(\lambda) b(\lambda) c(-\lambda)^{-1} \exp i\lambda(\log h)\, d\lambda \tag{6.9}$$

then

$$\varphi_b(h) = \sum_\mu \exp -(\rho + \mu)(\log h) \cdot B_\mu(h). \tag{6.10}$$

Now, using the recursion relation (4.11) for Γ_μ and the formula (4.15) for $c(\lambda)$, it is possible to show that if $\xi + i\eta$ is such that $\eta \in \Lambda^+ = \{\lambda | \langle \lambda, \alpha \rangle > 0$ for all $\alpha \in \sum\}$, then $\Gamma_\mu(\lambda)$ and $c(-\lambda)^{-1}$ are holomorphic at $\lambda = \xi + i\eta$. This, together with estimates on $\Gamma_\mu(\lambda)$ (which are rational functions), enables one to shift the integration in (6.9) into the complex domain (using the fact that b is entire) and one proves, as in the classical Paley–Wiener theorem, that $B_\mu(h) = 0$ if $|h| > R$. See, e.g., Ref. [*11c*].

Thus, provided one can justify the interchange of summation and integration leading to it, Eq. (6.10) shows that $\varphi_b(h) = 0$ if $|h| > R$, proving Lemma 6.1.

In [*11c*], Helgason was able to show that if either G is complex or rank $(G|K) = 1$, then the functions Γ_μ enjoys the following estimate:

$$|\Gamma_\mu(\lambda)| \leqq C\, m(\mu)^{C'}, \qquad \lambda = \xi + i\eta, \quad \xi \in \Lambda_0, \quad \eta \in \Lambda_0^+ \tag{6.11}$$

where $m(\mu)$ is the level of μ. Because of the factor $\exp -\mu(\log h)$ in (6.9), this estimate immediately implies that in these cases the term-by-term integration in (6.9) is valid, and the lemma is proved for these cases.

The results in Ref. [*4*] show that the estimate (6.11) is valid even if rank $(G/K) > 1$. The basic idea is the following: Fix λ and consider $\mu \to \Gamma_\mu(\lambda)$ as a function on the semilattice L. We want to estimate this function. If rank $(G|K) > 1$ (since Helgason's method runs into trouble in estimating this function), we resort to smoothing this function as a function on the semilattice L. Namely instead of the function $\Gamma(\lambda)\colon \mu \to \Gamma_\mu(\lambda)$ consider, for a suitable function d on L, the convolution of $\Gamma(\lambda)$ and d. Let $a(\lambda)$ be the convolution of $\Gamma(\lambda)$ and d as functions on L. Thus $a(\lambda)\colon \mu \to a_\mu(\lambda)$ where $a_\mu(\lambda) = \sum_{\nu+\sigma=\mu} d_\nu \Gamma_\sigma(\lambda)$. Now the idea is to choose d suitably so as to be able to estimate a_μ, and then to estimate Γ_μ from its relation to a_μ. It is, a priori, hard to guess what d should be. However, if we introduce the function $D(h) = \sum_{\mu\in L} d_\mu \exp -\mu(\log h)$ formally, then we see that $D(h)\, \Phi_\lambda(h) = \sum_\mu a_\mu(\lambda) \exp[(i\lambda - \rho - \mu)(\log h)]$. Thus, smoothing of Γ_μ by d "corresponds" to multiplication of $\Phi_\lambda(h)$ by $D(h)$, and the smoothed coefficients $\Gamma_\mu(\lambda)$ are those of the function $\Phi_\lambda(h)\, D(h)$.

The trick is to choose $D(h)$ so that it is possible to estimate these coefficients a_μ more sharply than was possible for Γ_μ. We note that the main tool for the estimation of Γ_μ is the recurrence relation (4.11), which in turn comes from the differential equation (4.10) for Φ_λ. Thus $D(h)$ should be so chosen that the differential equation for $\psi_\lambda = \Phi_\lambda D$ has simpler structure than the differential equation for Φ_λ. Indeed one sees that if $D(h) = \Delta^{1/2}(h)$ where $\Delta(h) = \prod_{\alpha\in\rho_+} (\exp \alpha(\log h) - \exp -\alpha(\log h))$, then ψ_λ satisfies a differential equation of the following sort:

$$\left(\sum_{i=1}^{l} H_i^2 + g\right)\psi_\lambda = -(\langle\lambda, \lambda\rangle + \langle\rho, \rho\rangle)\psi_\lambda \tag{6.12}$$

where g is a function that involves only the structure of $(\mathfrak{g}_0, \mathfrak{k}_0)$. The point is that this differential equation has no first derivatives in it. This equation now leads to a recurrence for $a_\mu(\lambda)$ analogous to (4.11). The absence of first derivatives in (6.12) simplifies the structure of this recurrence relation and enables one to get sharp estimates.

$$|a_\mu(\lambda)| \leqq C_1\, m(\mu)^{C'(\lambda)} \tag{6.13}$$

for a_μ. Now, $\Phi_\lambda = \Delta^{-1/2}\psi_\lambda$. Thus if $\Delta^{-1/2}(h) = \sum_\mu e_\mu \exp -\mu(\log h)$, we have $\Gamma(\lambda) = c * a(\lambda)$, i.e., $\Gamma_\mu(\lambda) = \sum_{\nu+\sigma=\mu} c_\nu a_\sigma(\lambda)$. However, $\Delta^{-1/2}$ is such that the coefficients c_ν enjoy an estimate $|c_\nu| \leqq M\, m(\nu)^{M'}$. So finally one gets $|\Gamma_\mu(\lambda)| \leqq C\, m(\mu)^{C'}$. The details of this are in [*4*].

The device used above is reminiscent of a similar device used by Liouville in the classical theory of hypergeometric functions. Namely, by multiplying a given function φ (which solves $L\varphi = \lambda\varphi$, say) by a suitable function D, one produces a function $\psi = D\varphi$ such that the differential equation for ψ has no first derivatives in it. This form of the hypergeometric equation is known as Liouville's normal form. See, e.g., [*3*].

We are now in a position to state the Paley–Wiener theorem.

Theorem 6.2: Suppose $f \in C_c^\infty(K\backslash G/K)$ and that the support of f is contained in the ball of radius R. Then $f(\lambda)$ is an entire W-invariant holomorphic function of $\lambda \in \Lambda$, and given $N > 0$, we can find $C_N > 0$ so that

$$\hat{f}|(\xi + i\eta)| \leqq C_N(1 + (|\xi|^2 + |\eta|^2))^{-N} \exp R|\eta|, \qquad \xi, \eta \in \Lambda_0. \tag{6.14}$$

Conversely, suppose b is an entire W-invariant holomorphic function on Λ which satisfies a condition analogous to (6.14). Then $b(\lambda) = \hat{f}(\lambda)$ for a unique $f \in C_c^\infty(K\backslash G/K)$ and the support of f is contained in the ball of radius R.

The first half is evident. To see the second half, note that the remarks made above, together with Lemma 6.1, show that if we write $f = \varphi_b$, then $f \in C_c^\infty(K\backslash G/K)$, the support of f is in the ball of radius R, and $\hat{f} = b$. Thus only the uniqueness requires proof. But if f' were another function in $C_c^\infty(K\backslash G/K)$ such that $\hat{f}' = b$, then $h = f - f' \in C_c^\infty(K\backslash G/K)$ and $h = 0$ on Λ_0. But since h is obviously in $L_1(K\backslash G/K)$, this means that $h \equiv 0$ by Proposition 2.5. This proves that f is unique.

Theorem 6.3: Thus we have for $f \in C_c^\infty(K\backslash G/K)$ that the formulas

$$f(\lambda) = \int f(x)\varphi_\lambda(x)\, dx, \qquad \lambda \in \Lambda_0 \tag{6.15}$$

$$f(x) = w^{-1} \int_{\Lambda_0} \hat{f}(\lambda)\varphi_\lambda(x)|c(\lambda)|^{-2}\, d\lambda, \qquad x \in G \tag{6.16}$$

hold literally. Moreover we have the Plancherel–Parseval relation

$$\int_G |f(x)|^2\, dx = w^{-1} \int_{\Lambda_0} |\hat{f}(\lambda)|^2 |c(\lambda)|^{-2}\, d\lambda. \tag{6.17}$$

From this it follows in the usual way that the map $f \to \hat{f}$ can be extended to a unitary transformation of $L_2(K\backslash G/K)$ onto a closed subspace of $I_2(\Lambda_0, w^{-1}|c(\lambda)|\, d\lambda)$. To see that the map is actually onto $I_2(\Lambda_0, w^{-1}|c(\lambda)|^{-2}\, d\lambda)$, one has only to note that if $b \in \mathscr{I}(\Lambda_0)$, then $\hat{\varphi}_b = b$. Thus $\mathscr{I}(\Lambda_0)$ is contained in the range of the map $f \to \hat{f}$. It follows that the range is all of the space $I_2(\Lambda_0, w^{-1}|c(\lambda)|^{-2}\, d\lambda)$. We should note that incidentally all this means that the transform $f \to F_f$ is one-to-one on $\mathscr{C}(K\backslash G/K)$ and maps $\mathscr{C}(K\backslash G/K)$ onto $\mathscr{I}(A)$ and $C_c^\infty(K\backslash G/K)$ onto $I_c^\infty(A)$. Finally one has the above formula (6.17) for suitable f,

$$\int_G |f(x)|^2\, dx = w^{-1} \int_{\Lambda_0} |\hat{f}(\lambda)|^2 |c(\lambda)|^{-2}\, d\lambda.$$

Note that this means that the Plancherel measure $d\mu(\lambda)$ on Λ_P is concentrated on the subset Λ_0 of Λ_P. (Abstractly, it is only known to be supported by Λ_P.)

7. More about L_1 $(K\backslash G/K)$

In this section, we shall discuss some questions about the harmonic analysis of $L_1(K\backslash G/K)$.

Because of Proposition 2.5, one knows that an $f \in L_1(K\backslash G/K)$ is determined by $\hat{f}(\lambda)$, $\lambda \in \Lambda_B$. In general, for an arbitrary f in $L_1(K\backslash G/K)$, getting a characterization for its Fourier transform would seem impossible in view of the well-known intractability of this problem even on **R**! However, it is possible to ask if for f in a suitable dense subspace of $L_1(K\backslash G/K)$, $\hat{f}$ can be characterized, and if an inversion formula like (5.2) holds.

There is a natural candidate for such a space. Namely the analogue of $\mathscr{C}(K\backslash G/K)$ for $L_1(K\backslash G/K)$. Without going into details, we define the space $\mathscr{C}_1(K\backslash G/K)$ to be the space of those C^∞ spherical functions of such that if $D \in \mathscr{G}$ and $k \geqq 0$ is any integer, then $\mathrm{Sup}_{x\in G}\, (1 + \sigma(x))^k(\varphi_0(x))^{-2}\, |Df(x)| < \infty$. Then it is easy to show that $\mathscr{C}_1 \subset L_1$. Moreover for any ε, $0 \leqq \varepsilon \leqq 1$, we can define $\mathscr{C}_\varepsilon(K\backslash G/K)$ analogously, by the condition $\sup_{x\in G}$ (1 +

$\sigma(x))^k(\varphi_0(x))^{-\varepsilon-1}|(Df)(x)| < \infty$, and topologize these spaces in the obvious way via the above seminorms.

One can show easily that $\mathscr{C}_\varepsilon(K\backslash G/K) \subset L_{2-\varepsilon}(K\backslash G/K)$, $\quad 0 \leqq \varepsilon \leqq 1$. The problem would be to characterize the Fourier transforms of these spaces. Consider for example the space $\mathscr{C}_1(K\backslash G/K)$. If $f \in \mathscr{C}_1(K\backslash G/K)$, then its Fourier transform $\hat{f}$ is holomorphic in $\Lambda_B = \Lambda_0 + iC_\rho$. Moreover one can show that $\hat{f}(\xi + i\eta)$ is a rapidly decreasing function of ξ. Indeed $\hat{f}(\xi + i\eta)$ is rapidly decreasing in ξ, uniformly for η in C_ρ. This leads to the following possible theorem.

Question 7.1: Let F be a function on Λ_B such that (i) F is holomorphic in the interior of Λ_B, (ii) F is W-invariant, and (iii) $F(\xi + i\eta)$ is rapidly decreasing as a function of ξ, uniformly in $\eta \in C_\rho$; then is F the Fourier transform of a function f in $\mathscr{C}_1(K\backslash G/K)$?

The obvious thing to do is of course to *define* f by

$$f(x) = w^{-1} \int F(\lambda)\varphi_\lambda(x)|c(\lambda)|^{-2}\, d\lambda, \qquad x \in G \tag{7.1}$$

and show that this f is in $\mathscr{C}_1(K\backslash G/K)$, using the holomorphy of F to shift the integration in the above formula into the complex domain. However, the term-by-term technique of Section 6 gives in this case estimates on the behavior of f only at regular points of A_+. This turns out to be not good enough for our purposes.

Nevertheless, I believe I have a proof which shows that the answer to this question is yes. The idea of the proof is to average the various terms in the expansion

$$\sum_{\mu \in L} \Gamma_\mu(\lambda) \exp[(i\lambda - \mu - \rho)(\log h)] \tag{7.2}$$

for $\Phi_\lambda(h)$ over orbits of the form $[s\lambda;\, s \in W']$ where W' is a suitable subgroup of W and to estimate the functions so obtained. However all the details of this proof have not been checked, so I cannot claim with certainty that the above formulation is correct. If G is $SL(2, R)$, then the theorem is a consequence of results of Ehrenpreis and Mautner [2]. If G is complex, or if rank $(G|K) = 1$, then the methods of Section 5 work and the answer to the above questions is yes. Of course, similar theorems are probably true for $\mathscr{C}_\varepsilon(K\backslash G/K)$. Here instead of the tube $\Lambda_B = \Lambda_0 + iC_\rho$, one would have to work with the tube $\Lambda_{B,\varepsilon} = \Lambda_0 + iC_{\varepsilon\rho}$.

Note added in proof: Since this was written, this question has been completely answered by Trombi and Varadarajan. See Ref. [*22*].

As was remarked above, the algebra $L_1(K\backslash G/K)$ is a commutative Banach algebra which is not a regular Banach algebra. Indeed the only function in $L_1(K\backslash G/K)$ whose Fourier transform has compact support in Λ_B is the zero function. Via the Fourier transform, the algebra $L_1(K\backslash G/K)$ becomes an algebra of holomorphic functions with continuous boundary values on the tube Λ_B. Thus we might ask the question:

Question 7.2: What is the ideal theory of this algebra?

For example, when the rank of $G|K$ is 1, the ideal theory of this algebra seems to have many interesting analogies with the work of Beurling on outer functions for the disk algebra.

Specifically one asks the usual questions of Tauberian and spectral synthetic type. For example:

Question 7.3: Suppose $f \in L_1(K\backslash G/K)$ and $\hat{f}$ never vanishes on Λ_B. Under what further conditions, if any, can we assert that the ideal generated by f is dense in $L_1(K\backslash G/K)$?

The solution to the problem seems to depend on the solution of 7.1 in some way. If one assumes that 7.1 has an affirmative solution, then one can derive from it a number of interesting results about 7.3.

Also, one should note that Ehrenpreis and Mautner [*2*] have interesting answers to this question which make it quite clear that in general one must demand that $\hat{f}$ decay at infinity no faster than at a certain rate, in order that the ideal generated by f be dense in $L_1(K\backslash G/K)$.

There is also

Question 7.4: What kind of closed sets in Λ_B are sets of spectral synthesis?

I know nothing about this question.

It is evident from these remarks that while the L_2-theory is now reasonably complete, no more than a beginning has been made on the theory of $L_1(K\backslash G/K)$. I hope that the exposition in these notes will make accessible the basic formulations on this topic and will prove useful to workers who may wish to investigate this fascinating subject.

References

[*1*] Bhanu-Murthy, T.S.
(*a*) Plancherel's measure for the factor space *SL*(*n*, *R*) / *SO*(*n*). *Dokl. Akad. Nauk SSSR*, **133** (1960), 503–506.
(*b*) The asymptotic behavior of zonal spherical functions on the Siegel upper half-plane. *Dokl. Akad. Nauk SSSR*, **135** (1960), 1027–1030.

[*2*] Ehrenpreis, L., and Mautner, F.I.
Some properties of the fourier transform on semisimple Lie groups, I. *Ann. of Math.*, **61** (1955), 406–439.

[*3*] Erdelyi *et al.*
Higher Transcendental functions, McGraw-Hill, New York, 1953.

[*4*] Gangolli, R.
On the Plancherel formula and the Paley–Wiener theorem for spherical functions semisimple Lie groups. To appear in *Ann. of Math.*

[*5*] Gelfand, I.M.
Spherical functions on Riemannian symmetric spaces. *Dokl. Akad. Nauk SSSR*, **70** (1950), 5–8.

[*6*] Gelfand, I.M., and Naimark, M.
Unitary representations of a unimodular group containing the identity representation of the unitary subgroup. *Trudy Moskov. Mat. Obsc.*, **1** (1952), 423–475.

[*7*] Gelfand, I.M., and Raikov, D.A.
Irreducible unitary representations of locally compact groups. *Mat. Sb.*, **13** (1943), 301–316.

[*8*] Gindikin, S., and Karpelevic, F.
Plancherel measure for Riemannian symmetric spaces of nonpositive curvature. *Soviet Math. Dokl.*, **3** (1962), 962–965.

[*9*] Godement, R.
(*a*) A theory of spherical functions. I. *Trans. Amer. Math. Soc.*, **73** (1952), 496–556.
(*b*) Introduction à travaux de A. Selberg, *Seminaire Bourbaki*, 1957.

[*10*] Harish-Chandra
(*a*) Representations of semisimple Lie groups, I. *Trans. Amer. Math. Soc.*, **75** (1953), 185–243.
(*b*) Representations of semisimple Lie groups, II. *Trans. Amer. Math. Soc.*, **76** (1954), 26–65.
(*c*) Representations of semisimple Lie groups, III. *Trans. Amer. Math. Soc.*, **76** (1954), 234–253.
(*d*) Spherical functions on a semisimple Lie goup, I. *Amer. J. Math.*, **80** (1958), 241–310.
(*e*) Spherical functions on a semisimple Lie goup, II. *Amer. J. Math.*, **80** (1958), 553–613.
(*f*) Discrete series for semisimple Lie groups, II. *Acta Math.*, **116**, (1966), 1–111.

(*g*) On the Plancherel formula for right invariant functions on a semisimple Lie group. *Proc. Nat. Acad. Sci. U.S.A.*, **40** (1954), 200–204.

[*11*] Helgason, S.

(*a*) *Differential Geometry and Symmetric Spaces*, Academic Press, New York, 1962.

(*b*) Fundamental solutions of invariant differential operators on symmetric spaces. *Amer. J. Math.*, **86** (1964), 565–601.

(*c*) An analogue of the Paley–Wiener theorem for the fourier transform on certain symmetric spaces. *Math. Ann.*, **165** (1966), 297–308.

(*d*) Duality and Radon transform for symmetric spaces. *Amer. J. Math.*, **85** (1963), 667–692.

(*e*) *Lie Groups and Symmetric Spaces. Battelle Rencontres, 1967*, W.A. Benjamin, New York, 1968.

(*f*) A duality for symmetric spaces with applications to group representations. *Advances in Math.*, **5** (1970), 1–154.

[*12*] Helgason, S., Johnson, K.
The bounded spherical functions on symmetric spaces. *Advances in Math.*, **3** (1969), 586–593.

[*13*] John, F.
Plane Waves and Spherical Means Applied to Partial Differential Equations, Interscience, New York, 1955.

[*14*] Kostant, B.

(*a*) On the existence and irreducibility of certain series of representations. *Bull. Amer. Math. Soc.*, **75** (1969), 627–642.

(*b*) Unpublished notes of lectures at Bowdoin College.

[*15*] Kunze, R., and Stein, E.
Uniformly bounded representations, II. *Amer. J. Math.*, **83** (1961), 723–786.

[*16*] Lipsman, R.
An explicit realization of Kostant's complementary series with applications to uniformly bounded representations. (Preprint)

[*17*] Rader, C.
Thesis, University of Washington, Seattle, Washington, 1970.

[*18*] Segal, I.

(*a*) Hypermaximality of certain operators on Lie groups. *Proc. Amer. Math. Soc.*, **3** (1952), 13–15.

(*b*) An extension of Plancherel's formula to separable unimodular groups. *Ann. of Math.*, **52** (1950), 272–292.

[*19*] Selberg, A.
Harmonic analysis and discontinuous subgroups in weakly symmetric Riemannian spaces with applications to dirichlet series. *J. Indian Math. Soc.*, **20** (1956), 47–87.

[*20*] Takahashi, R.
Sur les representations unitaires des groupes de Lorenz généralisees. *Bull. Soc. Math. France*, **91** (1963), 289–433.

[*21*] Tamagawa, T.
On Selberg's trace formula. *J. Fac. Sci. Univ. Tokyo*, **8** (1960), 363–386.

[*22*] Trombi, P., and Varadarajan, V.S.
Spherical transforms on semisimple Lie groups. To appear.

[*23*] Wallach, N.

(*a*) Cyclic vectors and irreducibility for principal series representations. To appear.

(*b*) Application of higher osculating spaces to the spherical principal series. To appear.

[*24*] Warner, G.
Representations of Semisimple Lie groups. To appear.

SPECTRA OF DISCRETE UNIFORM SUBGROUPS OF SEMISIMPLE LIE GROUPS

RAMESH GANGOLLI

University of Washington
Seattle, Washington 98105

Preface

This paper, like the preceding one, is based on the lectures which I gave at Washington University in the Spring of 1970. All the prefatory comments in the preceding paper also apply to this one. Indeed, the material here presented is much more tentative. There is not much that is new here, but I have tried to give an account of the basic formulations and a couple of results on spectra of discrete uniform subgroups. The deep ideas contained in Selberg's paper [*7*], which provided the impetus for this subject, have indeed been very fruitful. But in the uniform case, apart from the results of Langlands [*6*] and Gelfand [*2*], few positive results exist. My own interest in this subject is expressed in a paper [*1*] which not unnaturally forms the bulk of the material here presented. I believe I am not far wrong in stating that the uniform case is at present not too well understood in spite of the ease with which the basic results can be proved.

I once again give thanks to the members of my audience.

1. The Spectrum of a Discrete Uniform Subgroup

Let G be a connected semisimple Lie group with finite center and K a maximal compact subgroup of G. Let Γ be a discrete subgroup of G such that $\Gamma\backslash G$ is compact. G acts on $\Gamma\backslash G$ by right translation, and there exists a positive measure dX on $\Gamma\backslash G$, unique up to a scalar factor, which is invariant under this action. Now fix a Haar measure dx on G. For any $f \in C_c(G)$, define the function F on $\Gamma\backslash G$ as follows: $F(X) = \sum_{\gamma\in\Gamma} f(\gamma x)$ where x is any element of G which belongs to the coset $X \in \Gamma\backslash G$. Then we normalize the measure dX in such a way that for each $f \in C_c(G)$, we have

$$\int_G f(x)\, dx = \int_{\Gamma\backslash G} \Big(\sum_{\gamma\in\Gamma} f(\gamma x) \Big)\, dX \tag{1.1}$$

In the future, we shall write for the measure dX the symbol dx even though this is ambiguous in some cases.

If f is a function on G such that $f(\gamma x) = f(x)$ for all $\gamma \in \Gamma$, $x \in G$, then f induces a function on $\Gamma\backslash G$. Ignoring the risk of confusion, we shall call this function f also. For any $p > 0$, let $L_p(\Gamma\backslash G)$ be the space of measurable functions f on G such that (i) $f(\gamma x) = f(x)$, $\gamma \in \Gamma$, $x \in G$, and (ii) $\int_{\Gamma\backslash G} |f(x)|^p\, dx < \infty$. We write $\|f\|_{p,\Gamma} = (\int_{\Gamma\backslash G} |f(x)|^p\, dx)^{1/p}$.

In particular, $L_2(\Gamma\backslash G)$ is a separable Hilbert space. G acts on $L_2(\Gamma\backslash G)$ by right translation, and this action gives rise to a unitary representation

R of G on $L_2(\Gamma\backslash G)$. Indeed, if $x \in G$, $f \in L_2(\Gamma\backslash G)$, then

$$(R_x f)(y) = f(yx), \qquad y \in G. \tag{1.2}$$

The basic problem is to analyze the representation R.

In order to analyze this representation, it is convenient to extend R so that it becomes a representation of various group algebras of functions on G, such as $L_1(G)$, $C_c^\infty(G)$, etc. For example, suppose that $g \in L_1(G)$; then we put $R_g = \int_G g(x)R_x\,dx$. It is obvious that R_g is a well-defined bounded operator on the Hilbert space $L_2(\Gamma\backslash G)$ and $\|R_g\| \leqq \|g\|_1$. If $f \in L_2(\Gamma\backslash G)$ and $g \in L_1(G)$, then

$$\begin{aligned}(R_g f)(y) &= \int_G g(x)(R_x f)(y)\,dx \\ &= \int_G f(yx)g(x)\,dx \\ &= \int_G f(yx)g(x^{-1})\,dx \\ &= (f * \tilde{g})(y) \qquad \text{where} \quad \tilde{g}(x) = g(x^{-1}).\end{aligned} \tag{1.3}$$

Thus the operator R_g indicates merely right convolution by $\tilde{g}$.

Clearly $g \to R_g$ is a representation of the convolution algebras $C_c^\infty(G)$, $L_1(G)$. It is obviously continuous with respect to the usual topologies on these spaces. We shall study $L_2(\Gamma\backslash G)$ as a $C_c^\infty(G)$-module.

Lemma 1: If $g \in C_c^\infty(G)$, then R_g is a compact operator on $L_2(\Gamma\backslash G)$.

Proof: First let $f \in L_2(\Gamma\backslash G)$ be continuous. Then

$$\begin{aligned}(R_g f)(y) &= (f * \tilde{g})(y) \\ &= \int_G f(x)\,\tilde{g}(x^{-1}y)\,dx \\ &= \int_G f(x)g(y^{-1}x)\,dx \\ &= \int_G \Big(\sum_{\gamma\in\Gamma} f(\gamma x)g(y^{-1}\gamma x)\Big)\,dx \qquad \text{by (1.1)} \\ &= \int_G f(x)\left(\sum_{\gamma\in\Gamma} g(y^{-1}\gamma x)\right)\,dx, \qquad \because f(\gamma x) = f(x)\end{aligned}$$

$$= \int_G f(x)\, G(x, y)\, dx \tag{1.4}$$

where $G(x, y) = \int_{\gamma\in\Gamma} g(y^{-1}\gamma x)$.

Consider $G(x, y) = \sum_{\gamma\in\Gamma} g(y^{-1}\gamma x)$. If $(x, y) \in U \times U$ where U is any compact subset of G, then $y^{-1}\gamma x \in \operatorname{supp}(g)$ only if $\gamma \in U \operatorname{supp}(g)U^{-1}$. Since supp($g$) is compact, and Γ is a discrete subgroup, there will be only a finite number of $\gamma \in \Gamma$ such that $\gamma \in U \operatorname{supp}(g)U^{-1}$. It follows that the series $\sum_{\gamma\in\Gamma} g(y^{-1}\gamma x)$ is a finite sum when $(x, y) \in U \times U$. Hence $G(x, y)$ is a continuous function of (x, y) on $G \times G$. Moreover, $G(\gamma_1 x, \gamma_2 y) = G(x, y)$ for all $\gamma_1, \gamma_2 \in \Gamma$ and $x, y \in G$. Thus $G(x, y)$ is a continuous function on $\Gamma\backslash G \times \Gamma\backslash G$. Now (1.4) shows that R_g is an integral operator on $L_2(\Gamma\backslash G)$ whose kernel is $G(x, y)$. Since $\Gamma\backslash G$ is a finite measure space and $G(x, y)$ is continuous, it follows from the classical theory of integral operators that R_g is compact.

Now consider the algebra of operators $\mathfrak{a} = \{R_g; g \in C_c^\infty(G)\}$. It is obviously closed under the operation of taking adjoints, because $R_g = R_{g^*}$ where $g^*(x) = \overline{g(x^{-1})}$. Moreover, let $\{g_n, n = 1, 2, \ldots\}$ be a sequence of functions in $C_c^\infty(G)$ such that (i) $g_n \geqq 0$, (ii) $\int_G g_n(x)\, dx = 1$, and (iii) if $U_n = \operatorname{supp} g_n$, then $U_{n+1} \subset U_n$ and $\cap_{n\geqq 1} U_n = \{e\}$. Then it is easily seen that for each $f \in L_2(\Gamma\backslash G)$, $\|f * g_n - f\|_{2,\Gamma} \to 0$ as $n \to \infty$. This means that $R_{g_n} \to$ identity, strongly on $L_2(\Gamma\backslash G)$. Thus $\mathfrak{a}$ is an algebra of compact operators on $L_2(\Gamma\backslash G)$ closed under adjoints, and the strong closure of $\mathfrak{a}$ contains the identity. We now have the following theorem.

Theorem 2: Let $\mathfrak{a}$ be an algebra of compact operators on a separable Hilbert space H such that (i) $T \in \mathfrak{a} \Rightarrow T^* \in \mathfrak{a}$, and (ii) the strong closure of $\mathfrak{a}$ contains the identity operator on H.

Then, H is the orthogonal sum of a countable number of minimal $\mathfrak{a}$-stable closed subspaces H_i, $i \geqq 0$. Moreover, given any $i \geqq 0$, there exists only a finite number of subspaces H_j in this family such that H_j is isomorphic to H_i as an $\mathfrak{a}$-module.

Proof: Let T be a self-adjoint element of $\mathfrak{a}$. Then by the spectral theorem for compact operators, the spectrum of T is discrete, say 0, $\lambda_1(T)$, $\lambda_2(T)$, . . . , etc., and

$$H = H_{T,0} + \sum_{k\geqq 1} H_{T,\lambda_k(T)} \tag{1.5}$$

where the sum is orthogonal, and $H_{T,0}$ is the null-space of T and $H_{T,\lambda k(T)}$ is the subspace of H consisting of eigenvectors of T with eigenvalue $\lambda_k(T)$. Moreover $H_{T,\lambda k(T)}$ is finite dimensional for each $k \geqq 1$.

Let S be the collection of all closed $\mathfrak{a}$-stable subspaces L of H such that $L \cap H_{T,\lambda_1(T)} \neq \{0\}$. Then, since $\dim H_{T,\lambda_1(T)} < \infty$, there exists a member H' of S, such that $H' \cap H_{T,\lambda_1(T)}$ has minimum dimension. Fix such an H' and put $H'_{T,\lambda_1(T)} = H' \cap H_{T,\lambda_1(T)}$.

Now let $\mathfrak{G}$ be the collection of all the closed $\mathfrak{a}$-stable subspaces L of H such that $L \cap H_{T,\lambda_1(T)} = H_{T,\lambda_1(T)}$, and let $H_0 = \cap_{L\in\mathfrak{F}} L$. Then $H_0 \in \mathfrak{G}$. We claim that H_0 is irreducible under $\mathfrak{a}$. Indeed suppose $H_{01} + H_{02} = H_0$ with H_{01}, H_{02} closed and $\mathfrak{a}$-stable. Then because $H_0 \in \mathfrak{G}$,

$$H'_{T,\lambda_1(T)} = H_0 \cap H_{T,\lambda_1(T)} = H_{01} \cap H_{T,\lambda_1(T)} + H_{02} \cap H_{T,\lambda_1(T)}. \quad (1.6)$$

If both members on the right were nontrivial, then $0 < \dim(H_{01} \cap H_{T,\lambda_1(T)}) < \dim H'_{T,\lambda_1(T)}$. This would mean that H_{01} is a closed $\mathfrak{a}$-stable subspace of H whose intersection with $H_{T,\lambda_1(T)}$ has smaller dimension than that of $H'_{T,\lambda_1(T)}$. This is impossible by the choice of $H'_{T,\lambda_1(T)}$. Hence one of the two terms on the right side must be trivial. Suppose $H_{02} \cap H_{T,\lambda_1(T)} = \{0\}$. Then $H_{01} \cap H_{T,\lambda_1(T)} = H'_{T,\lambda_1(T)}$. So, $H_{01} \in \mathfrak{G}$. But $H_0 = \cap_{L\in L} L$. Hence $H_0 \subset H_{01}$ and hence $H_0 = H_{01}$. This proves that H_0 is irreducible.

Let $H_0^\perp$ be the orthogonal complement of H_0 in H. Then, since $\mathfrak{a}$ is self-adjoint, $H_0^\perp$ is $\mathfrak{a}$-stable. If $H_0^\perp = \{0\}$, we are finished. So suppose $H_0^\perp \neq \{0\}$. Then we claim that there exists a self-adjoint element $A \in \mathfrak{a}$ such that the restriction of A to $H_0^\perp$ is not zero. Suppose the contrary, and let $x \in H_0^\perp$, $x \neq 0$. Then $Ax = 0$ for every self-adjoint $A \in \mathfrak{a}$. Now, because of hypothesis (ii), there exists a sequence $T_n \in \mathfrak{a}$ such that $T_n x \to x$ as $n \to \infty$. But then $\frac{1}{2}(T_n + T_n^*)x \to x$ also. But since $(T_n + T_n^*)/2$ is a self-adjoint element of $\mathfrak{a}$, the above means that $((T_n + T_n^*)/2)x = 0$. Hence $x = 0$ which is a contradiction. This proves the claim.

We can now use the same argument on the Hilbert space $H_0^\perp$ and the algebra of restrictions of $\mathfrak{a}$ to $H_0^\perp$ that we have used on H and $\mathfrak{a}$. We will then get a closed $\mathfrak{a}$-stable irreducible subspace H_1 of $H_0^\perp$. Putting $H_1^\perp$ equal to the orthogonal complement of H_1 in $H_0^\perp$, we get $H = H_0 + H_1 + H_1^\perp$.

Proceeding in this way, we see that we get closed $\mathfrak{a}$-stable irreducible subspaces $\{H_i\}_{i\geqq 0}$, such that $H = \oplus_{i\geqq 0} H_i$.

Next, fix an index i, $i = 0$ say, and consider the subspaces H_{i_k}, $k = 1, \ldots, m$ (possibly ∞), that H_{i_k} is isomorphic to H_0 as an $\mathfrak{a}$-module. We claim that the number of such indices is finite, i.e., that $m < \infty$. For let T be an element in $\mathfrak{a}$ such that $T|_{H_0} \neq 0$. Then there is a $v_0 \neq 0 \in H_0$ such that $Tv_0 = \lambda v_0$ and $\lambda \neq 0$. Now since H_{i_k} is isomorphic to H_0 as an $\mathfrak{a}$-module, there exists a vector $v_k \neq 0 \in H_{i_k}$ such that $Tv_k = \lambda v_k$. But then $v_1, v_2, \ldots,$ are linearly independent, and belong to the eigenspace of T corresponding to the eigenvalue λ. But since T is compact, this eigenspace is finite dimensional. Hence m must be finite. This concludes the proof of the theorem.

As a consequence of this theorem, it is obvious that there exist integers n_i, $i = 0, 1, 2, \ldots$ and closed subspaces H_{ij}, $1 \leqq j \leqq n_i$, $i \geqq 0$ of $L_2(\Gamma/G)$ such that

(a) Each H_{ij} is stable under the representation R.
(b) $L_2(\Gamma\backslash G) = \sum_{i \leqq 0} \sum_{1 \leqq j \leqq n_i} H_{ij}$, orthogonal direct sum.
(c) H_{ij} is irreducible under R; moreover, for fixed i, the restrictions of R to H_{ij}, $1 \leqq j \leqq n_i$ are equivalent, while if $i \neq i'$ the restrictions of R to H_{ij} and $H_{i'j'}$ are not equivalent.

Let $\hat{G}$ be the set of equivalence classes of irreducible unitary representations of G. As we saw above, for any $i \geqq 0$, the restrictions of R to H_{ij}, $i \leqq j \leqq n_i$, give rise to irreducible unitary representations of G belonging to the same equivalence class. We shall denote this class by α_i.

Definition 3: By the spectrum of Γ we mean the sequence of ordered pairs (α_i, n_i), $i \geq 0$, where α_i is a class occurring in the above, and n_i is the multiplicity which it occurs.

Obviously the spectrum of Γ contains important information about Γ itself. Two important problems can now be formulated. First, one may ask: Given Γ and G, which classes $\alpha \in \hat{G}$ occur in the spectrum of Γ, and with what multiplicity? As an example of the sense in which this question may be interpreted, let $G = KAN$ be the Iwasawa decomposition and let M be the centralizer of A in K. Let $B = MAN$. Then given a class $\delta \in \hat{M}$ and a unitary character $\lambda \in \hat{A}$, the pair (δ, λ) gives a representation $\xi_{\delta,\lambda}$ of B, and one can construct a unitary representation $\pi_{\delta,\lambda}$ of G by inducing up from $\xi_{\delta,\lambda}$. One would like to know for which pairs δ, λ the class of $\pi_{\delta,\lambda}$ occurs in the spectrum of Γ. Moreover, one would like to determine the multiplicity with which it occurs, in terms of the data δ, λ.

Ths problem is very far from being solved at the present state of our knowledge. There is an important special case in which it has been solved by Langlands. We will describe his result in Section 3.

The second problem is as follows: To what extent does the spectrum of

Γ determine Γ? For example, suppose Γ and Γ' have the same spectra. Are they then isomorphic as abstract groups? Once again, very little is known in answer to this problem in general. However, we shall see below that at least in one special case this problem has a surprisingly nice answer.

We shall now consider the subspace $L_2(\Gamma\backslash G/K)$ of $L_2(\Gamma\backslash G)$. This subspace consists of those functions f in $L_2(\Gamma\backslash G)$ such that $f(xk) = f(x)$ for all $k \in K$, $x \in G$. This means that $L_2(\Gamma\backslash G/K) = \{f | f \in L_2(\Gamma\backslash G);\ R_k f = f$ for all $k \in K\}$. If $\mathbf{1} \in \hat{K}$ is the class of the trivial representations of K, then $L_2(\Gamma\backslash G/K)$ is just the subspace of vectors in $L_2(\Gamma\backslash G)$ that transform under R according to $\mathbf{1}$. If $R_{\mathbf{1}} = \int_K R(k)\,dk$, it is obvious that $R_{\mathbf{1}}$ is the orthogonal projection sending $L_2(\Gamma\backslash G)$ onto $L_2(\Gamma\backslash G/K)$. Now suppose $f \in L_2(\Gamma\backslash G/K)$; then obviously

$$f = \sum_{i \geqq 0} \sum_{1 \leqq j \leqq n_i} f_{ij} \qquad f_{ij} \in H_{ij}. \tag{1.7}$$

But $f = R_{\mathbf{1}} f$ so

$$f = \sum_{i \geqq 0} \sum_{1 \leqq j \leqq n_i} R_{\mathbf{1}} f_{ij}. \tag{1.8}$$

But obviously $R_{\mathbf{1}} H_{ij} \subset H_{ij}$, so $R_{\mathbf{1}} f_{ij} \in H_{ij}$. Hence comparing (1.7) and (1.8), we conclude

$$R_{\mathbf{1}} f_{ij} = f_{ij}, \qquad i \geqq 0, \quad i \leqq j \leqq n_i. \tag{1.9}$$

Now recall that the restriction of R to H_{ij} induces on H_{ij} a representation of G belonging to $\alpha_i \in \hat{G}$. If the class α_i is of class 1 with respect to K, then H_{ij} will contain a vector $v_{ij} \neq 0$ such that $R_{\mathbf{1}} v_{ij} = v_{ij}$. Otherwise $R_{\mathbf{1}} H_{ij} = \{0\}$. Thus (1.9) means that $f_{ij} = 0$ for all those indices i for which α_i is not of class $\mathbf{1}$ with respect to K. On the other hand, if α_i *is* of class $\mathbf{1}$ with respect to K, then $R_{\mathbf{1}} H_{ij}$ is of dimension 1. Let φ_{ij} be a unit vector in $R_{\mathbf{1}} H_{ij}$. Then we see that

$$f = \sum_{i \geqq 0}{}' \sum_{1 \leqq j \leqq n_i}{}' a_{ij} \varphi_{ij} \tag{1.10}$$

where $\sum'$ means that the sum is over those i for which α_i is of class $\mathbf{1}$ with respect of K. Since we shall be concerned only with $L_2(\Gamma\backslash G/K)$ in what follows, we shall restrict ourselves to these classes. Thus we will omit the prime in (1.10), and write it

$$f = \sum_{i \geqq 0} \sum_{1 \leqq j \leqq n_i} a_{ij} \varphi_{ij} \tag{1.11}$$

with i running only over those indices for which $\alpha_i \in \hat{G}_0$ where $\hat{G}_0$ consists of those $\alpha \in \hat{G}$ that are of class **1**. This implies that we have done some renumbering of the indices appearing in (1.7)–(1.8). Note that the functions φ_{ij} are unique up to a scalar factor of absolute value 1.

The expansion (1.11) should be regarded as the Fourier series for f. The functions φ_{ij} play the role in the Fourier analysis of $L_2(\Gamma\backslash G/K)$ analogous to that of representative functions in harmonic analysis on compact Lie groups. In particular, the functions φ_{ij} have a number of nice properties with respect to convolution and differentiation. These are developed in the next theorem.

Theorem 4: (a) Each function φ_{ij} is an eigenfunction of every operator of the form R_g, $g \in C_c(K\backslash G/K)$. Thus $R_g\varphi_{ij} = \chi_{ij}(g)\varphi_{ij}$. The mapping χ_{ij}: $g \to \chi_{ij}(g)$ is a nonzero homomorphism of the convolution algebra $C_c(K\backslash G/K)$ into the complex numbers.

(b) Each $\dot{\varphi}_{ij}$ is C^∞ and is an eigenfunction of each differential operator $D \in \mathcal{D}(G/K)$. Thus $D\varphi_{ij} = \chi_{ij}(D)\varphi_{ij}$. The mapping χ_{ij}: $D \to X_{ij}(D)$ is a homomorphism of $\mathcal{D}(G/K)$ into the complex numbers.

(c) Each φ_{ij} is an analytic function on G/K, and $\varphi_{ij}(e) \neq 0$.

(d) Let $\varphi_{ij}^{\#}(x) = \int_K \varphi_{ij}(kx)\, dk/\varphi_{ij}(e)$. Then $\varphi_{ij}^{\#}(x)$ is an elementary spherical function; for any $D \in \mathcal{D}(G/K)$, $D(\varphi_{ij}^{\#}) = \chi_{ij}(D)\varphi_{ij}^{\#}$.

(e) For a fixed i, the elementary spherical functions $\varphi_{ij}^{\#}$, $i \leqq j \leqq n_i$, are all the same. Indeed $\varphi_{ij}^{\#} = \varphi_i$, where φ_i is the unique elementary positive definite spherical function associated with the class $\alpha_i \in \hat{G}_0$. In particular, the homomorphisms χ_{ij}, $i \leqq j \leqq n_i$, are all the same, equal to χ_i, say.

(f) If $i \neq i'$, then $\chi_i \neq \chi_{i'}$.

Proof: (a) Since $\varphi_{ij} \in H_{ij}$ and $R_x H_{ij} \subset H_{ij}$ for $x \in G$, it is obvious that $R_g\varphi_{ij} \in H_{ij}$ for any $g \in C_c(G)$. If moreover, $g \in C_c(K\backslash G/K)$, then

$$\begin{aligned} R_k R_g \varphi_{ij} &= R_k(\varphi_{ij} * \tilde{g}) \\ &= \varphi_{ij} * R_k\tilde{g} \\ &= \varphi_{ij} * \tilde{g} \\ &= R_g\varphi_{ij}, \qquad k \in K. \end{aligned} \tag{1.12}$$

Hence $R_g\varphi_{ij}$ is a vector in H_{ij} that is fixed by each R_k, $k \in K$. But the space of such vectors in H_{ij} is spanned by the element φ_{ij}. Hence $R_g\varphi_{ij} = \chi_{ij}(g)$ φ_{ij} with $\chi_{ij}(g) \in C$. It is obvious that χ_{ij} is a homomorphism of the convolution algebra $C_c(K\backslash G/K)$. Moreover, χ_{ij} is not zero. For if it were, then this would mean that $R_g\varphi_{ij} = 0$ for all $g \in C_c(K\backslash G/K)$, i.e., that $\varphi_{ij} * \tilde{g} = 0$ for all $\tilde{g} \in C_c(K\backslash G/K)$. It follows by an approximation argument that if m is the Haar measure of K, regarded as a measure with compact support on G, then $\varphi_{ij} * m = 0$ but $\varphi_{ij} * m = \varphi_{ij}$ because $\varphi_{ij}(xk) = \varphi_{ij}(x)$ for all $k \in K$. Hence we would get $\varphi_{ij} = 0$. This contradicts the fact that $\|\varphi_{ij}\|_{2,\Gamma} = 1$. Hence $\chi_{ij} \neq 0$.

(b) Let $g \in C_c(K\backslash G/K)$. Then $\varphi_{ij} * \tilde{g} = \chi_{ij}(g)\varphi_{ij}$. Moreover, the above argument shows that $\exists\ g \in C_c^\infty(K\backslash G/K)$ such that $\chi_{ij}(g) \neq 0$. Hence $\varphi_{ij} = (\varphi_{ij} * \tilde{g})/X\int_j(g)$, for this g. Now let $D \in \mathscr{D}(G/K)$. Then

$$\begin{aligned} D\varphi_{ij} &= D(\varphi_{ij} * \tilde{g}/\chi_{ij}(g)) \\ &= \varphi_{ij} * D\tilde{g}/\chi_{ij}(g). \end{aligned} \tag{1.13}$$

But since $g \in C_c^\infty(K\backslash G/K)$ and $D \in \mathscr{D}(G/K)$, it is obvious that $Dg \in C_c^\infty (K\backslash G/K)$. Hence $\varphi_{ij} * D\tilde{g} = \chi_{ij}((D\tilde{g})^\sim)\varphi_{ij}$. Hence

$$D\varphi_{ij} = \frac{\chi_{ij}((D\tilde{g})^\sim)}{\chi_{ij}(g)}\,\varphi_{ij} \tag{1.14}$$

so (b) is true with $\chi_{ij}(D) = \chi_{ij}((D\tilde{g})^\sim)/\chi_{ij}(g)$. It is obvious that $\chi_{ij}(D)$ does not depend on the choice of g.

(c) Since the Laplace–Beltrami operator of G/K is in $\mathscr{D}(G/K)$ and is elliptic, it follows that φ_{ij} is analytic by Bernstein's theorem. Now suppose $\varphi_{ij}(e) = 0$. Then $(D\varphi_{ij})(e) = \chi_{ij}(D)\varphi_{ij}(e) = 0$ for all $D \in \mathscr{D}(G|K)$. Thus as a function on $G|K$, φ_{ij} and all its "derivatives" vanish at e. By Taylor's theorem, we conclude $\varphi_{ij} = 0$, since it is analytic. This is a contradiction, so $\varphi_{ij}(e) \neq 0$.

(d) Since $\varphi_{ij}^\#(xk) = \varphi_{ij}^\#(x)$, it is obvious that $\varphi_{ij}^\#(k'xk) = \varphi_{ij}^\#(x)$ for all $k', k \in K$. Moreover, $\varphi_{ij}^\#(e) = \int_K \varphi_{ij}(k)\,dk/\varphi_{ij}(e) = 1$, because $\varphi_{ij}(k) = \varphi_{ij}(ek) = \varphi_{ij}(e)$. Now let $D \in \mathscr{D}(G/K)$. Then since D is a left-invariant differential operator, we see that

$$\begin{aligned} (D(\varphi_{ij}^\#))(x) &= (1/\varphi_{ij}(e)) \int_K (D\varphi_{ij})(kx)\,dk \\ &= \chi_{ij}(D)\left(\int_K \varphi_{ij}(kx)\,dk\right)\Big/\varphi_{ij}(e) \\ &= \chi_{ij}(D)\varphi_{ij}^\#(x). \end{aligned} \tag{1.15}$$

Thus $\varphi_{ij}^\#$ is an eigenfunction of each $D \in \mathscr{D}(G|K)$. It follows that $\varphi_{ij}^\#$ is an elementary spherical function.

(e) Fix i, j. Consider the function $\psi_{ij}(x) = \langle R_x\varphi_{ij}, \varphi_{ij}\rangle$ where $\langle , \rangle$ is the inner product in $L_2(\Gamma\backslash G)$. It is obvious that $\psi_{ij}(x)$ is positive definite. Since $R_k\varphi_{ij} = \varphi_{ij}$ for all $k \in K$, it is clear that $\psi_{ij}(k_1xk_2) = \psi_{ij}(x)$. Thus ψ_{ij} is spherical; also $\psi_{ij}(e) = 1$. Now, the restriction of R to H_{ij} is an irreducible unitary representation of G, of class **1** with respect to K. It follows that ψ_{ij} is the unique positive definite spherical function associated with this irreducible unitary representation. Moreover, for a fixed i, the representations in the different H_{ij}, $1 \leqq j \leqq n_i$ are all equivalent and are all members of the class α_i. Hence their elementary spherical functions are all the same. Thus all these ψ_{ij} are the same, equal to φ_i, the elementary positive definite spherical function associated with the class α_i.

We shall now identify ψ_{ij} with $\varphi_{ij}^{\#}$. This will prove (e). We have

$$\begin{aligned}\varphi_{ij}(x) &= \langle R_x\varphi_{ij}, \varphi_{ij}\rangle \\ &= \int_{\Gamma\backslash G} \varphi_{ij}(yx)\,\overline{\varphi_{ij}(y)}\,dy.\end{aligned} \tag{1.16}$$

If X is any left-invariant differential operator on G, we can regard it as a differential operator on $\Gamma\backslash G$. It is easily seen, since $\Gamma\backslash G$ is compact and φ_{ij} is C^∞, that

$$(X\psi_{ij})(x) = \int_{\Gamma\backslash G} (X\varphi_{ij})(yx)\,\overline{\varphi_{ij}(y)}\,dy. \tag{1.17}$$

Now if $D \in \mathscr{D}(G/K)$, we have $D\varphi_{ij} = \chi_{ij}(D)\varphi_{ij}$ so

$$\begin{aligned}(D\psi_{ij})(x) &= \int_{\Gamma\backslash G} \chi_{ij}(D)\,\varphi_{ij}(yx)\,\overline{\varphi_{ij}(y)}\,dy \\ &= \chi_{ij}(D)\psi_{ij}(x).\end{aligned} \tag{1.18}$$

This shows that the elementary spherical functions $\varphi_{ij}^{\#}$ and ψ_{ij} both have the same system of eigenvalues with respect to $\mathscr{D}(G/K)$. This means that $\psi_{ij} = \varphi_{ij}^{\#}$. Since all the $\psi_{ij} = \varphi_i$ for $1 \leqq j \leqq n_i$, (e) is proved.

(f) If $X_i = X_{i'}$, this would mean that the elementary spherical functions φ_i, φ_i' would have the same system of eigenvalues with respect to $\mathscr{D}(G/K)$. They would then have to be identically equal. However, if $i \neq i'$, the classes α_i, $\alpha_{i'}$ are distinct classes in $\hat{G}_0$. Hence their elementary spherical functions φ_i, $\dot{\varphi}_{i'}$ cannot be equal. This proves (f).

The functions $\dot{\varphi}_{ij}$ may be called the elementary automorphic functions. We say that the functions φ_{ij}, $1 \leqq j \leqq n_i$ *lie over* the elementary spherical function φ_i.

Let χ be a homomorphism of $\mathscr{D}(G/K)$ into C. We denote by $A(\chi)$ the set of functions $f \in L_2(\Gamma\backslash G/K)$ such that $Df = \chi(D)f$ for all $D \in \mathscr{D}(G|K)$. $A(\chi)$ is the space of Γ-automorphic functions on G/K which belong to the homomorphism χ.

Corollary 5: $A(\chi) = \{0\}$ unless $\chi = \chi_i$ for some i. Moreover the functions φ_{ij}, $1 \leqq j \leqq n_i$, are a basis of $A(\chi_i)$. In particular, $\dim A(\chi_i) = n_i < \infty$.

Proof: Suppose $f \in A(\chi)$. Then

$$f = \sum_{i \geqq 0} \sum_{1 \leqq j \leqq n_i} a_{ij}\varphi_{ij}.$$

Now let $D \in \mathscr{D}(G|K)$. Then $Df \in L_2(\Gamma\backslash G/K)$ and

$$\begin{aligned} Df &= \sum_i \sum_j a_{ij}\, D\varphi_{ij} \\ &= \sum_i \sum_j a_{ij}\chi_i(D)\varphi_{ij} \end{aligned}$$

But $Df = \chi(D)f$ so we get

$$a_{ij}(\chi_i(D) - \chi(D)) = 0 \qquad \text{for all } i, j.$$

Thus if $\chi_i \neq \chi$, for any i, we conclude that $a_{ij} = 0$ for all i, j. Thus $A(\chi) = \{0\}$ if $\chi \neq \chi_i$ for any i.

On the other hand if $\chi = \chi_i$, then the above argument shows that $a_{i'j'} = 0$ for all $i' \neq i$ and any $1 \leqq j' \leqq n_{i'}$. Therefore $f \in A(\chi_i) \Rightarrow f = \sum_{1 \leqq j \leqq n_i} a_{ij}\varphi_{ij}$. Hence $\{\varphi_{ij}, 1 \leqq j \leqq n_i\}$ span $A(\chi_i)$. Since they are evidently linearly independent, we conclude that they form a basis of $A(\chi_i)$. This proves the corollary.

Corollary 6: The dimension of $A(\chi_i)$ is equal to the multiplicity of the class α_i in the decomposition of the representation R.

Indeed the multiplicity is n_i.

Of course, the homomorphism χ_i of $\mathscr{D}(G/K)$ determines (cf. the paper preceding this one) the elementary spherical function φ_i and hence also a class $\alpha_i \in \hat{G}$. Obviously, α_i is of class **1** with respect to K. Thus the data

$(\alpha_i, n_i)_{i \leq 0}$ is a piece of the full spectrum of Γ, consisting of those classes α_i that are of class **1** with respect to K. We may call this the **1**-spectrum of Γ, denoting it by spec(Γ, **1**).

Consider the elementary spherical function φ_i. We know that there is some point of Λ_P, say $-\lambda_i$, such that $\varphi_{-\lambda_i} = \varphi_i$. Thus $\chi_i(D) = \Gamma(D)(-\lambda_i)$, in the notation of the last chapter. It is obvious that $-\lambda_i$ is unique modulo action of W on Λ_P.

Theorem 7. Let $g \in C_c(K\backslash G/K)$. Then

$$R_g\varphi_{ij} = \hat{g}(\lambda_i)\varphi_{ij}, \qquad 1 \leqq j \leqq n_i \tag{1.19}$$

where

$$\hat{g}(\lambda_i) = \int_G g(y)\varphi_{\lambda i}(y^{-1})\, dy \tag{1.20}$$

is the Fourier transform of g.

Proof: We know that $R_g\varphi_{ij}$ is a constant multiple of φ_{ij}, say $R_g\varphi_{ij} = C_{ij}\varphi_{ij}$. It is obvious that

$$\begin{aligned} C_{ij} &= (R_q\varphi_{ij})(e)/\varphi_{ij}(e) \\ &= \left(\int_G \varphi_{ij}(y)g(y)\, dy\right)/(\varphi_{ij}(e)). \end{aligned} \tag{1.21}$$

Now $\int_G \varphi_{ij}(y)g(y)\, dy = \int_G (\int_K \varphi_{ij}(ky)\, dk)g(y)\, dy$ because $g(ky) = g(y)$ for all k. Hence

$$\begin{aligned} C_{ij} &= \int_G \left(\left(\int \varphi_{ij}(ky)\, dk\right)\Big/(\varphi_{ij}(e))\right)g(y)\, dy \\ &= \int_G \varphi_i(y)g(y)\, dy \\ &= \int_G \varphi_{-\lambda_i}(y)g(y)\, dy \\ &= \int_G g(y)\varphi_{\lambda_i}(y^{-1})\, dy \\ &= \hat{g}(\lambda_i). \qquad \text{Q.E.D.} \end{aligned} \tag{1.22}$$

Corollary 8: Suppose $f \in C_c(K\backslash G/K)$ and let $g = f * f^*$ where

$f^*(x) = f(x^{-1})$. Then the operator R_g is of trace class and

$$\text{Trace } R_g = \int_G \sum_{\gamma \in \Gamma} g(x^{-1}\gamma x)\, dx$$

$$= \sum_{i=0} \hat{g}(\lambda_i) n_i. \tag{1.23}$$

Proof: $R_g = R_f R_f^*$; hence R_g is a positive operator. It is obvious that R_f is a Hilbert–Schmidt operator. Hence R_g is of trace class. We saw above that the φ_{ij} are the eigenfunctions of R_g with eigenvalue $\hat{g}(\lambda_i)$. Since there are n_i of them corresponding to this eigenvalue, the corollary follows.

Formula (1.23) which expresses the trace of R_g in two different ways is known as the Selberg Trace Formula. The ideas behind it are contained in Selberg's paper [*7*]. (See also papers [*6*] and [*8*]).

While the formula (1.23) was proved above for $g \in C_c(K\backslash G/K)$, it can easily be envisaged for a wider class of functions g than the class $C_c(K\backslash G/K)$. For example if g is any continuous spherical function in $L_1(K\backslash G/K)$, and if $\sum_{\gamma \in \Gamma} g(y^{-1}\gamma x) = G(x, y)$ defines a continuous function on $\Gamma\backslash G \times \Gamma\backslash G$, then the above argument can be substantially repeated and there results a trace formula for R_g. We will not give the details here but will content ourselves with stating a definition and theorem.

Definition 9: A Continuous spherical function g is said to be admissible if the series $\sum_{\gamma \in \Gamma} g(y^{-1}\gamma x)$ converges absolutely, uniformly for (x, y) in compact subset of $G \times G$.

It is clear that if g is admissible, then the function $G(x, y) = \sum_{\gamma \in \Gamma} g(y^{-1}\gamma x)$ is a continuous function on $G \times G$, and further $G(\gamma x, \gamma' x) = G(x, y)$ for all $(x, y) \in G \times G$ and $(\gamma, \gamma') \in \Gamma \times \Gamma$.

We now state:

Theorem 10 (Selberg [*7*], Tamagawa [*8*]): Suppose g is an admissible function in $L_1(K\backslash G/K)$ and let $\hat{g}(\lambda)$ be the Fourier transform of g. Suppose that the series $\sum_i \hat{g}(\lambda_i) n_i$ converges absolutely. Then the operator R_g has a trace, and

$$\text{Trace } R_g = \int_{\Gamma\backslash G} \sum_{\gamma \in \Gamma} g(x^{-1}\gamma x)\, dx$$

$$= \sum_i \hat{g}(\lambda_i) n_i.$$

We note here that if g is of the form $f * f^*$ where $f^*(x) = f(x^{-1})$, then $R_g = R_f R_{f^*}$. Thus R_g is a positive operator. Hence by Mercer's theorem, it is automatic that R_g has a trace and that $\sum \hat{g}(\lambda_i) n_i$ converges absolutely (note that $\hat{g} \geqq 0$ in this case).

The point of Theorem 10 is that the class $C_c^\infty(K\backslash G/K)$ does not contain some of the functions to which we wish to apply the trace formula, as we shall see below. The class of admissible functions is wider.

Though the class of admissible functions is wider, it is often a difficult matter to decide whether a given function g in $L_1(K\backslash G/K)$ is admissible or not. This is in fact a major difficulty in using the trace formula.

2. Asymptotic Behavior of spec (Γ, 1)

Let α_i be the classes occurring in the decomposition of $L_2(\Gamma\backslash G/K)$ under $\{R_g,\ g \in C_c^\infty(K\backslash G/K)\}$ with multiplicities n_i. As we saw above, each α_i is of class **1**, with elementary spherical function φ_i, say. Further, there is a point $\lambda_i \in \Lambda$, unique modulo the action of W on Λ so that $\varphi_i = \varphi_{-\lambda_i}$. Ideally, we would like to be able to say, in terms of the structure of Γ, precisely which these λ_i are, and moreover, we would like to express n_i in terms of the λ_i. However, this problem is very far from being solved even in the simplest cases. Not being able to get precise information about the integers n_i, one has to content oneself with some sort of gross information about their distribution.

Let ω be the Laplace–Beltrami operator of G/K; clearly $\omega \in \mathscr{D}(G|K)$, and ω is elliptic: Moreover each $\varphi_{-\lambda_i}$ is an eigenfunction of ω; indeed $\omega\varphi_{-\lambda_i} = -(\langle\lambda_i, \lambda_i\rangle + \langle\rho, \rho\rangle)\varphi_{-\lambda_i}$, where ρ is the half-sum of the positive roots of (G, K). Moreover it is not hard to see that, since $\varphi_{-\lambda_i}$ is positive definite, so that its absolute value has a local maximum at e, the eigenvalue $-(\langle\lambda_i, \lambda_i\rangle + \langle\rho, \rho\rangle)$ must be nonpositive. Thus $\langle\lambda_i, \lambda_i\rangle + \langle\rho, \rho\rangle$ is nonnegative.

The eigenvalue $\langle\lambda_i, \lambda_i\rangle + \langle\rho, \rho\rangle$ provides a rough measure of the "size" of λ_i and we now introduce for a real $r > 0$, the function

$$N(r) = \sum_{\{i;\ <\lambda_i\lambda_i> + <\rho,\rho> \leqq r\}} n_i. \tag{2.1}$$

Thus $N(r)$ is the sum of the integers n_i corresponding to those λ_i whose "size" (as measured by the above eigenvalue of ω) is less than r. It is not clear at the moment that $N(r)$ is finite valued. However we shall see below that there are only a finite number of λ_i such that $\langle\lambda_i, \lambda_i\rangle + \langle\rho, \rho\rangle \leqq r$, so that $N(r)$ is indeed finite valued. Granting this, $N(r)$ is a very rough measure of the "size" of spec(Γ, **1**). It is the asymptotic distribution of $N(r)$ as $r \to \infty$ that we wish to discuss now.

This question was first formulated by Gelfand [*2*] in his address at the International Congress at Stockholm in 1962. In fact, Gelfand announced without proof the following formula:

$$N(r) \sim C_G \operatorname{Vol}(\Gamma\backslash G)\, r^{n/2} \qquad \text{as} \quad r \to \infty \tag{2.2}$$

where C_G is a constant depending only on G, $\operatorname{Vol}(\Gamma\backslash G)$ is the total measure of $\Gamma\backslash G$, and $n = \dim G/K$.

In their book, Gelfand, Graev, and Pyatetskii-Shapiro [*3*] discuss the special cases $G = SL(2, R)$ and $G = SL(2, \mathbf{C})$. However, even for these, the proof of (2.2) is not completed there. It would seem that while (2.2) is almost certainly correct in general, there are unexpected technical difficulties in its proof. Hence one must regard (2.2) as an extremely plausible conjecture.

I will now describe briefly a method which has proved successful in verifying Gelfand's formula (2.2) when G is a complex semisimple group. Full details of this method can be found in Gangolli [*1*]. It is different from the method of Ref. [*3*].

In order to study the asymptotic behavior of $N(r)$ as $r \to \infty$, one introduces the Laplace transform

$$L(t) = \int_0^\infty e^{-tr}\, dN(r) \tag{2.3}$$

and tries to study the behavior of $L(t)$ as $t \to 0$. A Tauberian argument then yields the behavior of $N(r)$ as $r \to \infty$.

Assuming for the moment that $N(r)$ is finite valued, one sees that

$$L(t) = \sum_i n_i \exp -(\langle \lambda_i, \lambda_i\rangle + \langle \rho, \rho\rangle)t. \tag{2.4}$$

Note here that if we can somehow show that the right side is finite for some positive t, it would automatically imply that $N(r)$ is finite valued.

Recall that on the space H_{ij}, $1 \leqq j \leqq n_i$, the operator ω acts via the scalar $-(\langle \lambda_i, \lambda_i\rangle + \langle \rho, \rho\rangle)$, cf. Theorem 4 of Section 1. Thus, formally, the right side of (2.4) is the trace of the operator $e^{t\omega}$ on $L_2(\Gamma\backslash G/K)$. Thus if we can attach some sense to the operator $e^{t\omega}$ and show that its trace is finite and equals the right side of (2.4), then we will have accomplished an important step in the study of $N(r)$.

The formal operators $e^{t\omega}$ can be given a perfectly good meaning on

$L_2(G/K)$. Namely, they form the semigroup of the heat equation $\omega u = \partial u/\partial t$ on G/K. One knows from generalities that if g_t is the fundamental solution of this heat equation, then right convolution by g_t "is" the operator $e^{t\omega}$. Thus if R_{g_t} is the operator of right convolution by g_t on $L_2(G/K)$, it is not unreasonable to hope to prove that it will induce an operator R_{g_t} on $L_2(\Gamma\backslash G/K)$ which would be of trace class there, and that its trace there would be the right side of (2.4).

Notice also that the right side of (2.4) is very reminiscent of Eq. (1.24) of Theorem 10 of Section 1. Indeed that equation suggests that if g_t were an admissible function with Fourier transform $\hat{g}_t(\lambda) = \exp -(\langle\lambda, \lambda\rangle + \langle\rho, \rho\rangle)t$, then the trace of R_{g_t} would be finite and would equal the right side of (2.4). Thus the crux of the matter is defining g_t and showing that R_{g_t} is of trace class, i.e., that g_t is admissible.

The function $\exp -(\langle\lambda, \lambda\rangle + \langle\rho, \rho\rangle)t$ is for each $t > 0$ a rapidly decreasing W-invariant function on Λ_0. Thus by the inversion formula described in the preceding paper, the way g_t ought to be defined is clear. We define

$$g_t(x) = w^{-1} \int_{\Lambda_0} \exp -(\lambda, \lambda\rangle + \langle\rho, \rho\rangle)t \ \varphi_\lambda(x)|c(\lambda)^{-2} \, d\lambda. \tag{2.5}$$

With this definition, the main properties of g_t can be summarized in the following theorem.

Theorem 1: The functions g_t, $t > 0$ have the following properties:
(i) $g_t \in \mathscr{C}(K\backslash G/K)$ for each $t > 0$.
(ii) $\hat{g}_t(\lambda) = \exp -(\langle\lambda, \lambda\rangle + \langle\rho, \rho\rangle)t$.
(iii) $g_t * g_s = g_{t+s}$.
(iv) $\omega g_t = \partial g_t/\partial t$.
(v) If $f \in C_c^\infty(K\backslash G/K)$, then $f * g_t$ is a solution of $\partial u/\partial t = \omega u$, and $\|f * g_t - f\|_2 \to 0$ as $t \to 0$.

Proof: (i) and (ii) are given by the inversion formula for the Fourier transform, cf. the preceding paper. To get (iii), note that $\hat{g}_s\hat{g}_t = \hat{g}_{t+s}$ by (ii) and also $g_t * g_s = \hat{g}_t\hat{g}_s$. Thus $g_t * g_s = \hat{g}_{t+s}$. But $f \to \hat{f}$ is one-to-one on $\mathscr{C}(K\backslash G/K)$. So (iii) follows. Property (iv) follows immediately upon using the definition of g_t, recalling that $\omega\varphi_\lambda = -(\langle\lambda, \lambda\rangle + \langle\rho, \rho\rangle)\varphi_\lambda$ and differentiating under the integral sign in the equation defining g_t. Finally, if $u_t = f * g_t$, then by (iv), u_t solves $\omega u = \partial u/\partial t$. Also $\|u_t - f\|_2^2 = \int_{\Lambda_0}|(\hat{u}_t - \hat{f})(\lambda)|^2|c(\lambda)|^{-2} d\lambda$ by the Plancherel Theorem, but $\hat{u}_t(\lambda) = \hat{f}(\lambda)g_t(\lambda)$, so $\|u_t - f\|^2 = \int_{\Lambda_0} |\hat{f}(\lambda)|^2 | \exp -t(\langle\lambda, \lambda\rangle + \langle\rho,\rho\rangle) - 1|^2 |c(\lambda)|^{-2} d\lambda$. As $t \to 0$, the integrand approaches 0, and by dominated convergence, we conclude that $\|u_t - f\|_2 \to 0$.

The last part of this theorem shows that convolution by g_t does indeed solve the heat equation. Thus g_t is indeed the fundamental solution of the heat equation. Standard methods now show that $g_t \geqq 0$. Moreover, $\{g_t\}_{t \geqq 0}$ is an approximate identity for $L_2(K\backslash G/K)$. It follows that if U is any spherical neighborhood of e in G, then $\int_{U^c} g_t(x)\, dx \to 0$ as $t \to 0$, where U^c is the complement of U. This fact will be very useful to us.

The next step is to show that g_t is admissible. Let us note here that g_t cannot possibly have compact support, because, being a solution of a parabolic equation, it is analytic as a function of x. Thus one has to establish ad hoc that g_t is admissible. The point is that we only know the Fourier transform of g_t, and not g_t itself, explicitly. On the other hand the conditions of Definition 9 of Section 1 are in terms of g_t, not in terms of $\hat{g}_t$. It seems a reasonably difficult problem to give a criterion on $\hat{f}$ in order that f be admissible.

However, when G is a complex group, one can compute g_t explicitly. This will enable us to show that g_t is admissible. This is done via the following proposition. For any $r > 0$, let $U_r(x)$ be the set of $x' \in G$ such that the distance of $x'K$ from xK in G/K is less than or equal to r.

Proposition 2 (cf. [1, p. 165]): Let G be a complex group. Then given $r > 0$, there exists $t_0 > 0$ such that for all $0 < t < t_0$ and for all x in the complement of $U_r(e)$, we have $(\omega g_t)(x) \geqq 0$. Moreover, given $t_0 > 0$, there exists an $r > 0$ such that for all x in the complement of $U_r(e)$ and for all t such that $0 < t < t_0$, we have $(\omega g_t)(x) \geqq 0$.

The explicit formula for g_t is used in this proposition.

The point of Proposition 2 is that for certain ranges of values of t, and for x "away" from e, g_t is subharmonic at x. Thus one can dominate $g_t(x)$ by means of its average over a small neighborhood of x. This proves to be decisive for us. Indeed we have

Proposition 3 (cf. [1, p. 170]): Suppose G is a complex group. Then, given $r > 0$, there exists $t_0 > 0$ and a constant $C_r > 0$ so that for any x in the complement of $U_{3r}(e)$ and for any t such that $0 < t < t_0$, we have

$$g_t(x) \leqq C_r \int_{U_r(x)} g_t(z)\, dz. \tag{2.6}$$

Further, given any $t_0 > 0$, there exists $r > 0$ and a constant $C_r > 0$ such that for each x in the complement of $U_{3r}(e)$, and for any t such that $0 < t \leqq t_0$, we have

$$g_t(x) \leqq C_r \int_{U_r(x)} g_t(z)\, dz. \tag{2.7}$$

We can now show that g_t is admissible.

Proposition 4: For each $t > 0$, g_t is an admissible function.

Proof: Let P be a compact subset of G whose image under the projection map covers $\Gamma\backslash G$. It is enough to show that $\sum_\gamma g_t(y^{-1}\gamma x)$ converges uniformly for $(x, y) \in P \times P$. Fix $r > 0$ so that if $y' \notin U_{3r}(e)$, then $g_t(y') \leqq C_r \int_{U_r(y')} g_t(z)\, dz$. Now consider the series $\sum_{\gamma\in\Gamma} g_t(y^{-1}\gamma x)$, for $(x, y) \in P \times P$. It is easily seen that there is a finite subset F_r of Γ such that if $\gamma \in \Gamma - F_r$, then $y^{-1}\gamma x \notin U_{3r}(e)$ for any $(x, y) \in P \times P$. Thus if $\gamma \in \Gamma - F_r$, we have

$$g_t(y^{-1}\gamma x) \leqq C_r \int_{U_r(y^{-1}\gamma x)} g_t(z)\, dz. \tag{2.8}$$

Put $\Gamma' = \Gamma - F_r$. Now suppose we are given $\varepsilon > 0$ and choose R so large that if $V = U_R(e)$, we have $\int_{V^c} g_t(z)\, dz < \varepsilon$. This can be done because $\int_G g_t$ is finite, and $g_t \geqq 0$. We now claim that there is a finite subset F' of Γ' such that if $\gamma \notin F'$, then the set $U_r(y^{-1}\gamma x)$ lies in V^c for all $(x, y) \in P \times P$. Indeed, if $U_r(y^{-1}\gamma x) \cap V \neq \phi$, let $z \in U_r(y^{-1}rx) \cap V$. Then the distance of zK from eK is at most R and the distance of $y^{-1}\gamma xK$ from zK is at most r. Hence $y^{-1}\gamma x \in U_{R+r}(e)$, and $\gamma \in yU_{R+r}(e)x^{-1} \subset PU_{R+r}(e)P^{-1}$. This last set is compact, and so Γ intersects it in a finite set F. Putting $F' = \Gamma' \cap F$, we see that our claim follows. Now let $\gamma \in \Gamma' - F'$. Then we have

$$\sum_{\gamma\in\Gamma'-F'} g_t(y^{-1}\gamma x) \leqq C_r \sum_{\gamma\in\Gamma'-F'} \int_{U_r(y^{-1}\gamma x)} g_t(z)\, dz. \tag{2.9}$$

Now one can show that there is an integer M such that for any $(x, y) \in P \times P$, a given point of G can belong to at most M of the sets $\{U_t(y^{-1}\gamma x);\ \gamma \in \Gamma\}$. Thus, the right side of (2.9) is majorized by $M\, C_r \int_{V^c} g_t(z)\, dz$ since all the sets $\{U_r(y^{-1}\gamma x),\ \gamma \in \Gamma' - F'\}$ are contained in V^c. Hence, since V was chosen carefully above, we have for all $(x, y) \in P \times P$,

$$\sum_{\gamma\in\Gamma'-F'} g_t(y^{-1}\gamma x) \leqq M\,C_{r\,\lambda}. \tag{2.10}$$

This shows that the series $\sum_{\gamma\in\Gamma} g_t(y^{-1}\gamma x)$ converges uniformly for $(x, y) \in P \times P$. Since $g_t \geqq 0$, this means that g_t is admissible. Q.E.D.

The formula for g_t shows that $g_t(x^{-1}) = g_t(x)$. Since g_t is real, this means that $g_t{}^* = g_t$. Moreover, $g_t = g_{t\backslash 2}{}^* g_{t\backslash 2}$. Hence combining the above proposition with Theorem 10 of Section 1 and the remark following it, we get

Corollary 5. The series $\sum_{\gamma\in\Gamma} g_t(y^{-1}\gamma x)$ converges uniformly for (x, y) in compact subsets of $G \times G$, and defines a continuous function $G_t(x, y)$ on $\Gamma\backslash G \times \Gamma\backslash G$. The operator R_{g_t} is an integral operator with kernel $G(x, y)$ on $L_2(\Gamma\backslash G)$. It leaves $L_2(\Gamma\backslash G/K)$ invariant and is zero on the orthocomplement of $L_2(\Gamma\backslash G/K)$ in $L_2(\Gamma\backslash G)$. It is of trace class on $L_2(\Gamma\backslash G/K)$ and the trace of R_{g_t} is given by

$$\begin{aligned}\text{Trace } R_{g_t} &= \int_{\Gamma\backslash G} \sum g_t(x^{-1}\gamma x)\,dx \\ &= \sum_i (\exp -(\langle\lambda_i, \lambda_i\rangle + \langle\rho, \rho\rangle)t)n_i.\end{aligned} \tag{2.11}$$

In particular the right side is finite for each $t > 0$. Hence for each r there are only finitely many λ_i such that $\langle\lambda_i\lambda_i\rangle + \langle\rho,\rho\rangle \leqq r$, so that $N(r)$ is finite-valued. The right side of (2.11) is the Laplace transform $L(t)$ of $N(r)$.

One now has to analyze the behavior of $L(t)$ as $t \to 0$. Here we use the other half of the trace formula, viz. the expression $\int_{\Gamma\backslash G} \sum_{\gamma\in\rho} g_t(x^{-1}\gamma x)\,dx$. We will describe the results briefly, referring to [*1*] for details.

Since g_t is spherical, it is obvious that the function $x \to \sum_\gamma g_t(x^{-1}\gamma x)$ is invariant under right translations by $k \in K$. Hence the integral $\int_{\Gamma\backslash G} \sum_\gamma g_t(x^{-1}\gamma x)\,dx$ may be written $\int_{\Gamma\backslash G\backslash K} \sum_\gamma g_t(x^{-1}\gamma x)\,dx$. Now let F be a measurable fundamental domain for Γ in G/K. Then since $g_t \geqq 0$, we have by Fubini's theorem that this integral equals $\sum_{\gamma\in\Gamma} \int_F g_t(x^{-1}\gamma x)dx$. Now the elements of Γ fall into three classes Γ_C, Γ_E, Γ_H, defined as follows. Γ_C is just those elements of Γ that are in the center of G. Obviously Γ_C is finite, say of cardinality c. Γ_E is those points $\gamma \in \Gamma$ that have fixed points for their action on G/K. These are the so-called elliptic points in Γ. It is easy to see that if $\gamma \in \Gamma_E$ and if γ' is another element of Γ which is conjugate to γ in G, then $\gamma' \in \Gamma_E$. Thus Γ_E is a union of G-conjugacy classes of Γ. It can be shown that this is a finite union. Thus $\Gamma_E = \cup_{i=E} E_i$ where each E_i is a class of elements of Γ that are conjugate in G. Finally Γ_H is those elements of Γ which act without fixed points on G/K. These are the hyperbolic elements in Γ. The sum $L(t) = \sum_{\gamma \in \Gamma_c} \int_F g_t(x^{-1}\gamma x)\,dx$ can be now broken into three pieces $J_C(t)$, $J_E(t)$, $J_H(t)$, where $J_C(t) = \sum_{r \in \Gamma} \int_F g_t(x^{-1}\gamma x)\,dx$, and J_E, J_H are defined similarly. For $r \in \Gamma_C, g_t(x^{-1}\gamma x) = g_t(\gamma)$. But since the center of G is contained in K, and g_t is spherical, $g_t(\gamma) = g_t(e)$. Thus

Proposition 6: $J_C(t) = cg_t(e)\ \mathrm{Vol}(F) = cg_t(e)\ \mathrm{Vol}(\Gamma\backslash G)$ where c is the cardinality of Γ_C.

We next consider $J_H(t)$. Here we have

Proposition 7: $J_H(t) \to 0$ as $t \to 0$.

Proof: We will sketch the proof. Let P be a compact set in G whose image under projection covers $\Gamma\backslash G$. Then clearly $\sum_{\gamma \in \Gamma_H} \int_F g_t(x^{-1}\gamma x)\,dx \leqq \sum_{\gamma \in \Gamma_H} \int_P g_t(x^{-2}\gamma x)\,dx$. Now, it is a consequence of the compactness of $\Gamma\backslash G$ and the definition of Γ_H that if $A = \{x^{-1}\gamma x | \gamma \in \Gamma_H,\ x \in G\}$, then A is closed and $A \cap K = \phi$. Hence there is an $r > 0$ such that $U_{3r}(e) \cap A = \phi$. Pick this r and choose $t_0 > 0$ so small that for $0 < t \leqq t_0$, and $\gamma \in \Gamma_H$, we have $g_t(x^{-1}\gamma x) \leqq C_r \int_{U_r(x^{-1}\gamma x)} g_t(z)\,dz$, as we may because of Proposition 3.

Thus

$$\sum_{\gamma \in \Gamma_H} g_t(x^{-1}\gamma x) \leqq C_r \sum_{\gamma \in \Gamma_H} \int_{U_r(x^{-1}\gamma x)} g_t(z)\,dz. \tag{2.12}$$

It is easy to see that the sets $\{U_r(x^{-1}\gamma x),\ \gamma \in \Gamma_H\}$ all lie in the complement of $U_r(e)$, since $x^{-1}\gamma x \notin U_{3r}(e)$. Further there is an integer M such that for any $x \in P$, any point of G lies in at most M of the sets $\{U_r(x^{-1}\gamma x),\ \gamma \in \Gamma\}$. It follows that $\sum_{\gamma \rho \Gamma_H} g_t(x^{-1}\gamma x) \leqq C_r M \int_{G - U_r(e)} g_t(z)\,dz$. Now as $t \to 0$, $\int_{G - U_r(e)} g_t(z)\,dz \to 0$ because $g_t(z)$ is an approximate identity in $L_1(K\backslash G/K)$ and $U_r(e)$ is a spherical neighborhood of e (cf. the remark following Theorem 1). This shows that $\sum_{\gamma \in \Gamma_H} g_t(x^{-1}\gamma x) \to 0$ as $t \to 0$, uniformly for $x \in P$. It follows that $J_H(t) \to 0$ as $t \to 0$.

Finally one has to deal with the term $J_E(t)$. Here we recall that Γ_E is a *finite* union of G-conjugacy classes E_i, $i = 1, \ldots, k$. So

$$J_E(t) = \sum_{i=1}^{k} \int_{\Gamma\backslash G} \sum_{\gamma\in E_i} g_t(x^{-1}\gamma x)\, dx. \tag{2.13}$$

Following Selberg one writes $\gamma_E(t)$ as

$$J_E(t) = \sum_{i=1}^{k} \mathrm{vol}\,(\Gamma_{\gamma_i}\backslash G_{\gamma_i}) \int_{G_{\gamma_i}/G} g_t(x^{-1}\gamma_i x)\, dx. \tag{2.14}$$

Here γ_i is some element in E_i, G_{γ_i} is the centralizer of γ_i in G, and $\Gamma_{\gamma_i} = \Gamma \cap G_{\gamma_i}$. $d\bar{x}_i$ is the invariant measure on $G_{\gamma_i}\backslash G$. It can be checked that the summand is independent of the choice of γ_i in E_i and depends only on the class E_i.

The integrals $\int_{G_\gamma\backslash G} f(x^{-1}\gamma x)\, d\bar{x}$ occur in Harish-Chandra's work on invariant integrals on G. Using Harish-Chandra's results and some rather technical observations, one can show that $t^{n/2} \int_{G\gamma_i\backslash G} g_t(x^{-1}\gamma_i x)\, d\bar{x}_i \to 0$ as $t \to 0$. To do this one has again to use the hypothesis that G is a complex group. Thus we have (cf. [*1*])

Proposition 8: $t^{n/2} J_E(t) \to 0$ as $t \to 0$, $n = \dim G/K$.

Note that we have used the hypothesis that G is complex most crucially only in Propositions 2 and 8. Propositions 3, 4, Corollary 5, Propositions 6 and 7 follow from Proposition 2 without the use of this hypothesis.

If Γ acts without fixed points on $G|K$, then Γ_E is empty and Proposition 8 is unnecessary. In this case, the only use of this hypothesis is in Proposition 2.

Due to Propositions 6, 7, and 8, we see that the behavior of $L(t)$ as $t \to 0$ is the same as that of $g(e)$ as $t \to 0$. However

$$g_t(e) = w^{-1} \int_{\Lambda_0} \exp -t(\langle\lambda, \lambda\rangle + \langle\rho, \rho\rangle)\, |c(\lambda)|^{-2}\, d\lambda \tag{2.15}$$

and we know an explicit formula for $c(\lambda)$. This makes it possible to get a pretty good idea of the behavior of $g_t(e)$ near $t = 0$. Indeed let, for $r > 0$, $M(r)$ be defined by

$$M(r) = w^{-1} \int_{\{\lambda, \langle\lambda,\lambda\rangle + \langle\rho,\rho\rangle \leq \gamma\}} |c(\lambda)|^{-2}\, d\lambda. \tag{2.16}$$

Then we have by (2.15), (2.16)

$$g_t(e) = \int_0^\infty e^{-tr}\, dM(r). \tag{2.17}$$

On the other hand, using the formula of Gindikin–Karplevic for the c-function (cf. the preceding paper), one can show that $M(r) \sim C_G r^{n/2}$ as $r \to \infty$, where C_G is a constant depending only on G, and $n = \dim G|K$. Thus, the Tauberian theorem for the Laplace transform shows that $g_t(e) \sim C'_G t^{-n/2}$ as $t \to 0$ where C'_G depends only on the group G. We have therefore

Proposition 9: As $t \to 0$, $t^{n/2}L(t)$ has a limit. Indeed $\lim_{t\to 0} t^{n/2}L(t) = cC'_G \operatorname{Vol}(\Gamma\backslash G)$, where $n = \dim G|K$, C'_G is a constant depending only on the structure of G, and c is the order of Γ_C.

One has immediately

Corollary 10: $N(r)/r^{n/2} \sim cC''_G \cdot \operatorname{Vol}(\Gamma\backslash G)$ where C''_G is a constant depending only on G.

This is precisely Gelfand's conjecture for complex G.

The reader should note that if Propositions 2 and 8 can be upheld without the hypothesis that G is complex, then all this analysis can be carried out in general. If rank $(G/K) = 1$, it is an unpublished result due to the author that Proposition 2 is valid. Thus, if in addition Γ_E is empty, Gelfand's conjecture is seen to be true if rank $(G/K) = 1$. For rank $G/K > 1$, I do not at present know how to proceed. Thus:

Proposition 11: Suppose rank $(G/K) = 1$ and Γ_E is empty. Then $N(r)/r^{n/2} \to cC_G \operatorname{Vol}(\Gamma\backslash G)$ as $r \to \infty$.

3. Concluding Remarks

Proposition 11 of Section 2 allows us to answer a question posed above in Section 1 in an important special case. Namely, let R be a compact Riemann surface of genus $g \geq 2$. Then the covering surface of R is the upper half-plane H, and $R = \Gamma\backslash H$ where Γ is the fundamental group of R. One knows that $H = SL(2, R)/SO(2)$ and Γ is a discrete subgroup of

$SL(2, R)$ with compact quotient. Thus the above applies with $G = SL(2, R)$, $K = SO(2)$. Clearly rank $(G|K) = 1$, and moreover, since R is a manifold, Γ acts without fixed points on G/K. Hence Proposition 11 of Section 2 is applicable and we see that $N(r)/r^{n/2} \to C''_G \operatorname{Vol}(\Gamma\backslash G) = C''_G \operatorname{Vol}(R)$ where $\operatorname{Vol}(R)$ is computed in the obvious measure. Recall that $R = \Gamma\backslash G/K$. Now there is a classical formula of Hurwitz which determines $\operatorname{Vol}(R)$ in terms of the genus g. Hence it follows that the asymptotic behavior of $N(r)$ determines the genus g, and hence the topological type of R. Moreover, since Γ is the fundamental group of R, it follows that $N(r)$ determines Γ up to isomorphism. We have

Theorem 1: Suppose $G = SL(2, R)$, $K = SO(2)$, and suppose Γ, Γ' are discrete subgroups such that $\Gamma\backslash G$, $\Gamma'\backslash G$ are compact. Suppose Γ, Γ' act without fixed points on G/K, and that $\operatorname{spec}(\Gamma, 1) = \operatorname{spec}(\Gamma', 1)$. Then Γ and Γ' are isomorphic.

Note added in proof: After this was written, the author came to know that this result is also contained in the paper of S. Tanaka [9].

Note that we only used the fact that the functions $N(r)$ and $N'(r)$ corresponding to Γ, Γ' are the same and did not use the full force of the hypothesis $\operatorname{spec}(\Gamma, \mathbf{1}) = \operatorname{spec}(\Gamma', \mathbf{1})$. A natural question that arises is whether Γ, Γ' are conjugate in G under this hypothesis.

Thus we see that in this special case a very small piece of the spectrum of Γ already determines the isomorphism class of Γ. This gives more interest to the general problem: How far does $\operatorname{spec}(\Gamma)$ determine Γ?

In the case of the Riemann surface R above, we saw that the topology of R is determined by its spectrum. The question arises: To what extent, if any, does the spectrum determine the conformal structure of R? Very little seems to be known about this problem.

One should probably also investigate the relations between the spectrum of R and the cohomology of various natural sheaves on R. Presumably this would lead to some understanding of the connection between representation theory and the theorem of Riemann–Roch.

In the same spirit one might ask if there is a similar theory for the case of elliptic curves. Here the genus is 1, and the covering surface is just $\mathbf{C}$ (equal to R^2). As G one may take the group of all Euclidean motions and as K we can take $SO(2)$. Then Γ is the lattice that one has to factor out from $\mathbf{C}$ to get the elliptic curve. The question arises, what can we say about Γ from a knowledge of its spectrum. I do not know of any results in this direction.

Finally, in connection with the question "which classes occur in spec Γ and with what multiplicity?," there is the lovely result of Langlands, which we now describe very briefly. Let G/K be such that G/K is a Hermitian symmetric space. Such spaces were classified by Cartan. For these, rank G = Rank K and K has a 1-dimensional center. In this case a theorem of Harish-Chandra says that G has a discrete series. That is, there exist irreducible unitary representations π of G on Hilbert spaces H that are square integrable, i.e., such that every matrix-element of π is square-integrable. As we saw in the preceding paper, such π cannot be of class **1** with respect to K.

In his paper [*4*], Harish-Chandra showed how to construct such a representation π_λ, starting from a suitable character $\lambda \in \hat{T}$. Here, T is a maximal torus in K. Moreover, Harish-Chandra also gave a sufficient condition in order that π_λ be square-integrable on G. In his paper [*6*] Langlands showed that if π_λ is an integrable (and not merely square-integrable) representation whose class occurs in spec(Γ), then one can compute the multiplicity with which it occurs in terms of λ. However, it is not known precisely which λ correspond to integrable representations. This result of Langlands is the only result to date in which something so precise is said about spec(Γ). Note that even here one does not know "which" π_λ appear in spec(Γ).

It will be clear to the reader that the contents of the subject described by the title of this paper are at present very tentative and much remains to be done. I hope that my exposition has not obscured the essentially scanty nature of the results.

References

[*1*] Gangolli, R., Asymptotic behaviour of spectra of compact quotients of certain symmetric spaces, *Acta Math.*, **121** (1968), 151–192.

[*2*] Gelfand, I., Automorphic functions and the theory of representation, *Proc. Intern. Cong. Mathematicians*, Stockholm, 1962. Almquist & Wiksell, Stockholm, 1963.

[*3*] Gelfand, I.M., Graev, M., and Pyatetskii-Shapiro, I.I., Generalized functions, Vol. VI, Moscow, 1966.

[*4*] Harish-Chandra, Representations of semisimple Lie groups, VI, *Amer. J. Math.*, **78** (1956), 564–628.

[*5*] Harish-Chandra, Invariant eigendistributions on a semisimple Lie group, *Trans. Amer. Math. Soc.*, **119** (1965), 457–508.

[6] Langlands, R.P., The dimension of spaces of holomorphic forms, *Amer. J. Math.*, **85** (1963), 99–125.

[7] Selberg, A., Harmonic analysis and discontinuous groups in weakly symmetric Riemannian spaces with applications to dirichlet series. *J. Indian Math. Soc.*, **20** (1956), 47–87.

[8] Tamagawa, T., On Selberg's trace formula. *J. Fac. Sci. Univ. Tokyo*, **8** (1960), 363–386.

[9] Tanaka S., Selberg's trace formula and spectrum. *Osaka J. Math, 3* (1966) 205–216.

FOURIER DECOMPOSITIONS OF CERTAIN REPRESENTATIONS

*K. I. GROSS** and *R. A. KUNZE*[†]

Dartmouth College
Hanover, New Hampshire

University of California
Irvine, California

*Research supported in part by National Science Foundation Grant GP–11910.
†Research supported in part by National Science Foundation Grant GP–21081.

Introduction

The results in this paper were suggested by a number of technical questions which arose in connection with work involving a new construction of holomorphic discrete series representations for $U(n, n)$ and similar groups. Our intention was to give a construction similar to that used by Shale [1] and Weil [2] in other contexts and analogous to the procedure employed by Tanaka [3] for the construction of discrete series for $SL(2, K)$, K being an arbitrary nondiscrete locally compact field. In general, the representations constructed by this method are far from irreducible, and it is necessary to decompose them in order to get the desired (irreducible) discrete series. Although the simple results presented here were specifically developed to achieve this reduction, they seem to be of independent interest.

For our work on discrete series, in particular, to prove square-integrability, we also need a suitable generalization of the usual Bessel functions J_k of integral order. These generalized Bessel functions are matrix valued, and in the case of $U(n, n)$, they are indexed by appropriate unitary representations of $U(n)$. These functions and their basic properties will be discussed in a future paper.

1. Unitary Representations of Compact Transformation Groups on L^2 Spaces

Let (U, X) be a topological transformation group in which U and X are Hausdorff spaces, the group U is compact, and X is a locally compact space with a regular U-invariant Baire measure dx. For our purpose, it is convenient to make the additional assumption that

$$\int_X f(x)\,dx > 0 \tag{1.1}$$

whenever $f \in C_c^+(X)$ and $f \neq 0$.

The first assumptions, exclusive of (1.1), guarantee that equation

$$L(u)f(x) = f(u^{-1}x), \qquad u \in U, \quad x \in X \tag{1.2}$$

defines a (continuous) unitary representation L of U on the space $L^2(X)$ of functions on X square-integrable with respect to dx.

Let λ be an irreducible unitary matrix representation of U. If T is any (continuous) unitary representation of U, there is a canonical subspace of the representation space $\mathscr{H}(T)$ consisting of the vectors that transform under T according to λ [4]. If d_λ is the degree of λ, the orthogonal projection of $\mathscr{H}(T)$ on the subspace which transforms according to λ is given by the equation

$$P_\lambda = d_\lambda \int_U \operatorname{tr} \bar{\lambda}(u)\, T(u)\, du \tag{1.3}$$

in which du denotes Haar measure on U. A vector lies in the range of P_λ if and only if its translates under T span a finite dimensional subspace in which T decomposes into irreducible representations equivalent to λ.

Our goal is to give a simple and fairly explicit realization of the range of P_λ in the special case that T is the representation L defined by (1.2). This will be accomplished by means of a generalized Fourier–Plancherel transform for the pair (U, X). This Fourier transform defines a unitary map from $L^2(X)$ onto another space $L^2(X)^\wedge$; it diagonalizes L in the sense that it converts "translations" into "multiplications," and maps each subspace $P_\lambda L^2(X)$ onto a space $L^2(X, \lambda)$ which is easy to describe. In fact, $L^2(X, \lambda)$ may be viewed as the space of square-integrable sections of a certain Hilbert bundle on $U\backslash X$ that is canonically associated with λ.

To motivate the definition of the spaces $L^2(X, \lambda)$, we recall the fact that the projections P_λ may be defined more generally for Banach space representations of U [4]. In the present context, this suggests that one should consider pointwise averages of the form

$$(E_\lambda f)(x) = d_\lambda \int_U \operatorname{tr} \bar{\lambda}(u) f(u^{-1}x)\, du \tag{1.4}$$

where f is an arbitrary continuous function on X. We may write

$$(E_\lambda f)(x) = d_\lambda \operatorname{tr} \int_U f(u^{-1}x)\, \bar{\lambda}(u)\, du$$

and this in turn suggests consideration of the matrix-valued function

$$x \to \int_U f(u^{-1}x)\, \bar{\lambda}(u)\, du, \qquad x \in X.$$

We use the notation $\mathfrak{C}^{m \times n}$ to denote the space of complex $m \times n$ matrices equipped with the standard inner product

$$(A|B) = \operatorname{tr}(B^* A).$$

Lemma 1. If $f \in C(X)$, the equation

$$\hat{f}(\lambda)(x) = \int_U f(u^{-1}x)\,\bar{\lambda}(u)\,du, \qquad x \in X \tag{1.5}$$

defines a continuous function

$$\hat{f}(\lambda) : X \to \mathfrak{C}^{d\lambda \times d\lambda}$$

such that

$$\hat{f}(\lambda)(ux) = \bar{\lambda}(u)\hat{f}(\lambda)(x) \tag{1.6}$$

for all (u, x) in $U \times X$, and if $f \in C_c(X)$, so does $\hat{f}(\lambda)$.

Proof. Let $f \in C(X)$. Then the integrand in (1.5) is a continuous function on the compact group U. Hence (1.5) gives a well-defined map of X into $\mathfrak{C}^{d\lambda \times d\lambda}$. Since

$$\begin{aligned}\|\hat{f}(\lambda)(x) - \hat{f}(\lambda)(y)\| &\leqq \int_U |f(u^{-1}x) - f(u^{-1}y)|\;\|\bar{\lambda}(u)\|\,du\\ &= d_\lambda^{\frac{1}{2}} \int_U |f(u^{-1}x) - f(u^{-1}y)|\,du,\end{aligned}$$

the continuity of $\hat{f}(\lambda)$ follows by a simple compactness argument from the joint continuity of $f(u^{-1}x)$ as a function of u and x. If f is supported by the compact set K, it is immediate that $\hat{f}(\lambda)$ is supported by the compact set UK. The remaining property (1.6) follows from the invariance of Haar measure.

Definitions. A function $\varphi\colon X \to \mathfrak{C}^{d\lambda \times d\lambda}$ will be called *λ-covariant* if

$$\varphi(ux) = \bar{\lambda}(u)\,\varphi(x)$$

for all (u, x) in $U \times X$. The vector space of continuous λ-covariant functions will be denoted by $C(X, \lambda)$, and $L^2(X, \lambda)$ will denote the space of all λ-covariant Baire functions φ such that

$$\int_X \|\varphi(x)\|^2\,dx < \infty.$$

Modulo null functions $L^2(X,\lambda)$ is a Hilbert space with respect to the inner product

$$(\varphi|\psi)_\lambda = d_\lambda \int_X (\varphi(x)|\psi(x))\,dx. \tag{1.7}$$

It follows from (1.1) that $\|\varphi\|_\lambda > 0$ whenever

$$\varphi \in C(X, \lambda) \cap L^2(X, \lambda)$$

and $\varphi \neq 0$. Moreover, standard arguments involving the regularity properties of the measure dx show that the space $C_c(X, \lambda)$ of continuous λ-covariant functions of compact support is dense in $L^2(X, \lambda)$.

It should be noted that at this stage we are not asserting the existence of any λ-covariant functions $\neq 0$. Indeed, the integrals (1.5) may all reduce to 0; we show later that $L^2(X, \lambda) \neq 0$ if and only if λ occurs in the reduction of L.

Definitions. Let $\hat{U}$ be a set of representatives, fixed once and for all, for the equivalence classes of continuous irreducible unitary matrix representations of U. Let

$$L^2(X)^\wedge = \bigoplus_{\lambda \in \hat{U}} L^2(X, \lambda),$$

i.e., the set of all functions F on $\hat{U}$ such that $F(\lambda) \in L^2(X, \lambda)$ for every λ and

$$\textstyle\sum_\lambda \|F(\lambda)\|^2 < \infty.$$

The Hilbert space structure of $L^2(X)^\wedge$ is given by the formula

$$(F|G) = \textstyle\sum_\lambda (F(\lambda)|G(\lambda))_\lambda \tag{1.8}$$

for the inner product of the elements F and G of $L^2(X)^\wedge$. The (generalized) *Fourier transform* of a function f in $C(X)$ is the function

$$\hat{f} : \hat{U} \to \cup_{\lambda \in \hat{U}} C(X, \lambda)$$

whose value at a point λ is defined by (1.5).

Having introduced sufficient terminology, we may now state our first result.

Theorem 1. The Fourier transform $f \to \hat{f}$ maps $C(X) \cap L^2(X)$ into $L^2(X)^\wedge$, and its restriction to $C(X) \cap L^2(X)$ extends uniquely to a unitary transformation $\mathscr{T}$ of $L^2(X)$ onto $L^2(X)^\wedge$ which maps $P_\lambda L^2(X)$ onto $L^2(X, \lambda)$. In $L^2(X, \lambda)$ the representation

$$\hat{L}(\cdot) = \mathscr{T}L(\cdot)\mathscr{T}^{-1}$$

is given by

$$(\hat{L}(u)\,\varphi)(x) = \varphi(x)\,\lambda(u)^t. \tag{1.9}$$

In the proof we use several lemmas.

Lemma 2. The set $U\backslash X$ of orbits of U in X equipped with the quotient topology is a locally compact Hausdorff space on which there exists a unique regular Baire measure dx such that

$$\int_X f(x)\,dx = \int_{U\backslash X} dx \int_U f(ux)\,du$$

for all nonnegative Baire functions f.

Proof. See Bourbaki [5].

Lemma 3. If $f \in C(X) \cap L^2(X)$, then $\hat{f}(\lambda) \in L^2(X, \lambda)$ for every λ in $\hat{U}$, and

$$\int_X |f(x)|^2\,dx = \sum_{\lambda \in \hat{U}} \|\hat{f}(\lambda)\|_\lambda^2.$$

Proof. Let $f \in C(X) \cap L^2(X)$. Then by Lemma 2 and the invariance properties of Haar measure on compact groups, it follows that

$$\int_X |f(x)|^2\,dx = \int_{U\backslash X} dx \int_U |f(u^{-1}x)|^2\,du.$$

Fix $x \in X$, let $g(u) = f(u^{-1}x)$, and set

$$\hat{g}(\lambda) = \int_U g(u)\bar{\lambda}(u)\,du.$$

Then, by the Peter–Weyl theorem

$$\int_U |g(u)|\hat{g}^2 du = \sum_{\lambda \in \hat{U}} \|\hat{g}(\lambda)\|^2 d_\lambda.$$

Hence, $\int_U |f(u^{-1}x)|^2\, du = \sum_\lambda \|\hat{f}(\lambda)(x)\|^2 d_\lambda$. By λ-covariance, the function $x \to \|\hat{f}(\lambda)(x)\|$ is constant on orbits of U in X. Therefore

$$\int_X |f(x)|^2 dx = \sum_\lambda d_\lambda \int_{U\backslash X} \|\hat{f}(\lambda)\|^2\, dx.$$

Since $f \in L^2(X)$, this implies

$$\int_{U\backslash X} \|\hat{f}(\lambda)(x)\|^2 dx < \infty$$

for every λ in $\hat{U}$. On the other hand, Lemma 2 shows that

$$\int_X \|\hat{f}(\lambda)(x)\|^2\, dx = \int_{U\backslash X} dx \int_U \|\hat{f}(\lambda)(ux)\|^2\, du = \int_{U\backslash X} \|\hat{f}(\lambda)(x)\|^2\, dx,$$

it being tacitly assumed here and throughout that the Haar measure of U is 1. The desired formula now follows from (1.7).

Lemma 4. If $\varphi \in C(X, \lambda)$, the equation

$$f(x) = d_\lambda \operatorname{tr} \varphi(x), \qquad x \in X \tag{1.10}$$

defines a function f in $C(X)$ such that $\hat{f}(\lambda) = \varphi$ and $\hat{f}(\gamma) = 0$ whenever $\gamma \in \hat{U}$ and $\gamma \neq \lambda$.

Proof. The function f is clearly continuous, and

$$\begin{aligned}
\hat{f}(\gamma)(x) &= d_\lambda \int_U \operatorname{tr}(\varphi(u^{-1}x))\bar{\gamma}(u)\, du \\
&= d_\lambda \int_U \operatorname{tr}(\varphi(x)\, \lambda(u)^t)\bar{\gamma}(u)\, du \\
&= d_\lambda \int_U \left(\textstyle\sum_{ij} \varphi_{ij}(x)\, \lambda_{ij}(u)\right)\bar{\gamma}(u)\, du \\
&= \textstyle\sum_{ij} \varphi_{ij}(x) d_\lambda \displaystyle\int_U \lambda_{ij}(u)\bar{\gamma}(u)\, du
\end{aligned}$$

where $\varphi_{ij}(x)$ and $\lambda_{ij}(u)$ are the ij entries of $\varphi(x)$ and $\lambda(u)$, respectively. When γ is not equivalent to λ, i.e., when $\gamma \in \hat{U}$ and $\gamma \neq \lambda$, the orthogonality relations imply that

$$\int_U \lambda_{ij}(u)\bar{\gamma}(u)\,du = 0$$

for all i and j. On the other hand,

$$d_\lambda \int_U \lambda_{ij}(u)\bar{\lambda}(u)\,du = E_{ij},$$

where E_{ij} is the standard matrix unit whose ij entry is 1. Thus

$$\hat{f}(\lambda)(x) = \textstyle\sum_{ij}\varphi_{ij}(x)E_{ij} = \varphi(x).$$

Lemma 5. The linear map

$$E_\lambda : C(X) \to C(X)$$

given by (1.4) is idempotent and coincides on $C_c(X)$ with the orthogonal projection P_λ of $L^2(X)$ on the subspace which transforms under L according to λ.

Proof. Let $f \in C(X)$. Then by (1.4) and (1.5)

$$(E_\lambda f)(x) = d_\lambda \operatorname{tr} \hat{f}(\lambda)(x). \tag{1.11}$$

Thus

$$E_\lambda(E_\lambda f)(x) = d_\lambda \operatorname{tr} (E_\lambda f)^\wedge(\lambda)(x).$$

This equation may be written in the form

$$(P_\lambda f | g) = (E_\lambda f | g)$$

because $E_\lambda f$ is a continuous function of compact support and hence an element of $L^2(X)$. Since $C_c(X)$ is dense in $L^2(X)$, it follows that

$$P_\lambda f = E_\lambda f \qquad \text{a.e.}\,.$$

Lemma 6. For each λ in $\hat{U}$,

$$(E_\lambda C_c(X))^\wedge = C_c(X, \lambda).$$

Proof. Let $f \in C_c(X)$. By Lemma 1, $\hat{f}(\lambda) \in C_c(X, \lambda)$, and by (1.10)

$$(E_\lambda f)(x) = d_\lambda \operatorname{tr} \hat{f}(\lambda)(x).$$

It follows from Lemma 4, that

$$(E_\lambda f)(\gamma) = \begin{cases} \hat{f}(\lambda) & \text{when } \gamma = \lambda \\ 0 & \text{when } \gamma \neq \lambda. \end{cases}$$

Therefore, $(E_\lambda C_c(X))^\wedge \subset C_c(X, \lambda)$. Conversely, if $\varphi \in C_c(X, \lambda)$, then the function

$$f(x) = d_\lambda \operatorname{tr} \varphi(x)$$

lies in $C_c(X)$, and

$$\hat{f}(\gamma) = \begin{cases} \varphi & \text{when } \gamma = \lambda \\ 0 & \text{when } \gamma \neq \lambda \end{cases}$$

by Lemma 4. Thus

$$\begin{aligned} f(x) &= d_\lambda \operatorname{tr} \hat{f}(\lambda)(x) \\ &= (E_\lambda f)(x) \end{aligned}$$

and consequently, $(E_\lambda C_c(X))^\wedge = C_c(X, \lambda)$.

Proof of Theorem 1. Since $C(X) \cap L^2(X)$ is dense in $L^2(X)$, it follows from Lemma 3 that the restricted Fourier transform $f \to \hat{f}$, $f \in C(X) \cap L^2(X)$, extends uniquely to an isometric linear mapping $\mathcal{T}$ of $L^2(X)$ into $L^2(X)$. Because $C_c(X)$ is dense in $L^2(X)$ and P_λ is continuous, $P_\lambda C_c(X)$ is dense in $P_\lambda L^2(X)$. At the same time, $C_c(X, \lambda)$ is dense in $L^2(X, \lambda)$. Thus, to show that

$$(P_\lambda L^2(X))^\wedge = L^2(X, \lambda),$$

it is sufficient, in view of Lemma 5, to show that $(E_\lambda C_c(X))^\wedge = C_c(X, \lambda)$. But this is exactly the assertion made in Lemma 6.

Now suppose that F is an arbitrary member of $L^2(X)^\wedge$. Then for a given λ in $\hat{U}$, $F(\lambda) \in L^2(X, \lambda)$, and by what we have just shown, there is a function f_λ in $P_\lambda L^2(X, \lambda)$ such that

$$\hat{f}_\lambda(\gamma) = \begin{cases} F(\lambda) & \text{when } \gamma = \lambda \\ 0 & \text{when } \gamma \neq \lambda \end{cases}$$

and

$$\int_X |f_\lambda(x)|^2 \, dx = \|F(\lambda)\|_\lambda^2.$$

Since $\sum_\lambda \|F(\lambda)\|_\lambda^2 < \infty$ and $(f_\gamma | f_\lambda) = 0$ whenever $\gamma \neq \lambda$, the series $\sum_\lambda f_\lambda$ converges in $L^2(X)$ to a well defined element f such that $\hat{f} = F$. Thus $\mathscr{T}$ maps $L^2(X)$ onto $L^2(X)^\wedge$.

To prove (1.9), it is sufficient to consider functions φ of the form $\hat{f}(\lambda)$ where $f \in C_c(X)$ and $E_\lambda f = f$. For any such φ and any v in U,

$$\begin{aligned}\hat{L}(v)\varphi(x) &= \int_U f(v^{-1}u^{-1}x)\bar{\lambda}(u)\,du \\ &= \int_U f(u^{-1}x)\,\bar{\lambda}(uv^{-1})\,du \\ &= \varphi(x)\,\lambda(v)^t.\end{aligned}$$

Next, we briefly consider the question of obtaining criteria for determining whether a given λ in $\hat{U}$ actually occurs in the reduction of L.

Necessary conditions are easily obtained. For, by the theorem, $P_\lambda \neq 0$ if and only if $L^2(X, \lambda) \neq 0$, and that is true if and only if $C_c(X, \lambda) \neq 0$. If φ is any λ-covariant function and x is an arbitrary point in X, the covariance condition implies

$$\bar{\lambda}(v)\varphi(x) = \varphi(x) \tag{1.12}$$

for all v in the subgroup of U fixing x.

Definitions. Let V_x be the stability group of the point x, let

$$F(x, \lambda) = \{a \in \mathfrak{C}^{d\lambda \times 1} : \lambda(v)a = a, \forall v \in V_x\},$$

and $m_\lambda(x)$ be the dimension of $F(x, \lambda)$. Finally, let $M(x, \lambda)$ denote the $m_\lambda(x)d_\lambda$ dimensional subspace of $\mathfrak{C}^{d\lambda \times d\lambda}$ consisting of all matrices whose columns are complex conjugates of elements of $F(x, \lambda)$.

If φ is any λ-covariant function on X, it follows from (1.12) that $\varphi(x) \in M(x, \lambda)$ for every x in X; hence, it is necessary that $m_\lambda(x) > 0$ for all x such that $\varphi(x) \neq 0$.

2. The Homogeneous Case

Although the homogeneous case, in which U acts transitively on X, is well known, in fact, classical [6, 7], it is of some interest to see how our methods apply to that situation.

Suppose then that U acts transitively on X. Let x_0 be an arbitrary point of X, fixed once and for all, and $V = V_{x_0}$. Then $uV \to ux_0$ is a homeomorphism of U/V onto X, and X automatically has an essentially unique measure invariant under U. In this context, Theorem 1 admits the following refinement.

Theorem 2. For each λ in $\hat{U}$, $L^2(X, \lambda)$ is a finite dimensional space of continuous functions isomorphic to $M(x_0, \lambda)$ via the map

$$\varphi \to \varphi(x_0).$$

In addition, $\|\varphi\|^2 = d_\lambda \|\varphi(x_0)\|^2$, and the number of times that λ occurs in the reduction of L is precisely the dimension m_λ of the space $F(x_0, \lambda)$.

Proof. Since $X = Ux_0$, every λ-covariant function on X is determined by its value at x_0. Thus, the map $\varphi \to \varphi(x_0)$ is an isomorphism of the vector space of all λ-covariant functions into $M(x_0, \lambda)$. On the other hand, given an arbitrary element A of $M(x_0, \lambda)$,

$$\bar{\lambda}(u_1)A = \bar{\lambda}(u_2)A$$

whenever $u_1 x_0 = u_2 x_0$. Since $X = Ux_0$, it follows that there is one and only one function φ on X such that

$$\varphi(ux_0) = \bar{\lambda}(u)A$$

for all u in U, and φ is evidently λ-covariant. Therefore, the space of λ-covariant functions is isomorphic to $M(x_0, \lambda)$. Moreover, a λ-covariant function φ is necessarily continuous. For if $x \in X$ and $\varepsilon > 0$, there is a neighborhood N of the identity in U on which

$$\|\varphi(x) - \bar{\lambda}(u)\varphi(x)\| < \varepsilon.$$

Since $\bar{\lambda}(u)\varphi(x) = \varphi(ux)$ and Nx is a neighborhood of x, this implies the continuity of φ at x. In addition

$$\int_X \|\varphi(x)\|^2 \, dx = \int_U \|\varphi(ux_0)\|^2 \, du = \|\varphi(x_0)\|^2$$

by Lemma 2. The restriction of L to $P_\lambda L^2(X)$ is a certain multiple m_λ of λ. Since $P_\lambda L^2(X)$ is isomorphic to $M(x_0, \lambda)$, its dimension is $m_\lambda(x_0)\cdot d_\lambda$ where $m_\lambda(x_0)$ is the dimension of $F(x_0, \lambda)$; hence $m_\lambda = m_\lambda(x_0)$.

The computation of the multiplicity of λ in L is easily done in more explicit terms. For it follows from Theorem 1 and the above that in

$P_\lambda L^2(x)$, the translation operator $L(u)$ $(u \in U)$ is unitarily equivalent to the multiplication operator

$$R(u) : A \to A\lambda(u)^t$$

in $M(x_0, \lambda)$. If $a \in F(x_0, \lambda)$ and $a \neq 0$, the map $b \to ab^t$ is an isomorphism of $\mathfrak{C}^{d\lambda \times 1}$ onto an R-invariant subspace S_a of $M(x_0, \lambda)$ in which R is equivalent to λ. If $\{a_1, \ldots, a_m\}$ is an orthonormal base for $F(x_0, \lambda)$, it is easy to check that $M(x_0, \lambda)$ is the orthogonal direct sum of $S_{a_1}, \ldots, S_{a_m}$.

In the important classical example in which U is the orthogonal group $O(n)$ acting linearly on the unit sphere in $\mathfrak{R}^{n \times 1}$, it is well known that m_λ is either 0 or 1. The proof of this fact given in [8] suggests the following result.*

Theorem 3. If the double cosets of V in U are symmetric, that is, invariant under the map $u \to u^{-1}$ $(u \in U)$, then each representation λ in $\hat{U}$ occurs with multiplicity 0 or 1 in the reduction of L, and the space $P_\lambda L^2(X)$ is closed under complex conjugation.

Proof. Suppose that $VuV = Vu^{-1}V$ for each u in U. Let $\lambda \in \hat{U}$ and suppose a and b are orthogonal elements of $F(x_0, \lambda)$ with $a \neq 0$. Then, since λ is irreducible, there exist scalars c_j and elements u_j in U such that

$$b = \textstyle\sum_j c_j \lambda(u_j)a.$$

Hence, $(\lambda(u)b|a) = \sum_j c_j(\lambda(uu_j)a|a)$.

On the other hand, for each u in U there exist elements v_1, v_2 in V such that $u = v_1 u^{-1} v_2$. If c and d are vectors in $F(x_0, \lambda)$, it follows that

$$\begin{aligned}(\lambda(u)d|c) &= (\lambda(u^{-1})\lambda(v_2)d|\lambda(v_1{}^{-1})c)\\ &= (\lambda(u^{-1})d|c)\\ &= \overline{(\lambda(u)c|d)}.\end{aligned}$$

Therefore

$$\begin{aligned}\int_U (\lambda(u)b|a)\,\overline{(\lambda(u)b|a)}\,du &= \int_U (\lambda(u)b|a)\,(\lambda(u)a|b)\,du\\ &= \textstyle\sum_j c_j \displaystyle\int_U (\lambda(u)a|b)\,\overline{(\lambda(uu_j)a|a)}\,du.\end{aligned}$$

*C. Moore has shown us a rough draft of a manuscript in which he and J. Wolf have proved a more comprehensive result along these lines.

Since λ is unitary and irreducible, the orthogonality relations

$$\int_U (\lambda(u)c|c')\,\overline{(\lambda(u)d|d')}\,du = d_\lambda^{-1}(c|d)\,(c'|d')$$

hold for all c, c', d, d' in $\mathfrak{C}^{d_\lambda \times 1}$.
Thus

$$\begin{aligned} d_\lambda^{-1}(b|b)\,(a|a) &= \textstyle\sum_j c_j(b|\lambda(u_j)a)\,\overline{(b|a)} \\ &= 0 \end{aligned}$$

so that $b = 0$; hence m_λ is 0 or 1.

Suppose $m_\lambda = 1$, and let a be a nonzero vector in $F(x_0, \lambda)$. Then since λ is irreducible, it follows from Theorem 1, Theorem 2, and Lemma 4, that $P_\lambda L^2(X)$ is finite dimensional and spanned by the translates of the function f such that

$$f(u\,x_0) = d_\lambda\,\mathrm{tr}(\bar{\lambda}(u)\bar{a}a^t).$$

By a simple computation

$$\mathrm{tr}(\bar{\lambda}(u)\bar{a}a^t) = (\lambda(u)a)^*a = (a|\lambda(u)a)$$

and above we showed that $(a|\lambda(u)a)$ is real. Therefore, f and its translates $L(u)f$ $(u \in U)$ are real valued, and $P_\lambda L^2(X)$ is invariant under complex conjugation.

3. A Decomposition for the Regular Representation of Abelian-Compact Semidirect Products

Let G be a locally compact group and dx a left Haar measure on G. Let L denote the left regular representation of G on the space $L^2(G)$ of Baire functions square-integrable with respect to the measure dx. Specifically, L is the continuous unitary representation of G defined for y and x in G by

$$L(y)f(x) = f(y^{-1}x), \qquad f \in L^2(G). \tag{3.1}$$

Our problem is to find a simple natural decomposition of L into primary representations, analogous to the decomposition achieved in Section 1, given special assumptions about the structure of G. For the present purpose, a primary representation may be defined as a finite multiple of an irreducible representation.

We assume that G is the semidirect product of a closed normal Abelian subgroup A and a compact subgroup K. Then

$$(a, k) \to ak$$

is a homeomorphism of $A \times K$ onto G. If da is a Haar measure on A, elementary arguments show that da is invariant under the transformations

$$a \to kak^{-1}, \qquad k \in K.$$

From this it follows that if dk is a Haar measure on K, then the product $da\,dk$ is a two-sided invariant measure on G. Thus G is unimodular, and with the proper normalizations of the measures,

$$\int_G f(x)\,dx = \int_K \left(\int_A f(ak)\,da \right) dk \tag{3.2}$$

for all nonnegative Baire functions [*5*] or [*9*, Lemma 12]. Since G is unimodular, L is unitarily equivalent to the right regular representation R whose definition is analogous to that of L; in fact, if $Jf(x) = f(x^{-1})$, then

$$R(y) = JL(y)J^{-1}. \tag{3.3}$$

It seems easier technically and results in no essential loss to focus our attention on the equivalent problem of decomposing R rather than L.

The requisite primary representations may all be described in terms of a slight variant of the usual Mackey theory [*10*]. For this purpose, let V be the dual of A, i.e., the group of continuous (unitary) characters of A equipped with the compact open topology. If $k \in K$ and $v \in V$, the equation

$$(kv)(a) = v(k^{-1}ak), \qquad a \in A \tag{3.4}$$

defines a character kv in V. It is straightforward to check that the map

$$(k, v) \to kv$$

defines a continuous action of K on V. Given an element v in V, let

$$U_v = \{u \in K : uv = v\}. \tag{3.5}$$

Suppose λ is a continuous irreducible unitary matrix representation of U_v of degree d_λ. Since U_v acts on K, we may form the space $L^2(K, \lambda)$ as in Section 1; here, however, it is convenient to make a minor change of notation, so that the functions in $L^2(K, \lambda)$ transform according to

$$\varphi(uk) = \lambda(u)\varphi(k), \qquad u \in U_v, \quad k \in K. \tag{3.6}$$

Let the columns of $\varphi(k)$ be denoted by $\varphi_1(k), \varphi_2(k), \ldots$. Then each φ_j $(1 \leqq j \leqq d_\lambda)$ is a member of the space $H(K, \lambda)$ of square-integrable maps

$$\psi : K \to \mathfrak{C}^{d\lambda \times 1}$$

which satisfy (3.6). It is convenient to define the norm in $H(K, \lambda)$ by

$$\|\psi\|^2 = d_\lambda \int_K \|\psi(k)\|^2 \, dk.$$

If $\psi \in H(K, \lambda)$ and $a \in A$, it is easy to check that the function $k \to v(kak^{-1})$ $\psi(k)$ is also in $H(K, \lambda)$ and that it has the same norm as ψ. Since $H(K, \lambda)$ is invariant under right translations, we may define a continuous unitary representation $T(\cdot, v, \lambda)$ of G on $H(K, \lambda)$ by setting

$$T(ab, v, \lambda)\, \psi(k) = v(kak^{-1})\psi(kb) \tag{3.7}$$

for $a \in A$ and $b \in K$. The Mackey theory shows that these representations are all irreducible and that given certain assumptions about $K\backslash V$, every infinite dimensional irreducible unitary representation of G is equivalent to some $T(\cdot,v,\lambda)$. If v_1 and v_2 lie in distinct orbits of V under K, the representations $T(\cdot, v_1, \lambda_1)$ and $(T\cdot, v_2, \lambda_2)$ are always disjoint, i.e., not equivalent, but this need not be the case if v_1 and v_2 are in the same orbit. For given an element v in V and c in K, it is evident that $U_{cv} = cU_v c^{-1}$; hence, if λ is a representation of U_v, we may define a representation $c\lambda$ of U_{cv} by setting

$$(c\lambda)(u) = \lambda(c^{-1}uc), \qquad u \in U_{kv}. \tag{3.8}$$

The representations $T(\cdot, v, \lambda)$ and $T(\cdot, cv, c\lambda)$ are unitarily equivalent. In fact, as is readily verified, the left translation operator

$$(L(c)\psi)(k) = \psi(c^{-1}k)$$

defines a unitary map $L(c) : H(K, \lambda) \to H(K, c\lambda)$ such that

$$L(c)T(ab, v, \lambda)L(c)^{-1} = T(ab, cv, c\lambda) \tag{3.9}$$

for a in A and b in K.

Since $L^2(X, \lambda)$ is obviously the direct sum of d_λ copies of $H(K, \lambda)$, the primary representation

$$R(\cdot, \nu, \lambda) = d_\lambda T(\cdot, \nu, \lambda)$$

is given quite explicitly on $L^2(X, \lambda)$ by

$$R(ab, \nu, \lambda)\, \varphi(k) = \nu(kak^{-1})\varphi(kb). \tag{3.10}$$

Definitions. For each ν in V let $\hat{U}_\nu$ be a set of representatives fixed once and for all for the equivalence classes of continuous irreducible unitary matrix representations of U. Let P be the set of all pairs (ν, λ) in which $\nu \in V$ and $\lambda \in \hat{U}_\nu$. For each f in $C_c(G)$ and each $\rho = (\nu, \lambda)$ in P, let $\tilde{f}(\rho) = \tilde{f}(\nu, \lambda)$ denote the function on K defined by the integral

$$\tilde{f}(\nu, \lambda)\,(k) = \int_{U\nu}\int_A f(auk)\, \nu(a^{-1})\, \lambda(u^{-1})\, da\, du. \tag{3.11}$$

Here, du is the element of Haar measure on U_ν.

The definition of $\tilde{f}(\rho)$ can also be given in a slightly different fashion. For this let $C_\nu = AU_\nu$. Then C_ν is a closed subgroup of G, and $dc = da\, du$ is a two-sided invariant measure on C_ν. Each $\rho = (\nu, \lambda)$ in P defines a matrix representation

$$\rho : au \to \nu(a)\, \lambda(u) \qquad (a \in A,\, u \in U_\nu) \tag{3.12}$$

of C_ν, and from the fact that ν is fixed by U_ν, it follows that

$$\tilde{f}(\rho)\,(k) = \int_{C_\nu} f(ck)\, \rho(c^{-1})\, dc. \tag{3.13}$$

As the first and major step in showing that the representations (3.10) are the desired "primary components" of the right regular representation R, we prove the following result.

Theorem 4. For each f in $C_c(G)$ and each pair (ν, λ) in P, the matrix-valued function $\tilde{f}(\nu, \lambda)$ is continuous on K and has the property that

$$\tilde{f}(\nu, \lambda)\,(uk) = \lambda(u)\tilde{f}(\nu, \lambda)\,(k) \tag{3.14}$$

for all (u, k) in $U_\nu \times K$; hence $\tilde{f}(\nu, \lambda) \in L^2(K, \lambda)$. If y is any element of G, then

$$(R(y)f)^{\sim}(\nu, \lambda) = R(y, \nu, \lambda)\,\tilde{f}(\nu, \lambda) \tag{3.15}$$

and with the proper normalizations of the Haar measures on A, K, and V

$$\int_G |f(x)|^2\,dx = \int_V \left(\sum_{\lambda \in \hat{U}_v} \|\tilde{f}(v, \lambda)\|_\lambda^2 \right) dv \tag{3.16}$$

for every f in $C_c(G)$.

Proof. The continuity of $\tilde{f}(v, \lambda)$ and (3.14) follow easily from (3.11) or (3.13). To prove (3.15), let $y = ab$ with a in A, b in K, and assume that ρ is given by (3.12). Then by (3.13)

$$\begin{aligned}(R(ab)f)^{\sim}(\rho)\,(k) &= \int_{C_v} f(c\,k\,ab)\,\rho(c^{-1})\,dc \\ &= \int_{C_v} f(c\,k\,ak^{-1}kb)\,\rho(c^{-1})\,dc \\ &= \int_{C_v} f(c\,k\,b)\,\rho(kak^{-1}c^{-1})\,dc \\ &= v(kak^{-1})\,\tilde{f}(\rho)\,(kb)\end{aligned}$$

and this is the desired equation.

The final Eq. (3.16) is obtained in the following way. By (3.2),

$$\int_G |f(x)|^2\,dx = \int_K \left(\int_A |f(ak)|^2\,da \right) dk.$$

Now for v in V and f in $C_c(G)$, let

$$(Ff)\,(v, k) = \int_A f(ak)\,v(a^{-1})\,da.$$

Then Ff is a bounded continuous function on $V \times K$, and by the Plancherel theorem the Haar measure dv on V may be normalized so that

$$\int_V |(Ff)\,(v, k)|^2\,dv = \int_A |f(ak)|^2\,da$$

for all f in $C_c(G)$. Therefore

$$\begin{aligned}\int_G |f(x)|^2\,dx &= \int_K \left(\int_V |(Ff)\,(v, k)|^2\,dv \right) dk \\ &= \int_V \left(\int_K |(Ff)\,(v, k)|^2\,dk \right) dv.\end{aligned}$$

On the other hand, if we fix v and set

$$g(k) = (Ff)(v, k),$$

it follows from Theorem 1, in particular, from Lemma 3, that

$$\int_K |g(k)|^2 \, dk = \sum_{\lambda \in \hat{U}_v} \|\hat{g}(\lambda)\|_\lambda^2.$$

It is easy to see that this also holds if λ is replaced by $\bar{\lambda}$. Since

$$\tilde{f}(v, \lambda) = \hat{g}(\bar{\lambda})$$

(cf. (1.5) and (3.11)), it follows that

$$\int_K |Ff(v, k)|^2 \, dk = \sum_{\lambda \in \hat{U}_v} \|\tilde{f}(v, \lambda)\|_\lambda^2.$$

Integrating both sides of this equation with respect to v, we obtain (3.16).

With each f in $C_c(G)$ we may associate the function

$$\tilde{f} : \rho \to \tilde{f}(\rho), \qquad \rho \in P$$

on P. The collection $C_c(G)^\sim$ of all $\tilde{f}$ with $f \in C_c(G)$ is a complex vector space with respect to the usual pointwise operations. By (3.16), the map $f \to \tilde{f}$, which is evidently linear, is an isomorphism of $C_c(G)$ onto $C_c(G)^\sim$. This being the case, the following result is immediate.

Corollary. On $C_c(G)^\sim$, there is a unique inner product such that

$$(\tilde{f} | \tilde{g}) = (f | g) \tag{3.17}$$

which is also given by

$$(\tilde{f} | \tilde{g}) = \int_V \left(\sum_{\lambda \in \hat{U}_v} (\tilde{f}(v, \lambda) | \tilde{g}(v, \lambda))_\lambda \right) dv. \tag{3.18}$$

Let $L^2(G)^\sim$ denote the completion of $C_c(G)^\sim$. Then the map $f \to \tilde{f}$ ($f \in C_c(G)$) extends to a unitary transformation of $L^2(G)$ onto $L^2(G)^\sim$, and the representation $\tilde{R}$ of G on $L^2(G)^\sim$ which corresponds to R on $L^2(G)$ is such that

$$(\tilde{R}(y)\tilde{f})(\rho) = R(y, \rho)\tilde{f}(\rho) \tag{3.19}$$

for all y in G, f in $C_c(G)$, and ρ in P.

From the conventional abstract point of view, the Fourier transform $\hat{f}$ of a function f in $C_c(G)$ at a point (ν, λ) of P is the operator

$$\hat{f}(\nu, \lambda) = \int_G f(x)\, T(x^{-1}, \nu, \lambda)\, dx \tag{3.20}$$

where $T(\cdot, \nu, \lambda)$ is the irreducible unitary representation of G on $H(K, \lambda)$ defined by (3.7). It is of some interest to determine the relation between $\hat{f}(\nu, \lambda)$ and $\tilde{f}(\nu, \lambda)$.

Theorem 5. Let f be a continuous function of compact support on G and (ν, λ) an element of P. If ψ is any continuous function in $H(K, \lambda)$, then

$$(\hat{f}(\nu, \lambda)\psi)(k) = \int_K M_f(k, b; \nu, \lambda)\, \psi(b)\, db \tag{3.21}$$

where the matrix-valued kernel M_f is given by

$$M_f(k, b; \nu, \lambda) = (L(b)f)^{\sim}(\nu, \lambda)(k). \tag{3.22}$$

In addition,

$$\int_G |f(x)|^2\, dx = \int_V \Big(\sum_{\lambda \in \hat{U}_\nu} d_\lambda \int_K \int_K \| M_f(k, b; \nu, \lambda) \|^2\, dk\, db \Big)\, d\nu. \tag{3.23}$$

Proof. Suppose ψ is a continuous function in $H(K, \lambda)$. Then standard arguments, which we shall omit, show that

$$\begin{aligned}
\hat{f}(\nu, \lambda)\psi(k) &= \int_G f(x^{-1})\, (T(x, \nu, \lambda)\, \psi)(k)\, dx \\
&= \iint f(b^{-1}a^{-1})\, \nu(kak^{-1})\, \psi(kb)\, da\, db \\
&= \iint f(b^{-1}k^{-1}a^{-1}k)\, \nu(a)\, \psi(kb)\, da\, db \\
&= \int f(b^{-1}a^{-1}k)\, \nu(a)\, \psi(b)\, da\, db \\
&= \int_{U_\nu \backslash K} \Big(\int_{U_\nu} \int_A f(b^{-1}u^{-1}a^{-1}k)\, \nu(a)\, \lambda(u)\, da\, db \Big)\, \psi(b)\, d\dot{b} \\
&= \int_{U_\nu \backslash K} \Big(\int_{U_\nu} \int_A f(b^{-1}a^{-1}u^{-1}k)\, \nu(a) \lambda(u)\, da\, db \Big)\, \psi(b)\, d\dot{b}.
\end{aligned}$$

Now defining $M_f(k, b; v, \lambda)$ by (3.22), we obtain the formula

$$(\hat{f}(v, \lambda)\, \psi)\, (k) = \int_{U_v \backslash K} M_f(k, b;\, v, \lambda)\, \psi(b)\, d\dot{b}.$$

Of course, this integral only makes sense when the integrand is constant on the cosets $U_v b$ $(b \in K)$. That this is true follows from the fact that

$$M_f(k, ub;\, v, \lambda) = M_f(k, b;\, v, \lambda)\, \lambda(u^{-1}),$$

which is readily verified by change of variables. Since the integrand above is constant on the cosets $U_v b$, $(\hat{f}(v, \lambda)\psi)\,(k)$ may also be computed as in (3.21). To prove (3.23), first note that

$$d_k \int_K \|M_f(k, b;\, v, \lambda)\|^2\, dk = \|L(b)f)^{\sim}(v, \lambda)\|_\lambda^2.$$

From this it follows that

$$\begin{aligned}
&\int_V \left(\sum_{\lambda \in U_v} d_\lambda \int_K \|M_f(k, b;\, v, \lambda)\,\|^2\, dk \right) dv \\
&\quad = \int_V \sum_{\lambda \in \hat{U}_v} \|\,(L(b)f)^{\sim}\;(v, \lambda)\,\|_\lambda^2\, dv \\
&\quad = \int_G |f(b^{-1}x)|^2\, dx \quad \text{(by (3.16))} \\
&\quad = \int_G |f(x)|^2\, dx.
\end{aligned}$$

Now integrate both sides of this equation with respect to b to obtain (3.23).

References

[1] D. Shale, "Linear symmetries of free boson fields," *Trans. Amer. Math. Soc.*, **102** (1962), 149–167.

[2] A. Weil, "Sur certains groupes d'operateurs unitaries," *Acta Math.*, **111** (1964), 143–211.

[3] S. Tanaka, "On irreducible unitary representations of some special linear groups of the second order *I*," *Osaka J. Math.*, **3** (1966), 217–227.

[4] R. Godement, "A theory of spherical functions 1," *Trans. Amer. Math. Soc.*, **73** (1952), 496–556.

[5] N. Bourbaki, *Livre VI*, Hermann, Paris, 1963, Chapitre 7, §1, no. 4.

[6] E. Cartan, *Euvres Completes*, Gauthier-Villars, Paris, 1939.

[7] H. Weyl, "Harmonics on homogeneous manifolds," *Ann. of Math.*, **35** (1934), 486–499.
[8] R. Coifman and G. Weiss, "Representations of compact groups and spherical hormonics," *Enseignement Math., 14* (1968), 121–173.
[9] R. Kunze and E. Stein, "Uniformly bounded representations III," *Amer. J. Math.*, **89**, No. 2 (1967), 385–442.
[10] G. Mackey, "Infinite-dimensional group representations," *Bull. Amer. Math. Soc.*, **69** (1963), 628–686.

CONICAL DISTRIBUTIONS AND GROUP REPRESENTATIONS

SIGURDUR HELGASON

Department of Mathematics
Massachusetts Institute of Technology
Cambridge, Massachusetts

From the point of view of the duality between points and horocycles in a symmetric space, the counterparts to the spherical functions on the symmetric space are the conical distributions on the manifold of horocycles. In this paper, we give a sketch of the proof of Theorem 5.4 which deals with the explicit determination of these distributions. We shall also indicate some roles they play in representation theory.

As customary we use the notation **Z**, **R**, and **C** for the set of integers, real numbers, and complex numbers, respectively; $\mathbf{Z}^+$ will denote the set of nonnegative integers. We also use Schwartz' notation $\mathscr{E}$ and $\mathscr{D}$ for the space of C^∞ functions and the space of C^∞ functions of compact support, respectively. The dual spaces $\mathscr{D}'$ and $\mathscr{E}'$ are then identified with the space of distributions and the space of distributions of compact support, respectively.

1. The Space of Horocycles

Let X be a symmetric space of the noncompact type, that is a coset space $X = G/K$ where G is a noncompact, connected semisimple Lie group with finite center and K a maximal compact subgroup. Let $G = KAN$ be an Iwasawa decomposition of G, the groups A and N being Abelian and nilpotent, respectively. A *horocycle* in X is by the definition [*6*] an orbit in X of a subgroup of G conjugate to N. Since N is closed in G and K is compact, the horocycles are closed submanifolds of X. They are related to the "horyspheres" considered in Gelfand [*1*, p. 79] but are however not the same, because while the horocycles in G/K are always permuted transitively by G this is not the case for the horyspheres in G/K. For the simplest case when $G = \mathbf{SU}(1, 1)$, the group of conformal maps of the unit disk $|z| < 1$, the two notions coincide and are the circles tangential to the boundary $|z| = 1$.

Let Ξ denote the set of horocycles in X. If o is the origin $\{K\}$ in X and ξ_0 the horocycle $N \cdot o$, the subgroup of G which leaves ξ_0 invariant equals MN where M is the centralizer of A in K. Thus we have the identifications

$$X = G/K, \qquad \Xi = G/MN,$$

the latter of which turns the set Ξ into a manifold. Using the Iwasawa decomposition of G, it is readily seen that the map $(kM, a) \to \xi = ka \cdot \xi_0$ is a diffeomorphism of $K/M \times A$ onto Ξ. Here the coset kM is called the *normal* to ξ and a the *complex distance* from o to ξ. Let $B = K/M$; given $x \in X$, $b \in B$ there exists exactly one horocycle through x with normal b.

Let us consider again the case when X is the unit disk. If ξ is a horocycle and x a point on ξ there exists exactly one other horocycle through x tangential to ξ. Thus in the analogy between horocycles in X and hyperplanes in $\mathbf{R}^n$ (both are perpendicular to a family of parallel geodesics), the hyperplanes should be considered as oriented hyperplanes.

For a general X let M' be the normalizer of A in K and W the Weyl group M'/M. For $s \in W$ select a representative $m_s \in M'$ mod M. Corresponding to the Iwasawa decompositions $G = KA(m_s N m_s^{-1})$, the horocycles $m_s N m_s^{-1} \cdot o$ have the same tangent space at o. In general if $\xi \in \Xi$ and $x \in \xi$, there exist precisely $|W|$ distinct horocycles through x tangential to ξ at x. (Here $|W|$ denotes the order of W.)

2. Heuristic Basis for Theorem 5.4

Let $\mathbf{D}(X)$ and $\mathbf{D}(\Xi)$ denote the algebras of differential operators on X and Ξ, respectively, which are invariant under the action of G. We recall that a *spherical function* f on X is by definition a K-invariant eigenfunction of all $D \in \mathbf{D}(X)$ satisfying $f(o) = 1$. By analogy we make the following definition.

Definition. A distribution Ψ on Ξ is called *conical* if it is: (1) MN-invariant; (2) an eigendistribution of each $D \in \mathbf{D}(\Xi)$.

The analogy is precise because a "spherical distribution" would automatically be a function. For the applications to representation theory, it is important to consider not just conical functions but also conical distributions; while the former are closely related to certain finite-dimensional representations of G, the latter will have applications to the principal series of infinite-dimensional representations of G.

Of course there is no guarantee in advance that any nonconstant conical distributions exist. Let us first try to make some rough guesses on the basis of the analogy with spherical functions.

As is well known, each $g \in G$ can be written $g = k_1 a k_2$ ($k_1, k_2 \in K$, $a \in A$); here a is uniquely determined modulo an operation of W. This can be expressed in the formula

$$K\backslash G/K \cong A/W. \tag{2.1}$$

Next let $\mathfrak{g} = \mathfrak{k} + \mathfrak{a} + \mathfrak{n}$ be the direct Lie algebra decomposition which corresponds to the decomposition $G = KAN$. If $g \in G$, let $H(g) \in \mathfrak{a}$, $k(g) \in K$, $n(g) \in N$ be determined by $g = k(g) \exp H(g) n(g)$. Let $\mathfrak{a}^*$ (respectively

$\mathfrak{a}_c^*$) denote the space of real-valued (respectively complex-valued) linear forms on $\mathfrak{a}$ and for each $H \in \mathfrak{a}$ let $\rho(H)$ denote half the trace of the linear transformation $X \to [H, X]$ of $\mathfrak{n}$. If $\lambda \in \mathfrak{a}_c^*$ the function

$$\varphi_\lambda(gK) = \int_K \exp[(i\lambda - \rho)(H(gk))]\, dk$$

(dk is Haar measure) is a spherical function, and according to Harish-Chandra each spherical function is obtained in this way. Moreover $\varphi_\lambda \equiv \varphi_\mu$ if and only if λ and μ are conjugate under W. Denoting by S the set of spherical functions on X, we express this result in the form

$$S \cong \mathfrak{a}_c^*/W. \tag{2.2}$$

Now there is an analog of (2.1) for the horocycle space. According to the Bruhat lemma (see Harish-Chandra [*4*]), each $g \in G$ can be written (with obvious notation) $g = m_1 n_1 a m' m_2 n_2$, where $m' \in M'$ is unique (mod M). The element $a \in A$ is also unique [*6*, p. 674], so we can state

$$MN \backslash G / MN \cong A \times W. \tag{2.3}$$

If we now compare (2.1) and (2.3), the identification (2.2) suggests that the set C of conical distributions would somehow be parametrized by the product set $\mathfrak{a}_c^* \times W$. Theorem 5.4 below shows that this is essentially the case.

3. Reformulation of the Definition

In order to understand better the action of the operators in $\mathbf{D}(\Xi)$, we consider (by Section 1) Ξ as a trivial fiber bundle with base K/M, the fiber over the point $kM \in K/M$ being $\{kaMN \mid a \in A\}$. If $g \in G$, we write $gk = k(gk)a(gk)n(gk)$ by the Iwasawa decomposition. Then since A normalizes N,

$$g \cdot kaMN = k(gk)a(gk)aMN,$$

which shows that the action of g on the bundle is: (1) fiber preserving; (2) the mapping of the fiber over kM onto the fiber over $k(gk)M$ is a translation by $a(gk)$. Now let $\mathbf{D}(A)$ denote the set of translation-invariant differential operators on A. If $U \in \mathbf{D}(A)$, we can define a differential operator D_U on Ξ by

$$(D_U\varphi)(ka \cdot \xi_0) = U_a(\varphi(ka \cdot \xi_0)), \tag{3.1}$$

the subscript a on the right referring to the variable on which U acts. Properties (1) and (2) above show that D_U is invariant under the action of G on Ξ.

Theorem 3.1. The mapping $U \to D_U$ is an isomorphism of $\mathbf{D}(A)$ onto $\mathbf{D}(\Xi)$.

This is proved in Ref. [6]; the "onto" statement requires some Lie algebra computations but the remaining statements are clear from the above.

If $a \in A$, then the map $\sigma(a)$: $gMN \to gaMN$ is a well-defined diffeomorphism of Ξ. From (3) and Theorem 3.1, it is easy to see that $\varphi \in \mathscr{E}(\Xi)$ is an eigenfunction of the operators $D \in \mathbf{D}(\Xi)$ if and only if

$$\varphi \circ \sigma(a) = \chi(a)\varphi, \qquad a \in A \tag{3.2}$$

where χ: $A \to \mathbf{C}-\{0\}$ is a homomorphism. This criterion can be extended to distributions. If $\Psi \in \mathscr{D}'(\Xi)$, let $\Psi^{\sigma(a)}$ denote the distribution $\varphi \to \Psi(\varphi \circ \sigma(a))$ on Ξ. Then Ψ is an eigendistribution of each $D \in \mathbf{D}(\Xi)$ if and only if

$$\Psi^{\sigma(a)} = \chi(a)\Psi, \qquad a \in A \tag{3.3}$$

for some χ as above. Now if $\lambda \in \mathfrak{a}_c^*$, we define

$$\mathscr{D}'_\lambda(\Xi) = \{\Psi \in \mathscr{D}'(\Xi) | \Psi^{\sigma(a)} = \exp[-(i\lambda + \rho)(\log a)]\Psi, \, a \in A\}.$$

Then each conical distribution lies in a unique space $\mathscr{D}'_\lambda$; more precisely the conical distributions are the MN-invariants in the spaces $\mathscr{D}'_\lambda$. This homogeneity under the group MNA leads us to consider the orbits of this group on Ξ. With the notation from Section 1, let $\xi_s = m_s MN$, Ξ_s the orbit $MNA \cdot \xi_s$. Then the Bruhat lemma implies

$$\Xi = \bigcup_{s \in W} \Xi_s \qquad \text{(disjoint union).}$$

Let θ denote the (Cartan) involution on G which has K as a fixed point set and put $\overline{N} = \theta N$. If we put $N'_s = N \cap (m_s \overline{N} m_s^{-1})$, then the map $(n'_s, a) \to n'_s a \cdot \xi_s$ is a diffeomorphism of $N'_s \times A$ onto Ξ_s. Carrying the product measure $dn'_s \, da$ over on Ξ_s by means of this diffeomorphism, we obtain a measure dv on Ξ_s. Postponing convergence questions, we ask what properties a function ψ on Ξ_s should have in order that the functional

$$\varphi \to \int_{\Xi_s} \varphi(\xi)\psi(\xi) \, dv(\xi)$$

should be an MN-invariant element in $\mathscr{D}'_\lambda$. We are then led to the relations

$$\psi(n'_s a \cdot \xi_s) = \psi(a \cdot \xi_s) = \exp[(is\lambda + s\rho)(\log a)]\, \psi(\xi_s),$$

so if $a(\xi) \in A$ is given by $\xi = n'_s a(\xi) \cdot \xi_s$, we define the functional $\Phi_{\lambda,s}$ by

$$\Phi_{\lambda,s}(\varphi) = \int_{\Xi_s} \varphi(\xi) \exp[(is\lambda + s\rho)(\log a(\xi))] dv(\xi) \tag{3.4}$$

on the set of $\varphi \in \mathscr{D}(\Xi)$ for which the integral converges absolutely. Although φ has compact support in Ξ, the restriction to Ξ_s may not have compact support in Ξ_s since the orbit Ξ_s is not in general closed; thus there is no guarantee that (3.4) will converge. However a rather simple computation relates the integral (3.4) to the integral

$$\int_{\bar{N}_s} \exp[-(i\lambda + \rho)(H(\bar{n}_s))]\, d\bar{n}_s, \tag{3.5}$$

where $\bar{N}_s = \bar{N} \cap m_s^{-1} N m_s$, for which a convergence criterion was given by Gindikin and Karpelevič [2]. Thus we obtain

Lemma 3.2. The integral (3.4) converges absolutely for all $\varphi \in \mathscr{D}(\Xi)$ if and only if

$$\mathrm{Re}(\langle i\lambda, \alpha \rangle) > 0 \qquad \text{for} \quad \alpha \in \Sigma^+ \cap s^{-1}\Sigma^-. \tag{3.6}$$

Here Re denotes real part, $\langle\ ,\ \rangle$ is the inner product on $\mathfrak{a}_c^*$ induced by the Killing form, $\langle\ ,\ \rangle$ of $\mathfrak{g}$, Σ^+ (respectively Σ^-) denotes the set of positive (respectively, negative) restricted roots.

If (3.6) is satisfied then $\Phi_{\lambda,s}$ is a conical distribution in $\mathscr{D}'_\lambda$.

The problem is now for each $s \in W$ to continue the distribution-valued function $\lambda \to \Phi_{\lambda,s}$ defined in the tube (3.6) to a meromorphic function on $\mathfrak{a}_c^*$ with values in $\mathscr{D}'(\Xi)$. Since the MN-invariance and the homogeneity (3.3) would remain valid under such an extension, the values would all be conical distributions.

It will be convenient later to express the elements in $\mathscr{D}'_\lambda$ in terms of the product representation $\Xi = (K/M) \times A$. The following proposition follows rather easily from the definition of $\mathscr{D}'_\lambda$.

Proposition 3.3. The space $\mathscr{D}'_\lambda(\Xi)$ consists precisely of the distributions

Ψ on Ξ given by

$$\Psi(\varphi) = \int_{K/M}\left(\int_A \varphi(ka\cdot\xi_0) \exp[(i\lambda + \rho)(\log a)]\, da\right) dS(k_M)$$

where S is an arbitrary distribution on K/M.

Let $T_{\lambda,s}$ denote the distribution on K/M which via this proposition corresponds to $\Phi_{\lambda,s}$. Then

$$T_{\lambda,s}(F) = \int_{\bar{N}_s} F(m_s k(\bar{n}_s)M) \exp[-(i\lambda + \rho)(H(\bar{n}_s))]\, d\bar{n}_s, \qquad F \in \mathscr{E}(K/M). \tag{3.7}$$

4. A Reduction Principle. The Meromorphic Continuation in the Rank-One Case

In the Iwasawa decompositions $G = KAN$, $\mathfrak{g} = \mathfrak{k} + \mathfrak{a} + \mathfrak{n}$, the Lie algebra $\mathfrak{n}$ is given by

$$\mathfrak{n} = \sum_{\alpha\in\Sigma^+} \mathfrak{g}_\alpha,$$

where Σ^+ is the set of positive restricted roots and

$$\mathfrak{g}_\alpha = \{X \in \mathfrak{g} \mid [H, X] = \alpha(H)X \text{ for all } H \in \mathfrak{a}\}.$$

It is well known that if α, $\beta \in \Sigma^+$ are proportional, the proportionality factor is $\frac{1}{2}$, 1, or 2. Again let θ denote the Cartan involution of G with fixed point set K.

Theorem 4.1. Suppose α and 2α lie in Σ^+. Select arbitrarily $X_\alpha \neq 0$ in $\mathfrak{g}_\alpha$ and $X_{2\alpha} \neq 0$ in $\mathfrak{g}_{2\alpha}$. Then the subalgebra $\mathfrak{g}^*$ of $\mathfrak{g}$ generated by X_α, $X_{2\alpha}$, θX_α, $\theta X_{2\alpha}$ is isomorphic to $\mathfrak{su}(2, 1)$, the Lie algebra of $\mathbf{SU}(2, 1)$.

The proof is quite elementary (see [*8*], Chapter III). Let G^* denote the analytic subgroup of G with Lie algebra $\mathfrak{g}^*$. Then G^* has Iwasawa, Cartan, and Bruhat decompositions compatible with those of G, that is

$$G^* = K^*A^*N^*, \qquad G^* = K^*A^*K^*, \qquad G^* = \bar{N}^*M^*A^*N^* \cup M^*A^*N^*, \tag{4.1}$$

where the starred groups are the intersections of G^* with the corresponding subgroups of G. Thus the components of an element $g \in G$ in the three decompositions can be found by explicit matrix computations in the group $\mathbf{SU}(2, 1)$, provided g is contained in a group of the form G^*. An application of this principle is the following proposition which was proved independently by G. Schiffmann in a different manner.

If $\alpha \in \Sigma^+$, the dimension $m_\alpha = \dim \mathfrak{g}_\alpha$ is called the multiplicity of α. The Bruhat decomposition of G shows that $\bar{N}MAN$ is dense in G; moreover if $g \in \bar{N}MAN$, its components in each subgroup are unique. Let $\bar{n}(g)$ denote the component in $\bar{N}$.

Proposition 4.2. Suppose the symmetric space G/K has rank one and put $|Z|^2 = -\langle Z, \theta Z\rangle$ for $Z \in \mathfrak{g}$. Let $m^* \in K$ induce the nontrivial Weyl group element t, α the simple root. Then if $\bar{n} = \exp X \exp Y$ $(X \in \mathfrak{g}_{-\alpha}, Y \in \mathfrak{g}_{-2\alpha})$

$$\exp[\rho(H(\bar{n}))] = [(1 + c|X|^2)^2 + 4c|Y|^2]^{(m_\alpha + 2m_{2\alpha})/4}$$

and

$$\exp[\rho(H(\bar{n}(m^*\bar{n})) - H(\bar{n}))] = [c^2|X|^4 + 4c|Y|^2]^{-(m_\alpha + 2m_{2\alpha})/4}.$$

The constant c is given by $c^{-1} = 4(m_\alpha + 2m_{2\alpha})$.

In the remainder of this section, we assume G/K of rank one and keep the notation of Proposition 4.2. Our problem is now to give a meromorphic continuation of the functions $\lambda \to \Phi_{\lambda,e}$ and $\lambda \to \Phi_{\lambda,t}$ in Section 3. The first is clearly holomorphic on the entire space $\mathfrak{a}_c^*$; in fact condition (3.6) is vacuous in agreement with the orbit Ξ_e being closed. In accordance with Proposition 3.3, we have two distributions $T_{\lambda,e}$, $T_{\lambda,t}$ on K/M such that

$$\Phi_{\lambda,e}(\varphi) = \int_{K/M}\left(\int_A \varphi(ka\cdot\xi_0)\exp[(i\lambda + \rho)(\log a)]\,da\right)dT_{\lambda,e}$$

$$\Phi_{\lambda,t}(\varphi) = \int_{K/M}\left(\int_A \varphi(ku\cdot\xi_0)\exp[(i\lambda + \rho)(\log a)]\,da\right)dT_{\lambda,t}. \qquad (4.2)$$

Here $T_{\lambda,e}$ is the delta-distribution at the origin in K/M whereas by (3.7),

$$T_{\lambda,t}(F) = \int_{\bar{N}} F(m^*k(\bar{n})M)\exp[-(i\lambda + \rho)(H(\bar{n}))]\,d\bar{n}, \qquad F \in \mathscr{E}(K/M). \qquad (4.3)$$

Now K/M is canonically identified with G/P (where $P = MAN$) and under this identification if $\bar{n} \neq e$,

$$m^*k(\bar{n})M = m^*\bar{n}P = \bar{n}(m^*\bar{n})P.$$

The mapping $\bar{n} \to \bar{n}(m^*\bar{n})$ is a diffeomorphism of $\bar{N} - \{e\}$ onto itself. By computing its Jacobian, we obtain from (4.3)

$$T_{\lambda,s}(F) = \int_{\bar{N}} F(\bar{n}P) \exp[-2\rho(H(\bar{n}))] \exp[-(i\lambda - \rho)(H(\bar{n}(m^*\bar{n})))] \, d\bar{n}. \tag{4.4}$$

Now the mapping $(X, Y) \to \exp X \exp Y$ is a diffeomorphism of $\mathfrak{g}_{-\alpha} \times \mathfrak{g}_{-2\alpha}$ onto $\bar{N}$ under which the product measure $dX\, dY$ corresponds to a constant multiple of $d\bar{n}$. Let $z \in \mathbf{C}$ be determined by $\lambda = z\rho$. Using now the latter formula of Proposition 4.2, the integral (4.4) is transformed to an integral of the form

$$\int_{\mathfrak{g}_{-\alpha} \times \mathfrak{g}_{-2\alpha}} F^*(X,Y) (|X|^4 + |Y|^2)^{-p(1-iz)} \, dX \, dY \tag{4.5}$$

where F^* is a C^∞ function and p is determined by $4p = m_\alpha + 2m_{2\alpha}$. Using polar coordinates first in $\mathfrak{g}_{-\alpha}$ and in $\mathfrak{g}_{-2\alpha}$ and again in the resulting double integral, the expression (4.5) reduces to an integral

$$\int_0^\infty \rho^{2piz-1} \varphi_z(\rho) \, d\rho, \qquad Re(iz) > 0, \tag{4.6}$$

where the C^∞ function $\varphi_z(\rho)$ depends holomorphically on z. This integral extends to a meromorphic function of z with, at most, simple poles when $-2piz$ is a nonnegative integer. This proves the following result where $\alpha_0 = \alpha/\langle\alpha, \alpha\rangle$ and Γ is the usual gamma function.

Theorem 4.3. The function

$$\lambda \to \Gamma(\langle i\lambda, \alpha_0\rangle)^{-1} \Phi_{\lambda,t} = \Psi_{\lambda,t}$$

is an entire distribution-valued function on $\mathfrak{a}_c^*$.

If one looks more carefully at (4.5), one finds that the conical distribution $\Psi_{\lambda,t}$ is $\neq 0$ for all λ. Moreover the residues of (4.6) at the poles are certain derivatives of φ_z at the origin. It follows that if λ_0 is a pole for

$\Phi_{\lambda,t}$, then

$$\operatorname*{Residue}_{\lambda=\lambda_0} \Phi_{\lambda,t} = D\Phi_{\lambda,e}, \tag{4.7}$$

where D is a certain differential operator on Ξ transversal to the orbit Ξ_e.

5. The General Case. Use of the Radon Transform

We now go back to considering G/K of arbitrary rank and shall give the meromorphic continuation of each of the distribution-valued functions $\lambda \to \Phi_{\lambda,s}$. Since the coset space $B = K/M$ is compact, it will be more convenient to deal with the functions $\lambda \to T_{\lambda,s}$ which correspond to $\Phi_{\lambda,s}$ by means of Proposition 3.3; in fact, any two distributions on K/M have a well-defined convolution $\times$ inherited from the convolution product of distributions on K.

We recall that each $s \in W$ can be written $s = s_{\alpha_1} \ldots s_{\alpha_p}$ where each α_i is a simple restricted root and s_{α_i} the corresponding Weyl reflexion. This expression for s is called *reduced* if p is as small as possible.

Theorem 5.1. Let $s \in W$ be arbitrary and $s = s_{\alpha_1} \ldots s_{\alpha_p}$ any reduced expression of s. Put $s^{(q)} = s_{\alpha_{q+1}} \ldots s_{\alpha_p}$ $(1 \leqq q \leqq p-1)$, $s^{(p)} = e$. Then for a suitable normalization of Haar measures

$$T_{\lambda,s} = T_{s'\lambda,s_{\alpha_1}} \times \cdots \times T_{s(P)\lambda,s_{\alpha_p}} \tag{5.1}$$

for all λ in the tube

$$\operatorname{Re}(\langle i\lambda, \alpha\rangle) > 0 \qquad \text{for} \quad \alpha \in \Sigma^+ \cap s^{-1}\Sigma^-.$$

The proof starts from the integral (3.7) and reduces it to a product integral over the groups $\bar{N}_{s_{\alpha_i}}$; to some extent this works as if these nilpotent groups were Abelian. For $F \equiv 1$, this reduction was done by Gindikin and Karpelevič [*2*] in the proof of their product formula for Harish-Chandra's **c**-function. For general F, such a reduction was done by Schiffmann [*10*, Theorem 1].

Now let β denote one of the simple roots α_i. Let G_β denote the analytic subgroup of G whose Lie algebra is the subalgebra of $\mathfrak{g}$ generated by $\mathfrak{g}_\beta$ and $\mathfrak{g}_{-\beta}$. Then G_β is semisimple and the orbit $G_\beta \cdot o$ in G/K is a symmetric space G_β/K_β of rank one. The orthogonal complement $\mathfrak{a}_\beta$ of the hyperplane $\beta = 0$ in $\mathfrak{a}$ plays the role of $\mathfrak{a}$ for the space G_β/K_β. The centralizer M_β of exp $\mathfrak{a}_\beta$ in K_β satisfies $M_\beta = M \cap K_\beta$ so K_β/M_β is in a natural way a submanifold of K/M. If $F \in \mathscr{E}(K/M)$, let $\bar{F}$ denote its restriction to K_β/M_β. If μ is a linear function on $\mathfrak{a}$, let $\bar{\mu}$ denote its restriction to $\mathfrak{a}_\beta$. Let $T^\beta_{\bar{\mu},s_\beta}$ denote the distribution T_{λ,s_β} constructed for the rank one space G_β/K_β. Then one proves without difficulty,

$$T_{\mu,s_\beta}(F) = T^\beta_{\bar{\mu},s_\beta}(\bar{F}). \tag{5.2}$$

Now using Theorem 4.3, we deduce that the function $\mu \to \Gamma(\langle i\mu, \beta_0\rangle)^{-1}T_{\mu,s_\beta}$ extends to an entire distribution-valued function on $\mathfrak{a}_c^*$. We apply this now to the individual factors in (5.1). It is not hard to show that the roots $s^{(j)^{-1}}\alpha_j$ $(1 \leqq j \leqq p)$ constitute precisely the elements in $\sum_0^+ \cap s^{-1}\sum_0^-$, where $\sum_0^+ = \{\alpha \in \sum^+ | \frac{1}{2}\alpha \notin \sum^+\}$ and $\sum_0^-$ is the set of negatives in $\sum_0^+$. This gives the following meromorphic continuation of the function $\lambda \to \Phi_{\lambda,s}$.

Theorem 5.2. Let $s \in W$ be arbitrary and define the function $\boldsymbol{d}_s$ on $\mathfrak{a}_c^*$ by

$$\boldsymbol{d}_s(\lambda) = \prod_{\alpha \in \Sigma_0^+ \cap s^{-1}\Sigma_0^-} \Gamma(\langle i\lambda, \alpha_0\rangle) 2^{-\langle i\lambda, \alpha_0\rangle}, \tag{5.3}$$

where for any $\alpha \in \Sigma^+$ we put $\alpha_0 = \alpha/\langle\alpha, \alpha\rangle$. Then the map

$$\lambda \to \boldsymbol{d}_s(\lambda)^{-1}\Phi_{\lambda,s}$$

extends from the tube (3.6) to a holomorphic function, denoted $\Psi_{\lambda,s}$, on $\mathfrak{a}_c^*$ with values in $\mathscr{D}'(\Xi)$. For each $\lambda \in \mathfrak{a}_c^*$, $\Psi_{\lambda,s}$ is a conical distribution in $\mathscr{D}'_\lambda(\Xi)$.

The residues of $\Phi_{\lambda,s}$, that is the values of $\Psi_{\lambda,s}$ at the singular points $\lambda = \lambda_0$ for $\Phi_{\lambda,s}$, have the following geometric interpretation: the closure of the orbit Ξ_s in Ξ is a union of Ξ_s and some other orbits Ξ_t where the $t \in W$ are obtained from the expression $s = s_{\alpha_1} \ldots s_{\alpha_p}$ by deleting some factors. The residue $\mathrm{Res}_{\lambda=\lambda_0}\Phi_{\lambda,s}$ consists of a linear combination of certain transversal derivatives of the various $\Phi_{\lambda,t}$ constructed for these other orbits [cf. (4.7)]. This is clear from the product (5.1) because at a singularity $\lambda = \lambda_0$ some of the factors in (5.1) are replaced by derivatives of the delta function on K/M.

In accordance with the discussion in Section 2, one would now like to prove that all the conical distributions on Ξ can be constructed from the distributions $\Psi_{\lambda,s}$. To this end, we use the Radon transform on X which to each function $f \in \mathscr{D}(X)$ associates the function $\hat{f} \in \mathscr{D}(\Xi)$ given by

$$\hat{f}(\xi) = \int_{\xi} f(x)\, ds(x), \qquad \xi \in \Xi, \tag{5.4}$$

ds being the volume element on ξ (induced by the Riemannian structure of X). If $x \in X$, the isotropy subgroup $K_x \subset G$ at x acts transitively on the set $\check{x}$ of horocycles passing through x. Thus $\check{x}$ has a unique K_x-invariant measure $d\mu$ such that $\check{x}$ has total measure 1. For $\varphi \in \mathscr{E}(\Xi)$, we put

$$\check{\varphi}(x) = \int_{\check{x}} \varphi(\xi)\, d\mu(\xi), \qquad x \in X$$

and call $\check{\varphi}$ the *dual Radon transform* of φ. For suitably normalized G-invariant measures dx on X, $d\xi$ on Ξ, we have

$$\int_X f(x)\check{\varphi}(x)\, dx = \int_{\Xi} \hat{f}(\xi)\varphi(\xi)\, d\xi \tag{5.5}$$

for f and φ of compact support. Accordingly, we define for distributions $S \in \mathscr{E}'(X)$, $\Sigma \in \mathscr{D}'(\Xi)$,

$$\hat{S}(\varphi) = S(\check{\varphi}), \qquad \check{\Sigma}(f) = \Sigma(\hat{f})$$

for $f \in \mathscr{D}(X)$, $\varphi \in \mathscr{D}(\Xi)$. The following lemma now plays a basic role.

Lemma 5.3. For each $\lambda \in \mathfrak{a}_c^*$ and each $s \in W$, the distribution $(\Psi_{\lambda,s})^\vee$ on X is a C^∞ function given by

$$(\Psi_{\lambda,s})^\vee(naK) = \boldsymbol{e}_s(\lambda) \exp[(is\lambda + \rho)(\log a)], \qquad n \in N, \qquad a \in A, \tag{5.6}$$

where $\boldsymbol{e}_s(\lambda)$ is given by

$$\boldsymbol{e}_s(\lambda)^{-1} = \prod_{\alpha \in \Sigma_0^+ \cap s^{-1}\Sigma_0^-} \Gamma\left(\tfrac{1}{2}\left(\frac{m_\alpha}{2} + 1 + \langle i\lambda, \alpha_0\rangle\right)\right)\Gamma\left(\tfrac{1}{2}\left(\frac{m_\alpha}{2} + m_{2\alpha} + \langle i\lambda, \alpha_0\rangle\right)\right).$$

This lemma comes from the above mentioned relationship between $\Psi_{\lambda,s}$ and the integral (3.5) which by the formulas of Harish-Chandra [*5*, p. 303] and Gindikin–Karpelevic [*2*] equals $\boldsymbol{d}_s(\lambda)\boldsymbol{e}_s(\lambda)$. In case s is the Weyl group element s^* which interchanges Σ^+ and Σ^-, the group $\overline{N}_s$ equals $\overline{N}$ and the integral (3.5) equals Harish-Chandra's function $\boldsymbol{c}(\lambda)$,

[5]. While this function plays a basic role in the theory of spherical functions on X, the factors $\boldsymbol{d}_s(\lambda)$ and $\boldsymbol{e}_s(\lambda)$ have their individual importance for the conical distributions. We shall also see that the factor $\boldsymbol{e}_{s^*}(\lambda)$ is intimately related to representation theory.

In order to determine all the conical distributions $\Psi \in \mathscr{D}'_\lambda$, we consider the dual Radon transform $\check{\Psi}$. Since Ψ is an eigendistribution of the operators in $\mathbf{D}(\Xi)$, one finds that $\check{\Psi}$ is an eigendistribution of the operators in $\mathbf{D}(X)$, hence an *eigenfunction*. Since the dual Radon transform preserves the MN-invariance, we are led to look for the MN-invariant functions on X which are eigenfunctions of all the operators in $\mathbf{D}(X)$. This problem can be solved completely, giving for suitable constants $a_s \in \mathbf{C}$ the formula

$$\check{\Psi}(naK) = \sum_{s\in W} a_s \exp[(is\lambda + \rho)(\log a)], \qquad n \in N, \quad a \in A,$$

at least if λ is regular, that is $s\lambda \neq \lambda$ for all $s \neq e$ in W. This leads to the following main result.

Theorem 5.4. Let $\lambda \in \mathfrak{a}_c^*$. Assume $s\lambda \neq \lambda$ for all $s \neq e$ in W and $\boldsymbol{e}_{s^*}(\lambda) \neq 0$. Then the linear combinations

$$\sum_{s\in W} c_s \Psi_{\lambda,s}, \qquad c_s \in \mathbf{C}$$

constitute all the conical distributions in $\mathscr{D}'_\lambda$.

Which of these distributions are conical functions? It is clear from considering supports that at most the distributions Ψ_{λ,s^*} could be given by C^∞ functions on Ξ. In order to see what conditions on λ have to be imposed, one first proves that for $G = \mathbf{SL}(2, \mathbf{R})$ with the standard Iwasawa decomposition the conical functions are the constant multiples of the functions

$$\Psi\left(\begin{pmatrix} x_{11} & x_{12} \\ x_{21} & x_{22} \end{pmatrix} \cdot \xi_0\right) = x_{21}^{2r}, \tag{5.7}$$

where r is a nonnegative integer. If $\Psi = \Psi_{\lambda,s^*}$ is a C^∞ function on Ξ, we construct for each $\alpha \in \Sigma^+$ a subgroup $G_\alpha \subset G$ locally isomorphic to $\mathbf{SL}(2, \mathbf{R})$; expressing that Ψ has a C^∞ restriction to the corresponding space Ξ_α gives by (5.7) the condition

$$\frac{\langle i\lambda - \rho, \alpha\rangle}{\langle \alpha, \alpha\rangle} \in \mathbf{Z}^+. \tag{5.8}$$

On the other hand, if this condition is satisfied for each $\alpha \in \Sigma^+$, one can by means of representation theory construct a conical function in $\mathscr{D}'_\lambda$. This gives the following result.

Theorem 5.5. The conical functions on Ξ are the constant multiples of the functions Ψ_{λ,s^*} where

$$\frac{\langle i\lambda - \rho, \alpha \rangle}{\langle \alpha, \alpha \rangle} \in \mathbf{Z}^+ \qquad \text{for } \alpha \in \Sigma^+.$$

For the generalized Lorentz groups, the conclusion of Theorem 5.4 holds without any restriction on λ [*8*, Chapter III, Section 4].

Theorem 5.6. Suppose $G = \mathbf{SO}_0(1, n)$, $n > 2$. For each $\lambda \in \mathfrak{a}_c^*$, the linear combinations

$$\sum_{s \in W} c_s \Psi_{\lambda,s} \qquad c_s \in W$$

constitute all the conical distributions in $\mathscr{D}'_\lambda$.

In order for the theorem to remain valid also for $n = 2$, a minor change in the definition of conical distributions is required.

6. Conical Distributions and Representations

Finally we describe briefly some connections between conical distributions and representations of the group G. For a more complete treatment see Refs. [*7, 8*]. The distribution space $\mathscr{D}'_\lambda$ defined in Section 3 is a locally convex space in the topology induced by the strong topology of $\mathscr{D}'(\Xi)$. The natural action of G on $\mathscr{D}'(\Xi)$ gives a continuous representation π_λ of G on $\mathscr{D}'_\lambda$. The locally square-integrable functions in $\mathscr{D}'_\lambda$ form a dense invariant subspace $\mathscr{H}_\lambda$ and the representation π_λ^0 on this Hilbert space is the representation induced by the representation $man \to \exp[i\lambda(\log a)]$ of MAN. However, since we are considering λ in the complex dual $\mathfrak{a}_c^*$ (and not just the real dual $\mathfrak{a}^*$), there are certain advantages in passing from the Hilbert space $\mathscr{H}_\lambda$ to the distribution space $\mathscr{D}'_\lambda$. A vector $\Psi \in \mathscr{D}'_\lambda$ is called an *extreme weight vector* if there exists a $\mu \in \mathfrak{a}_c^*$ such that

$$\pi_\lambda(an)\Psi = \exp[\mu(\log a)]\Psi, \qquad a \in A,\ n \in N.$$

Theorem 6.1. Let $\lambda \in \mathfrak{a}_c^*$ satisfy the assumptions of Theorem 5.4. Then the scalar multiples of the vectors $\Psi_{\lambda,s}$ $(s \in W)$ are precisely the extreme weight vectors in $\mathscr{D}'_\lambda$.

This result follows from the proof of Theorem 5.4.

Theorem 6.2. The representation π_λ of G on $\mathscr{D}'_\lambda$ is irreducible if and only if

$$e_{s^*}(\lambda)e_{s^*}(-\lambda) \neq 0.$$

This can be derived by means of Kostant's criterion [*9*, p. 631] for the algebraic irreducibility of the representation $d\pi_\lambda^0$ of $\mathfrak{g}$ on the space of K-finite vectors in $\mathscr{H}_\lambda$, combined with Harish-Chandra's Theorem 5 in [*3*] relating this irreducibility to that of π^0; see Ref. [*12*] for complex G. However, a more direct proof is also possible, cf. [*7*, §4] and [*8*, p. 133].

Let $S_{\lambda,s} \in \mathscr{D}'(K/M)$ correspond to the conical distribution $\Psi_{\lambda,s}$ under the bijection $S \to \Psi$ of $\mathscr{D}'(K/M)$ onto $\mathscr{D}'_\lambda$ given by Proposition 3.3. Let $\times$ denote the convolution product of distributions on K/M, induced by convolution product on K.

Theorem 6.3. Assume π_λ irreducible. Let $s \in W$. Then the convolution $S \to S \times S_{s\lambda,s^{-1}}$ on $\mathscr{D}'(K/M)$ induces (by Proposition 3.3) a bijection $A_{\lambda,s}$ of $\mathscr{D}'_\lambda$ onto $\mathscr{D}'_{s\lambda}$ which intertwines π_λ and $\pi_{s\lambda}$, that is

$$A_{\lambda,s}\pi_\lambda(g) = \pi_{s\lambda}(g)A_{\lambda,s}, \qquad g \in G.$$

In particular π_λ and $\pi_{s\lambda}$ are equivalent.

This theorem is proved in [*8*, Chapter III, Section 6] where the relationship with work of F. Bruhat, A. Knapp, R. Kunze, G. Schiffmann, E. Stein, and D. Zhelobenko is also explained. In particular, Schiffmann [*11*] constructed independently intertwining operators for $d\pi_\lambda^0$ and $d\pi_{s\lambda}^0$ giving infinitesimal equivalence between π_λ^0 and $\pi_{s\lambda}^0$.

References

[*1*] I. M. Gelfand, "Automorphic functions and the theory of representations," *Proc. Inter. Cong. Math. Stockholm 1962*, pp. 74–85.

[*2*] S. G. Gindikin and F. I. Karpelevič, "Plancherel measure of Riemannian symmetric spaces of nonpositive curvature," *Soviet Math. Dokl.*, **3** (1962), 962–965.

[*3*] Harish-Chandra, "Representations of a semisimple Lie group on a Banach space I," *Trans. Amer. Math. Soc.*, **75** (1953), 185–243.

[*4*] ———, "On a lemma of Bruhat," *J. Math. Pures Appl.* **35** (1956), 203–210.

[*5*] ———, "Spherical functions on a semisimple Lie group I," *Amer. J. Math.*, **80** (1958), 241–310.

[*6*] S. Helgason, "Duality and Radon transform for symmetric spaces," *Amer. J. Math.*, **85** (1963), 667–692.

[*7*] ———, "Applications of the Radon transform to representations of semisimple Lie groups," *Proc. Nat. Acad. Sci. U.S.A.*, **63** (1969), 643–647.

[*8*] ———, "A duality for symmetric spaces with applications to group representations," *Advances in Math.*, **5** (1970), 1–154.

[*9*] B. Kostant, "On the existence and irreducibility of certain series of representations," *Bull. Amer. Math. Soc.*, **75** (1969), 627–642.

[*10*] G. Schiffmann, "Intégrales d'entrelacement," *Compt. Rend. Acad. Sci. Paris*, **266** (1968), 47–49.

[*11*] ———, "Sur les intégrales d'entrelacement de R. A. Kunze et E. M. Stein," Thèse, Paris, 1969.

[*12*] K. R. Parthasarathy, R. Ranga Rao, and V. S. Varadarajan, "Representations of complex semisimple Lie groups and Lie algebras," *Ann. of Math.*, **85** (1967), 383–429.

GEOMETRIC IDEAS IN LIE GROUP HARMONIC ANALYSIS THEORY*

ROBERT HERMANN

Department of Mathematics
Rutgers University
New Brunswick, New Jersey

*Work on this paper was done under Contr. N00014–69–C–0265, NR 041–378. Reproduction in whole or in part is permitted for any purpose of the United States Government.

Preface

The aim of this report is to present qualitative, geometric ideas relating the differential geometry of homogeneous spaces to certain problems in Lie group representation theory, particularly the problem of "asymptotic behavior of matrix elements of Lie group representations," and to study the interrelation between the theory of harmonic analysis of functions on a Lie group and the compactifications of homogeneous spaces on which the Lie group acts.

Many of these ideas are scattered through my own work cited in the bibliography. I hope that presenting a more unified exposition will make them more accessible and lead to further applications. For example, in addition to the evident application to group representation theory, there are, I believe, close relations and analogies with many problems in mathematical physics. In particular, refer to Ref. [*6*] for an exposition of many problems of this type, as well as an overall viewpoint towards functional analysis and generalized function theory that is intermediate between that developed—in rather complete isolation from each other's problems and ideas—by mathematicians and physicists.

In addition to this background, the formalism of differential and integral geometry on manifolds will be needed. As a basic reference in this area, my book [*5*] will be used.

1. Hilbert and Dirac Spaces and Generalized Functions

1.1. Introduction

In this chapter, we will rely on the ideas developed in Ref [*6*] to give a general framework for our work. In particular, we will work throughout with incomplete Hilbert spaces, using a formalism [*6*] that is intermediate between the mathematician's and the physicist's view of abstract functional analysis.

Let H be a complex vector space, with a real-bilinear form: $H \times H \to C$, denoted by $(\psi_1, \psi_2) \to \langle\psi_1|\psi_2\rangle$. We suppose that this form is Hermitian symmetric, i.e., that

$$\begin{aligned} &\langle\psi_1|\psi_1\rangle > 0 \quad \text{if} \quad \psi_1 \neq 0, \\ &\langle\psi_1|\psi_2\rangle = \langle\psi_2|\psi_1\rangle^*, \\ &\langle\psi_1|c\psi_2\rangle = c\langle\psi_1\psi_2\rangle = \langle c^*\psi_1|\psi_2\rangle, \end{aligned} \tag{1.1.1}$$

for $\psi_1, \psi_2 \in H$, $c \in C$. (* denotes the complex conjugate of the complex number.)

We will follow the custom that is inherent in the physics literature and call H, with the inner-product structure $\langle | \rangle$, a *Hilbert space*. (Note that the "completeness" axiom, necessary for the mathematicians's version of a "Hilbert space," has been left out.)

Now, let D denote the space of complex-linear mappings: $H \to C$. Denote a typical element of D, i.e., a linear functional on H, by "α". D, as a space of linear functionals, is obviously a real vector space. It will be made into a *complex* vector space, using the following rule:

$$(c\alpha)(\psi) = c^*\alpha(\psi) \tag{1.1.2}$$

for $c \in C$, $\alpha \in D$, $\psi \in H$.

With this definition of D as a complex vector space, the Hermitian symmetric inner product (1.1) defines an imbedding of $H \to D$ that is a complex linear map. Namely, associate with $\psi \in H$ the linear functional α_ψ defined as follows:

$$\alpha_\psi(\psi_1) = \langle \psi | \psi_1 \rangle. \tag{1.1.3}$$

In Dirac's language [*2*], the elements of H are "ket" vectors, while the elements of D are "bra" vectors. In order to keep as close as possible to Dirac's notation, we will usually use the following notational convention:

$$\alpha(\psi) = \langle \alpha | \psi \rangle$$

for $\alpha \in D$, $\psi \in H$.

A key question is always the extension of a linear operator $A\colon H \to H$ to a linear operator: $D \to D$ that is compatible with the imbedding (1.3) of H into D. This can be done in a very natural way for such A's that have adjoint operators $A^\dagger\colon H \to H$ which satisfy the following condition:

$$\langle A^\dagger \psi_1 | \psi_2 \rangle = \langle \psi_1 | A \psi_2 \rangle. \tag{1.1.4}$$

In fact, if such an adjoint exists, extend A to D using the following formula:

$$\langle A\alpha | \psi \rangle = \langle \alpha | A^\dagger \psi \rangle \tag{1.1.5}$$

for $\alpha \in D$, $\psi \in H$.

If such an adjoint does not exist, the extension to D must be made by various approximation procedures, about which very little is known in general.

In this paper, we will mainly be concerned with convergence of curves $t \to \alpha_t$, $0 \leqq t < \infty$, in D. We will consider "weak convergence," namely, such a curve is said to *converge* to an element $\alpha_\infty \in D$ if

$$\lim_{t \to \infty} \langle \alpha_t | \psi \rangle = \langle \alpha_\infty | \psi \rangle \tag{1.1.6}$$

for all $\psi \in H$. However, we will not become involved with the various functional-analysis technicalities concerning these limits. Instead, we will mainly deal with specific examples and situations where such limits may be discussed less abstractly.

1.2. The Projective Spaces Associated with the Denote Spaces

As we will see when those general ideas are specialized, one often encounters curves $t \to \alpha(t)$ in a Dirac space D associated with a Hilbert space H which do not converge, even "weakly," but do have the following property:

> There is a curve $t \to \lambda(t)$ in C such that the curve $t \to \lambda(t)\alpha(t)$ does converge weakly in D.

To understand the abstract features of this situation, it is convenient to introduce the projective space, PD, associated with the vector space D. To do this, let C' denote the multiplicative group of nonzero complex numbers. Denote a typical element of C' by λ. Then, C' acts on D in the obvious way:

$$(\lambda, \psi) \to \lambda\psi$$

for $\lambda \in C'$, $\psi \in D$. PD is now defined as the orbit space of C' acting in this way on the nonzero elements of D.

There is then a mapping $D - (0) \to PD$ which assigns to each nonzero $\psi \in D$ the orbit of C' on which it lies. We will refer to this as the projection of $D - (0)$ on PD.

Now, we face an interesting problem in functional analysis: Discuss various general notions of convergence in PD. In fact, there seems to be very little work along these lines in the literature on which to base our discussion. However, we will proceed as if one did have such a notion available. For example, the most naive point of view might be described as follows:

> A curve $t \to \alpha'(t)$ in PD converges, as $t \to \infty$, to a point $\alpha' \in PD$ if there is a curve $t \to \alpha(t)$ in $D - (0)$ with projects into the curve

$t \to \alpha'(t)$, such that:

$$\lim_{t \to \infty} \alpha(t) = \alpha \qquad \text{(weakly)}, \qquad \alpha \text{ projects into } \alpha'.$$

2. Generalized Functions Defined Using the Lebesgue Dominated Convergence Theorem

2.1. Introduction

In Section 1, we have briefly discussed an abstract framework for consideration of various sorts of asymptotic-limit problems that arise in mathematical physics and Lie group representation theory problems. There are two directions which should be developed to understand the general mathematical structure and nature of those problems: Develop first the general geometric techniques needed to understand the important classes of examples, then abstract from this the general functional analysis framework in which one can imbed these problems. In this report, we are concentrating on the first problem. Accordingly, we now descend from the very general framework of Section 1 to consider various situations with more geometric structure and content.

We will always deal with "Hilbert spaces" H that are composed of C^∞ cross sections of vector bundles on manifolds, with the Hermitian inner product $\langle | \rangle$ given by integration over the manifold. In evaluating limits of the form

$$\lim_{t \to \infty} \langle \alpha(t) | \psi \rangle,$$

one encounters problems of finding limits of integrals on functions or manifolds that depend on parameters. The most powerful tool available for attacking these problems is the Lebesgue dominated convergence theorem. Accordingly, we will now review it, in the context of the integration theory on manifolds that was sketched in [*5*].

2.2. The Lebesgue Dominated Convergence Theorem

Let us use the book by Riesz and Nagy [*17*] as the basic reference for measure theory and the associated ideas of functional analysis. We will not state the theorem for the most general sort of measure space, but will restrict attention to manifolds, using [*5*] for basic facts and notations concerning differential and integral geometry on manifolds.

Let M be an orientable manifold, with "dp" a volume element differential form on M. Let $F_0(M)$ denote the C^∞ compact support real valued functions on M. The integral

$$\int_M f\,dp \tag{2.2.1}$$

can be defined for $f \in F_0(M)$, as explained in [5]. This is basically the "Riemann integral." Following the pattern of extending the Riemann integral to the Lebesgue integral for the case: $M = R$, the integral (2.2.1) can be used to define a measure on Borel subsets of M, and one then knows what is meant by a "measurable, integrable real valued function" on M. If f is such a function, its integral can be defined, and will also be denoted by (2.2.1).

Suppose now that $t \to f_t$, $0 \leqq t < \infty$, is a one-parameter family of complex-valued, measurable functions. Suppose also that h is a real-valued, measurable, and *integrable* function such that:

$$|f_t(p)| \leqq h(p) \tag{2.2.2}$$

for all t, and all $p \in M$, except possibly for a set of measure zero. Suppose, finally, that f is a complex-valued function such that

$$\lim_{t\to\infty} f_t(p) = f(p) \tag{2.2.3}$$

for all $p \in M$, except possibly for a set of measure zero.

The Lebesgue theorem now asserts that

$$\lim_{t\to\infty} \int f_t\,dp = \int f\,dp, \tag{2.2.4}$$

i.e., limits and integrals can be interchanged.

We now turn to studying situations where these hypotheses can be readily applied, namely the study of limits of curves in generalized function spaces generated by flows on M.

2.3. Generalized Functions Definited by Flows

Let M and dp be as defined in Section 2.2. Denote by $F(M)$ the C^∞, real-valued functions on M. Let H be the C^∞, compact-support, complex-valued functions on M. Define the Hermitian inner product $\langle | \rangle$ satisfying (2.1.1) by the following formula:

$$\langle \psi_1 | \psi_2 \rangle = \int_M \psi_1(p)^* \psi_2(p)\,dp. \tag{2.3.1}$$

Define D as the "Dirac space" associated with H, as explained in Section 2.1. Also, imbed H as a complex subspace of D, as explained in Section 1 and in [6].

Let φ be a diffeomorphism: $M \to M$. If $\psi \in H$, let $\varphi(\psi)$ be the function

$$p \to \psi(\varphi^{-1}(p)) \tag{2.3.2}$$

on M. Let $j \in F(M)$ be the Jacobian function associated with φ, defined by

$$\int \varphi(\psi)\, dp = \int j\psi\, dp \tag{2.3.3}$$

for $\psi \in H$. Note that j can also be defined by the following formula:

$$\varphi^*(dp) = j\, dp \tag{2.3.4}$$

where $\varphi^*(dp)$ is the "pull-back" of differential forms defined in Ref. [5]. (Due to a conflict between the "standard" notations used in mathematics and in physics literature, the reader must understand from context which meaning of * is used—complex conjugate or pull-back map.)

Using (2.3.3), and the rule (1.1.5), we can extend φ, considered as a linear map: $H \to H$, to a linear map: $D \to D$. First, let us compute the adjoint map $\varphi\dagger$: $H \to H$.

$$\begin{aligned}\langle \psi_1 | \varphi\psi_2 \rangle &= \int \psi_1(p)^* \psi_2(\varphi^{-1}p)\, dp \\ &= \int \psi_1(\varphi^{-1}(\varphi(p)))^* \psi_2(\varphi^{-1}p)\, dp \\ &= \int \psi_1(\varphi(p))^* \psi_2(p) j(p)\, dp \qquad \text{using (2.3.3),} \\ &= \langle j\varphi^{-1}(\psi_1) | (\psi_2 \rangle,\end{aligned}$$

i.e.,

$$\varphi\dagger = j\varphi^{-1}. \tag{2.3.5}$$

Hence, for $\alpha \in D$, $\psi \in H$,

$$\langle \varphi(\alpha) | \psi \rangle = \langle \alpha | \varphi^\dagger(\psi) \rangle = \langle \alpha | j\varphi^{-1}(\psi) \rangle. \tag{2.3.6}$$

Now, a *flow* on M is a one-parameter family $t \to \varphi_t$, $0 \leqq t < \infty$, of diffeomorphisms of M. We also require that the map $M \times [0, \infty] \to M$

defined by $(p, t) \to \varphi_t(p)$ be C^∞. Our general aim can be described as follows:

Given $\alpha \in D$, and a curve $t \to \lambda(t)$ in R, find sufficient conditions that the curve

$$t \to \alpha_t = \lambda(t)\varphi_t(\alpha) \tag{2.3.7}$$

converge to an element $\alpha_\infty \in D$ according to the rule (1.1.6). (In fact what we want to do is to write down "sufficient conditions" involving the Lebesgue theorem discussed in Section 2.2.)

In this section, we will only treat the case where α is of the form α_{ψ_0}, where $\psi_0 \in H$. Then, using (3.6), we have, for $\psi \in H$,

$$\begin{aligned} \langle \alpha_t | \psi \rangle &= \lambda(t)\langle \varphi_t(\alpha) | \psi \rangle \\ &= \lambda(t)\langle \alpha | j\varphi_t^{-1}(\psi) \rangle \\ &= \lambda(t)\langle \psi_0 | j_t\varphi_t^{-1}(\psi) \rangle \\ &= \lambda(t) \int \psi_0(p)^* j_t(p)\psi(\varphi_t(p))\, dp \\ &= \int \psi_0(p)^*(\lambda(t)j_t(p)\psi(\varphi_t(p)))\, dp. \end{aligned} \tag{2.3.8}$$

We can now see that formula (2.3.8) exhibits a form of $\langle \alpha_t | \psi \rangle$ to which the Lebesgue theorem can be readily applied, *if* one has geometric information about the limits, as $t \to \infty$, of the "streamline" curves $t \to \varphi_t(p)$ of the flow. For example, the following result holds.

Theorem 3.1. Suppose that, for all $p \in M$ except for a set of measure zero, the following conditions are satisfied:

(a) $\lim_{t\to\infty} \lambda(t) j_t(p) = 1$,
(b) $\lim_{t\to\infty} \varphi_t(p) = p_0$, where p_0 is a fixed point of M,
(c) $\int \psi_0(p)^*\, dp = 1$.

Then,

$$\lim_{t\to\infty} \alpha_t = \delta_{p_0}, \tag{2.3.9}$$

where δ_{p_0} is the Dirac delta function associated with the point p_0.

Of course, δ_{p_0} is defined as an element of D via the following formula:

$$\langle \delta_{p_0} | \psi \rangle = \psi(p_0) \tag{2.3.10}$$

for all $\psi \in H$.

A class of flows (generated by gradient vector fields on Riemannian manifolds) satisfying condition (b) has been given in Ref. [8]. Rather than repeat this discussion, we will just give some simpler examples, for which needed hypotheses can be readily verified.

EXAMPLE 1. $M = R^n$, $p_0 = 0$. Denote a point of M by: $x = (x_1, \ldots, x_n)$. Set

$$\varphi_t(x) = e^{-t}x.$$

Then, using (2.3.4),

$$j_t = e^{-nt}.$$

Thus, if $\lambda(t) = e^{nt}$, α_t converges to the Dirac delta function $\delta(x)$.

EXAMPLE 2. $M = R^p \times R^q$. Denote a point of R^p and R^q by $x = (x_1, \ldots, x_p)$, $y = (y_1, \ldots, y_q)$. Thus, a point of M can be denoted by (x, y). Now set:

$$\varphi_t(x, y) = (e^{-t}x, y).$$

Then

$$j_t = e^{-pt}.$$

Choose

$$\lambda(t) = e^{pt}.$$

Then, α_t converges to the delta function concentrated on the surface $x = 0$ of M, i.e., the delta function $\delta(x_1)\delta(x_2) \ldots \delta(x_p)$.

3. Generalized Functions Definied by Functions with Critical Points

3.1. Introduction

Let M be an orientable manifold, with volume element differential form dp. Let f be a C^∞, real-valued function on M. Let N be the subset of points of $p \in M$ such that $f(p) = 0$.

If 0 is a regular value of f, i.e., if $df \neq 0$ at each point of N, then N is a submanifold of M. A volume element form can be defined on N, as follows:

Choose a vector field X in a neighborhood of N such that

$$X(f) = 1. \tag{3.1.1}$$

Then, $X \lrcorner dp$ restricted to N is a volume element form for N. This form can be used to define a "generalized function" $\delta(f)$ on M, as follows:

$$\langle \delta(f) | \psi \rangle = \int_N \psi(X \lrcorner dp). \tag{3.1.2}$$

In (3.1.2), ψ is a C^∞, complex-valued, compact-support function on M. (See [*3*, *6*] for the background to formula (3.1.2).)

Our aim in this section is to consider a definition for $\delta(f)$ in case 0 is a critical value of f. (This problem was suggested on the last page of [*6*].) However, we shall only consider the simplest case, that of a "nondegenerate critical value" in the sense of Morse theory [*15*].

3.2. The Definition of $\delta(f)$ by Principal Value When N Contains a Critical Value of f

Now, let us suppose that p_0 is a point of N such that:

(a) p_0 is a critical point of f, i.e.,

$$df(p_0) = 0;$$

(b) p_0 is the only point of N which is a critical point of f.

As a result of these assumptions, we can find a vector field X in a neighborhood of $N - (p_0)$ such that

$$X(f) = 1. \tag{3.2.1}$$

Let H be the space of C^∞, compact-support, complex-valued functions on M. Let us define $\delta(f)$ as a linear functional: $H \to C$ by means of the following formula:

$$\delta(f)|\psi = \text{Cauchy principal value of the integral } \int_{N-(p_0)} \psi(X \lrcorner dp). \tag{3.2.2}$$

To define the right-hand side, proceed as follows: Let ε be a small real positive parameter. Let N_ε be a family of open submanifolds of $N - (p_0)$, which fill up $N - (p_0)$ as $\varepsilon \to 0$. Now, define the right-hand side of (3.2.2) as follows:

$$\lim_{\varepsilon \to 0} \int_{N_\varepsilon} \psi(X \lrcorner\, dp). \tag{3.2.3}$$

To see that this makes sense in a simple case, let us consider a simple example.

EXAMPLE. $M = R^2$; $f = \frac{1}{2}(x^2 - y^2)$; $p_0 = (0, 0)$. N is then the cone

$$x^2 = y^2.$$

(x, y denote Euclidean coordinates on R^2.)

Set

$$dp = dx \wedge dy$$

$$X = \frac{1}{x^2 + y^2}\left(x\frac{\partial}{\partial x} - y\frac{\partial}{\partial y}\right).$$

Notice then that (3.2.1) is satisfied. Then,

$$X \lrcorner\, dp = [1/(x^2 + y^2)]\,(x\,dy + y\,dx).$$

Let N_ε be the set of points $(x, y) \in N$ such that $|x| > \varepsilon$ Then,

$$\int_{N_\varepsilon} \psi X \lrcorner\, dp = \int_\varepsilon^\infty \frac{\psi(t, t)}{t^2}\, t\, dt + \int_{-\infty}^{-\varepsilon} \frac{\psi(t, t)}{t^2}\, t\, dt - \int_\varepsilon^\infty \frac{\psi(t, -t)}{t^2}\, t\, dt - \int_{-\infty}^{-\varepsilon} \frac{\psi(t, -t)}{t^2}\, t\, dt$$

$$= \int_\varepsilon^\infty \frac{\psi(t, t) - \psi(-t, -t)}{t}\, dt - \int_\varepsilon^\infty \frac{\psi(t, -t) - \psi(-t, t)}{t}\, dt. \tag{3.2.4}$$

Since ψ is C^∞ on N—in particular at p_0—the integrals on the right-hand side of (3.2.3) do indeed exist, and define $\delta(x^2 + y^2)$ as a generalized function.

3.3. $\delta(f)$ When p_0 Is a Nondegenerate Critical Point of Hyperbolic Type

Continue with the notations of Section 3.2. Suppose that p_0 is the only point of N at which $df = 0$, and that

$$\dim M = m.$$

Choose indices as follows:

$$0 \leqq \mu, \nu \leqq m - 1,$$

$$1 \leqq i, j \leqq m - 1.$$

We can suppose, without loss in generality, that

$$f(p_0) = 0.$$

Definition. p_0 is a nondegenerate hyperbolic critical value for f if there is a coordinate system (x_μ) for M valid in a neighborhood of p_0 such that

$$f = \tfrac{1}{2}(x_0^2 - x_i x_i). \tag{3.3.1}$$

Let us now proceed to calculate the limit on the right-hand side of (3.2.3). Set

$$r^2 = x_0^2 + x_i x_i, \tag{3.3.2}$$

$$X = (1/r^2)(x_0 \partial_0 - x_i \partial_i),$$

with

$$\partial_\mu = \partial/\partial x_\mu.$$

Then, one readily calculates that

$$X(f) = 1. \tag{3.3.3}$$

Now, let dx denote the form $dx_0 \wedge \cdots \wedge dx_{m-1}$. Suppose that

$$dp = j\, dx, \tag{3.3.4}$$

where j is a positive, real-valued function on a neighborhood of p_0.

We must then calculate:

$$\lim_{\varepsilon \to 0} \int_{N_\varepsilon} \psi(x)\, j(x)\, (X \lrcorner dx), \tag{3.3.4a}$$

where ψ is a function on M that vanishes outside a small coordinate neighborhood of p_0. This enables us to suppose $M = R^m$, and to suppose that $(N - p_0)$ is the submanifold of R^m defined by the following relations:

$$x_0^2 = \mathbf{x}^2, \qquad x \neq 0. \tag{3.3.5}$$

($\mathbf{x}$ denotes the element (x_i) of R^{m-1}. x denotes the vector (x_μ) of R^m. $\mathbf{x}^2$ denotes $x_i x_i$.)

Let N^+ and N^- be the set of points of $N - (p_0)$ at which $x_0 > 0$ and $x_0 < 0$. On $N - (p_0)$

$$0 = x_0\, dx_0 - x_i\, dx_i \qquad \text{or} \qquad dx_0 = \frac{x_i}{x_0}\, dx_i. \tag{3.3.6}$$

Set

$$\mathbf{dx} = dx_1 \wedge \cdots \wedge dx_{m-1}. \tag{3.3.7}$$

Then,

$$dx = dx_0 \wedge \mathbf{dx}, \tag{3.3.8}$$

$$\begin{aligned} X \lrcorner\, dx &= X(x_0)\, \mathbf{dx} - dx_0 \wedge (X \lrcorner\, \mathbf{dx}) \qquad (3.3.9)\\ &= X(x_0)\mathbf{dx} - \frac{x_i}{x_0}\, dx_i \wedge (X \lrcorner\, \mathbf{dx}) \qquad \text{using (3.3.6)},\\ &= X(x_0)\, \mathbf{dx} - \frac{x_i}{x_0}\, X(x_i)\, \mathbf{dx}\\ &= \frac{1}{2x_0{}^2}\left(x_0 + \frac{\vec{x}^2}{x_0}\right) \mathbf{dx}\\ &= \frac{\mathbf{dx}}{x_0}. \qquad (3.3.10)\end{aligned}$$

(Notice, as a side point, that, in the case $m = 4$, which is the case of physical interest, the right-hand side of (3.3.10) is the Lorentz invariant volume element on $N - (p_0)$, which is the "light cone.")

Now, let φ be diffeomorphism of $N - (p_0) \to N - (p_0)$ defined as follows:

$$\varphi(x) = -x. \tag{3.3.11}$$

Let N_ε be the subset of points $(x_0, \mathbf{x}) \in N$ such that $\mathbf{x}^2 > \varepsilon$. Then,

$$\varphi(N_\varepsilon) = N_\varepsilon.$$

Hence, since φ reverses the orientation of N_ε,

$$\begin{aligned}\int_{N_\varepsilon} \psi j(X \lrcorner\, dx) &= -\int_{N_\varepsilon} \varphi^*(\psi j)\, \varphi^*(X \lrcorner\, dx)\\ &= -\int_{N\varepsilon} \varphi^*(\psi j)\, \frac{\mathbf{dx}}{x_0}\\ &= -\int_{N\varepsilon} \psi(-x)\, j(-x)\, \frac{\mathbf{dx}}{d_0}.\end{aligned}$$

Then,

$$\int_{N_\varepsilon} \psi j(X \lrcorner\, dx) = \tfrac{1}{2}\int_{N_\varepsilon} \frac{(\psi(x)j(x) - \psi(-x)j(-x))}{x_0}\, d\mathbf{x}. \tag{3.3.12}$$

The integral on the right-hand side of (3.3.12) now obviously converges, because ψj is C^∞ at $x = 0$.

3.4. Definition of $\delta(f)$ and Their Derivatives at a Nondegenerate Critical Point by Means of Renormalization Techniques

Continue with M, N, f, p_0 as in Section 3.2, 3.3. Let us suppose now that p_0 is a nondegenerate critical point for f, in the sense of Morse theory [*15*]. This means that there is a coordinate system (x_μ) for M, valid in a neighborhood of p_0, such that

$$x_\mu(p_0) = 0, \qquad f = \tfrac{1}{2} g_{\mu\nu} x_\mu x_\nu, \tag{3.4.1}$$

where $(g_{\mu\nu})$ is a constant, symmetric, nonsingular matrix. Now, we attack the general problem of defining $\delta(f)$ in a natural way. However, in order to avoid various special assumptions, we make the following assumption:

$$M \text{ is compact.} \tag{3.4.2}$$

(Of course, cases where M is noncompact are of primary importance in physics. These can often be reduced to (3.4.2) by suitable compactifications, a technique we will discuss in later sections.)

Let H continue to denote the vector space of C^∞, complex-valued functions on M.

Theorem 4.1. There is a linear functional $\delta(f)\colon H \to C$ defined by the following formula:

$$\langle\, \delta(f)|\psi\, \rangle = \int_{N-(p_0)} (\psi - \psi(p_0))\, (X \lrcorner\, dp) \tag{3.4.2a}$$

for $\psi \in H$.

Proof. In view of the assumption (3.4.2)that M is compact, it suffices—in order to show that the integral (3.4.2a) makes sense—to work in the coordinate neighborhood of p_0 defined by the coordinate system (3,4.1). Set

$$dx = dx_0 \wedge \cdots \wedge dx_{m-1}, \qquad dp = j\, dx. \tag{3.4.2b}$$

We can suppose, without any loss in generality, that

$$g_{\mu\nu}g_{\nu\bar{\omega}} = \delta_{\mu\bar{\omega}}. \tag{3.4.3}$$

(The choice (3.4.3) amounts to normalizing the coordinate system (x_μ).) Set

$$r^2 = x_\mu x_\mu. \tag{3.4.4}$$

Then, X is readily calculated as follows:

$$X = (g_{\mu\nu}x_\mu/r^2)\partial_\nu. \tag{3.4.5}$$

Set

$$\theta_\mu = \partial_\mu \lrcorner\ dx. \tag{3.4.6}$$

Then (3.4.2) can be written as follows:

$$\langle\ \delta(f)|\psi\ \rangle = \int_{N-(p_0)} (\psi(x) - \psi(0)g_{\mu\nu}x_\mu)\,/\,r^2\ \theta_\nu. \tag{3.4.7}$$

We see now that the integral (3.4.2) is well defined, since the function $x \to [(\psi(x) - \psi(0)\,)/r^2]g_{\mu\nu}x_\mu$ is continuous at $x = 0$.

Now, let us concern ourselves with the "derivatives" $\delta^1(f), \delta^2(f), \ldots$ defined, in case N has no critical points of f, in [3] or [6]. In [6], the definition was given in the following form:

$$\langle\delta^1(f)|\psi\rangle = -\langle\delta(f)|X(\psi)\rangle - \langle\delta(f)|(\text{div } X)\psi\rangle, \tag{3.4.8}$$

for $\psi \in H$, where (div X) is the function such that

$$X(dp) = (\text{div } X)dp. \tag{3.4.9}$$

(The left-hand side of (3.4.9) denotes the Lie derivative of the form dp by the vector field X.)

Suppose we then try to carry over (3.4.8) naively to the case where f has a critical point at p_0. Then, (3.4.8) takes the following form:

$$\langle\ \delta^1(f)|\psi\ \rangle = -\int_{N-(p_0)} X(\psi)\ (X \lrcorner\ dp) - \int_{N-(p_0)} \psi(\text{div } X)\ (X \lrcorner\ dp). \tag{3.4.10}$$

Again, the integrals on the right-hand side of (3.4.10) have trouble converging in a neighborhood of p_0.

However, it is readily verified that the integrals defining the right-hand side of (3.4.9) do indeed converge if ψ vanishes to the first order at p_0, i.e., if ψ and its first derivatives vanish at p_0. Let H_1 denote the subspace of H consisting of the $\psi \in H$ which vanish to the first order at p_0. Let π_1 be a projection map: $H \to H_1$. Then define

$$\langle \delta^1(f)|\psi\rangle = -\langle \delta(f)|X(\pi_1(\psi))\rangle - \langle \delta(f)|(\text{div } X)\pi_1(\psi)\rangle. \quad (3.4.11)$$

This gives one possible definition of $\delta^1(f)$ as an element of the Dirac space associated with H. The higher derivatives $\delta^2(f)$, $\delta^3(f)$, . . . are handled similarly, but involve projections of H on the space of functions that vanish to the second, third, . . . order at p_0.

4. Asymptotic Behavior of Matrix Elements of Group Representations

4.1. Introduction

Let H be a complex vector space, with a Hermitian symmetric inner product $\langle|\rangle$, mapping $H \times H$ into C. Let G be a Lie group, and let $g \to \rho(g)$, $g \to \rho'(g)$ be representations of G by linear operators mapping H into itself. Let **G** be the Lie algebra of G.

Definition. The representations ρ, ρ' are *adjoint* to each other if the following condition is satisfied:

$$\langle \psi_1|\rho(g)\psi_2\langle = \langle \rho'(g^{-1})\psi_1|\psi_2\rangle \quad (4.1.1)$$

for $\psi_1, \psi_2 \in H$, $g \in G$.

In the sequel, let us suppose that ρ and ρ' are indeed adjoints of each other.

Let D be the Dirac space associated with H. Extend ρ and ρ' to be representations of G by linear operators on D by using the following formulas:

$$\langle \rho(g)\alpha|\psi\rangle = \langle \alpha|\rho'(g^{-1})\psi\rangle, \qquad \langle \rho'(g)\alpha|\psi\rangle = \langle \alpha|\rho(g)\psi\rangle \quad (4.1.2)$$

for $\alpha \in D$, $\psi \in H$.

Now, let $t \to g(t) = \exp(tX)$, $X \in G$, be a one parameter subgroup of G; let a be a real number and let α, α_∞ be elements of D.

Definition. α_∞ is the *asymptotic limit of strength a* of the element α, relative to the one parameter group defined by X, if the following relation holds:

$$\lim_{t\to\infty} e^{at} \langle \rho(g(t))\alpha|\psi \rangle = \langle \alpha_\infty|\psi \rangle \qquad (4.1.3)$$

for all $\psi \in H$. Of course, the element α_∞ in (4.1.3) really is a function of α, a, and X. In order to keep the notation as simple as possible, we will not indicate this explicitly, except when it is necessary to avoid complete confusion.

Our aim in this section is to describe general geometric methods for studying relations of the form (4.1.3). We will not be, for the moment, studying abstractly given representations of G, but will restrict ourselves to the case where G is a real semisimple Lie group of noncompact type, with the representations ρ and ρ' obtained "geometrically" by letting G act on cross sections of homogeneous vector bundles of coset spaces of G. H will consist of cross sections of these vector bundles.

In order to orient ourselves further, it is most convenient to study the case where: $G = SL(2, R)$, and where H is the space of C^∞, complex-valued functions on the unit circle in R^2, which is of course a coset space of G. This situation has been described briefly in Parts V and VI of the series of papers titled "Analytic continuation of group representations" [*10*]. We will now go over this material again, emphasizing[1] certain geometric features that appear in the case of much more general groups G.

4.2. Representations of $G = SL(2,R)$

Let us suppose that the manifold M is the unit circle in R^2. Let H be $F_c(M)$, the C^∞, complex-valued functions on M. We will parametrize M in the usual way, with an angular variable θ, $0 \leqq \theta \leqq 2\pi$. Then, H consists of the complex-valued, C^∞ functions $\theta \to \psi(\theta)$ of θ which are periodic, i.e.,

$$\psi(\theta + 2\pi) = \psi(\theta) \qquad \text{for} \quad \theta \in R. \qquad (4.2.1)$$

[1]The reader who is not familiar with the history of Lie group and group representation theory should note that understanding how general ideas specialize to this group has often been the key to making progress in problems involving more complicated and general groups. True, a given problem often involves considerably more technical detail for a general semisimple Lie group G than for the case $G = SL(2, R)$, but one can usually discern a sense in which the ideas carry over. In a way this is the beauty of Lie group theory—there are usually techniques available to generalize relations found by explicit calculations for the groups of simplest type to vastly more complicated and interesting situations. Thus, I feel it is more important to test out general ideas on simple situations, rather than to attempt to build a general theory immediately.

The inner product $\langle | \rangle$ in H is chosen in the usual way:

$$\langle \psi_1 | \psi_2 \rangle = \int_0^{2\pi} \psi_1(\theta)^* \, \psi_2(\theta) \, d\theta. \tag{4.2.2}$$

Now, let **G** be the Lie algebra of $SL(2, R)$, with a basis consisting of elements X, Y, Z such that

$$[Z, X] = Y, [Z, Y] = -X, \qquad [X, Y] = -Z. \tag{4.2.3}$$

Define a representation ρ, ρ' of **G** by differential operators in H as follows:

$$\begin{aligned} \rho(Z) &= d/d\theta, \\ \rho(X) &= \sin\theta \, d/d\theta - \lambda \cos\theta, \\ \rho(Y) &= \cos\theta \, d/d\theta + \lambda \sin\theta, \end{aligned} \tag{4.2.4}$$

$$\begin{aligned} \rho'(Z) &= d/d\theta, \\ \rho'(X) &= \sin\theta \, d/d\theta + (\lambda^* + 1) \cos\theta, \\ \rho'(Y) &= \cos\theta \, d/d\theta - (\lambda^* + 1) \sin\theta. \end{aligned} \tag{4.2.5}$$

(In (4.2.4)–(4.2.5), λ is a free complex parameter labeling the representation.)

Let G denote the Lie group $SL(2, R)$. Now, one can show that the representations (4.2.4)–(4.2.5) of G by operators on H can be integrated to give representations of G, again denoted by ρ, ρ'. We will not actually need the specific formulas for $\rho(G)$, $\rho'(G)$ and hence will not write them down. In addition, one can show that ρ and ρ' are adjoints to each other, in the sense defined in Section 4.1, hence both representations can be extended to be representations on the Dirac space D associated with H.

Now, let $t \to g(t)$ be the one parameter subgroup of G generated by the element X of G. Our goal is to compute the "asymptotic behavior"—in the sense described by (4.1.3)—of the orbits of the one parameter group $t \to \rho(g(t))$ on elements of D. To do this, let us work with an idea mentioned in [*10*]: namely, write $\rho(X)$ in the following alternate form to (4.2.5):

$$\rho(X) = (\sin\theta)^{\lambda} \rho_0(X) \, (\sin\theta)^{-\lambda}, \tag{4.2.6}$$

with

$$\rho_0(X) = \sin\theta \, d/d\theta. \tag{4.2.7}$$

Formula (4.2.6) enables us to give an explicit formula for the one parameter group $\rho(g(t))$. Let $t \to \varphi(t)$ be the one parameter group of diffeomorphisms generated by the vector field on M given by the right-hand side of (4.2.7). Namely,

$$\varphi(-t)^*(\psi) = \exp(t\,\rho_0(X))\,(\psi) \tag{4.2.8}$$

for $\psi \in H$.

(Here, $\varphi(-t)^*$ denotes the pull-back map on functions induced by the diffeomorphism: $M \to M$, i.e., this is the differential-geometric use of the starsymbol.) Note that the explicit formulas for $\varphi(t)(\theta)$ are given in [*10*, Part V]. However, we will not need them. Combining (4.2.6) and (4.2.8), we see that, for $\psi \in H$,

$$\rho(g(t))(\psi)(\theta) = (\sin\theta)^\lambda \varphi(-t)^*((\sin\theta)^{-\lambda}\psi(\theta)),$$

or

$$\rho(g(t))(\psi)(\theta) = \sin^\lambda\theta\,\sin^{-\lambda}(\varphi(-t)(\theta))\psi(\varphi(-t)\theta). \tag{4.2.9}$$

Hence, combining (4.2.9) with (4.2.2) gives the following formula:

$$\langle\rho(g(t))(\psi)|\psi_1\rangle^* = \int_0^{2\pi} \sin^\lambda\theta\,\sin^{-\lambda}(\varphi(-t)(\theta))\psi(\varphi(-t)\theta)\psi_1(\theta)^*\,d\theta \tag{4.2.10}$$

for $\psi_1 \in H$. Note now that there are two qualitative points to keep in mind when examining the asymptotic behavior as $t \to \infty$ of the right-hand side of (4.2.10). First, $t \to \varphi(t)$ is a one parameter group of diffeomorphisms of M, with the fixed points, namely $\theta = 0$ and $\theta = \pi$. One of these fixed points is "attractive," the other "repulsive." As $t \to -\infty$, i.e., if p is a point of M which is not the "repulsive" fixed point, then

$$\lim_{t\to\infty} \varphi(-t)\,(p) = \text{the attractive fixed point.} \tag{4.2.11}$$

(It does not really matter for our purposes—since it is not our aim to do explicit calculations here—which of $\theta = 0, \pi$ is the attractive one.)

Second, the function $(\sin\theta)$ which appears in (4.2.10) is a C^∞ function on M which has zeros at precisely the fixed points of the group $t \to \varphi(t)$.

Then, depending on the value of λ, the Lebesgue dominated convergence theorem can be used to calculate the limit as $t \to \infty$ of the right-hand side of (4.2.10). However, for certain values of λ, the integral will not, in fact, be dominated by an integrable function as $t \to \infty$. Then, to calculate the limit, certain "subtractions"—similar to the subtractions used in quantum field renormalization theory—must be used.

Rather than proceed further with this example, it is convenient to indicate a more general setting for this procedure. As a bonus, we will see general ideas that may be applied in two directions—determining asymptotic behavior of orbits of more general groups on Dirac spaces, and, more speculatively, perhaps obtaining new algebraic and intuitive insights into the renormalization problem for Feynman integrals in quantum field theory.

4.3. Asymptotic Behavior under a Set of Geometric Assumptions

Let us now proceed to work with the following hypotheses. M is a compact, orientable manifold, with a volume element form denoted by dp. H consists of the C^∞, complex-valued functions on M, with the inner product $\langle | \rangle$ defined as follows:

$$\langle \psi_1 | \psi_2 \rangle = \int_M \psi_1(p)^* \psi_2(p)\, dp \qquad \text{for} \quad \psi_1, \psi_2 \in H. \tag{4.3.1}$$

Let $t \to \varphi(t)$, $0 \leqq t < \infty$, be a one parameter semigroup of diffeomorphisms of M. Suppose that $p_0 \in M$ is an attractive fixed point for this semigroup, i.e., suppose there exists an open submanifold M' of M containing p_0 such that

$$\lim_{t \to \infty} \varphi(t)p = p_0 \tag{4.3.2}$$

for all $p \in M'$.

M-M' has measure zero with respect to the measure on M determined by the form dp. (4.3.3)

Using (4.3.3), we can then rewrite (4.3.1) as follows:

$$\langle \psi_1 | \psi_2 \rangle = \int_{M'} \psi_1(p)^* \psi_2(p)\, dp. \tag{4.3.4}$$

Suppose now that $t \to \rho(t)$ is a one parameter semigroup of linear transformations of H, such that, for $\psi_1, \psi_2 \in H$,

$$\langle \psi_1 | \rho(t) \psi_2 \rangle = \int_{M'} \psi_1^* |f|^\lambda |\varphi(t)^*(f)|^{-\lambda} \varphi(t)^*(\psi_2)\, dp. \tag{4.3.5}$$

In (4.3.5), f is a real-valued, C^∞ function on M which has a zero on M' only at the point p_0, and λ is a complex number. (Of course, f and λ depend only on ρ, not on ψ_1 or ψ_2.)

Now, let us suppose that M' has a coordinate system $(x_1, \ldots, x_m)$ of functions such that:

$$\begin{aligned} \varphi(t)^*(x_1) &= e^{-b_1 t}x_1 \\ \varphi(t)^*(x_m) &= e^{-b_m t}x_m \\ 0 = x_1(p_0) &= \cdots = x_m(p_0). \end{aligned} \tag{4.3.6}$$

In (4.3.6), $b_1, \ldots, b_m$ are positive real numbers, such that $b_1 \geqq b_2 \geqq \ldots \geqq b_m$. Then, f has a Taylor expansion about $x = 0$ in this coordinate system.

$$f = a_1 x_1 + \cdots + a_m x_m + \cdots. \tag{4.3.7}$$

Thus, combining (4.3.6) and (4.3.7) gives the following result:

$$\lim_{t \leftarrow \infty} e^{b_1 t}\, \varphi(t)^*\, (f) = f_1, \tag{4.3.8}$$

where f is a new function on M', readily computable using (4.3.7) and (4.3.8). (For example, if

$$b_1 = b_2 > b_3 \geqq \ldots, \qquad \text{then} \quad f_1 = a_1 x_1 + a_2 x_2.)$$

Of course, we see from (4.3.8) that f_1 is not C^∞ on all of M—it will have a "pole" at the points of M–M' which are fixed points of the semigroup $t \to \varphi(t)$.

Now, in favorable circumstances (to be prescribed below) Eq. (4.3.8) enables us to compute the asymptotic behavior as $t \to \infty$ of the left-hand side of (4.3.5):

$$\lim_{t \to \infty} e^{b_1 t} \langle \psi_1 | \rho(t) \psi_2 \rangle = \int_{M'} \psi_1^* \; |f|^\lambda \, |f_1|^{-\lambda}\, \psi_2(p_0)\, dp. \tag{4.3.9}$$

Now, in order to derive (4.3.9), we must be able to apply the Lebesgue dominated convergence theorem to the integrals on the right-hand side of (4.3.5). A glance at our hypotheses will convince the reader that this can be applied if the following condition is satisfied:

$$\int_{M'} |f|^\lambda |f_1|^{-\lambda}\, dp < \infty. \tag{4.3.10}$$

Then, for the values of λ for which (4.3.10) is satisfied, the following result holds:

$$\begin{gathered}\rho(t)(\psi_2) \text{ converges asymptotically (in the sense of Section 4.1)} \\ \text{with strength } b_1 \text{ to the element of} \\ D \text{ determined by the function } |f|^{\lambda}|f_1|^{-\lambda}\psi_2(p_0).\end{gathered} \tag{4.3.11}$$

Now, we must examine what happens if (4.3.10) is not satisfied. What must be done is to "improve" the convergence of the integral (4.3.10) by a "subtraction." This can be done as follows, using (4.3.5) and (4.3.8):

$$\begin{aligned} & e^{b_1 t}|\psi_1|\rho(t)\psi_2\rangle - e^{b_1 t}\langle\psi_1(p_0)|\rho(t)(\psi_2)\rangle \\ &= \int_{M'} (\psi_1{}^* - \psi_1(p_0)^*)|f|^{\lambda}|e^{b_1 t}\varphi(t)^*(f)|^{-\lambda}\varphi(t)^*(\psi_2)\, dp. \end{aligned} \tag{4.3.12}$$

Suppose, for example, that the real part of λ is positive, so that the difficulty in the convergence of (4.3.10) is in the neighborhood of p_0. Now, the integral on the right-hand side will converge if

$$\int_{M'} |\psi_1 - \psi_1(p_0)|\, |f_1|^{-\lambda}\, dp < \infty. \tag{4.3.13}$$

Let us suppose that the singularity of f_1 on the boundary M–M' does not give any trouble, i.e., we can concentrate on the neighborhood of p_0. Now, of course, (4.3.13) may converge for a wider range of values than (4.3.10), because the function $\psi - \psi_1(p_0)$ is zero at p_0, hence it can eliminate part of the singularity of $|f_1|^{-\lambda}$ at p_0. If so, the limit $t \to \infty$ of the right-hand side of (4.3.10) will be simply (4.3.13) times $\psi_2(p_0)$. If not, the convergence can again be "improved" by more subtractions.

To see what is involved here more explicitly, let us return to the case considered in Section 4.2 where M is the unit circle in R^2, with $G = SL(2, R)$. Denote a point of R^2 by a complex number z, with $z = e^{i\theta}$ denoting a point of M. Suppose that

$$\psi_1 = z^m, \qquad \psi_2 = z^{-n}. \tag{4.3.14}$$

Set

$$t = \log(\,(1 + s)/(1 - s)\,). \tag{4.3.15}$$

Then, in [*10*, Part V, p. 182], one will find the following formula:

$$i(s^2 - 1)\, \langle \psi_1 | \rho(t) \psi_2 \rangle$$
$$= \int \left(\frac{z}{z^2 s + z(s^2 + 1)ts} \right)^{-\lambda} \left(\frac{sz + 1}{z + s} \right)^n z^{-m+1}\, dz. \qquad (4.3.16)$$

(The integral in (4.3.16) is over the unit circle in the z-plane.) If the real part of λ is negative, one can now obviously take the limit as $s \to 1$ under the integral sign. However, for other values of λ, the singularity of the denominator for $s = 1$, $z = -1$ obviously gives trouble. However, this difficulty can be ameliorated by performing a subtraction:

$$i(s^2 - 1) < \psi_1 - \psi_1(1) | \rho(t) \psi_2 >$$
$$= \int \left(\frac{z}{z^2 s + z(s^2 + 1) + s} \right)^{-\lambda} \left(\frac{sz + 1}{z + s} \right)^n (z^{-m+1} - 1)\, dz. \qquad (4.3.17)$$

The range of λ for which the Lebesgue dominated convergence theorem can be applied has obviously been increased by this subtraction. If this is not sufficient, further subtractions can be made.

5. Compactification of Real Vector Spaces by Means of a Quadratic Form

5.1. Introduction

Let V be a real finite dimensional vector space, with $\beta : V \times V \to R$ a nonsingular, real bilinear, symmetric form. In this section, we shall indicate a method for "compactifying" V, i.e., imbedding V as an open subset of a compact manifold M in such a way that $M - V$ is a union of submanifolds of lower dimension.

This method will be a generalization of compactifications known for two special cases:

(A) β is positive definite, with $\dim V = n$. M will then be an n-sphere S^n and the imbedding $V \to M$ is essentially the compactification of V obtained by adding to V a "point at infinity."

(B) $\dim V = 4$, β is a form defining the Lorentz metric, i.e., V is Minkowski space. The compactification $V \to M$ will then be the one known and applied for various purposes in mathematical physics [*13*, *16*].

The method is based on imbedding V as a linear subspace of a real vector space V', with

$$\dim V' = \dim V + 2.$$

M is then a quadratic hypersurface of the real projective space defined by V'.

5.2. The Construction of the Compactification

Let us now start over again, from a different point of view. Suppose that V' is a real vector space, with

$$\dim V' = n + 2. \tag{5.2.1}$$

Suppose that β is a nonsingular, symmetric, real bilinear form: $V' \times V' \to R$. Suppose also that V is a real subspace of V' such that

$$\dim V = n, \tag{5.2.2}$$

$$\beta \text{ restricted to } V \text{ is nonsingular,} \tag{5.2.3}$$

$$\beta \text{ restricted to the orthogonal complement of } V \text{ in } V' \text{ has isotropic vectors.} \tag{5.2.4}$$

More explicitly, assumption (5.2.4) means that there is a nonzero vector $v_0 \in V'$ such that

$$\beta(v_0, V) = 0 \tag{5.2.5}$$

$$\beta(v_0, v_0) = 0. \tag{5.2.6}$$

Let π be the linear orthogonal projection map: $V' \to V$. In other words

$$\pi^2 = \pi, \tag{5.2.7}$$

$$\beta(\,(1 - \pi)(V'), \pi(V')\,) = 0. \tag{5.2.8}$$

Let $V'(\beta)$ denote the isotropic vectors of V' with respect to β, i.e.,

$$V'(\beta) = \text{the set of nonzero } v' \in V' \text{ such that } \beta(v', v') = 0. \tag{5.2.9}$$

Let $V'(\beta, v_0)$ be the subset of $V'(\beta)$ consisting of the vectors $v' \in V'(\beta)$ such that

$$\beta(v', v_0) \neq 0. \tag{5.2.10}$$

Then, $V'(\beta, v_0)$ is an open subset of $V'(\beta)$. Define a map $\phi : V'(\beta, v_0) \to V$ by the following formula:

$$\phi(v') = \frac{\pi(v')}{\beta(v', v_0)}. \tag{5.2.11}$$

Theorem 5.2.1 If v', v_1 are elements of $V'(\beta, v_0)$ such that

$$\phi(v') = \phi(v'_1), \tag{5.2.12}$$

then there is a real scalar λ such that

$$v' = \lambda v'_1, \tag{5.2.13}$$

i.e., v' and v'_1 lie in the same one-dimensional subspace of V'.

Proof. We may suppose, without loss in generality, that

$$\beta(v', v_0) = \beta(v'_1, v_0) = 1. \tag{5.2.14}$$

Then, from (5.2.12), we deduce that

$$\pi(v') = \pi(v'_1). \tag{5.2.15}$$

Now, let π' be the projection map $(1 - \pi)$. Then,

$$v' = \pi(v') + \pi'(v'),$$
$$v_1 = \pi(v'_1) + \pi'(v'_1).$$

Now, v' and v'_1 are elements of $V(\beta)$, i.e.,

$$0 = \beta(v', v') = \beta(\pi(v'), \pi(v')) + \beta(\pi'(v'), \pi'(v')),$$
$$0 = \beta(v'_1, v'_1) = \beta(\pi(v'_1), \pi(v'_1)) + \beta(\pi'(v'_1), \pi'(v'_1)).$$

In view of (5.2.15), we deduce that

$$\beta(\pi'(v'_1), \pi'(v'_1)) = \beta(\pi'(v'), \pi'(v')). \tag{5.2.16}$$

Also, from (5.2.14), we have

$$\beta(\pi'(v'), v_0) = \beta(\pi'(v'_1), v_0),$$

or

$$\beta(\pi'(v') - \pi'(v'_1), v_0) = 0.$$

But, using (5.2.1) and (5.2.2), we see that $\pi'(V')$ is 2-dimensional, hence any vector in it perpendicular to v_0 must be a multiple of v_0, i.e.,

$$\pi'(v') - \pi'(v'_1) = cv_0, \qquad \text{with} \quad c \in R. \tag{5.2.17}$$

Thus,

$$\beta(\pi'(v'), \pi'(v')) = \beta(\pi'(v_1'), \pi'(v_1')) + 2c\beta(\pi'(v_1'), v_0).$$

Using (5.2.16) and (5.2.14), we see that

$$c = 0,$$

which implies (5.2.13) (with $\lambda = 1$, in this case) and finishes the proof.

Now, let M be the orbit space of $V'(\beta)$ under the action of the multiplicative group of nonzero real numbers: $v \to \lambda v$. Let M_0 be the orbits that touch $V'(\beta, v_0)$. Then, M_0 is an open subset of M. Now, ϕ is obviously constant on these orbits, hence passes to the quotient to define a map: $M_0 \to V$, that we also denote by ϕ. It is obviously onto. Theorem 5.2.1 says that it is one–one. It is now readily seen that ϕ is a diffeomorphism of M_0 with V. The inverse map

$$\phi^{-1} : V \to M_0$$

thus is the desired compactification of V.

5.3. Computation of the Compactifying Map ϕ^{-1}

Continue with the notations of Section 5.2. Then, one can find a basis

$$(v_0, v_0') \text{ of } \pi'(V') \text{ such that}$$

$$\beta(v_0, v_0) = 0 = \beta(v_0', v_0'), \tag{5.3.1}$$

$$\beta(v_0, v_0') = 1.$$

Suppose now that $v \in V$. Consider $v' \in V$ of the following form:

$$v' = v + v_0' + av_0, \qquad \text{with} \quad a \in R.$$

Then,

$$\beta(v', v_0) = 1.$$

We must find the condition that

$$v' \in V(\beta).$$

Then,

$$\beta(v', v') = 0 = \beta(v, v) + 2a.$$

Thus, we can define a map $\phi' : V \to V(\beta)$ by the formula

$$\phi'(v) = v + v_0' - (\beta(v, v)/2)v_0. \tag{5.3.2}$$

Clearly, ϕ', followed by the projection map: $V(\beta) \to M$, is the inverse map ϕ^{-1}.

Let us compute the complement:

$$M - \phi^{-1}(V).$$

Suppose $v' \in V'(\beta)$. Then, the projection of v' in M lies in $M - \phi^{-1}(V)$ if and only if

$$\beta(v', v_0) = 0. \tag{5.3.3}$$

In view of (5.3.1) and (5.3.3), such an element must be of the following form:

$$v' = \pi(v') + av_0. \tag{5.3.4}$$

The condition that $v' \in V'(\beta)$ is then that

$$\beta(\pi(v'), \pi(v')) = 0. \tag{5.3.5}$$

Thus, we have proved:

Theorem 5.3.1 The image in M of a point $v' \in V(\beta)$ lies on the boundary of $\phi^{-1}(V)$ if and only if v' is the sum of an isotropic vector in V and a multiple of v_0.

Corollary 5.3.2 If β is positive definite on V, then $M - \phi^{-1}(V)$ consists of a single point; namely the image in M of v_0.

Proof. If β is positive definite on V, condition (5.3.5) implies that $\pi(v') = 0$, which shows, via (5.3.4), that v' is a scalar multiple of v.

Remark. One can further show that in this case, M is the n-sphere S^n, and that the imbedding $\phi^{-1} : V \to S^n$ is the usual one obtained by "stereographic projection."

Let us now return to the general situation. Let $V(\beta)$ denote the space of isotropic vectors of β restricted to V. Define a map $\gamma : V(\beta) \times R \to M$ as follows:

$$\gamma(v, a) = \text{image under the map } V'(\beta) \to M \quad \text{of the element } v + av_0. \tag{5.3.6}$$

Reversing the analysis given above, we see that

$$\gamma(V(\beta) \times R) = M - \phi^{-1}(V). \tag{5.3.7}$$

Let us examine the fibers of γ. Suppose

$$\gamma(v, a) = \gamma(v_1, a_1), \qquad \text{with} \quad v, v_1 \in V(\beta), \quad a, a_1 \in R.$$

Then, there is a nonzero $\lambda \in R$ such that

$$v + av_0 = \lambda(v_1 + a_1 v_0),$$

i.e.,

$$a = \lambda a_1, \tag{5.3.8}$$

$$v = \lambda v_1. \tag{5.3.9}$$

This suggests that we split up $M - \phi^{-1}(V)$ into three pieces, $M_1 \cup M_2 \cup M_3$. Namely, set

$$\gamma_1(v) = \gamma(v, 0); \qquad M_1 = \gamma_1(V(\beta));$$

$$\gamma_2(v) = \gamma(v, 1); \qquad M_2 = \gamma_2(V(\beta));$$

$$M_3 = \text{image in } M \text{ of the point } v_0.$$

This completes our analysis of the structure of the boundary of V in M, when it is compactified by identification with $\phi^{-1}(V)$.

Now, following (5.3.8)–(5.3.9), γ_2 is one–one and hence defines $V(\beta)$ as a submanifold of M. γ_1 is invariant under the action $v \to \lambda v$ of R on $V(\beta)$. Let N be the orbit space of the action of the multiplicative group of real numbers on $V(\beta)$. Then, γ passes to the quotient to define a diffeomorphism between N and M_1.

Now, let us study the limit properties of sequences of points of V as they approach points of $M - \phi^{-1}(V)$.

5.4. Asymptotic Properties of Sequences of Points of V

Continue with the notations of Sections 5.2 and 5.3. Suppose that $v^1, \ldots$ is a sequence of points of V, no subsequence of which converges within

V. Our aim now is to calculate the limit of these points when imbedded in M via ϕ^{-1}.

Suppose that $\mu^1, \mu^2, \ldots$ is a sequence of points in V which converges to a vector $\mu \in V$, such that

$$v^n = \lambda_n \mu^n, \qquad n = 1, 2, \ldots, \tag{5.4.1}$$

where $\lambda_n \in R$.

Suppose that

$$\phi^{-1}(v^n) = v^n + v_1 - (\beta(v^n, v^n)/2v_0, \tag{5.4.2}$$

where (v_1, v_0) are the basis elements of the orthogonal complement of V in V' which were introduced in Section 5.1.

Then, combining (5.4.1) and (5.4.2),

$$\phi^{-1}(v^n) = \lambda_n \mu^n + v_1 - (\lambda_n^2/2)\beta(\mu_n, \mu_n)v_0.$$

CASE 1.

$$\beta(\mu, \mu) \neq 0. \tag{5.4.3}$$

Then,

$$\frac{\phi^{-1}(v^n)}{\lambda_n^2} \to -\tfrac{1}{2}\beta(\mu, \mu)v_0 \qquad \text{as} \quad n \to \infty. \tag{5.4.4}$$

Thus, in terms of the classification (5.3.10) of the boundary of V in M, the sequence $v', v^2, \ldots$ converges to the point M_0.

CASE 2.

$$\beta(\mu, \mu) = 0; \qquad \lim_{n\to\infty} \lambda_n \beta(\mu_n, \mu_n) = 0. \tag{5.4.5}$$

In this case, we see that v^n converges, as $n \to \infty$, to a point on the submanifold M_1; namely, to the point corresponding to the point $\mu \in V(\beta)$.

CASE 3.

$$\beta(\mu, \mu) = 0: \qquad \lim_{n\to\infty} \lambda_n \beta(\mu_n, \mu_n) \neq 0. \tag{5.4.6}$$

Then, v^n converges to the point of the submanifold M_2 (described in

(5.3.12)) of the form

$$\mu + av_0, \qquad \text{with:} \qquad a = -\lim_{n\to\infty} \frac{\lambda_n\beta(\mu_n, \mu_n)}{2}. \tag{5.4.7}$$

These three cases exhaust the classification of the limit points reached by sequences of points in V.

Now, let us look at the above computations in a slightly different way. A *line* in V is a one-dimensional submanifold of V which is the translation of a one-dimensional linear subspace of the vector space V. An *oriented line* is such a submanifold with a fixed orientation. Let W denote the space of such oriented lines of V.

W can be given a manifold structure by exhibiting it as a homogeneous space. In fact, it is acted on transitively by the "affine linear group of V," which is the semidirect product of the group of all linear transformations of V with the group of translations of V.

Suppose now that

$$t \to \mu' + t\mu, \qquad -\infty < t < \infty \tag{5.4.8}$$

is such an oriented line of V, with v', $\mu \in V$. Set

$$v^t = \mu' + t\mu, \qquad \mu^t = \frac{v^t}{t} = \frac{\mu'}{t} + \mu. \tag{5.4.9}$$

Then, we are in Case 1, i.e., (5.4.6) holds, if $\beta(\mu, \mu) \neq 0$.

Suppose now that

$$\beta(\mu, \mu) = 0. \tag{5.4.10}$$

Then $t\beta(\mu^t, \mu^t) = (\beta(\mu', \mu')/t) + 2\beta(\mu', \mu)$. Thus, we are in situation (5.4.5) if

$$\beta(\mu', \mu) = 0,$$

and in situation (5.4.6) if

$$\beta(\mu', \mu) \neq 0.$$

We can now define a mapping

$$W \to M - \phi^{-1}(V) \tag{5.4.11}$$

by assigning to a line of form (5.4.8) the limit as $t \to \infty$ of v^t. This map clearly intertwines the action of the group of linear transformations that preserve the form β.

6. Projective Deformations of Lie Algebras and Their Representations

6.1. Introduction

In Refs. [*4, 6, 11*], we have explained how the theory of deformations of Lie algebras and their representation play a role in understanding various asymptotic limit relations concerning Lie group representations and the associated Plancherel expansions. In this work, there was one typical example—the deformation of the Poincaré into the Galilean group—whose theory was not well-understood from a general point of view. In this section, we will present a more satisfactory framework in which one may understand these phenomenon and indicate an even more general framework that covers other physical situations.

6.2. Deformation of the Poincaré into the Galilean Group

We refer to [*11*, part III], for the background of the example to be presented here.

Let **G** be the three dimensional vector space spanned by three elements, that we label "X," "E," and "p." For each real number $\lambda > 0$, we define a Lie algebra structure $[\ ,\ ]_\lambda$ on **G** by means of the following rules:

$$[E, p]_\lambda = 0, \qquad [X, E]_\lambda = p, \qquad [X, p]_\lambda = E/\lambda. \tag{6.2.1}$$

(With λ identified with c^2, with c the velocity of light, (6.2.1) is the Lie algebra of the Poincaré group in one space and one time dimension.) Then, as $\lambda \to \infty$, the bracket $[\ ,\ ]_\lambda$ goes over smoothly to the bracket $[\ ,\ ]_\infty$ defined as follows:

$$[X, p]_\infty = 0, \qquad [X, E]_\infty = p, \qquad [X, p]_\infty = 0. \tag{6.2.2}$$

Notice in fact this limit Lie algebra is the Lie algebra of the Galilean group of one space and one time variable. (It is also the Lie algebra of the Heisenberg group of particle quantum mechanics.)

We shall now present a family ρ_λ of representations of the λth Lie algebra by linear operators. Let F be the space of C^∞, real-valued functions of a single real variable v, i.e., $F = F(R)$. (Physically, v is the velocity of a particle traveling in one dimension.)

Now set

$$\rho_\lambda(X) = (1 - (v^2/\lambda))\, d/dv,$$
$$\rho_\lambda(p) = mv/(1 - v^2/\lambda)^{1/2}, \qquad (6.2.3)$$
$$\rho_\lambda(E) = m\lambda/(1 - v^2/\lambda)^{1/2}$$

(m is a free constant; physically, it is the mass). Notice that $\rho_\lambda(\mathbf{G})$ lies in $D_1(F, F)$, the space of first order, linear, differential operators: $F \to F$. It was shown in [*11*, part II], that (6.2.3) provides a representation of the commutation relations (6.2.1), when $D_1(F, F)$ is considered a Lie algebra via the operator commutator as bracket.

Let us now examine the limit as $\lambda \to \infty$ of ρ_λ. Notice that $\rho_\lambda(p)$ and $\rho_\lambda(X)$ have limits:

$$\rho_\infty(X) = d/dv = \text{the nonrelativistic "acceleration,"}$$
$$\rho_\infty(p) = mv = \text{the nonrelativistic "momentum."}$$

However, $\rho_\lambda(E)$ does not have a limit. Instead, its singularity must be "renormalized."

To do this, set

$$\rho'_\lambda(E) = \rho_\lambda(E) - m\lambda,$$
$$\rho'_\lambda(p) = \rho_\lambda(p), \qquad (6.2.4)$$
$$\rho'_\lambda(X) = \rho_\lambda(X).$$

Let us regard the assignments (6.2.4) of operators to elements of $\mathbf{G}$ as a linear mapping from $\mathbf{G}$ to $D_1(F, F)$. This assignment is *not* a Lie algebra homomorphism from $\mathbf{G}_\lambda$ to $D_1(F, F)$. Instead, it is a "ray representation" or a "representation up to a factor." For the reader's convenience, we will give a formal definition of this concept.

Definition. Let $\mathbf{G}$ be a real Lie algebra and let V be a real vector space. Let ρ be a linear mapping from $\mathbf{G}$ to the space of linear operators on V. For $X, Y \in \mathbf{G}$, set

$$\omega(X, Y) = \rho([X, Y]) - \rho(X)\rho(Y) - \rho(Y)\rho(X). \qquad (6.2.5)$$

Then, ρ is said to be a *ray representation* of $\mathbf{G}$ if, for each $X, Y \in \mathbf{G}$, $\omega(X, Y)$ is a scalar multiple of the identity operator.

The name "ray representation" comes from the following geometric point of view: Let $P(V)$ be the "projective space" associated with V, i.e., a "point" of $P(V)$ is a one-dimensional linear subspace of V. Let us suppose that G is a group whose Lie algebra is **G**. Associate to $y = \exp(X) \in G$, with $X \in G$, the linear transformation: $V \to V$ defined as follows:

$$\rho(g) = \exp(\rho(X)).$$

Since $\rho(g)$ is a linear transformation, it acts on linear subspaces of V, hence passes to the quotient to define a transformation $\rho_P(G)$ in $P(V)$.

EXERCISE. Show that (6.2.5) is the condition that the assignment $g \to \rho_P(g)$ defines a transformation group action in $P(V)$, i.e., ρ_P is a homomorphism of G into the group of transformations on $P(V)$. Return now to the consideration of ρ_λ. The reader will readily verify that (6.2.1) defines ρ'_λ as a ray representation of G_λ. The key fact is now that the operator $\rho'_\lambda(G)$ converge as $\lambda \to \infty$. To see this, we must examine only the first term on the right-hand side of (6.2.4), using (6.2.3):

$$\rho'_\lambda\,(E) = \frac{m\lambda}{(1 - v^2/\lambda)^{1/2}} - m\lambda = \frac{m(1 - (1 - v^2/\lambda)^{1/2})}{(1 - v^2/\lambda)^{1/2}}.$$

In particular,

$$\lim_{\lambda \to \infty} \rho'_\lambda\,(E) = \tfrac{1}{2}mv^2. \tag{6.2.6}$$

Notice that the right-hand side of (6.2.6) is the nonrelativistic expression for "energy."

Thus, we may say that by "renormalizing" the representation $\rho_\lambda(G_\lambda)$ to a ray representation $\rho'_\lambda(G_\lambda)$, we have arranged convergence, as $\lambda \to \infty$, of ρ'_λ to give a ray representation of the limit Lie algebra G_∞, the Galilean group.

In summary, these remarks give a clearer version of the explanation given in [*11*, part III] of the algebraic mechanism for obtaining the representations of the Galilean group as limits of those of the Poincaré group. This version is also better suited to permit understanding of the underlying generalities. In the rest of this section, we will undertake the task of putting this construction in its general setting.

6.3. Renormalization of Representation Deformations Using Lie Algebra Extensions

Again, we shall consider deformations parametrized by a real parameter

λ. However, it will be more convenient to suppose that λ ranges over $0 < \lambda \leqq 1$, with $\lambda = 0$ corresponding to $\lambda = \infty$ in the previous work.

Let **G** be a real vector space. Suppose a given family $[\ ,\ ]_\lambda$ of Lie algebra brackets on **G**, defining **G** as a Lie algebra, which we denote by "$\mathbf{G}_\lambda$." Let us suppose that $[\ ,\ ]_0$ is a Lie algebra structure on **G** such that

$$[X, Y]_0 = \lim_{\lambda \to 0} [X, Y]_\lambda \qquad \text{for} \quad X, Y \in \mathbf{G}. \tag{6.3.1}$$

Suppose given now a Lie algebra **A** and a family $\lambda \to \rho_\lambda$ of Lie algebra homorphisms: $\mathbf{G}_\lambda \to \mathbf{A}$. Alternately, ρ_λ can be considered as a linear map: $\mathbf{G} \to \mathbf{A}$ such that

$$\rho_\lambda([X, Y]_\lambda) = [\rho_\lambda(X), \rho_\lambda(Y)] \qquad \text{for} \quad X, Y \in G. \tag{6.3.2}$$

We suppose that ρ_λ is defined only for $0 < \lambda \leqq 1$, but that the limit as $\lambda \to 0$ does not exist.

Now, let $\mathbf{A}'$ be an ideal of **A**, and let $\mathbf{A}''$ be the quotient Lie algebra $\mathbf{A}/\mathbf{A}'$. Let ρ'_λ be the quotient Lie algebra homomorphism

$$\mathbf{G}_\lambda \to \mathbf{A}''.$$

Definition. The representation-deformation $\lambda \to \rho$ is *renormalizable modulo* $\mathbf{A}'$ if there is a map $\rho''_0 : \mathbf{G} \to \mathbf{A}''$ such that

$$\rho_0''(X) = \lim_{\lambda \to 0} \rho_\lambda''(X) \qquad \text{for all} \quad X \in \mathbf{G}. \tag{6.3.3}$$

Remark. It follows from (6.3.3) that ρ''_0 is also a Lie algebra homomorphism from the Lie algebra $\mathbf{G}_0$ to the Lie algebra $\mathbf{A}''$.

6.4. Projective Lie Algebra Deformations

Let us review what is involved in the previous sections. One is given a real vector space **G**, whose typical elements are denoted by X, Y, Z. One is also given a one-parameter family $[\ ,\ ]_\lambda$, $0 < \lambda \leqq 1$, of Lie algebra structures on **G**, together with a one parameter family $\lambda \to \rho_\lambda$, $0 < \lambda \leqq 1$, of linear representations of these Lie algebras on a fixed vector space. In earlier work, we have assumed that the Lie algebra structure $[\ ,\ ]_\lambda$ is smooth about $\lambda = 0$, i.e., can be expanded in a Taylor series, but that the operators $\rho_\lambda(X)$ have singularity at $\lambda = 0$. However, we find in certain key examples that this singularity can be "renormalized" by passing to the projective space associated with the vector space.

A natural generalization is then to consider possible singularities in the Lie algebra structures $[\ ,\ ]_\lambda$ also. We shall now consider a general framework for discussing analogous "renormalization" techniques for this sort of deformations.

Let us begin again, starting with **G** as a real vector space. For technical simplicity, we will assume that **G** is finite dimensional, although the most interesting physical application (in quantum field theory) occurs when **G** is infinite dimensional.

Let **L(G)** denote the set of Lie algebra structures defined on the vector space **G**. Thus, each element of **L(G)** is a bilinear map, typically denoted by $[\ ,\ ]$, of $\mathbf{G} \times \mathbf{G} \to \mathbf{G}$, which satisfies the following conditions:

$$[X, Y] = -[Y, X], \tag{6.4.1}$$

$$[X, [Y, Z]] = [[X, Y], Z] + [Y, [X, Z]] \quad \text{for} \quad X, Y, Z \in G.$$

Notice now that if $[\ ,\ ]$ satisfies (6.4.1), so does $[\ ,\ ]'$ defined as follows:

$$[X, Y]' = t[X, Y] \quad \text{for} \quad X, Y \in G, \tag{6.4.2}$$

where t is a real number.

Thus, the multiplicative group of nonzero real numbers acts, via formula (6.4.2), on the set **L(G)**. Denote by **PL(G)** the orbit space under this action; call it the space of *projective Lie algebra structures* on **G**.

Consider now a curve $\lambda \to [\ ,\ ]_\lambda$ in $L(G)$, defined for $0 < \lambda \leqq 1$. We want to allow the possibility of a "singularity" of this "deformation," at $\lambda = 0$. However, let us suppose that the projection of this curve in **PL(G)** is smooth in the neighborhood of $\lambda = 0$. Let us denote by $\sigma(\lambda)$ the projection in **PL(G)** of this curve in $L(G)$. Then, we suppose that

$$\sigma(0) = \lim_{\lambda \to 0} \sigma(\lambda)$$

exists. Let $[\ ,\ ]_0$ denote an element of **L(G)** which projects into $\sigma(0)$. Thus, there is a curve $\lambda \to t(\lambda)$ in R such that

$$[X, Y]_0 = \lim_{\lambda \to 0} t(\lambda)\, [X, Y]_\lambda . \tag{6.4.3}$$

Thus, formula (6.4.3) allows for the possibility of the deformation $\lambda \to [X, Y]_\lambda$ developing a singularity at $\lambda = 0$, but allows this singularity to be "renormalized" to obtain a new Lie algebra $[\ ,\ ]_0$.

The next step in the program should be to develop a cohomology-type algebraic theory to calculate the relation between the Lie algebra structures

$[\ ,\ ]_0$ and those of $[\ ,\ ]_\lambda$ for $\lambda \neq 0$, generalizing the now classic theory of Gerstenhaber, Nijenhuis, and Richardson for "smooth" deformations. However, such a study would take us too far afield here and will be deferred to another work.

7. Rapidly Decreasing Functions on Manifolds Defined by Compactifications

7.1. Introduction

A key question in the application of distribution-generalized function theory to such disciplines as mathematical physics, partial differential equations, and harmonic analysis on Lie groups is the proper choice of a class of "test functions" for each broad class of problems. For example, the class of "rapidly decreasing functions on R^n," defined by Schwartz, is ideally suited to studying constant-coefficient partial differential equations, using Fourier transform theory. One will find in the "Generalized functions" treatises by Gel'fand and collaborators extensive discussions of this point of view.

However, it seems that little thought has been given to the geometric principles underlying a possible general theory of generalized functions on manifolds. The aim of this section is to begin such a discussion, based on the idea that classes of test functions on a manifold may be defined by means of compactifications of the manifold.

7.2. General Principles

Let M be a manifold. Adopt the differential-geometric conventions of [5]. In particular, the C^∞, real-valued functions on M are denoted by $F(M)$.

Definition. Given a point $p \in M$ and an $f \in F(M)$, f is said to *vanish to the infinite order at p* if

$$D(f)(p) = 0 \tag{7.2.1}$$

for each differential operator $D : F(M) \to F(M)$.

Now, suppose that N is a submanifold of M. We suppose, for simplicity, at least, that N is *regularly imbedded* in M in the sense that the topology N inherits as a subset of M is the same as its "own" manifold topology. Let B be the boundary of N in M, i.e., the set of points of M that do not lie in N but that can be approached as a limit of sequences of points in N.

Definition. A function $f \in F(N)$ is *rapidly decreasing* (relative to the imbedding of $N \to M$) if it is the restriction to N of a function in $F(M)$ which vanishes to the infinite order at each point in B. Denote the set of these rapidly decreasing functions by $S(N)$. If the dependence on M must be explicitly displayed, we will denote this by $S_M(N)$.

Certain properties of $S(N)$ are obvious from this definition. First, the sum and product of two elements of $S(N)$ lie in $S(N)$, i.e., $S(N)$ forms an algebra over the real numbers.

Further let D' be a differential operator $F(N) \to F(N)$. We now ask: When is

$$D'(S(N)) \subset S(N).$$

Definition. The differential operator $D' : F(N) \to F(N)$ is the *restriction to N* of a *differential operator* $D'' : F(M) \to F(M)$ if

$$D''(f)|_N = D'(f|_N) \qquad \text{for all} \quad f \in F(M). \tag{7.2.2}$$

(Here, the symbol $|_N$ denotes the restriction of the function to the subset N.) D' is said to be *extendable to M* if there is a differential operator D'' which satisfies (7.2.2).

Now, the following result is obvious from the above definitions, but it is an important property nonetheless.

Theorem 7.2.1. If D' is a differential operator: $F(N) \to F(N)$ that is extendable to M, then

$$D'(S(N)) \subset S(N). \tag{7.2.3}$$

An example of such an extendable operator is available using geometric reasoning. Suppose that $t \to g(t)$, $0 \leqq t < \infty$ is a differentiable one-parameter semigroup of diffeomcrphisms of M, each of which leaves N invariant. Let $D'' : F(M) \to F(M)$, $D' : F(N) \to F(N)$ be the first-order linear differential operators which are the infinitesimal generators of these are parameter semigroups, i.e.,

$$D'(f) = \frac{\partial}{\partial t} g(t)^* (f)|_{t=0} \qquad \text{for} \quad f \in F(N),$$

$$D''(f) = \frac{\partial}{\partial t} g(t)^* (f)|_{t=0} \qquad \text{for} \quad f \in F(M). \tag{7. 2. 4}$$

Then, it should be obvious from (7.2.4) that condition (7.2.2) is satisfied, i.e., that (7.2.3) is satisfied.

Now, let us suppose that a volume element differential form $d_N p$ is given on N.

Definition. The volume element form $d_N p$ is said to be *compatible* with the choice $S(N)$ of rapidly decreasing functions if the following condition is satisfied:

$$\int_N f\, d_N p < \infty \qquad \text{for all} \quad f \in S(N). \tag{7.2.5}$$

As we shall see later, many of the interesting and simple examples satisfy condition (7.2.5). It is the key abstract property needed to apply many of the standard ideas of Fourier transform and generalized function theory developed in the "classical" situation where $N = R^n$ to more complicated and sophisticated Lie group situations.

We can investigate condition (7.2.5) further under the following hypotheses, which are often realized:

$$M \text{ is compact}; \qquad N \text{ is an open subset of } M. \tag{7.2.6}$$

Let $d_M p$ be a volume element form for M. Then, there is a C^∞ function j on N such that

$$d_N p = j\, d_M p. \tag{7.2.7}$$

Thus,

$$\int_N f\, d_N p = \int_N jf\, d_M p. \tag{7.2.8}$$

Definition. Let us say that a function $j \in F(N)$ is *slowly increasing* if the following condition is satisfied:

For each function $f \in S(N)$, the function jf is bounded on N. (7.2.9)

Putting all this together, proves the following observation:

Theorem 7.2.2. With conditions (7.2.6) and (7.2.7) as hypotheses, let us suppose that the function j defined by (7.2.7) is slowly increasing in the sense of the above definition. Then, condition (7.2.5) is satisfied, i.e., the volume element $d_N p$ is compatible with the choice of $S(N)$ of rapidly decreasing functions on N.

The real point to these observations, in terms of the applications we have in mind, is that often a choice of volume element $d_N p$ for N is imposed by the nature of the special situation. For example, N might be a Riemannian manifold or a Lie group coset space. It is then an interesting problem to find, using differential-geometric techniques, conditions that the volume element be compatible with the compactification.

7.3. The Rapidly Decreasing Functions in the Case Where N is an Open Submanifold of a Compact M, with a Hypersurface B as Boundary

The case covered by the title of this section is an obviously important special case. As we will see later, it covers a case which is important in physics, namely the compactification of the "mass-shell" submanifold of R^4.

Explicitly, we suppose that the following conditions are satisfied:

(a) $\dim M = \dim N = n$. N is connected.
(b) M is a compact manifold.
(c) B, the point-set boundary of N in M, is a compact submanifold of dimension $(n-1)$.
(d) There is a C^∞ function h defined on N and a neighborhood of B such that

$$\begin{aligned} h &= 0 \quad \text{on } B, \\ h &> 0 \quad \text{on } N, \\ dh &\neq 0 \quad \text{at each point of } B. \end{aligned}$$

Our problem is to describe the elements of $S(N)$ among all C^∞ functions on N relative to the asymptotic behavior of f relative to all polynomials in h.

Theorem 7.3.1. If $f \in S(N)$, then for each integer $r \geqq 0$, f/h^r is continuous on $N \cup B$.

Proof. Let p be a point of B. By virtue of the above assumptions, (a)–(d), we can find a coordinate system of M, consisting of functions

(x_i), $1 \leqq i, j \leqq n$, valid in a neighborhood of p, such that

$$h = x_1, \qquad x_i(p) = 0.$$

Now, use the fact that all derivatives of f vanish at each point of the subset: $x_1 = 0$ to show that $f/x_1^r = f/h^r$ is continuous in this neighborhood of p. Since p may be any point of B, f/h^r is obviously continuous on $N \cup B$.

7.4. The Compactification of the Mass-Shells in R^4

As an illustration of the general situation considered in Section 7.3, let us consider the example mentioned in the title of this section. Denote a point of R^4 by $p = (p_\mu)$, $0 \leqq \mu, \nu \leqq 3$, with the Lorentz inner product denoted by

$$p^2 = p_0 p_0 - p_i p_i, \qquad 1 \leqq i, \ j \leqq 3. \tag{7.4.1}$$

Let N^+ denote the set of $p \in R^4$ such that

$$p^2 = m^2, \tag{7.4.2}$$

$$p_0 > 0. \tag{7.4.3}$$

Here, m is a positive real number.

Let V be a real-3-dimensional vector space, with a positive definite, symmetric inner product denoted by

$$v_1 \cdot v_2. \tag{7.4.4}$$

As usual, denote the length of a $v \in V$, based on the inner product (7.4.4), by

$$|v| = (v \cdot v)^{1/2}. \tag{7.4.5}$$

Let N be the "unit ball" in V, i.e., the set of vectors v such that

$$|v| < 1. \tag{7.4.6}$$

Define a map $\phi : N \to N^+$ as follows:

$$\phi(v) = (m/(1 - |v|^2)^{1/2})\,(1,v). \tag{7.4.7}$$

(In formula (7.4.7), identify R^4 with the direct sum vector space $R + V$.)

EXERCISE. Show that ϕ is a diffeomorphism of N with N^+.

Then, ϕ^{-1} identifies N^+ with the open subset N of V. The closure of B in V is compact. Equation (7.4.7) then defines a "compactification" of N^+.

Remark. This compactification is well-known geometrically, as the "Poincaré realization" of the Lobachewski space of constant negative curvature. This compactification also has certain natural "equivariance" properties with respect to the action of the Lorentz group, which we will not go into in any detail. We will only mention that $SO^+(1, 3)$ (the connected component of the Lorentz group) acting on N^+ goes over onto a group of transformations on B that can be extended to a neighborhood of the boundary. The group also acts on the boundary of B in V. Now, it is obvious from (7.4.6) that this boundary is the unit sphere in V, namely S^2. Now, S^2 is also diffeomorphic to $P^1(C)$, the complex projective space of one complex dimension. The simply connected covering group of $SO^+(1, 3)$ is $SL(2, C)$, which also acts in a natural way on $P^1(C)$. It can be shown that these identifications all agree, so that, group-theoretically speaking, the "compactification" of N^+ by identification with N is a very natural thing.

The space V also has a very natural physical interpretation. It is the space of "velocity vectors" of particles.

Using formula (7.4.7), it is now straightforward to compute the volume element $d_N p$ when it is pulled back (using ϕ) to a differential form on B. The result is readily seen to be given by the following formula:

$$\phi^*(d_N p) = \frac{m\,dv}{(1 - |v|^2)^2}. \tag{7.4.8}$$

In (7.4.8), "dv" denotes the Euclidean volume element form on $V = R^3$.

Now, using this compactification, we can define a suitable class of "rapidly decreasing test functions on N," denoted by $S(N^+)$.

Definition. A function $f \in F(N)$ is *rapidly decreasing if* f can be extended in a C^∞ way to be a function in a neighborhood of the closure of B in V which vanishes to an infinite order on the boundary of B in V. A function $f \in F(N^+)$ is *rapidly decreasing* if $\phi^*(f)$ is rapidly decreasing. Denote the space of rapidly decreasing functions on N and N^+ by $S(N)$, $S(N^+)$.

Theorem 7.4.1. Each function $f \in S(N)$ is integrable with respect to the Lorentz invariant measure $d_N p$ on N.

Proof. We have, using (7.4.8),

$$\int_N f\, d_N p = \int_{|v| <} \frac{mf(\phi(v))}{(1 - |v|^2)^2}\, dv. \tag{7.4.9}$$

Now, if f vanishes to infinite order on the boundary of B, the function

$$v \to \frac{f(v)}{(1 - |v|^2)^2}$$

is continuous, hence bounded, on the closure of B in V (which is compact). Thus, the integral (7.4.9) is finite, since B has finite measure with respect to the Euclidean measure on B.

Now, we turn to the question of the relation between this class $S(N^+)$ of rapidly decreasing functions and the Schwartz class of rapidly decreasing functions $S(R^4)$ on R^4.

Definition. A function $f \in F(R^4)$ lies in $S(R^4)$ if, for each constant coefficient differential operator $D : F(R^4) \to F(R^4)$ and each polynomial function $p \to h(p)$ on R^4, the following relation holds:

$$\lim_{p \to \infty} h(p)\, D(f)\,(p) = 0. \tag{7.4.10}$$

Theorem 7.4.2. If $f \in S(R^4)$, then the restriction of f to the submanifold N^+ belongs to $S(N^+)$.

Proof. It suffices to show that

$$\lim_{|v| \to 1} \frac{f(\phi(v))}{(1 - |v|^2)^r} = 0 \qquad \text{for all} \quad r. \tag{7.4.11}$$

Now,

$$\phi(v) = \frac{m}{(1 - |v|^2)^{1/2}}\,(1, v).$$

Set

$$h(p) = p_0^{2r}.$$

Then,

$$h(\phi(v)) = \frac{m^{2r}}{(1 - |v|^2)^r}.$$

Using (7.4.10), we see that

$$0 = \lim_{p \to \infty} h(p) f(p) = \lim_{|v| \to 1} h(\phi(v)) f(\phi(v))$$
$$= m^{2r} \lim_{|v| \to 1} \frac{f(\phi(v))}{(1 - |v|^2)^r}.$$

This proves (7.4.11).

7.5. Generalized Function Solutions of the Klein–Gordon Equation

As an illustration of the utility of the definition of the class of rapidly decreasing functions on the mass-hyperboloid $p^2 = m^2$ in R^4, we will now present an application to the study of the Klein–Gordon equation of quantum mechanics. Although the results to be presented here do not really extend what is already presented in quantum field theory texts (for example, that by Bjorken and Drell [*1*]), the methods and ideas used here are useful in a wide variety of more general problems.

Let M be the manifold R^4. Denote a point of M by $x = (x_\mu)$, $0 \leqq \mu$, $\nu \leqq 3$. Let N' be the set of real four vectors $p = (p_\mu)$, such that

$$p^2 = m^2.$$

Then, N' is an orbit under the Lorentz group, $SO(1, 3)$. The isotropy subgroup at a point is $SO(3, R)$. Let $d_N p$ be the Lorentz invariant volume element on differential form N. Thus, as is well known, $d_N p$ takes the following form:

$$d_N p = \frac{dp_1 \wedge dp_2 \wedge dp_3}{|p_0|}. \tag{7.5.1}$$

Let us say that a real-valued function on N' is *rapidly decreasing* if its restriction to the disjoint open subsets

$$N^+ = \{p \in N' : p_0 > 0\},$$
$$N^- = \{p \in N' : p_0 < 0\}$$

is rapidly decreasing in the sense defined in Section 7.4. Let H' denote the complex vector space of complex-valued, rapidly decreasing functions on N'. Define a Hermitian-symmetric, positive-definite (but incomplete) inner product on H' as follows:

$$\langle \psi_1 | \psi_2 \rangle = \int_N \psi_1'(p)^* \psi_2'(p)\, d_N p \qquad \text{for} \quad \psi_1', \psi_2' \in H'. \tag{7.5.2}$$

(For the Hilbert-space type notations inherent in (7.5.2), we follow the physicist's notations, as explained in "Fourier analysis on groups and partial wave analysis" [*6*].)

Similarly, let H denote the rapidly decreasing, complex-valued functions on $M = R^4$, with inner product defined as follows:

$$\langle \psi_1 | \psi_2 \rangle = \int_{R^4} \psi_1(x)^* \psi_2(x) \, d^4x \tag{7.5.3}$$

(d^4x denotes the Euclidean volume element on R^4).

Now, let D, D' be the Dirac spaces associated with H and H'. Thus, an element $\alpha \in D$ is a complex-linear map: $H \to C$, with the product of α and a complex number defined as follows:

$$(c\alpha)(\Psi) = c^*\alpha(\Psi) \qquad \text{for} \quad \Psi \in H. \tag{7.5.4}$$

To carry over the Dirac bracket notations, we write

$$\alpha(\Psi) = \langle \alpha | \Psi \rangle.$$

An element $\Psi_1 \in H$ is then identified with the linear functional

$$\Psi \to \langle \Psi_1 | \Psi \rangle.$$

Define a linear mapping $F^\dagger : H \to H'$ as follows:

$$F^\dagger(\Psi)(p) = \int_{R^4} \Psi(x) e^{-ip.x} \, d^4x \qquad \text{for} \quad p \in N. \tag{7.5.5}$$

Using standard theorems of Fourier-transform-generalized function theory (see Gel'fand and Shilov [*3*]) and Theorem 7.4.2, one sees readily that (7.5.5) defines the function $p \to F^\dagger(\Psi)(p)$ as an element of H'. Now, define $F : D' \to D$ as follows:

$$\langle F(\alpha') | \Psi \rangle = \langle \alpha | F^\dagger(\Psi) \rangle \qquad \text{for} \quad \alpha \in D', \quad \Psi \in H. \tag{7.5.6}$$

Let Δ be the Klein–Gordon differential operator on R^4:

$$\Delta = \partial_0^2 - \partial_i \partial_i + m^2 \qquad \left(\partial_\mu = \frac{\partial}{\partial x_\mu}\right). \tag{7.5.7}$$

Then, Δ maps $H \to H$. Extend Δ to a map: $D \to D$ as follows:

$$\langle \Delta(\alpha) | \Psi \rangle = \langle \alpha | \Delta(\Psi) \rangle, \tag{7.5.8}$$

Theorem 7.5.1. If $\alpha' \in D'$, then

$$\Delta(F(\alpha')) = 0. \tag{7.5.9}$$

Proof.

$$\langle \Delta F(\alpha') | \Psi \rangle = \langle F(\alpha') | D\Psi \rangle = \langle \alpha' | F^{\dagger} D\Psi \rangle.$$

Now,

$$F^{\dagger} D\Psi = (p) = \int (D\Psi)(x) e^{-ip.x} \, d^4x \qquad \text{for} \quad p \in N. \tag{7.5.10}$$

Use (7.5.7), and integrate the right-hand side of (7.4.10) by parts, obtaining

$$F^{\dagger} D(\Psi)(p) = \int \Psi(x)(m^2 - p^2) e^{-ip.x} \, d^4x.$$

However, for $p \in N$, $p^2 = m^2$; hence this is zero, which proves (7.5.9).

Remark. In quantum field theory textbooks, one will find this result written in the following form:

$$\psi(x) = \int_{R^4} \alpha(p) \delta(p^2 - m^2) \, d^4p, \tag{7.5.11}$$

with the "function" $\psi(x)$ a solution of the Klein–Gordon equation $D\Delta = 0$. What we have done here is to put this in a framework where it can be readily extended to considerably wider situations.

7.6. Rapidly Decreasing Functions on Semisimple Lie Groups

In previous sections, we have defined a class of rapidly decreasing functions on the coset space $SO^+(1, 3)/SO(3, R)$ by means of a compactification of this coset space. Now, because of the geometric property of this compactification, the class of rapidly decreasing functions (defined using the general principles described in Section 7.3) is invariant under the action of the Lorentz group $SO^+(1, 3)$. This suggests two general problems:

(a) Is there a natural definition of a rapidly decreasing function on the group manifold of $SO^+(1, 3)$?

(b) Can one replace $SO^+(1, 3)$ with a more general group?

We shall now begin work on these two general problems. Suppose then that G is a connected, real semisimple Lie group. Let K be a maximal compact subgroup of G, and let N be the coset space G/K. Then, there is a mapping $\alpha : F(G) \to F(N)$ defined by integration over G. To define this map, identify a function on the coset space G/K with a function on G that is invariant under right multiplication by elements of K, i.e., satisfies the following condition:

$$f(gk) = f(g) \qquad \text{for} \quad k \in K, g \in G. \tag{7.6.1}$$

Now, define $\alpha : F(G) \to F(N)$ as follows:

$$\alpha(f)(g) = \int_K f(gk)\, dk \qquad \text{for} \quad f \in F(G), g \in G. \tag{7.6.2}$$

In (7.6.2), dk is the bi-invariant Haar measure volume element on K, such that K has total measure one. (Thus, $\alpha(f) = f$ if f already satisfies (7.6.1).)

Now, compactify $N = G/K$ as follows: Let G act on a compact manifold M such that there is a point $p_0 \in M$ at which the isotropy subgroup of G is K. Then, the orbit $N = Gp_0$, a submanifold of M, can be identified with G/K. This point of view can then be reversed, to consider the identification of G/K with the orbit N as the compactification of G/K. This compactification is *G-equivariant* in the sense that the transformations induced by G on N can be extended to transformations on M.

Now—in accordance with Section 7.2—define $S(N) = S(G(K))$, the *rapidly decreasing functions on* G/K, as the C^∞ functions which are the restriction to N of C^∞ functions on M which vanish to the infinite order on the boundary of N in M:

Definition. A function in G is *rapidly decreasing* if $\alpha(f)$, as defined above, is in $S(G/K)$. Denote the set of rapidly decreasing functions on G by $S(G)$. Thus, we have, by definition:

$$\alpha(S(G)) = S(G/K). \tag{7.6.3}$$

Theorem 7.6.1. The class $S(G)$ is invariant under left translations by elements of G.

Proof. Given $g_0 \in G$, let $L_{g_0} : F(G) \to F(G)$ denote the operation of left translation by g_0, i.e.,

$$L_{g_0}(f)(g) = f(g_0^{-1}g) \qquad \text{for} \quad g \in G, \quad f \in F(G).$$

Then, we see from (7.6.2) that

$$\alpha(L_{g_0}(f)) = L_{g_0}\alpha(f).$$

Now, left translation on a function satisfying (7.6.1) means translation under g_0 (via the natural action of G on the coset space $G(K)$ of the corresponding function on G/K). Since the chosen compactification of G/K is G-equivariant, translation by g_0 preserves $S(G/K)$. Putting these two remarks together with (7.5.3) proves that

$$L_{g_0}(S(G)) \subset S(G). \tag{7.6.4}$$

Now, let dg denote the left-invariant volume element form on G. One can prove that there is a G-invariant volume element form on G/K, denoted, say, by $d(g/k)$, such that

$$\begin{aligned}\int f(g)\,dg &= \int_{G/K} \alpha(f)\,d(g/k)\\ &= \int_{G/K} \left(\int_K f(gk)\,dk\right) d(g/k).\end{aligned} \tag{7.6.5}$$

In particular, note that $f \in F(G)$ is integrable with respect to dg if $\alpha(f)$ is integral with respect to $d(g/k)$. This proves the following result:

Theorem 7.6.2. Suppose that each $f \in S(G/K)$ is integrable with respect to the G-invariant measure $d(g/k)$. Then, each $f \in S(G)$ is integrable with respect to dg.

Now, this integrability property is the most important one if $S(G)$ is to serve as a family of "test functions" for the Fourier analysis of functions on G. (See "Fourier analysis on groups and partial wave analysis" [*6*].) This indicates the importance of a basic problem in compactification theory:

> Find G-equivariant compactifications of G/K such that, with $S(G/K)$ the rapidly decreasing function defined by means of this compactification, each $f \in S(G/K)$ is integrable with respect to the measure $d(g/k)$. (7.6.6)

For example, we have seen that, in the case $G = SO(1, 3)$, the natural compactification of G/K indeed has this property. Thus, there is a satisfactory class of rapidly decreasing test functions for the purposes of harmonic analysis on this group.

8. Some General Properties of Rapidly Decreasing Functions on Semisimple Lie Groups

In Section 7, we have introduced "geometric" ideas interrelating rapidly decreasing functions and compactifications of semisimple Lie groups. In this chapter, we will present more detail, emphasizing certain general geometric properties that will be useful in the final definitive theory.

8.1. Generalities about Semisimple Lie Algebras and Groups

In this section, we will describe briefly some features of semisimple group theory described in broad terms—with further precise references—in "Lie groups for physicists" [*4*]. In addition, many of the general facts about Lie groups theory and Riemannian geometry used here are explained in Ref. [*5*] and in Helgason's "Differential geometry and symmetric spaces."

Let $\mathbf{G}$ be a real semisimple Lie algebra. Let B be the Killing form, a symmetric, bilinear map: $\mathbf{G} \times \mathbf{G} \to R$, defined as follows:

$$B(X, Y) = \text{trace}(\text{Ad}\, X\, \text{Ad}\, Y) \qquad \text{for} \quad X, Y \in \mathbf{G}.$$

Definition. A decomposition of $\mathbf{G}$ as a direct sum $\mathbf{K} + \mathbf{P}$ of subspaces is a *Cartan decomposition* of $\mathbf{G}$ if

$$\begin{aligned}
&\text{(a)}\quad [\mathbf{K}, \mathbf{K}] \subset \mathbf{K}, \text{ i.e., } \mathbf{K} \text{ is a subalgebra of } \mathbf{G}.\\
&\text{(b)}\quad [\mathbf{P}, \mathbf{P}] \subset \mathbf{K},\ [K, \mathbf{P}] \subset \mathbf{P}.\\
&\text{(c)}\quad B(X, X) < 0 \text{ for } X \in \mathbf{K}. \qquad\qquad (8.1.1)\\
&\text{(d)}\quad B(X, X) > 0 \text{ for } X \in \mathbf{P}.\\
&\text{(e)}\quad B(\mathbf{K}, \mathbf{P}) = 0.
\end{aligned}$$

It can be proved that such decompositions exist and are unique in the sense that any two are related via an inner automorphism of $\mathbf{G}$. We shall suppose in the following work that one such decomposition is chosen and fixed.

Let G be the connected Lie group without center whose Lie algebra is isomorphic to $\mathbf{G}$. (It can be proved that G is unique up to isomorphism—

in fact, it is isomorphic to the connected identity component of the group of automorphisms of **G**.) Let K be the connected subgroup of G generated by the Lie algebra **K**. Let $\exp : \mathbf{G} \to G$ be the usual Lie group exponential map. Set

$$P = \exp(\mathbf{P}).$$

It is now known that

$$G = P \cdot K, \tag{8.1.2}$$

i.e., each element $g \in G$ can be written in one and only one way as a product of an element in P and one in K.

G/K is a noncompact Riemannian symmetric space of nonpositive sectional curvature which is homeomorphic to Euclidean space. (8.1.3)

Now, let G' be any other connected Lie group whose Lie algebra is isomorphic to G. There is then a Lie group homomorphism $\pi : G' \to G$ which is onto G and whose kernel is the center G of G'.

Let K' be a subgroup of G' such that

$$\pi(K') \subset K. \tag{8.1.4}$$

$$\dim K' = \dim K. \tag{8.1.5}$$

Then, π passes to the quotient to define a map $\pi' : G'/K' \to G/K$. In view of (8.1.5) and the general principles of Lie theory π' is a covering map. Because G/K is simply connected [(by (8.1.3)], π' is a diffeomorphism. In particular, π' is one–one, i.e.,

$$\text{center } G' \subset K'. \tag{8.1.6}$$

Let $\exp'$ be the exponential map: $\mathbf{G} \to G'$. Set

$$P' = \exp'(P). \tag{8.1.7}$$

Then,

$$\pi(\exp'(X)) = \exp(X), \qquad \text{for} \quad X \in \mathbf{G}, \tag{8.1.8}$$

$\exp'$ restricted to P is a one–one submanifold map. Also, using (8.1.6), one shows that

$$G' = P'K', \tag{8.1.9}$$

i.e., each element of G' can be written in a unique way as a product of an element of P' and an element of K'.

We can now sum up this discussion in the following form (after switching the notation used above, and replacing G by G'):

Theorem 8.1.1. Let G be a connected, semisimple Lie group with finite center. Let

$$\mathbf{G} = \mathbf{K} + \mathbf{P}$$

be a Cartan decomposition of its Lie algebra. Let K be the connected subgroup of G whose Lie algebra is $\mathbf{K}$. Then:

(a) K is a maximal compact subgroup of G.
(b) The exponential map exp, when restricted to $\mathbf{P}$, is a one–one submanifold map of $\mathbf{P}$ onto a closed subset P of G.
(c) $G = P \cdot K$, with the decomposition unique.
(d) center $G \subset K$.

8.2. Sequences Going to Infinity in Semisimple Lie Groups

Let G be a Lie group. A sequence of points $(g_1, g_2, g_3, \ldots)$ of G is said to *go to infinity* if there is no subsequence $g_{n(1)}, g_{n(2)}, \cdots$ with $n(1) < n(2) < \cdots$ which converges to a point of G.

Our aim in this section is to study such sequences of points of G that go to infinity in the case that G is connected semisimple with a finite center. Let $\mathbf{G} = \mathbf{K} + \mathbf{P}$ be a Cartan decomposition of $\mathbf{G}$. Let $\exp : \mathbf{G} \to G$ be the exponential map, with

$$P = \exp(\mathbf{P}).$$

Then,

$$G = P \cdot K, \tag{8.2.1}$$

with the decomposition of each $g \in G$ into a product pk unique. It follows from general Lie group principles that the map $P \times K \to G$ which assigns to $(p, k) \in P \times K$ the point $pk \in G$ is a diffeomorphism.

Suppose then that $g_1, g_2, \ldots$ is a sequence of points of G that goes to infinity. Then, following (8.2.1), each element admits a unique decomposition:

$$g_j = p_j k_j \qquad \text{with} \quad p_j \in P, \quad k_j \in K. \tag{8.2.2}$$

Since K is compact, we see that the sequence of points $p_j \in P = G/K$ must also go to infinity, in the sense that no subsequence converges.

We can decompose the p_j, $j = 1, 2, \ldots$, occurring in (8.2.2) further. Let A be a maximal Abelian subspace of $\mathbf{P}$, i.e., $\mathbf{A}$ is a linear subspace of the vector space $\mathbf{P}$, such that

$$[\mathbf{A}, \mathbf{A}] = 0 \tag{8.2.3}$$

and $\mathbf{A}$ is contained in no larger linear subspace of P which satisfies (8.2.3). Set

$$A = \exp(\mathbf{A}). \tag{8.2.4}$$

Then, it is known that any $p \in P$ can be written as

$$p = kak^{-1} \qquad \text{with} \quad k \in K, \quad a \in A. \tag{8.2.5}$$

Let us in particular apply this decomposition to the g_j's in (8.2.2):

$$g_j = k_j a_j k_j, \qquad \text{with} \quad k_j, k_j \in K, \qquad a_j \in A. \tag{8.2.6}$$

Let $\mathbf{A}^+$ be a Weyl chamber of $\mathbf{A}$. (See [*4*, p. 105]). We can even arrange that the a_j's in (8.2.6) to be in $\mathbf{A}^+ = \exp(A^+)$. These a_j in A^+ are then uniquely determined representatives of g_j. Since K is compact, we have proved:

Theorem 8.2.1. The sequence (g_j) goes to infinity in G if and only if the sequence (a_j) of points in A^+ defined by (8.2.6) goes to infinity.

8.3. Compactifications of G/K

Let G be any Lie group now, with a given transformation group action of G on a compact manifold M. Let K be a compact subgroup of G, and let p_0 be a point of M such that the isotropy subgroup of G at p_0 is precisely K. Then, the coset space G/K can be identified with the orbit Gp_0. This identification of G/K with the submanifold Gp_0 of M gives a compactification of G/K, from which one may construct classes $S(G)$ and $S(G/K)$ of rapidly decreasing functions on G and G/K.

A basic geometric property of this compactification—a result of the fact that the isotropy subgroup of p_0 is compact—can be described as follows:

If $g_1, g_2, \ldots$ is a sequence of elements of G such that the points

$g_1p, g_2p, \ldots$ of the orbit Gp_0 (where p is some point of Gp_0) converge to a point of Gp_0, then some subsequence of the sequence $g_1, g_2, \ldots$ converges to an element of G. (8.3.1)

Let us suppose now that f is a function in $F(M)$ that is zero on the boundary of $Gp_0 = G/K$ in M. Consider the function $f' \in F(G \times M)$ defined as follows:

$$f'(g, p) = f(gp). \tag{8.3.2}$$

Theorem 8.3.1. If $g_1, g_2, \ldots$ is a sequence of points of G that "goes to infinity," as defined in Section 8.2, then

$$\lim_{j\to\infty} f'(g_j, p) = 0 \qquad \text{for each} \quad p \in Gp_0. \tag{8.3.3}$$

Proof. Statement (8.3.1) implies that the sequence of points g_1p, $g_2p, \ldots$ has no subsequence of points converging to a point of Gp_0. (For, otherwise, by (8.3.1), there would be a subsequence of $g_1, g_2, \ldots$ convergent within G, contradicting the hypotheses.)

But, M is compact. Every convergent subsequence of $g_1p, g_2p, \ldots$ must go to a point on the boundary of Gp_0 in M. These two facts—and the assumption that f vanishes on the boundary of the orbit Gp_0—implies (8.3.3).

Now, let us study the following situation. Start off with a function $f \in S(G/K)$, i.e., a C^∞ function on M which vanishes to the infinite order on the boundary of Gp_0. Pick any element $g_0 \in G$. With this element fixed, construct a new function $f_0 \in F(G/K)$ as follows:

If $p = gp_0$, with $g \in G$, then

$$f_0(p) = \int_K f(gk\, g_0\, p_0)\, dk. \tag{8.3.4}$$

In (8.3.4), dk denotes the bi-invariant volume element on the compact group K.

Theorem 8.3.2. If $p_1, p_2, \ldots$ is a sequence of points that converge to a point on the boundary of Gp_0 in M, then

$$\lim_{j\to\infty} f_0(p_j) = 0. \tag{8.3.5}$$

Proof. We can then find a sequence $g_1, g_2, \ldots$ of group elements going to infinity in G, such that

$$p_j = g_j p_0, \qquad j = 1, 2, \ldots.$$

Then, using (8.3.4),

$$f_0(p_j) = \int_K f(g_j k g_0 p_0)\, dk.$$

By Theorem 8.3.1, for fixed k and g_0, we have

$$\lim_{j \to \infty} f(g_j k g_0 p_0) = 0. \tag{8.3.6}$$

Limit (8.3.5) then follows from the Lebesgue dominated convergence theorem.

Corollary to Theorem 8.3.2. If $f \in S(G)$, and if g_0 is an element of G, then the right-translate function

$$R_{g_0}(f) : g \to f(gg_0) \tag{8.3.7}$$

goes to zero on every sequence $g_1, g_2, \ldots$ of group elements which go to infinity.

Remark. It seems likely that the argument given above can be sharpened to prove that $R_{g_0}(f)$ also belongs to $S(G)$, i.e., $S(G)$ is also invariant under right translations by elements of G. This would be a property that would greatly extend the utility of $S(G)$ as a choice of "test functions" for the definition of the "Fourier transform" of functions on G [*6*].

References

[*1*] J. D. Bjorken and S. Drell, "Relativistic Quantum Fields," McGraw-Hill, New York, 1964.
[*2*] P. A. M. Dirac, "The Principles of Quantum Mechanics," Oxford, New York, 1947.
[*3*] I. M. Gel'fand and G. E. Silov, "Generalized Functions," Academic Press, New York, 1964.
[*4*] R. Hermann, "Lie Groups for Physicists," Benjamin, New York, 1966.
[*5*] ———, "Differential Geometry and the Calculus of Variations," Academic Press, New York, 1968.

[*6*] ———, "Fourier Analysis on Groups and Partial Wave Analysis," Benjamin, New York, 1969.
[*7*] ———, A Poisson kernel for certain homogeneous spaces, *Proc. Amer. Math. Soc.*, **69** (1963), 782–788.
[*8*] ———, Geometric aspects of potential theory in symmetric spaces, Part I, *Math. Ann.*, **148** (1962), 349–366; Part II, ibid., **181** (1963), 143–149; Part III, ibid., **152** (1964), 384–394.
[*9*] ———, Compactification of homogeneous spaces and contraction of Lie groups, *Proc. Nat. Acad. Sci. U.S.A.*, **51** (1964), 456–461.
[*10*] ———, Compactification of homogeneous spaces, Part I, *J. Math. Mech.*, **15** (1965), 655–678; Part II, ibid., **15**(1966), 667–689.
[*11*] ———, Analytic continuation of group representations, Part I, *Comm. Math. Phys.*, *2* (1966), 251–270; Part II ibid., **3** (1966), 53–74; Part III, ibid., **3** (1966), 75–97; Part IV, ibid., **5** (1967), 131–156; Part V, ibid., **5**(1967), 157–190; Part VI, **6** (1967), 205–225.
[*12*] ———, The Gell–Mann formula for representation of semisimple groups, *Comm. Math. Phys.*, **2** (1966), 155–164.
[*13*] ———, Compactification of Minkowski space and $SU(3)$ symmetry, *J. Math. Phys.*, **6** (1965), 817.
[*14*] ———, Differential equations of the matrix elements of the degenerate series, *Bull. Amer. Math. Soc.*, **73** (1967), 899–903.
[*15*] J. Milnor, "Morse Theory," Princeton Univ. Press, Princeton, New Jersey, 1963.
[*16*] R. Penrose, "Structure of Space-Time," 1967 Battelle Rencontres, Benjamin, New York, 1968.
[*17*] F. Riesz and B. S. Nagy, "Functional Analysis," Ungar, New York, 1955.

BOUNDED SYMMETRIC DOMAINS AND HOLOMORPHIC DISCRETE SERIES

A. W. KNAPP

Department of Mathematics
Cornell University
Ithaca, New York 14850

In this paper we shall study some special features of those symmetric spaces of noncompact type that are complex manifolds in a natural way. After discussing the case of the disk, we give Harish-Chandra's realization of all such spaces as bounded symmetric domains in $\mathbf{C}^n$, homogeneous under a connected group G, and we reproduce Harish-Chandra's construction of some irreducible infinite-dimensional group representations of G by means of this realization. Finally we show how the Hardy H^2 theory of the disk generalizes to yield further irreducible representations of G and to show that certain standard induced representations of G are reducible. A list of the material in each section follows.

1. Discrete series for $SU(1,1)$. Bargmann [*1*] classified the infinite-dimensional representations of $SU(1,1)$. The theorems mentioned at the beginning of this chapter were proved by him and then put in a more general setting by Harish-Chandra [*2*].
2. Bounded symmetric domains: examples, role of the center of K, structure of roots, Harish-Chandra decomposition. Much of this material appears in [*3*, pp. 354, 304–322], where it is treated from the point of view of differential geometry. The structure of the roots and the Harish-Chandra decomposition appear originally in [*2*, pp. 750, 756–761, 4, 590].
3. Holomorphic discrete series: definition, the "constant function," nonvanishing theorem, square-integrability, examples. See [*2*, pp. 6–13, 598–612].
4. Limits of homomorphic discrete series: case of $SU(1,1)$, a more general setting, statements of theorems. See [*4*].
5. Notes and references.

1. Discrete Series for $SU(1,1)$

Let

$$G = SU(1,1) = \left\{ \begin{pmatrix} \alpha & \beta \\ \bar{\beta} & \bar{\alpha} \end{pmatrix} \middle| \, |\alpha|^2 - |\beta|^2 = 1 \right\}.$$

The members of this group act as analytic automorphisms of the disk $\Omega = \{|z| < 1\}$ under $z \to zg = (\bar{\alpha}z + \beta)/(\bar{\beta}z + \alpha)$. [Here $(zg)h = z(gh)$.] The discrete series for G occurs in two parts:

(1) $$n \geqq 2,$$

$$\mathscr{H}_n = \{F(z) \text{ analytic in } \Omega,\ \|F\|^2 = \textstyle\int_\Omega |F(z)|^2\,(1 - |z|^2)^{n-2} dx\,dy < \infty\},$$

$$\mathscr{U}_n(g)F(z) = (\bar{\beta}z + \alpha)^{-n}\, F((\bar{\alpha}z + \beta)/(\bar{\beta}z + \alpha)).$$

This part will be called the holomorphic discrete series.

(2) The same, except that conjugates of the functions in $\mathcal{H}_n$ are used and $\mathcal{U}$ involves $(\overline{\beta z + \alpha})^{-n}$.

Some properties of $\mathcal{H}_n$ and $\mathcal{U}_n$ in the first case are the following:

(1) $\mathcal{H}_n$ is a Hilbert space, i.e., the analytic functions in

$$L^2(\Omega, (1 - |z|^2)^{n-2}\, dx\, dy)$$

form a closed subspace.

(2) $\mathcal{U}_n$ is a representation, i.e., $\mathcal{U}_n(gh) = \mathcal{U}_n(g)\mathcal{U}_n(h)$ and $\mathcal{U}_n(e) = I$.

(3) $\mathcal{U}_n$ is unitary, i.e., each $\mathcal{U}_n(g)$ is a linear isometry onto. (The proof consists of a change of variables, which uses the invariance of $(1 - |z|^2)^{-2}\, dx\, dy$ under $z \to zg$.)

(4) $\mathcal{U}_n$ is strongly continuous, i.e., $\lim_{g \to e} \|\mathcal{U}_n(g)f - f\| = 0$ for each f in $\mathcal{H}_n$.

(5) $\mathcal{U}_n$ is irreducible, i.e., there are no nontrivial closed invariant subspaces in $\mathcal{H}_n$.

(6) $\mathcal{U}_n$ is square integrable, i.e., $\int |(\mathcal{U}_n(g)\, \phi, \phi)|^2\, dg < \infty$ for some $\phi \in \mathcal{H}_n$. In the presence of (1)–(5), this implies that $\int |(\mathcal{U}_n(g)\, \phi, \psi)|^2\, dg < \infty$ for all $\phi, \psi \in \mathcal{H}_n$ and that $(\mathcal{H}_n, \mathcal{U}_n)$ is unitarily equivalent with a closed subspace of $L^2(G)$ with right translation as the operation.

We shall now describe the Harish-Chandra realization of the holomorphic discrete series for this G. Let $SL(2,\mathbf{C})$ be the group of 2-by-2 complex matrices of determinant 1 and let B be the lower triangular subgroup $\left\{\begin{pmatrix} a & 0 \\ c & a^{-1} \end{pmatrix}\right\}$.

Lemma. Every element of the set $BG \subseteq SL(2,\mathbf{C})$ has a unique decomposition as a product

$$\begin{pmatrix} 1 & 0 \\ \zeta & 1 \end{pmatrix}\begin{pmatrix} \gamma & 0 \\ 0 & \gamma^{-1} \end{pmatrix}\begin{pmatrix} 1 & z \\ 0 & 1 \end{pmatrix}, \qquad \zeta \in \mathbf{C}, \quad \gamma \in \mathbf{C} - \{0\}, \quad |z| < 1, \quad (*)$$

and every matrix (*) is in BG.

Indication of Proof. Since $B = \left\{\begin{pmatrix} 1 & 0 \\ \zeta & 1 \end{pmatrix}\begin{pmatrix} \gamma & 0 \\ 0 & \gamma^{-1} \end{pmatrix}\right\}$ is a group, existence of the decomposition follows from the identity

$$\begin{pmatrix}\alpha & \beta \\ \bar{\beta} & \bar{\alpha}\end{pmatrix} = \begin{pmatrix}1 & 0 \\ \bar{\beta}/\alpha & 1\end{pmatrix}\begin{pmatrix}\alpha & 0 \\ 0 & \alpha^{-1}\end{pmatrix}\begin{pmatrix}1 & \beta/\alpha \\ 0 & 1\end{pmatrix}.$$

The lemma suggests that we give BG the structure of a complex manifold by giving it the product structure. Now $SL(2,\mathbf{C})$ has its own complex structure, and the next lemma gives the connection.

Lemma. *BG is an open subset of $SL(2,\mathbf{C})$, and its product complex structure obtained from (*) is the same as what it inherits from $SL(2,\mathbf{C})$. In particular, right translation by $g \in G$ is a holomorphic automorphism of BG.*

The second statement in the lemma can be verified directly, without reference to the complex structure on $SL(2,\mathbf{C})$.

To construct the representation, let $n \geqq 2$ be an integer and let ξ_n be the holomorphic character of B given by $\xi_n\begin{pmatrix}a & 0 \\ c & a^{-1}\end{pmatrix} = a^{-n}$. Let

$$H_n = \left\{ F \;\middle|\; \begin{array}{ll} \text{(i)} & F(x) \text{ is holomorphic on } BG \\ \text{(ii)} & F(bx) = \xi_n(b)F(x) \quad \text{for} \quad b \in B, \quad x \in BG\ , \\ \text{(iii)} & \|F\|^2 = \int_G |F(g)|^2\, dg < \infty \end{array} \right\},$$

$$U_n(g)F(x) = F(xg).$$

If F satisfies (i), so does $U_n(g)F$ by the lemma above; thus it is clear that $U_n(g)$ preserves H_n and is unitary.

We shall give a correspondence between $\mathscr{H}_n$ and H_n and show that it is a unitary equivalence if the Haar measure dg is normalized suitably. Given $F(\zeta, \gamma, z) \in H_n$, define $f(z)$ on the disk Ω by

$$f(z) = F(0, 1, z).$$

For the inverse, if $f(z) \in \mathscr{H}_n$ is given, put

$$F(\zeta, \gamma, z) = \gamma^{-n} f(z).$$

Then F satisfies (i) and (ii) at least. To see that the correspondence $F \leftrightarrow f$ is an isometry, we note that it is possible to normalize dg in such a way that for any reasonable $h_0(z)$ on Ω the function $h(g) = h_0(0g)$ satisfies

$$\int_G h(g)\, dg = \int_\Omega h_0(z)(1-|z|^2)^{-2}\, dx\, dy. \qquad (**)$$

Let $F \in H_n$, $g = \begin{pmatrix} \alpha & \beta \\ \bar{\beta} & \bar{\alpha} \end{pmatrix}$. Then

$$F(g) = F(\bar{\beta}/\alpha, \alpha, \beta/\alpha) = \alpha^{-n}F(0, 1, \beta/\alpha) = \alpha^{-n}f(\beta/\alpha) = \alpha^{-n}f(0g)$$

since $\beta/\alpha = (\bar{\alpha}z + \beta)/(\bar{\beta}z + \alpha)$ for $z = 0$. Thus $|F(g)|^2 = |\alpha|^{-2n}|f(0g)|^2$. Since $|\alpha|^2 - |\beta|^2 = 1$, $1 - |0g|^2 = |\alpha|^{-2}$ and

$$|F(g)|^2 = (1 - |0g|^2)^n|f(0g)|^2.$$

If we put $h_0(z) = (1 - |z|^2)^n|f(z)|^2$ and $h(g) = |F(g)|^2$ in (**), we obtain

$$\|F\|^2 = \int_G |F(g)|^2\,dg = \int_\Omega |f(z)|^2(1 - |z|^2)^{n-2}\,dx\,dy = \|f\|^2,$$

and it follows that the correspondence is unitary.

To see that $F \leftrightarrow f$ commutes with the group actions, we compute that

$$U_n(g)F(\zeta,\gamma,z) = F\left(\zeta',\gamma(\bar{\beta}z + \alpha), \frac{\bar{\alpha}z + \beta}{\bar{\beta}z + \alpha}\right).$$

Thus

$$\begin{aligned} U_n(g)F(0,1,z) &= (\bar{\beta}z + \alpha)^{-n}F\left(0,1,\frac{\bar{\alpha}z + \beta}{\bar{\beta}z + \alpha}\right) \\ &= (\bar{\beta}z + \alpha)^{-n}f\left(\frac{\bar{\alpha}z + \beta}{\bar{\beta}z + \alpha}\right) = \mathcal{U}_n(g)f(z), \end{aligned}$$

and $F \leftrightarrow f$ is a unitary equivalence.

2. Bounded Symmetric Domains

2.1 Examples

These are examples of domains Ω in $\mathbf{C}^n$ of the form G/K, where G is noncompact semisimple and K is maximal compact in G, such that G operates holomorphically on Ω. These domains were classified by Cartan, and we use his numbering.

I. Let $m \leqq n$ and $\Omega = \{Z \in M_{nm}(\mathbf{C}) | I_m - Z^*Z > 0\}$.
Here M_{nm} refers to all n-by-m matrices, I_m is the identity of size m, and "> 0" means "is positive definite." Let

$$G = SU(n, m) = \left\{ g \in SL(n+m, \mathbf{C}) \middle| g^* \begin{pmatrix} I & 0 \\ 0 & -I \end{pmatrix} g = \begin{pmatrix} I & 0 \\ 0 & -I \end{pmatrix} \right\},$$

$$K = S(U(n) \times U(m)) = \left\{ \begin{pmatrix} A & 0 \\ 0 & D \end{pmatrix} \middle| A \in U(n),\ D \in U(m),\ \det = 1 \right\}.$$

The condition for g to be in G is that g preserve the Hermitian quadratic form $|z_1|^2 + \cdots + |z_n|^2 - |z_{n+1}|^2 - \cdots - |z_{n+m}|^2$. G operates holomorphically on Ω by

$$gZ = (AZ + B)(CZ + D)^{-1} \qquad \text{if} \quad g = \overset{\quad n \quad\ m}{\begin{pmatrix} A & B \\ C & D \end{pmatrix}}\begin{matrix} n \\ m \end{matrix}$$

[To see that $(CZ + D)^{-1}$ is defined and $g(\Omega) \subseteq \Omega$, write

$$\begin{aligned} (AZ+B)^*(AZ+B) - (CZ+D)^*(CZ+D) &= (Z^* \ I_m) g^* \begin{pmatrix} I & 0 \\ 0 & -I \end{pmatrix} g \begin{pmatrix} Z \\ I_m \end{pmatrix} \\ &= (Z^* \ I_m) \begin{pmatrix} I & 0 \\ 0 & -I \end{pmatrix} \begin{pmatrix} Z \\ I_m \end{pmatrix} \\ &= Z^*Z - I_m. \end{aligned}$$

Let $(CZ + D)v = 0$. Unless $v = 0$, we have

$$0 \leqq v^*(AZ + B)^*(AZ + B)v = v^*(Z^*Z - I_m)v < 0,$$

a contradiction. So $(CZ + D)^{-1}$ exists and

$$(gZ)^*(gZ) - I = (CZ + D)^{-1*}(Z^*Z - I_m)(CZ + D)^{-1} < 0.$$

Thus $g(\Omega) \subseteq \Omega$.

The isotropy subgroup at $Z = 0$ is K. One can show that G operates transitively on Ω, and thus $\Omega = G/K$. The Lie algebra of G is obtained from the definition of G and is

$$\mathfrak{g} = \left\{ \begin{pmatrix} a & b \\ \bar{b}^t & d \end{pmatrix} \in \mathfrak{sl}(n+m, \mathbf{C}) \middle| a = -a^*,\ d = -d^* \right\}$$

It is possible to show that G is the identity component of the group of all holomorphic automorphisms of Ω. This fact holds much more generally than for this Ω, and its proof is concealed in the proof of the first lemma in Section 2.3.

II. $\Omega = \{Z \in M_{nn}(\mathbf{C}) | I_n - Z^*Z > 0, Z = -Z^t\}$. Let

$$G = SO^*(2n) = \left\{ g \in SU(n,n) | g^t \begin{pmatrix} 0 & I \\ I & 0 \end{pmatrix} g = \begin{pmatrix} 0 & I \\ I & 0 \end{pmatrix} \right\}.$$

$$K = \left\{ \begin{pmatrix} A & 0 \\ 0 & \bar{A} \end{pmatrix} | A \in U(n) \right\}.$$

Then G acts in the same way as in type I, preserving Ω. The action is transitive with K the isotropy subgroup at 0. G has Lie algebra

$$\mathfrak{g} = \left\{ \begin{pmatrix} a & b \\ -\bar{b} & \bar{a} \end{pmatrix} | a = -a^*, b = -b^t \right\}.$$

III. $\Omega = \{z \in \mathbf{C}^n \mid |z^t z|^2 - 2z^*z + 1 < 0, |z^t z|^2 - 1 < 0\}$. The group $G = SO(n,2)$ has a transitive holomorphic action on Ω with isotropy subgroup $K = SO(n) \times SO(2)$.

IV. $\Omega = \{Z \in M_{nn}(\mathbf{C}) | I_n - Z^*Z > 0, Z = Z^t\}$. Let

$$G = Sp(n, \mathbf{R}) = \left\{ g \in SU(n,n) | g^t \begin{pmatrix} 0 & I \\ -I & 0 \end{pmatrix} g = \begin{pmatrix} 0 & I \\ -I & 0 \end{pmatrix} \right\},$$

$$K = \left\{ \begin{pmatrix} A & 0 \\ 0 & \bar{A} \end{pmatrix} | A \in U(n) \right\}.$$

(This is not the standard realization of $Sp(n, \mathbf{R})$ but is conjugate to it.) Then G acts in the same way as in type I, preserving Ω. The action is transitive with K the isotropy subgroup at 0. G has Lie algebra

$$\mathfrak{g} = \left\{ \begin{pmatrix} a & b \\ \bar{b} & \bar{a} \end{pmatrix} | a = -a^*, b = b^t \right\},$$

In addition to domains of types I to IV, there are two exceptional domains, and we can form products of any of these domains. In each case, we can verify that dim(center K) > 0. During the rest of this chapter, we shall be concerned with the following problems.

PROBLEM 1. Characterize those pairs (G, K) such that G/K admits an invariant complex structure. The answer for G simple will be that the dimension of the center of K is positive.

PROBLEM 2. When G/K has an invariant complex structure, exhibit G/K as a bounded domain in $\mathbf{C}^n$ in a group-theoretic way.

2.2 Almost Complex Manifolds

Let M be a C^∞-manifold, M_p the tangent space at p. A map $p \to J_p \in \operatorname{End}(M_p)$ is an *almost complex structure* on M if

(i) $J_p^2 = -I$ for all p,
(ii) JX is a smooth vector field whenever X is.

In this case, (M, J) is an *almost complex manifold.*

EXAMPLE. M = complex manifold of dimension m with charts (U_a, φ_a), where $\varphi_a : U_a \to \mathbf{C}^m$. M_p has a basis

$$\left(\frac{\partial}{\partial x_1}\right)_p, \left(\frac{\partial}{\partial y_1}\right)_p, \ldots, \left(\frac{\partial}{\partial x_m}\right)_p, \left(\frac{\partial}{\partial y_m}\right)_p,$$

defined relative to (U_a, φ_a). Put

$$J^\alpha\left(\frac{\partial}{\partial x_i}\right)_p = \left(\frac{\partial}{\partial y_i}\right)_p, \qquad J^\alpha\left(\frac{\partial}{\partial y_i}\right)_p = -\left(\frac{\partial}{\partial x_i}\right)_p.$$

Then $(J^\alpha)^2 = -I$. One can check that $J^\alpha = J^\beta$ if $p \in U_a \cap U_\beta$. Thus $J_p = J^\alpha = J^\beta$ is independent of the coordinates and the result J is an almost complex structure on M.

Let (M, J) and (M', J') be almost complex manifolds, $\Phi : M \to M'$ differentiable. We say Φ is *almost complex* if

$$d\Phi_p \circ J_p = J'_{\Phi(p)} \circ d\Phi_p \qquad \text{for} \quad p \in M, \qquad (*)$$

i.e., if Φ infinitesimally carries J into J'.

EXAMPLE. Let M and M' be complex manifolds, $\Phi : M \to M'$. Then Φ is holomorphic if and only if Φ is almost complex. In fact, Φ is holomorphic if and only if Φ is holomorphic in each variable (see Bochner-Martin, p. 33), if and only if Φ satisfies the Cauchy-Riemann equations in each variable, if and only if Φ satisfies (*).

2.3 Cartan Decomposition

Let

G = connected semisimple Lie group with finite center,
K = maximal compact subgroup,
$\mathfrak{g}, \mathfrak{k}$ = corresponding Lie algebras,

$$B(X, Y) = \mathrm{Tr}(\mathrm{ad}\, X\, \mathrm{ad}\, Y) = \text{Killing form, and}$$
$$\mathfrak{p} = \mathfrak{k}^{\perp} \text{ relative to } B.$$

Then $\mathfrak{g} = \mathfrak{k} \oplus \mathfrak{p}$ is a Cartan decomposition of $\mathfrak{g}$, in that

(1) B is negative definite on $\mathfrak{k}$, positive definite on $\mathfrak{p}$,

(2) $[\mathfrak{k}, \mathfrak{k}] \subseteq \mathfrak{k}$, $[\mathfrak{k}, \mathfrak{p}] \subseteq \mathfrak{p}$, $[\mathfrak{p}, \mathfrak{p}] \subseteq \mathfrak{k}$,

(3) the map $K \times \mathfrak{p} \to G$ given by $(k, X) \to k \exp X$ is a diffeomorphism onto G.

The tangent space $(G/K)_{eK}$ is naturally identified with $\mathfrak{p}$ in such a way that $dL_k = \mathrm{Ad}(k)|_{\mathfrak{p}}$, dL_k being the differential at eK of left translation by k. Suppose G/K is Hermitian, i.e., has an invariant complex structure. Form the corresponding almost complex structure, and let J be the structure on $(G/K)_{eK} \leftrightarrow \mathfrak{p}$. Since left translation by $k \in K$ is holomorphic, it is almost complex. That is,

$$dL_k \circ J = J \circ dL_k,$$

where $dL_k = \mathrm{Ad}(k)|_{\mathfrak{p}}$. Thus if G/K is Hermitian, then there is a $J \in \mathrm{Aut}(\mathfrak{p})$ with $J^2 = -I$ and with $J\,\mathrm{Ad}(k)|_{\mathfrak{p}} = \mathrm{Ad}(k)|_{\mathfrak{p}}J$ for $k \in K$.

Theorem. If G/K is Hermitian and if J is the associated automorphism of $\mathfrak{p}$, then

(i) there exists $V \in \mathrm{center}(\mathfrak{k})$ such that $\mathrm{ad}\, V|_{\mathfrak{p}} = J$,

(ii) if $V = V_1 + \cdots + V_n \in \mathfrak{g}_1 + \cdots + \mathfrak{g}_n$ is the decomposition according to the simple components of $\mathfrak{g}$, then $V_j \in \mathrm{center}(\mathfrak{k}_j)$ and $(\mathrm{ad}\, V_j)^2|_{\mathfrak{p}_j} = -I$. Consequently V_j defines a G_j-invariant almost complex structure on G_j/K_j.

Remarks. (ii) is immediate from (i) since the $\mathfrak{g}_j$ are ideals. (i) will be proved after the lemma below, which does not depend on Hermitian structures. The lemma contains the fact that G is the full connected component of the group of automorphisms of G/K. J-i. Hano remarked that if there were not independent interest in the lemma, then one could give a shorter proof of the theorem by using the fact that every derivation of $\mathfrak{g}$ is inner.

Lemma. Suppose $\mathfrak{g}$ has no compact factor. Let C be a K-invariant inner product on $\mathfrak{p}$ and let

$$\mathfrak{k}' = \left\{ M \in \mathrm{End}\, \mathfrak{p} \;\middle|\; \begin{array}{ll} \text{(i)} & C(MX,Y) + C(X,MY) = 0 \\ \text{(ii)} & [M, \mathrm{ad}[X,Y]|_{\mathfrak{p}}] = \mathrm{ad}[MX,Y]|_{\mathfrak{p}} \\ & \qquad + \mathrm{ad}[X,MY]|_{\mathfrak{p}};\, X, Y \in \mathfrak{p} \end{array} \right\}$$

Then $\mathfrak{k}'$ is a Lie subalgebra of End $\mathfrak{p}$ and the mapping $\mathfrak{k} \to \mathfrak{k}'$ given by $X \to \operatorname{ad} X|_{\mathfrak{p}}$ is an isomorphism onto.

Proof. Closure under brackets is straightforward. Also $\mathfrak{k}$ maps into $\mathfrak{k}'$ because (i) is just the $\mathfrak{k}$-invariance of C and (ii) is just the Jacobi identity for an element of $\mathfrak{k}$ and two elements of $\mathfrak{p}$. The map is one–one because $\{x \in \mathfrak{k} \,|\, \operatorname{ad} X|_{\mathfrak{p}} = 0\}$ is easily seen to be an ideal in $\mathfrak{g}$, and $\mathfrak{g}$ was assumed to have no compact factor.

We are to show the map is onto. Since it is one–one, we can regard $\mathfrak{g} = \mathfrak{k} + \mathfrak{p}$ as a subspace of the vector space $\mathfrak{g}' = \mathfrak{k}' + \mathfrak{p}$. Make $\mathfrak{g}'$ into a Lie algebra with $\mathfrak{g}$ as a subalgebra by defining

$$\begin{aligned} [X, Y]_{\text{new}} &= \operatorname{ad}[X, Y]|_{\mathfrak{p}} && \text{for } X, Y \in \mathfrak{p},\\ [T, X]_{\text{new}} &= -[X, T]_{\text{new}} = T(X) && \text{for } T \in \mathfrak{k}', X \in \mathfrak{p},\\ [S, T]_{\text{new}} &= ST - TS && \text{for } S, T \in \mathfrak{k}'. \end{aligned}$$

(Here $[\cdot\,,\,\cdot]_{\text{new}}$ is clearly skew, and the Jacobi identity is trivial for three vectors unless two of them are in $\mathfrak{p}$, the other in $\mathfrak{k}'$. In this case, the identity follows from (ii).) Since $\mathfrak{k} \supseteq [\mathfrak{p}, \mathfrak{p}]$, the proof will be complete if we show that $[\mathfrak{p}, \mathfrak{p}] = \mathfrak{k}'$.

Let $SO(\mathfrak{p})$ be the rotation group on $\mathfrak{p}$ relative to C, $\mathfrak{so}(\mathfrak{p})$ the Lie algebra. Then $\mathfrak{k}' \subseteq \mathfrak{so}(\mathfrak{p})$. Choose an $SO(\mathfrak{p})$-invariant inner product on $\mathfrak{so}(\mathfrak{p})$, restrict it to $\mathfrak{k}'$, and extend it by means of C to be defined on all of $\mathfrak{g}'$. The result is that $\mathfrak{g}'$ possesses a $\mathfrak{k}'$-invariant inner product. If $\mathfrak{z} =$ center $\mathfrak{g}'$, then $\mathfrak{z} \cap \mathfrak{k}' = 0$ since $\mathfrak{k}' \subseteq$ End $\mathfrak{p}$. From these two facts, it follows that the Killing form B of $\mathfrak{g}'$ is negative definite on $\mathfrak{k}'$ (cf. [*3*, p. 123]).

Since $[\mathfrak{k}', \mathfrak{k}'] \subseteq \mathfrak{k}'$ and $[\mathfrak{k}', \mathfrak{p}] \subseteq \mathfrak{p}$ and $[\mathfrak{p}, \mathfrak{p}] \subseteq \mathfrak{k}'$, $\mathfrak{k}'$ and $\mathfrak{p}$ are orthogonal with respect to B. From this fact and the fact that B is negative definite on $\mathfrak{k}'$, it follows that

$$\mathscr{I} = \{X \in \mathfrak{g}' \,|\, B(X, Y) = 0 \quad \text{for all} \quad Y \in \mathfrak{g}'\},$$

which is an ideal in $\mathfrak{g}'$, is contained in $\mathfrak{p}$. Thus $\mathscr{I}$ is an ideal in $\mathfrak{g}$ contained in $\mathfrak{p}$, and therefore $\mathscr{I} = 0$. Consequently B is nondegenerate on $\mathfrak{p}$.

Let $\mathfrak{v}$ be the orthogonal complement relative to B of $[\mathfrak{p}, \mathfrak{p}]$ in $\mathfrak{k}'$. Fix T in $\mathfrak{v}$ and let X and Y be arbitrary in $\mathfrak{p}$. Then

$$B([T, X], Y) = B(T, [X, Y]) = 0.$$

By the nondegeneracy on $\mathfrak{p}$, $T(X) = [T, X] = 0$ for all X. Thus $T = 0$, $\mathfrak{v} = 0$, and $[\mathfrak{p}, \mathfrak{p}] = \mathfrak{k}'$. This proves the lemma.

Proof of (i) in theorem. Without loss of generality, we can assume $\mathfrak{g}$ has no compact factor. Put $C(X, Y) = B(X, Y) + B(JX, JY)$ for $X, Y \in \mathfrak{p}$. Then C is a K-invariant inner product and J satisfies (i) in the definition of $\mathfrak{k}'$ in the lemma. In view of the lemma, the proof will be complete if we show that J satisfies

$$[J, \operatorname{ad}[X, Y]|_{\mathfrak{p}} = \operatorname{ad}[JX, Y]|_{\mathfrak{p}} + \operatorname{ad}[X, JY]|_{\mathfrak{p}}$$

for $X, Y \in \mathfrak{p}$. The left side is 0 since J commutes with ad $\mathfrak{k}|_{\mathfrak{p}}$. Complexify $\mathfrak{k}$, $\mathfrak{p}$, and C. Then the right side is a restriction of

$$-\operatorname{Im} \operatorname{ad}[X - iJX, Y - iJY],$$

where $J(X - iJX) = i(X - iJX)$, $J(Y - iJY) = i(Y - iJY)$. Changing notation, we see that it is enough to prove that $\operatorname{ad}[X, Y]|_{\mathfrak{p}^c} = 0$ if $JX = iX$ and $JY = iY$. Now

$$C(X, Y) = C(JX, CY) = C(iX, iY) = -C(X, Y),$$

so that

$$C(X, Y) = 0.$$

Let $Z \in \mathfrak{p}^{\mathbf{C}}$ and $T \in \mathfrak{p}^{\mathbf{C}}$ be arbitrary. Then

$$C([[Z, T], X], Y) = 0$$

since $J([[Z,T],X]) = i[[Z,T],X]$. On the other hand,

$$D(X, Y, Z, T) = C([[X, Y], Z], T)$$

is a quadrilinear form on $\mathfrak{p}^{\mathbf{C}}$ that

(i) is skew in X and Y,

(ii) is skew in Z and T (a small verification required here), and

(iii) satisfies the Jacobi identity in X, Y, and Z.

By Lemma 12.4 in [*3*, pp. 68–69],

$$C([[X, Y], Z], T) = C([[Z, T], X], Y),$$

and we have seen that the right side is 0. Now C is nondegenerate, being the complexification of an inner product, and thus $[[X, Y], Z] = 0$ for all $Z \in \mathfrak{p}^{\mathbf{C}}$. That is, $\operatorname{ad}[X, Y]|_{\mathfrak{p}^c} = 0$. The proof of the theorem is complete.

At the moment we settle for a weak converse, which will be used in the next section.

Proposition. If G is simple and $\mathfrak{c} = \text{center}(\mathfrak{k}) \neq 0$, then G/K has an invariant almost complex structure.

Lemma 1. $\mathfrak{k}$ is a maximal subalgebra of $\mathfrak{g}$.

Proof. Since $\mathfrak{g}$ is simple, ad $\mathfrak{k}$ acts irreducibly on $\mathfrak{p}$. (Namely, if $[\mathfrak{k},\mathfrak{p}_1] \subseteq \mathfrak{p}_1$ and $\mathfrak{p}_1 \subseteq \mathfrak{p}$, then $[\mathfrak{p},\mathfrak{p}_1] + \mathfrak{p}_1$ is an ideal in $\mathfrak{g}$.) If $\mathfrak{s} \subseteq \mathfrak{k}$ is a subalgebra of $\mathfrak{g}$, then $\mathfrak{s} = \mathfrak{k} + (\mathfrak{s} \cap \mathfrak{p})$. If $X \in \mathfrak{s} \cap \mathfrak{p}$ with $X \neq 0$, then iterated brackets of members of $\mathfrak{k}$ with X span $\mathfrak{p}$. So $\mathfrak{s} \cap \mathfrak{p} = \mathfrak{p}$ and $\mathfrak{s} = \mathfrak{g}$.

Corollary. K is a maximal Lie subgroup of G.

Proof. It is known that no other Lie subgroup of G has $\mathfrak{k}$ as Lie algebra.

Lemma 2. If $X \in \mathfrak{c}$ with $X \neq 0$, then the centralizer $Z_{\mathfrak{g}}(X)$ equals $\mathfrak{k}$. If $x \in \text{Ad}_G(\text{center } K)$ with $x \neq e$, then $Z_{\text{Ad } G}(x) = \text{Ad}_G K$.

Proof. $Z_{\mathfrak{g}}(X) \supseteq \mathfrak{k}$ and $Z_{\mathfrak{g}}(X)$ is a subalgebra. Since $\mathfrak{g}$ is semisimple, $Z_{\mathfrak{g}}(X) \neq \mathfrak{g}$. By Lemma 1, $Z_{\mathfrak{g}}(X) = \mathfrak{k}$. The second statement is similar.

Proof of Proposition. The identity component of $\text{Ad}_G(\text{center } K)$ is a torus and thus contains an element Ad(j) of order 4. By Lemma 2, Ad $K = Z_{\text{Ad}G}(\text{Ad } j^2)$. $\text{Ad}(j^2)$ has ± 1 as eigenvalues and is $+1$ on $\mathfrak{k}$. Since $Z_{\text{Ad}G}(\text{Ad } j^2) = \text{Ad } K$, $\text{Ad}(j^2) = -1$ on $\mathfrak{p}$. Thus $\text{Ad}(j)|_{\mathfrak{p}}$ provides the required automorphism of $(G/K)_e$. Translation of this by G gives the almost complex structure at the other points of G/K.

2.4 Compact Cartan Subgroups and Roots

In this section and the next, we sharpen the proposition of Section 2.3 to produce a complex structure on G/K and to exhibit G/K as a bounded domain in $\mathbf{C}_n$. In order not to complicate the notation, we assume that G is simple, rather than just semisimple. We therefore assume

$\mathfrak{g}$ = simple Lie algebra,
$\mathfrak{k}$ = maximal compact subalgebra,
$G \supset K$, corresponding subgroups, G having finite center,
$\mathfrak{c} = \text{center}(\mathfrak{k}) \neq 0$,

$\mathfrak{h}$ = maximal Abelian subalgebra of $\mathfrak{k}$, and
T^- = exp $\mathfrak{h}$, a subgroup.

Proposition. $\mathfrak{h}$ is a maximal Abelian subalgebra in $\mathfrak{g}$, and ad $\mathfrak{h}$ is fully reducible on $\mathfrak{g}^{\mathbf{C}}$. T^- is a (compact) torus, and T^- is maximal Abelian in G.

Proof. By Lemma 2, $Z_{\mathfrak{g}}(\mathfrak{h}) \subseteq Z_{\mathfrak{g}}(\mathfrak{c}) = \mathfrak{k}$. So $Z_{\mathfrak{g}}(\mathfrak{h}) = Z_{\mathfrak{k}}(\mathfrak{h}) = \mathfrak{h}$, and $\mathfrak{h}$ is maximal Abelian in $\mathfrak{g}$. Next, ad $\mathfrak{h}$ is skew relative to the positive definite form $B_\theta(X, Y) = -B(X, \theta Y)$ (where $\theta = 1$ on $\mathfrak{k}$ and $\theta = -1$ on $\mathfrak{p}$) and is therefore fully reducible. Since $\mathfrak{h}$ is maximal Abelian in $\mathfrak{k}$, $\overline{T^-} = T^-$; thus T^- is compact and is a torus. By Corollary 2.7 of [*3*, p. 247], T^- is maximal Abelian in K.

To see that T^- is maximal Abelian in G, let $z = k \exp X \in Z_G(T^-)$ with $k \in K$, $X \in \mathfrak{p}$. We want $X = 0$. We have $k(\exp X)^{-1} = \theta z \in Z_G(T^-)$, and thus $\exp 2X = (\theta z^{-1})z \in Z_G(T^-)$. That is, $h(\exp 2X)h^{-1} = \exp 2X$ for $h \in T^-$. Hence $\exp(\mathrm{Ad}(h)2X) = \exp 2X$, and $\mathrm{Ad}(h)2X = 2X$ because exp is one–one on $\mathfrak{p}$. Consequently $[\mathfrak{h}, X] = 0$ and $X \in Z_{\mathfrak{g}}(\mathfrak{h}) = \mathfrak{h}$. Since $X \in \mathfrak{p}$, $X \in \mathfrak{h} \cap \mathfrak{p} = 0$. Thus T^- is maximal Abelian in G.

It follows from the proposition that $\mathfrak{h}^{\mathbf{C}}$ is a Cartan subalgebra of $\mathfrak{g}^{\mathbf{C}}$. Form roots in the usual way: If $\alpha \in (\mathfrak{h}^{\mathbf{C}})'$, let

$$\mathfrak{g}_\alpha = \{X \in \mathfrak{g}^{\mathbf{C}} \mid [H,X] = \alpha(H)X \text{ for all } H \in \mathfrak{h}^{\mathbf{C}}\}$$

and call α a *root* if $\mathfrak{g}_\alpha \neq 0$. Let Σ be the set of nonzero roots. Then $\mathfrak{g}^{\mathbf{C}} = \mathfrak{h}^{\mathbf{C}} + \sum_{\alpha\in\Sigma} \mathfrak{g}_\alpha$. If $\alpha \in \Sigma$, it is known that $-\alpha \in \Sigma$, that $\dim_{\mathbf{C}} \mathfrak{g}_\alpha = 1$, and that $\frac{1}{2}$ is purely imaginary on $\mathfrak{h}$. Since $[\mathfrak{k}, \mathfrak{k}] \subset \mathfrak{k}$ and $[\mathfrak{k}, \mathfrak{p}] \subseteq \mathfrak{p}$, we have

$$[\mathfrak{h}^{\mathbf{C}}, \mathfrak{k}^{\mathbf{C}}] \subseteq \mathfrak{k}^{\mathbf{C}} \qquad \text{and} \qquad [\mathfrak{h}^{\mathbf{C}}, \mathfrak{p}^{\mathbf{C}}] \subseteq \mathfrak{p}^{\mathbf{C}}.$$

If $\alpha \in \Sigma$, it follows from the one-dimensionality of $\mathfrak{g}_\alpha$ that $\mathfrak{g}_\alpha \subseteq \mathfrak{k}^{\mathbf{C}}$ or $\mathfrak{g}_\alpha \subseteq \mathfrak{p}^{\mathbf{C}}$. We call α *compact* or *noncompact* accordingly. We have $\mathfrak{k}^{\mathbf{C}} = \mathfrak{h}^{\mathbf{C}} + \sum_{\alpha \text{ compact}} \mathfrak{g}_\alpha$ and $\mathfrak{p}^{\mathbf{C}} = \sum_{\alpha \text{ noncompact}} \mathfrak{g}_\alpha$.

Example. Let $\mathfrak{g} = \mathfrak{su}(m,n)$ as in 2.1. Then $\mathfrak{g}^{\mathbf{C}} = \mathfrak{sl}(m+n, \mathbf{C})$ and

$$\mathfrak{k}^{\mathbf{C}} = \left\{\begin{pmatrix} a & 0 \\ 0 & d \end{pmatrix}, \text{ trace } 0\right\} \quad \mathfrak{p}^{\mathbf{C}} = \left\{\begin{pmatrix} 0 & b \\ c & 0 \end{pmatrix}\right\}.$$

If $\mathfrak{h}$ is chosen as the subalgebra of diagonal matrices in $\mathfrak{g}$, then $\mathfrak{h}^{\mathbf{C}}$ consists of all diagonal matrices of trace 0. If e_i denotes evaluation of the ith diagonal entry, the roots are $e_i - e_j$, $i \neq j$. With $\alpha = e_i - e_j$, $\mathfrak{g}_\alpha = \mathbf{C}E_{ij}$, where E_{ij} is 1 in the i-jth entry and 0 elsewhere. The compact roots $e_i - e_j$ are thus the ones for which both i and j are $\leq m$ or both i and j are $\geq m + 1$.

Proposition. If $\alpha \in \Sigma$, the following are equivalent:
(i) α is compact,
(ii) α vanishes on $\mathfrak{c}$,
(iii) α vanishes on some $H \neq 0$ in $\mathfrak{c}$.

Proof. Let $H \in \mathfrak{c} \subseteq \mathfrak{h}$ with $H \neq 0$. Then $\alpha(H) = 0 \Leftrightarrow [H, X_\alpha] = 0$ for $X_\alpha \in \mathfrak{g}_\alpha \Leftrightarrow X_\alpha \in Z_{\mathfrak{g}^c}(H) = \mathfrak{k}^{\mathbf{C}} \Leftrightarrow \alpha$ is compact. The right side of this chain is independent of H, and the proposition follows.

Corollary. $\dim \mathfrak{c} = 1$.

Proof. Let α be any noncompact root. Such exist since $\mathfrak{p} \neq 0$. By the proposition, $0 = \dim(\mathfrak{c} \cap \ker \alpha) \geq \dim \mathfrak{c} - 1$ since α is a linear functional.

Corollary. G/K admits only two invariant complex structures.

Proof. In view of the previous corollary and of the theorem in 2.3, there are at most two possibilities for J. J determines the complex structure since an almost complex mapping between complex manifolds is holomorphic.

All roots are real on the real vector space $i\mathfrak{h}$. By choosing an ordered basis for $i\mathfrak{h}$, we can order the roots by saying that α is positive if it is positive on the first basis vector on which it is nonzero. Choose a *good ordering*, in that every noncompact positive root is larger than every compact root. (For instance, take a vector in $\mathfrak{c}$ as the first basis vector and appeal to the proposition.) Let

$$\mathfrak{p}^+ = \sum_{\substack{\alpha>0 \\ \alpha \text{ noncompact}}} \mathfrak{g}_\alpha , \qquad \mathfrak{p}^- = \sum_{\substack{\alpha<0 \\ \alpha \text{ noncompact}}} \mathfrak{g}_\alpha .$$

In the example $\mathfrak{su}(m, n)$, it is possible to choose a good ordering in such a way that the positive roots are $e_i - e_j$ with $i < j$. Then

$$\mathfrak{p}^+ = \left\{\begin{pmatrix} 0 & b \\ 0 & 0 \end{pmatrix}\right\} \text{ and } \mathfrak{p}^- = \left\{\begin{pmatrix} 0 & 0 \\ c & 0 \end{pmatrix}\right\}.$$

Proposition. $\mathfrak{p}^{\mathbf{C}} = \mathfrak{p}^+ \oplus \mathfrak{p}^-$, $[\mathfrak{k}^{\mathbf{C}}, \mathfrak{p}^+] \subseteq \mathfrak{p}^+$, $[\mathfrak{k}^{\mathbf{C}}, \mathfrak{p}^-] \subseteq \mathfrak{p}^-$, and $\mathfrak{p}^+$ and $\mathfrak{p}^-$ are Abelian subspaces.

Proof. Clearly $\mathfrak{p}^{\mathbf{C}} = \mathfrak{p}^+ \oplus \mathfrak{p}^-$. Let α, β, and $\alpha + \beta$ be in Σ with α compact and β noncompact. Then $[\mathfrak{g}_\alpha, \mathfrak{g}_\beta] \subseteq \mathfrak{g}_{\alpha+\beta}$ and β and $\alpha + \beta$ are both positive or both negative. Summing on α and β, we obtain $[\mathfrak{k}^{\mathbf{C}}, \mathfrak{p}^+] \subseteq \mathfrak{p}^+$ and $[\mathfrak{k}^{\mathbf{C}}, \mathfrak{p}^-] \subseteq \mathfrak{p}^-$.

The idea in the proof that $\mathfrak{p}^+$ and $\mathfrak{p}^-$ are Abelian is that they are the eigenspaces of the automorphism J of $\mathfrak{p}^{\mathbf{C}}$ that was considered in 2.3. However, we are given the existence of the center $\mathfrak{c}$, not of J, and J must be constructed. Thus by the proposition in 2.3, choose $J \in \operatorname{Aut} \mathfrak{p}$ with $J^2 = -I$ so that J commutes with Ad K. The proof of the theorem in 2.3 shows that $J = \operatorname{ad} V|_{\mathfrak{p}}$ for some $V \in \mathfrak{c}$. Suppose for definiteness that the positive noncompact roots are positive on $-iV$. If $\alpha > 0$ is noncompact and $X \in \mathfrak{g}_\alpha$, then $-i\,\alpha(V) > 0$ and

$$JX = (\operatorname{ad} V)X = \alpha(V)X = (-i\,\alpha(V))\,(iX).$$

Since $J^2 = -I$, J has eigenvalues $\pm i$, and it follows that $JX = iX$. If $\beta > 0$ is noncompact and $Y \in \mathfrak{g}_\beta$, then $JY = iY$ and the proof of the theorem in 2.3 shows that $\operatorname{ad}[X, Y]|_{\mathfrak{p}^c} = 0$. Since $\mathfrak{g}$ has no compact factor, $[X, Y] = 0$. Summing on α and β, we see that $[\mathfrak{p}^+, \mathfrak{p}^+] = 0$. Similarly $[\mathfrak{p}^-, \mathfrak{p}^-] = 0$.

Corollary (OF PROOF). All noncompact positive roots agree on $\mathfrak{c}$.

Proof. $-i\,\alpha(V) = 1$ for all positive noncompact roots.

2.5 Harish-Chandra Decomposition

We retain the notation of 2.4 but assume further that G is a matrix group. Define

$$\mathfrak{b} = \mathfrak{h}^{\mathbf{C}} + \sum_{\alpha<0} \mathfrak{g}_\alpha,$$

$\mathfrak{g}^{\mathbf{C}}$ = complexified algebra of matrices ($\mathfrak{g}$ being an algebra of matrices),
$G^{\mathbf{C}}$ = analytic subgroup of matrices with Lie algebra $\mathfrak{g}^{\mathbf{C}}$, and
$P^-, K^{\mathbf{C}}, P^+, B$ = subgroups of $G^{\mathbf{C}}$ corresponding to $\mathfrak{p}^-, \mathfrak{k}^{\mathbf{C}}, \mathfrak{p}^+, \mathfrak{b}$.

Each of these groups has a natural complex structure consistent with multiplication by i in $\mathfrak{g}^{\mathbf{C}}$. It is not hard to see that exp is a diffeomorphism on $\mathfrak{p}^+$ and $\mathfrak{p}^-$ and hence that P^+ and P^- are holomorphically isomorphic with $\mathbf{C}^n$ for a suitable n.

Theorem. Multiplication $P^- \times K^{\mathbf{C}} \times P^+ \to G^{\mathbf{C}}$ is one–one, holomorphic, and regular (with the image open in $G^{\mathbf{C}}$), BG is open in $G^{\mathbf{C}}$, and there exists a bounded open subset $\Omega \subseteq P^+$ such that

$$BG = P^-K^{\mathbf{C}}G = P^-K^{\mathbf{C}}\Omega.$$

Interpretation. G acts on Ω by holomorphic automorphisms under the definition $(w)g = p^+(wg)$, for $w \in \Omega$ and $g \in G$, where $p^+(\,\cdot\,)$ refers to the P^+ component of an element of $P^-K^{\mathbf{C}}P^+$. The formula $P^-K^{\mathbf{C}}G = P^-K^{\mathbf{C}}\Omega$ shows the action is transitive. For the isotropy subgroup at $1 \in \Omega$, we ask when $p^+(1g) = 1$ or $g \in P^-K^{\mathbf{C}}$. Thus the isotropy subgroup is $G \cap P^-K^{\mathbf{C}}$. On the Lie algebra level, one verifies that $\mathfrak{g} \cap (\mathfrak{p}^- + \mathfrak{k}^{\mathbf{C}}) = \mathfrak{k}$, and it follows that $G \cap P^-K^{\mathbf{C}} = K$ since no other Lie subgroup of G has $\mathfrak{k}$ as Lie algebra. Thus $\Omega = K\backslash G$, Ω is a bounded domain in $\mathbf{C}^n$, and G operates holomorphically. The theorem therefore completes the solution of the two problems at the end of 2.1.

For the proof of the theorem we shall use some further notation:

$$\mathfrak{n}_c^- = \sum_{\alpha<0} \mathfrak{g}_\alpha, \qquad \mathfrak{n}_c^+ = \sum_{\alpha>0} \mathfrak{g}_\alpha, \qquad \mathfrak{h}^+ = i\mathfrak{h},$$

N_c^-, T^+ = analytic subgroups of $G^{\mathbf{C}}$ with Lie algebras $\mathfrak{n}_c^-, \mathfrak{h}^+$.

Then
$$B = N_c^- T^+ T^-.$$

Lemma 1. $P^-K^{\mathbf{C}} \cap P^+ = \{1\}$ and $K^{\mathbf{C}}P^+ \cap P^- = \{1\}$.

Proof. Let $y \in P^-K^{\mathbf{C}} \cap P^+$. Since $y \in P^-K^{\mathbf{C}}$, $\mathrm{Ad}(y)\mathfrak{p}^- \subseteq \mathfrak{p}^-$. Choose $Y \in \mathfrak{p}^+$ so that $y = \exp Y$. Write $Y = \sum c_\alpha X_\alpha$, $X_\alpha \neq 0$ in $\mathfrak{g}_\alpha$, where the sum is over positive noncompact α. If $Y \neq 0$, let α_0 be the smallest α so that $c_\alpha \neq 0$. Then $[Y, X_{-\alpha_0}] \equiv c_{\alpha_0}[X_{\alpha_0}, X_{-\alpha_0}] \bmod \mathfrak{n}_c^+$, and so

$$\mathrm{Ad}(y)X_{-\alpha_0} \equiv X_{-\alpha_0} + c_{\alpha_0}[X_{\alpha_0}, X_{-\alpha_0}] \bmod \mathfrak{n}_c^+,$$

$$\mathrm{Ad}(y)X_{-\alpha_0} \equiv c_{\alpha_0}[X_{\alpha_0}, X_{-\alpha_0}] \bmod (\mathfrak{n}_c^+ + \mathfrak{p}^-).$$

Thus $\mathrm{Ad}(y)X_{-\alpha_0} \notin \mathfrak{p}^-$, contradiction. So $Y = 0$ and $P^-K^{\mathbf{C}} \cap P^+ = \{1\}$. The other identity is proved similarly.

Lemma 2. $(\mathfrak{n}_c^- + \mathfrak{h}^+) \cap \mathfrak{g} = 0$.

Reference. Corollary, [*2*, p. 2].

Lemma 3. Let $\mathfrak{g} = \mathfrak{h}_1 \oplus \mathfrak{h}_2$, and let G, H_1, and H_2 be corresponding analytic subgroups. Then multiplication $H_1 \times H_2 \to G$ is everywhere regular.

Reference. Lemma 5.2, [*3*, p. 235].

Lemma 4. BG is open in $G^{\mathbf{C}}$.

Proof. $BG = N_c^- T^+ T^- G = (N_c^- T^+)G$ and $(\mathfrak{n}_c^- + \mathfrak{h}^+) \cap \mathfrak{g} = 0$ by Lemma 2. Also $\dim_{\mathbf{R}} \mathfrak{g}^{\mathbf{C}} = \dim_{\mathbf{R}}(\mathfrak{n}_c^- + \mathfrak{h}^+) + \dim_{\mathbf{R}} \mathfrak{g}$ because

$$\begin{aligned}\dim_{\mathbf{R}} \mathfrak{g}^{\mathbf{C}} &= \dim_{\mathbf{R}} \mathfrak{g} + \dim_{\mathbf{C}} \mathfrak{g}^{\mathbf{C}} = \dim_{\mathbf{R}} \mathfrak{g} + \dim_{\mathbf{C}} \mathfrak{n}_c^- + \dim_{\mathbf{C}} \mathfrak{n}_c^+ + \dim_{\mathbf{C}} \mathfrak{h}^{\mathbf{C}} \\ &= \dim_{\mathbf{R}} \mathfrak{g} + \dim_{\mathbf{R}} \mathfrak{n}_c^- + \dim_{\mathbf{R}} \mathfrak{h}^+ = \dim_{\mathbf{R}} \mathfrak{g} + \dim_{\mathbf{R}}(\mathfrak{n}_c^- + \mathfrak{h}^+).\end{aligned}$$

So $\mathfrak{g}^{\mathbf{C}} = (\mathfrak{n}_c^- + \mathfrak{h}^+) \oplus \mathfrak{g}$, and the result follows from Lemma 3.

Lemma 5. Multiplication $P^- \times K^{\mathbf{C}} \times P^+ \to G^{\mathbf{C}}$ is one–one, holomorphic, and regular.

Proof. It is clearly holomorphic, and it is regular by two applications of Lemma 3. By Lemma 1, $(P^-K^{\mathbf{C}}) \cap P^+ = \{1\}$. Since $P^- \cap (K^{\mathbf{C}}P^+) = \{1\}$, $P^- \cap K^{\mathbf{C}} = \{1\}$. Thus 1 has only one preimage. But $p_1k_1q_1 = p_2k_2q_2$ implies $(p_2^{-1}p_1)(k_1k_2^{-1})(k_2(q_1q_2^{-1})k_2^{-1}) = 1$ and hence $p_1 = p_2$, $k_1 = k_2$, and $q_1 = q_2$.

An Iwasawa decomposition $G = KA_0N$ is developed from a maximal Abelian subspace $\mathfrak{a}_0$ of $\mathfrak{p}$. Then $G = KAK$ also, though not uniquely. We shall construct an $\mathfrak{a}_0$ with special properties.

Lemma 6. There exist positive noncompact roots $\gamma_1, \ldots, \gamma_s$ such that $\sum_{i=1}^{s} \mathbf{C}(X_{\gamma_j} + X_{-\gamma_j})$ is a maximal Abelian subspace of $\mathfrak{p}^{\mathbf{C}}$.

Reference. [*3*, pp. 314–315].

The Killing form is nondegenerate on $\mathfrak{h}^{\mathbf{C}}$, and we can use this form by duality to define $\langle \alpha, \beta \rangle$ for $\alpha, \beta \in (\mathfrak{h}^{\mathbf{C}})'$. If α is a nonzero root, then one can show that $\langle \alpha, \alpha \rangle > 0$. According to [*3*, p. 220], it is possible to choose vectors $X_\alpha \in \mathfrak{g}_\alpha$ in such a way that

$$X_\alpha - X_{-\alpha} \in \mathfrak{k} + i\mathfrak{p}, \qquad i(X_\alpha + X_{-\alpha}) \in \mathfrak{k} + i\mathfrak{p},$$
$$B(X_\alpha, X_{-\alpha}) = 2/\langle \alpha, \alpha \rangle. \qquad (*)$$

This normalization is arranged to connect $\mathfrak{g}^{\mathbf{C}}$ with $\mathfrak{sl}(2, \mathbf{C})$. Namely if $H = \begin{pmatrix} 1 & 0 \\ 0 & -1 \end{pmatrix}$, $X = \begin{pmatrix} 0 & 1 \\ 0 & 0 \end{pmatrix}$, $Y = \begin{pmatrix} 0 & 0 \\ 1 & 0 \end{pmatrix}$, then $\mathfrak{sl}(2, \mathbf{C}) = \mathbf{C}H + \mathbf{C}X + \mathbf{C}Y$ and the bracket relations are $[H, X] = 2X$, $[H, Y] = -2Y$, $[X, Y] = H$. The normalization (*) is arranged so that the mapping

$$H \to [X_\alpha, X_{-\alpha}], \qquad X \to X_\alpha, \qquad Y \to X_{-\alpha}$$

extends to an isomorphism of $\mathfrak{sl}(2, \mathbf{C})$ into $\mathfrak{g}^{\mathbf{C}}$ that carries $\mathfrak{su}(1,1)$ into $\mathfrak{g}$ and $\mathbf{R}H$ into $\mathfrak{h}^+$.

Corollary. Under the normalization (*), the positive noncompact roots $\gamma_1, \ldots, \gamma_s$ of Lemma 6 are such that

$$\mathfrak{a}_0 = \sum_{i=1}^{s} \mathbf{R}(X_{\gamma_i} + X_{-\gamma_i})$$

is a maximal Abelian subspace of $\mathfrak{p}$.

Proof. $X_{\gamma_i} + X_{-\gamma_i} \in \mathfrak{p}^{\mathbf{C}} \cap i^{-1}(\mathfrak{k} + i\mathfrak{p}) = \mathfrak{p}$. So $\mathfrak{a} \subset \mathfrak{p}$. The rest follows from Lemma 6.

Lemma 7. Let L be a Lie group with Lie algebra $\mathfrak{sl}(2, \mathbf{C}) = \mathbf{C}H + \mathbf{C}X + \mathbf{C}Y$ as above. Then

$$\exp t(X + Y) = \exp(\tanh t)Y \exp(\log(\cosh t))H \exp(\tanh t)X.$$

Proof. Since $SL(2, \mathbf{C})$ is simply connected, it is enough to verify this identity there, where it is the special case of the identity

$$\begin{pmatrix} \alpha & \beta \\ \bar{\beta} & \bar{\alpha} \end{pmatrix} = \begin{pmatrix} 1 & 0 \\ \bar{\beta}/\alpha & 1 \end{pmatrix} \begin{pmatrix} \alpha & 0 \\ 0 & \alpha^{-1} \end{pmatrix} \begin{pmatrix} 1 & \beta/\alpha \\ 0 & 1 \end{pmatrix}$$

with $\alpha = \cosh t$ and $\beta = \sinh t$.

Corollary. If $Z = \sum_{i=1}^{s} t_i(X_{\gamma_i} + X_{-\gamma_i}) \in \mathfrak{a}_0$, then

$$\exp Z = \exp Y_0 \exp H_0 \exp X_0$$

with

$$Y_0 = \sum(\tanh t_i)X_{-\gamma_i},\ H_0 = \sum(\log \cosh t_i)[X_{\gamma_i}, X_{-\gamma_i}], X_0 = \sum(\tanh t_i)X_{\gamma_i}.$$

Proof. Apply Lemma 7, using the commutativity of X_{γ_i} and $X_{\pm\gamma_j}$ for $i \neq j$.

In the corollary, we have $Y_0 \in \mathfrak{p}^-$, $H_0 \in \mathfrak{h}^+ \subseteq \mathfrak{k}^{\mathbf{C}}$, and $X_0 \in \mathfrak{p}^+$. The corollary therefore gives explicitly the imbedding of the group $A_0 = \exp \mathfrak{a}_0$ in $P^- K^{\mathbf{C}} P^+$.

Lemma 8. $BG \subseteq P^- K^{\mathbf{C}} G \subseteq P^- K^{\mathbf{C}} P^+$.

Proof. Since $\mathfrak{b} \subseteq \mathfrak{p}^- + \mathfrak{k}^{\mathbf{C}}$, we have $B \subseteq P^- K^{\mathbf{C}}$, and it is enough to prove $G \subseteq P^- K^{\mathbf{C}} P^+$. Since $G = KA_0K$ and K normalizes P^- and P^+, it is enough to show $A_0 \subseteq P^- K^{\mathbf{C}} P^+$. But this follows from the corollary above.

Proof of Theorem. Let $\mathfrak{b}_K = \mathfrak{h}^{\mathbf{C}} + \sum_{\substack{\alpha<0 \\ \alpha \text{ compact}}} \mathfrak{g}_\alpha$, B_K the corresponding analytic subgroup. B_K is known to be closed, and K is compact; thus $B_K K$ is closed. On the other hand, $B_K K$ is open in $K^{\mathbf{C}}$ by the same kind of argument as in Lemma 4. Thus $K^{\mathbf{C}} = B_K K$. Therefore

$$P^- K^{\mathbf{C}} G \subseteq BK^{\mathbf{C}} G = B\, B_K KG = BG.$$

That is, $BG = P^- K^{\mathbf{C}} G$ and so $BG = P^- K^{\mathbf{C}} \Omega$ with $\Omega \subseteq P^+$. Since BG is open in $G^{\mathbf{C}}$, Ω is open in P^+.

We are to show that Ω is bounded, i.e., that $p^+(G)$ is bounded. Choose a K-invariant inner product on $\mathfrak{p}^+$. Since $G = KA_0K$ with K normalizing P^- and P^+, we see that we are to prove that $\|\log p^+(A_0)\|$ is bounded. In view of the corollary to Lemma 7,

$$\|\log p^+(\exp \sum t_i(X_{\gamma_i} + X_{-\gamma_i})\| = \|\sum(\tanh t_i)X_{\gamma_i}\| \leqq \sum_{i=1}^{s} \|X_{\gamma_i}\|,$$

and the right side is bounded independently of the t_i. This proves the theorem.

3. Holomorphic Discrete Series

3.1 Construction of Representations

Throughout this chapter, we retain the notation of 2.5. We begin by summarizing some properties of irreducible unitary representations of the compact group K. Let τ be such a representation (necessarily finite-dimensional) on a complex vector space V. Since the subgroup T^- of K is Abelian, V is the orthogonal direct sum of subspaces on which each member of $\tau(T^-)$ operates as a scalar. If these subspaces are chosen as large as possible, they are called the *weight spaces* for τ. For the corresponding representation of $\mathfrak{k}$, each member of $\tau(\mathfrak{h})$ operates as a scalar on each weight space. If the weight space is held fixed and the member of $\mathfrak{h}$ is allowed to vary, the scalar value of the member of $\tau(\mathfrak{h})$ defines a complex-valued linear functional on $\mathfrak{h}$. The corresponding complex-linear functionals on $\mathfrak{h}^{\mathbf{C}}$ are called the *weights* of τ. Since τ is unitary, the weights are imaginary on $\mathfrak{h}$, and hence real on $\mathfrak{h}^+ = i\mathfrak{h}$. We have already defined an ordering on $(\mathfrak{h}^+)'$, and we can therefore speak of the *highest weight* of τ, which is the largest weight in this ordering.

Some known properties of the highest weight Λ of τ are as follows:

(1) Λ is integral, i.e., $\xi_\Lambda(h) = \exp \Lambda(\log h)$ is a well-defined character of $T^{\mathbf{C}} = \exp \mathfrak{h}^{\mathbf{C}}$.

(2) Λ is *dominant with respect to* $\mathfrak{k}$, i.e., $\langle \Lambda, \alpha \rangle \geqq 0$ for all positive compact roots α.

(3) The weight space for the weight Λ has complex dimension one.

(4) τ is determined up to unitary equivalence by Λ.

Conversely every integral linear form on $\mathfrak{h}^{\mathbf{C}}$, dominant with respect to $\mathfrak{k}$, is the highest weight of some irreducible unitary representation of K.

Let Λ be an integral form on $\mathfrak{h}^{\mathbf{C}}$ dominant with respect to $\mathfrak{k}$, and let ξ_Λ be the associated character of $T^{\mathbf{C}}$. We have $B = N_c^- T^{\mathbf{C}}$, and we extend ξ_Λ to a holomorphic character of B by setting $\xi_\Lambda(n) = 1$ for $n \in N_c^-$. Define

$$\Gamma(\Lambda) = \left\{ F : BG \to \mathbf{C} \;\middle|\; \begin{array}{ll} \text{(i)} & F \text{ holomorphic on } BG \\ \text{(ii)} & F(bx) = \xi(b)F(x) \text{ for } b \in B,\ x \in BG \end{array} \right. ,$$

$$H(\Lambda) = \left\{ F \in \Gamma(\Lambda) \middle| \ \|F\|^2 = \int_G |F(g)|\, dg < \infty \right\},$$

$$U_\Lambda(g)F(x) = F(xg) \qquad \text{for} \quad F \in \Gamma(\Lambda),\ g \in G,\ x \in BG.$$

Then $U_\Lambda(g)$ preserves both $\Gamma(\Lambda)$ and $H(\Lambda)$. These definitions are completely analogous to the ones discussed in Section 1 for $G = SU(1,1)$. The main result is as follows:

Theorem (Harish-Chandra). If Λ is an integral form on $\mathfrak{h}^{\mathbf{C}}$ dominant with respect to $\mathfrak{k}$, then $H(\Lambda)$ is a Hilbert space, and U is a continuous unitary representation on it. If furthermore

$$\langle \Lambda + \rho, \alpha \rangle < 0 \qquad \left(\rho = \tfrac{1}{2} \sum_{\beta > 0} \beta\right)$$

for every noncompact positive root α, then $H(\Lambda)$ is nonzero, U is irreducible, and the matrix coefficients of U are square integrable.

Example of $SU(1,1)$. Here

$$\mathfrak{h} = \begin{pmatrix} i\theta & 0 \\ 0 & -i\theta \end{pmatrix}, \qquad \mathfrak{h}^{\mathbf{C}} = \begin{pmatrix} a & 0 \\ 0 & -a \end{pmatrix}$$

$$\Lambda\begin{pmatrix} a & 0 \\ 0 & -a \end{pmatrix} = -na, \quad \alpha\begin{pmatrix} a & 0 \\ 0 & -a \end{pmatrix} = 2a, \quad \rho = \tfrac{1}{2}\alpha, \quad \rho\begin{pmatrix} a & 0 \\ 0 & -a \end{pmatrix} = a.$$

Λ integral means n is an integer; Λ is automatically dominant with respect to $\mathfrak{k}$ since the only positive root α is noncompact. Let

$$\Lambda_1\begin{pmatrix} 1 & 0 \\ 0 & -1 \end{pmatrix} = c_1, \qquad \Lambda_2 = \begin{pmatrix} 1 & 0 \\ 0 & -1 \end{pmatrix} = c_2.$$

Then $\langle \Lambda_1, \Lambda_2 \rangle = cc_1c_2$ for a certain constant $c > 0$. Thus

$$\langle \Lambda + \rho, \alpha \rangle = c(-n + 1)2,$$

and Harish-Chandra's condition $\langle \Lambda + \rho, \alpha \rangle < 0$ is the condition $n > 1$. Therefore

$$\xi_\Lambda\begin{pmatrix} a & 0 \\ 0 & a^{-1} \end{pmatrix} = a^{-n} \qquad \text{with} \quad n \geqq 2.$$

In this example, when the representations are viewed as operating on functions on the disk, the constant function plays a special role, in that the mean value property is expressed in terms of it. That is, we have

$$\frac{1}{2\pi}\int_0^{2\pi} f(e^{2i\theta}z)\,d\theta = f(0),$$

which we rewrite as

$$\frac{1}{2\pi}\int_0^{2\pi} e^{-in\theta}\, U_n\begin{pmatrix} e^{-i\theta} & 0 \\ 0 & e^{i\theta}\end{pmatrix} f(z)\,d\theta = f(0),$$

and we can regard the right side as equal to $f(0)1(z)$, where $1(z)$ is the constant function. Passing to Harish-Chandra's realization by putting $F(\zeta, \gamma, z) = \gamma^{-n} f(z)$, we obtain

$$\frac{1}{2\pi}\int_0^{2\pi} F\left(\begin{pmatrix} e^{i\theta} & 0 \\ 0 & e^{-i\theta}\end{pmatrix} x \begin{pmatrix} e^{-i\theta} & 0 \\ 0 & e^{i\theta}\end{pmatrix}\right) d\theta = F(1)\psi_n(x), \qquad x \in BG,$$

where $\psi_n(\zeta, \gamma, z) = \gamma^{-n}$ is the extension of the constant function. We begin the discussion of general G by defining the analog of ψ_n and by giving the analog of this mean value property.

Recall that $BG = P^- K^{\mathbf{C}}\Omega \subseteq P^+ K^{\mathbf{C}} P^+$. Let $\mu(x)$ be the middle (or $K^{\mathbf{C}}$) component of the member x of $P^- K^{\mathbf{C}} P^+$. Let τ be an irreducible unitary representation of K, and extend τ holomorphically to $K^{\mathbf{C}}$ (by passing from K to $\mathfrak{k}$ to $\mathfrak{k}^{\mathbf{C}}$ to $K^{\mathbf{C}}$). Let ϕ_Λ be a highest weight vector for τ_Λ and suppose $\|\phi_\Lambda\| = 1$. Define

$$\psi_\Lambda(x) = (\tau_\Lambda(\mu(x))\phi_\Lambda, \phi_\Lambda) \qquad \text{for} \quad x \in P^- K^{\mathbf{C}} P^+.$$

Lemma. The restriction to BG of ψ_Λ is in $\Gamma(\Lambda)$.

Proof. ψ_Λ is holomorphic, being the composition of the holomorphic maps $x \in P^- K^{\mathbf{C}} P^+ \to \mu(x) \in K^{\mathbf{C}}$, $k \in K^{\mathbf{C}} \to \tau_\Lambda(k) \in GL(n, \mathbf{C})$, $g \in GL(n, \mathbf{C}) \to (g\phi_1, \phi_2)$. To see that it transforms appropriately under B on the left, let $b \in B$ and $x \in BG$. Write $b = p^- k^{\mathbf{C}}$ with $p^- \in P^-$ and $k^{\mathbf{C}} \in K^{\mathbf{C}} \cap B$; then $\mu(bx) = \mu(k^{\mathbf{C}}x) = k^{\mathbf{C}}\mu(x)$ since $K^{\mathbf{C}}$ normalizes P^-. Thus

$$\psi_\Lambda(bx) = (\tau_\Lambda(\mu(bx))\phi_\Lambda, \phi_\Lambda) = (\tau_\Lambda(k^{\mathbf{C}})\tau_\Lambda(\mu(x))\phi_\Lambda, \phi_\Lambda).$$

Let $\phi_1 = \phi_\Lambda, \phi_2, \ldots, \phi_d$ be an orthonormal system of weight vectors arranged in order so that the corresponding weights decrease, and write

$$\tau_\Lambda(\mu(x)) = \sum c_j \phi_j$$

and $k^{\mathbf{C}} = hn$ with $h \in T^{\mathbf{C}} = \exp \mathfrak{h}^{\mathbf{C}}$ and $n \in N_c^- \cap K^{\mathbf{C}}$. Then the matrix of $\tau_\Lambda(n)$ can be seen to be lower triangular in the basis $\{\phi_j\}$, with ones on the diagonal, and so

$$\begin{aligned}\tau_\Lambda(k^{\mathbf{C}})\tau_\Lambda(\mu(x))\phi_\Lambda = \tau_\Lambda(h)\tau_\Lambda(n)\sum c_j\phi_j &= \tau_\Lambda(h)[c_1\phi_\Lambda + \sum_{j \geqq 2} d_j\phi_j] \\ &= c_1\xi_\Lambda(h)\phi_\Lambda + \sum_{j \geqq 2} d_j\xi_{\Lambda_j}(h)\phi_j.\end{aligned}$$

Thus

$$\psi_\Lambda(bx) = c_1\xi_\Lambda(h) = \xi_\Lambda(h)\,(\tau_\Lambda(\mu(x))\phi_\Lambda, \phi_\Lambda) = \xi_\Lambda(h)\psi_\Lambda(x) = \xi_\Lambda(b)\psi_\Lambda(x)$$

since $\xi_\Lambda(p^-) = \xi_\Lambda(n) = 1$.

Main Lemma. For every F in $\Gamma(\Lambda)$,

$$\int_{T^-} F(hxh^{-1})\, dh = F(1)\psi_\Lambda(x).$$

The main lemma gives the generalization of the mean value property. We shall give the idea of the proof in 3.3.

Note that the members of $\Gamma(\Lambda)$ are determined by their values on G. Since the functions in $\Gamma(\Lambda)$ are continuous, $\|F\| = 0$ implies $F = 0$.

Proposition. $|F(1)| \leqq \|\psi_\Lambda\|^{-1}\, \|F\|$ for all $F \in H(\Lambda)$.

Proof. We can assume $F(1) \neq 0$. Applying the main lemma to F and recalling that $T^- \subseteq B$, we have

$$\int_{T^-} F(xh)\overline{\xi_\Lambda(h)}\, dh = \int_{T^-} F(h^{-1}xh)\, dh = F(1)\psi_\Lambda(x).$$

Thus

$$\|F\|^2 = \int_G |F(x)|^2 dx = \int_G |F(xh)\overline{\xi_\Lambda(h)}|^2 dx \qquad \text{for all} \quad h \in T^-$$

$$= \int_G \int_{T^-} |F(xh)\overline{\xi_\Lambda(h)}|^2 \, dh \, dx$$

$$\geq \int_G \left| \int_{T^-} F(xh)\overline{\xi_\Lambda(h)} \, dh \right|^2 dx \qquad \text{by the Schwarz inequality}$$

$$= \int_G |F(1)|^2 \, |\psi_\Lambda(x)|^2 \, dx$$

$$= |F(1)|^2 \, \|\psi_\Lambda\|^2.$$

Corollary. $H(\Lambda) \neq 0$ if and only if $\|\psi_\Lambda\| < \infty$.

Proof. If $\|\psi_\Lambda\| < \infty$, then $\psi_\Lambda \in H(\Lambda)$. If $\|\psi_\Lambda\| = \infty$, then the proposition shows that $U_\Lambda(g)F(1) = 0$ for all $g \in G$ and $F \in H(\Lambda)$. Hence $F \equiv 0$.

Corollary. To each compact set $E \subseteq BG$ corresponds a constant $C_E < \infty$ such that

$$|F(x)| \leqq C_E \|F\|$$

for all $F \in H(\Lambda)$ and $x \in E$.

This is a simple consequence of the proposition since $\|U_\Lambda(g)F\| = \|F\|$ and since $|F(bx)| = |\xi_\Lambda(b)| \, |F(x)|$.

Corollary. $H(\Lambda)$ is complete.

Proof. Norm convergence in $H(\Lambda)$ implies uniform convergence on compact sets of BG, by the previous corollary. This shows that $H(\Lambda)$ is identified with a closed subspace of $L^2(G)$. Hence $H(\Lambda)$ is complete.

Clearly U_Λ is a unitary representation. It is continuous because $(H(\Lambda), U_\Lambda)$ is isomorphic with a subspace of $L^2(G)$, and right translation is continuous in $L^2(G)$.

3.2 Properties of the Representations

Proposition. If $H(\Lambda) \neq 0$, then U_Λ is irreducible.

Proof. Let $H \subseteq H(\Lambda)$ be a nonzero closed invariant subspace, and let

$F \neq 0$ be in H. After applying some $U_\Lambda(g)$, we can assume that $F(1) \neq 0$. The limit of a linear combination of translates of F given by

$$\int \overline{\xi_\Lambda(h)} U_\Lambda(h) F \, dh$$

is again in H. But the main lemma shows that this is $F(1)\psi_\Lambda$. Thus ψ_Λ is in H. If $H^\perp \neq 0$ also, then ψ_Λ is in $H^\perp$, which is a contradiction. Hence $H = 0$ or $H^\perp = 0$.

In the previous section we showed that $H(\Lambda) \neq 0$ if and only if $\|\psi_\Lambda\| < \infty$. We now turn to the evaluation of $\|\psi_\Lambda\|$. In $\mathfrak{a}_0$, it is possible to choose a closed cone $\mathfrak{a}_0^+$ that is a fundamental domain for the action of $\mathrm{Ad}(K)$ in the following sense: Any point of $\mathfrak{a}_0$ is conjugate by some member of K to a member of $\mathfrak{a}_0^+$, and no two points of the interior of $\mathfrak{a}_0^+$ are conjugate to each other. (In $SU(1,1)$, $\mathfrak{a}_0 = \left\{\begin{pmatrix} 0 & t \\ t & 0 \end{pmatrix}, t \in \mathbf{R}\right\}$, and $\mathfrak{a}_0^+ = \left\{\begin{pmatrix} 0 & t \\ t & 0 \end{pmatrix}, t \geq 0\right\}$. Let $A_0^+ = \exp \mathfrak{a}_0^+$, da = Haar measure on A_0. Then $G = KA_0^+K$. By [*3*, pp. 381–382], there is a continuous function $D(a) \geqq 0$ on A_0^+ such that

$$dx = D(a)\, dk_1\, da\, dk_2 \qquad \text{under} \quad x = k_1\, a\, k_2.$$

Lemma. $\|\psi_\Lambda\|^2 = d_\Lambda^{-2} \int_{A_0^+} \mathrm{Tr}(\tau_\Lambda(\mu(a))^2 D(a)\, da$ where d_Λ is the degree of τ_Λ.

Proof. Again choose an orthonormal basis of weight vectors $\phi_1 = \phi_\Lambda$, $\phi_2, \ldots, \phi_d$. If v is in the representation space of τ_Λ, then

$$\int_K |(\tau_\Lambda(k)v, \phi)|^2\, dk = d_\Lambda^{-1} \|v\|^2. \qquad (*)$$

[This is a consequence of the Schur orthogonality relations

$$\int_K (\tau_\Lambda(k)\phi_i, \phi_j)\, \overline{(\tau_\Lambda(k)\phi_{i'}, \phi_{j'})}\, dk = \delta_{ii'}\, \delta_{jj'}\, d_\Lambda^{-1}. \,]$$

Since $\mu(k'\, a\, k) = k'\mu\,(a)k$, we have

$$\|\psi_\Lambda\|^2 = \int_G |\psi_\Lambda(x)|^2\, dx = \int_{K \times A_0^+ \times K} |(\tau_\Lambda(k'\mu(a)k)\phi_\Lambda, \phi_\Lambda)|^2 D(a)\, dk'\, da\, dk$$

$$= d_\Lambda^{-1} \int_{A_0^+ \times K} \|\tau_\Lambda(\mu(a)\tau_\Lambda(k)\phi)\|^2 D(a)\, dk\, da \quad \text{by } (*).$$

Write $\tau_\Lambda(k)\phi_i = \sum_j \tau_{ij}(k)\phi_j$ and put $i = 1$. Then

$$\|\psi_\Lambda\|^2 = d_\Lambda^{-1}\int_{A_0^+ \times K} \sum_{j,l}(\tau_\Lambda(\mu(a))\phi_j, \tau_\Lambda(\mu(a))\phi_l)\tau_{j1}(k)\overline{\tau_{l1}(k)}D(a)\, dk\, da$$

$$= d_\Lambda^{-2}\int_{A_0^+} \sum_j \|\tau_\Lambda(\mu(a))\phi_j\|^2\, D(a)\, da.$$

In the corollary to Lemma 7 of Section 2.5, we saw that $\mu(a) \in T^+$. Therefore $\tau_\Lambda(\mu(a))$ is diagonal and self-adjoint. The lemma follows.

The evaluation of the integral in the statement of the lemma is complicated and will be omitted. In Section 2.3, we shall look at the integral more closely in two special cases as illustrations. We shall need the condition $\langle \Lambda + \rho, \alpha \rangle < 0$ for α positive and noncompact, in order to get a finite answer.

Proposition. If $\|\psi_\Lambda\| < \infty$, then $(U_\Lambda(x)\psi_\Lambda, \psi_\Lambda) = \psi_\Lambda(x)\|\psi_\Lambda\|^2$.

Remark. Thus when $\|\psi_\Lambda\| < \infty$, U_Λ has a square integrable matrix coefficient and is therefore square integrable.

Proof. Apply the main lemma to $U_\Lambda(y)\psi_\Lambda$ and get

$$\int_{T^-} \psi_\Lambda(hxh^{-1}y)\, dh = \psi_\Lambda(y)\psi_\Lambda(x).$$

Then

$$\psi_\Lambda(y)\|\psi_\Lambda\|^2 = \int_G \psi_\Lambda(y)\psi_\Lambda(x)\overline{\psi_\Lambda(x)}dx = \int_G\int_{T^-} \overline{\psi_\Lambda(x)}\psi_\Lambda(hxh^{-1}y)\, dh\, dx$$

$$= \int_{T^-}\int_G \overline{\psi_\Lambda(x)}\psi_\Lambda(hxh^{-1}y)\, dx\, dh = \int_{T^-}\int_G \overline{\psi_\Lambda(x)}\psi_\Lambda(xy)\, dx\, dh$$

$$= (U_\Lambda(y)\psi_\Lambda, \psi_\Lambda),$$

the next-to-last equality holding after the change of variables $x \to h^{-1}xh$ since $\psi_\Lambda(h^{-1}xh) = \psi_\Lambda(x)$.

3.3 Comments about the Main Lemma and $\|\psi_\Lambda\|$

Main Lemma. For every F in $\Gamma(\Lambda)$,

$$\int_{T^-} F(hxh^{-1})\, dh = F(1)\psi_\Lambda(x).$$

Idea of Proof. The idea is to show near $x = 1$ that the left side is $F(1)$ times a power series in x that is independent of F. The power series is evaluated as the series for $\psi_\Lambda(x)$ by putting $F = \psi_\Lambda$. Then since both sides are holomorphic on BG and equal in a neighborhood of 1, they are equal everywhere. Thus for $X \in \mathfrak{g}^{\mathbf{C}}$ and $g \in BG$, let

$$\tilde{X}f(g) = d/dt\, f(g \exp tX)\big|_{t=0}.$$

Since F is real analytic on BG, we can expand F in a power series about $1 \in G^{\mathbf{C}}$ (cf. [*3*, p. 95]):

$$F(\exp X) = \sum_{n\geq 0} \frac{1}{n!}(\tilde{X}^n F)(1).$$

Conjugate by $h \in T^-$ and integrate, letting L–S denote the left hand side:

$$L-S = \sum_{n\geq 0} \frac{1}{n!}(\{\int_{T^-} \mathrm{Ad}(h)\, \tilde{X}^n\, dh\}F)(1).$$

Choose a basis $\{X_1, \ldots, X_s\}$ of $\mathfrak{g}^{\mathbf{C}}$ over $\mathbf{C}$, write $X = z_1 X_1 + \cdots + z_s X_s$, and expand the integral into a sum of integrals of monomials. In this expansion, the various z_j's factor out because $\tilde{X}$ is complex-linear when applied to a holomorphic function. The integral of each monomial is T^--invariant. If $\{X_1, \ldots, X_s\}$ is chosen with the first vectors in $\mathfrak{n}_c^+$, the next ones in $\mathfrak{h}^{\mathbf{C}}$, and the last ones in $\mathfrak{n}_c^-$, then the integral of a monomial is 0 unless the monomial itself was T^--invariant. In the monomial, the $\mathfrak{n}_c^-$ vectors can be put at the right end, and any monomial that does have vectors in $\mathfrak{n}_c^-$ present contributes 0 to L-S since F is constant on N_c^-. It follows that after the integration, the only nonzero terms come from monomials with all vectors in $\mathfrak{h}^{\mathbf{C}}$. Such monomials have F as eigenfunction since $T^{\mathbf{C}} \subseteq B$, and consequently the value of these terms is $F(1)$ times an expression independent of F. The lemma follows.

In computing $\|\psi_\Lambda\|^2$, we are to look at the integral

$$\int_{A_0^+} \mathrm{Tr}\,(\tau_\Lambda(\mu(a))^2)D(a)\, da.$$

First consider $SU(1,1)$. Here we know from Lemma 7 in 2.5 that $a = \begin{pmatrix} \cosh t & \sinh t \\ \sinh t & \cosh t \end{pmatrix}$ and $\mu(a) = \begin{pmatrix} \cosh t & 0 \\ 0 & (\cosh t)^{-1} \end{pmatrix}$. From 3.1, we have $\tau_\Lambda\begin{pmatrix} \alpha & 0 \\ 0 & \alpha^{-1} \end{pmatrix} = \alpha^{-n}$, and one can show that $D(a) = \sinh 2t$, up to a constant factor. The integral is

$$\int_0^\infty (\cosh t)^{-2n} \sinh 2t \, dt = 2\int \frac{\sinh t \, dt}{(\cosh t)^{2n-1}} = \frac{1}{n-1} \quad \text{if} \quad n > 1.$$

The condition $n > 1$ agrees with Harish-Chandra's condition

$$\langle \Lambda + \rho, \alpha \rangle < 0.$$

Next consider $Sp(2, \mathbf{R})$. As in 2.1, we have

$$\mathfrak{sp}(2,\mathbf{R}) = \overset{\;\;2\quad\;2}{\begin{pmatrix} a & b \\ \bar{b} & \bar{a} \end{pmatrix}}\begin{matrix} 2 \\ 2 \end{matrix}, \qquad a = -a^*, \quad b = b^t$$

and

$$\mathfrak{k} = \begin{pmatrix} a & 0 \\ 0 & \bar{a} \end{pmatrix}, \quad a = -a^*; \qquad \mathfrak{k}^{\mathbf{C}} = \begin{pmatrix} a & 0 \\ 0 & -a^t \end{pmatrix}, \quad a \in \mathfrak{gl}(2,\mathbf{C}).$$

Choose

$$\mathfrak{h} = \begin{pmatrix} i\theta_1 & & & \\ & i\theta_2 & & \\ & & -i\theta_1 & \\ & & & -i\theta_2 \end{pmatrix}, \qquad \mathfrak{h}^{\mathbf{C}} = \begin{pmatrix} c_1 & & & \\ & c_2 & & \\ & & -c_1 & \\ & & & -c_2 \end{pmatrix}.$$

If e_i denotes evaluation of the ith diagonal entry of a member of $\mathfrak{h}^{\mathbf{C}}$, then the roots are

$$e_i - e_j,\ i \neq j, \qquad \text{compact,}$$
$$\pm(e_i + e_j),\ i < j, \qquad \text{noncompact.}$$

The positive roots can be taken as

$$e_1 - e_2; \qquad 2e_1,\ e_1 + e_2,\ 2e_2.$$

Choose $\mathfrak{g}_0$ to be the real matrices

$$\begin{pmatrix} 0 & \begin{pmatrix} t_1 & \\ & t_2 \end{pmatrix} \\ \begin{pmatrix} t_1 & \\ & t_2 \end{pmatrix} & 0 \end{pmatrix} = t_1 H_1 + t_2 H_2.$$

By the Corollary to Lemma 7 in 2.5,

$$\mu(\exp(t_1 H_1 + t_2 H_2)) = \begin{pmatrix} \cosh t_1 & & & \\ & \cosh t_2 & & \\ & & (\cosh t_1)^{-1} & \\ & & & (\cosh t_2)^{-1} \end{pmatrix}.$$

Let $\Lambda = -me_1 - ne_2$. Here $\rho = 2e_1 + e_2$ and so

$$\Lambda + \rho = -(m-2)e_1 - (n-1)e_2.$$

Λ integral says m and n are integers. Λ dominant with respect to $\mathfrak{k}$ says $\langle \Lambda, e_1 - e_2 \rangle \geq 0$ or $n \geq m$. $\langle \Lambda + \rho, \alpha \rangle < 0$ for α positive noncompact says

$$\langle \Lambda + \rho, 2e_1 \rangle < 0 \qquad \text{and} \qquad \langle \Lambda + \rho, 2e_2 \rangle < 0$$

or

$$m > 2 \qquad \text{and} \qquad n > 1.$$

Since $n \geq m$, Harish-Chandra's condition is $m > 2$. K is the 2-by-2 unitary group, and one computes easily that if $a = \exp(t_1 H_1 + t_2 H_2)$, then $\mathrm{Tr}(\tau_\Lambda(\mu(a))^2)$ equals

$$(\cosh t_1 \cosh t_2)^{-m-n} \left[\left(\frac{\cosh t_1}{\cosh t_2} \right)^{-m+n} + \cdots + \left(\frac{\cosh t_1}{\cosh t_2} \right)^{-(-m+n)} \right].$$

with the exponents decreasing one at a time. Also

$$D(a) = \sinh 2t_1 \sinh 2t_2 \sinh(t_1 + t_2) \sinh(t_1 - t_2),$$

and the integral in question is

$$\int_{t_1=0}^{\infty} \int_{t_2=0}^{t_1} \mathrm{Tr}(\tau_\Lambda(\mu(a))^2)\, D(a)\, dt_2\, dt_1.$$

When $n \geq m$, it is easy to see that this is finite if and only if $m > 2$, but it is not so easy to evaluate the integral. In [*2*, pp. 598–612], Harish-Chandra does the evaluation in the general case we have been considering.

4. Limits of Holomorphic Discrete Series

4.1 Case of $SU(1,1)$

Recall from Section 1 the two formulations of holomorphic discrete series for $G = SU(1,1)$. Let $n \geqq 2$ be an integer. In Bargmann's formulation

$$\mathscr{H}_n = \left\{ f \,\middle|\, \begin{array}{ll} \text{(i)} & f \quad \text{analytic in} \quad \Omega = \{|z| < 1\} \\ \text{(ii)} & \|f\|_n^2 = \int_\Omega |f(z)|^2 (1 - |z|^2)^{n-2}\, dx\, dy < \infty \end{array} \right\},$$

$$\mathscr{U}_n(g) f(z) = (\bar\beta z + \alpha)^{-n} f\left(\frac{\bar\alpha z + \beta}{\bar\beta z + \alpha}\right), \qquad g = \begin{pmatrix} \alpha & \beta \\ \bar\beta & \bar\alpha \end{pmatrix}.$$

In Harish-Chandra's formulation, put $\xi_n \begin{pmatrix} a & 0 \\ c & a^{-1} \end{pmatrix} = a^{-n}$. Then

$$H_n = \left\{ F \,\middle|\, \begin{array}{ll} \text{(i)} & F \quad \text{holomorphic on} \quad BG = P^- K^{\mathbf{C}} \Omega \\ \text{(ii)} & F(bx) = \xi_n(b) F(x),\ b \in B,\ x \in BG \\ \text{(iii)} & \|F\|^2 = \int_G |F(g)|^2\, dg < \infty \end{array} \right\}$$

$$U_n(g)\, F(x) = F(xg).$$

The unitary equivalance is given by

$$f(z) = F(0,1,z), \quad F(\zeta,\gamma,z) = \gamma^{-n} f(z).$$

We shall now consider the limiting case $n = 1$. Without some adjustment, the norm for $n = 1$ will lead to a null Hilbert space. But with the norm

$$\|f\|_1^2 = \lim_{n \downarrow 1} (n - 1) \int_\Omega |f(z)|^2 (1 - |z|^2)^{n-2}\, dx\, dy,$$

one finds that

$$\|f\|_1^2 = \lim_{r \uparrow 1} \tfrac{1}{2} \int_0^{2\pi} |f(re^{i\theta})|^2\, d\theta,$$

and the space $\mathscr{H}_1$ becomes the Hardy space $H^2(\Omega)$ defined by

$$H^2(\Omega) = \{ f \mid \text{analytic in } \Omega,\ \sup_{r<1} \int_0^{2\pi} |f(re^{i\theta})|^2\, d\theta < \infty \}.$$

This space is not 0, since all polynomials are in it. To each function f in $\mathscr{H}_1$, one can associate a boundary function $f(e^{i\theta}) \in L^2(\text{circle})$ and the formula for the norm becomes

$$\|f\|_1^2 = \tfrac{1}{2}\int_0^{2\pi} |f(re^{i\theta})|^2 \, d\theta.$$

This norm on $\mathscr{H}_1$, together with the group action $\mathscr{U}_1(g)$, gives an irreducible unitary representation of G.

There is another representation of interest here, with Hilbert space all of $L^2(\text{circle})$ and with action given by

$$\mathscr{V}(g)f(e^{i\theta}) = (\bar{\beta}e^{i\theta} + \alpha)^{-1} f\left(\frac{\bar{\alpha}e^{i\theta} + \beta}{\bar{\beta}e^{i\theta} + \alpha}\right).$$

This representation is of interest because it is unitarily equivalent to a member of what Bargmann called the principal continuous series. The principal series of $SU(1,1)$ consists of representations of $SU(1,1)$ induced from the one-dimensional unitary representations of the closed subgroup

$$MA^+N = \pm\begin{pmatrix} \cosh t - ixe^t & \sinh t - ixe^t \\ \sinh t + ixe^t & \cosh t + ixe^t \end{pmatrix}.$$

(This group is conjugate within $SL(2, \mathbf{C})$ to the group of real upper triangular matrices. A precise definition of induced representation will not be needed here; it suffices to know at present that there is a construction that associates a unitary representation of G in a natural way to a unitary representation of a closed subgroup.) The one-dimensional representations in question are of two kinds:

$$\pm(\text{as above}) \to e^{i\lambda t}, \qquad \lambda \in \mathbf{R} \text{ fixed},$$

$$\pm(\text{as above}) \to \pm e^{i\lambda t}, \qquad \lambda \in \mathbf{R} \text{ fixed}.$$

$\mathscr{V}(g)$ is equivalent with the induced representation of the second kind (in which the sign $\pm$ is retained) with $\lambda = 0$. The imbedding $H^2 \subseteq L^2$ commutes with the action of $SU(1,1)$ and exhibits $\mathscr{V}(g)$ as reducible. That this is an exceptional situation is indicated by the fact that all other members of the principal series of $SU(1,1)$ are irreducible.

We shall consider the representation $(\mathscr{H}_1, \mathscr{U}_1)$ in the context of the Harish-Chandra decomposition. (The imbedding $H^2 \subseteq L^2$ can also be treated in this context.) We take F holomorphic on BG, transforming under B according to ξ_1. The norm of $F(\zeta, \gamma, z)$ is to be roughly the norm of

$F(0, 1, e^{i\theta})$, and so we suppose that F is holomorphic in a neighborhood of the closure of BG. Put

$$u_1 = \begin{pmatrix} \sqrt{2}/2 & \sqrt{2}/2 \\ -\sqrt{2}/2 & \sqrt{2}/2 \end{pmatrix}, \qquad K = \left\{ k = \begin{pmatrix} e^{-i\theta} & 0 \\ 0 & e^{i\theta} \end{pmatrix} \right\}, \qquad f(e^{i\theta}) = F(0,1,e^{i\theta}).$$

Then

$$u_1 k = \begin{pmatrix} \sqrt{2}/2 & \sqrt{2}/2 \\ -\sqrt{2}/2 & \sqrt{2}/2 \end{pmatrix} \begin{pmatrix} e^{-i\theta} & 0 \\ 0 & e^{i\theta} \end{pmatrix} = \begin{pmatrix} \frac{\sqrt{2}}{2} e^{-i\theta} & \frac{\sqrt{2}}{2} e^{i\theta} \\ -\frac{\sqrt{2}}{2} e^{-i\theta} & \frac{\sqrt{2}}{2} e^{i\theta} \end{pmatrix}$$

$$= \begin{pmatrix} 1 & 0 \\ -1 & 1 \end{pmatrix} \begin{pmatrix} \frac{\sqrt{2}}{2} e^{-i\theta} & 0 \\ 0 & (-)^{-1} \end{pmatrix} \begin{pmatrix} 1 & e^{2i\theta} \\ 0 & 1 \end{pmatrix}.$$

Thus $u_1 k \in P^- K^{\mathbf{C}} \Omega = \overline{BG}$, and we have

$$F(u_1 k) = \left(\frac{\sqrt{2}}{2} e^{-i\theta} \right)^{-1} F \begin{pmatrix} 1 & e^{2i\theta} \\ 0 & 1 \end{pmatrix} = \sqrt{2} e^{i\theta} f(e^{2i\theta}).$$

Since

$$\int_K |F(u_1 k)|^2 \, dk = \frac{1}{\pi} \int_0^{2\pi} |f(e^{2i\theta})|^2 \, d\theta = \frac{1}{\pi} \int_0^{2\pi} |f(e^{i\theta})|^2 \, d\theta,$$

we can define

$$\|F\|^2 = \int_K |F(u_1 k)|^2 \, dk$$

and take the completion of the space of functions of finite norm as H_1. $U_1(g)$ is a right translation by g, and $(\mathscr{H}_1, \mathscr{U}_1)$ is unitarily equivalent with (H_1, U_1). Our intention in the next section is to generalize this construction and the imbedding $H^2 \subseteq L^2$.

4.2 General Case

We continue to use the notation of Section 2.5. In particular, G is a simple Lie group with $G \subseteq G^{\mathbf{C}}$, and center $\mathfrak{k}$ is not 0.

Let Λ be an integral form on $\mathfrak{h}^{\mathbf{C}}$ dominant with respect to $\mathfrak{k}$. Harish-Chandra's condition for nonvanishing $H(\Lambda)$ was $\langle \Lambda + \rho, \alpha \rangle < 0$ for all positive noncompact α, and we shall consider the limiting case of this condition in this section. One can show that the following three conditions are equivalent:

(i) $\langle \Lambda + \rho, \alpha \rangle \leq 0$ for all positive noncompact α with equality for some α,
(ii) $\langle \Lambda + \rho, \alpha_0 \rangle = 0$ for the largest root α_0, and $\langle \Lambda + \rho, \alpha \rangle < 0$ for all other positive noncompact α, and
(iii) $\langle \Lambda + \rho, \alpha_0 \rangle = 0$ for the largest root α_0.

Suppose that Λ satisfies these equivalent conditions.

We shall define shortly an element $u_1 \in G^{\mathbf{C}}$ and a closed subgroup $M_s \subseteq G$. Let

$$\Gamma_2(\Lambda) = \left\{ F \,\middle|\, \begin{array}{ll} \text{(i)} & F \text{ holomorphic in a neighborhood of } BG \\ \text{(ii)} & F(bx) = \xi_\Lambda(b)F(x) \text{ for } b \in B,\ x \in BG \\ \text{(iii)} & \|F\|^2 = \int_{M_s \times K} |F(u_1 mk)^2\, dm\, dk \end{array} \right\}$$

$$U_\Lambda(g)F(x) = F(xg) \qquad \text{for} \qquad F \in \Gamma_2(\Lambda), \quad g \in G, \quad x \in BG.$$

Here $\|F\|$ will make sense because we shall have $Bu_1G \subseteq \overline{BG}$. Also $\|F\| = 0$ will imply $F = 0$. Thus we define

$$H(\Lambda) = \text{completion of } \Gamma_2(\Lambda).$$

Theorem 1. If Λ is an integral form on $\mathfrak{h}^{\mathbf{C}}$, dominant with respect to $\mathfrak{k}$, and if $\langle \Lambda + \rho, \alpha_0 \rangle = 0$ for the largest root α_0, then U_Λ extends to a continuous unitary representation on $H(\Lambda)$, $H(\Lambda)$ is not 0, and U_Λ is irreducible.

We shall define u_1 and M_s while we establish notation for Theorem 2, which concerns the reducibility of certain induced representations. Though we shall give most of the definitions for general G, we shall concentrate on the examples $SU(1,1)$ and $Sp(2, \mathbf{R})$. Recall the earlier discussion of $Sp(2, \mathbf{R})$ in Section 3.3. Our definitions are set forth in the following table.

Item	*General*	$SU(1,1)$	$Sp(2,\mathbf{R})$
$\mathfrak{h}^{\mathbf{C}}$	(*see* 2.4)	$\begin{pmatrix} c & 0 \\ 0 & -c \end{pmatrix}$	$\begin{pmatrix} c_1 & & & \\ & c_2 & & \\ & & -c_1 & \\ & & & -c_2 \end{pmatrix}$

Compact roots > 0	(see 2.4)	none	$e_1 - e_2$
Noncompact roots > 0	(see 2.4)	$2e_1$	$2e_1, e_1 + e_2, 2e_2$
ρ	$\frac{1}{2} \sum_{\beta > 0} \beta$	e_1	$2e_1 + e_2$
α_0	largest root	$2e_1$	$2e_1$
Λ	—	$-ne_1$	$-me_1 - ne_2,\ n \geq m$
$\langle \Lambda + \rho, \alpha_0 \rangle = 0$	—	$n = 1$	$m = 2,\ n \geq m$
u_1	$\exp \frac{\pi}{4}(X_{\alpha_0} - X_{-\alpha_0})$ (see after 2.5, Lemma 6)	$\begin{pmatrix} \sqrt{2}/2 & \sqrt{2}/2 \\ -\sqrt{2}/2 & \sqrt{2}/2 \end{pmatrix}$	$\begin{pmatrix} \begin{pmatrix} \sqrt{2}/2 & 0 \\ 0 & 1 \end{pmatrix} & \begin{pmatrix} \sqrt{2}/2 & 0 \\ 0 & 0 \end{pmatrix} \\ \begin{pmatrix} -\sqrt{2}/2 & 0 \\ 0 & 0 \end{pmatrix} & \begin{pmatrix} \sqrt{2}/2 & 0 \\ 0 & 1 \end{pmatrix} \end{pmatrix}$
$\mathfrak{a}^+$	$\mathbf{R}(X_{\alpha_0} + X_{-\alpha_0})$	$\begin{pmatrix} 0 & t \\ t & 0 \end{pmatrix}$	$\begin{pmatrix} 0 & \begin{pmatrix} t & 0 \\ 0 & 0 \end{pmatrix} \\ \begin{pmatrix} t & 0 \\ 0 & 0 \end{pmatrix} & 0 \end{pmatrix}$
$\mathfrak{m}$	$(\mathfrak{a}^+)^\perp \cap Z_{\mathfrak{g}}(\mathfrak{a}^+)$	0	$\begin{pmatrix} 0 & 0 & 0 & 0 \\ 0 & a & 0 & b \\ 0 & 0 & 0 & 0 \\ 0 & \bar{b} & 0 & -a \end{pmatrix}$, a imaginary
M_0	corresponding group	1	$\begin{pmatrix} 1 & 0 & 0 & 0 \\ 0 & \alpha & 0 & \beta \\ 0 & 0 & 1 & 0 \\ 0 & \bar{\beta} & 0 & \bar{\alpha} \end{pmatrix}$, det $= 1$
M_s	commutator subgroup of M_0	1	$M_s = M_0 \cong SU(1,1)$
M	$Z_K(\mathfrak{a}^+) \cdot M_0$	$\begin{pmatrix} \pm 1 & 0 \\ 0 & \pm 1 \end{pmatrix}$	$\begin{pmatrix} \pm 1 & 0 & 0 & 0 \\ 0 & \alpha & 0 & \beta \\ 0 & 0 & \pm 1 & 0 \\ 0 & \bar{\beta} & 0 & \bar{\alpha} \end{pmatrix}$
A^+	$\exp \mathfrak{a}^+$	$\begin{pmatrix} \cosh t & \sinh t \\ \sinh t & \cosh t \end{pmatrix}$	$\begin{pmatrix} \cosh t & 0 & \sinh t & 0 \\ 0 & 1 & 0 & 0 \\ \sinh t & 0 & \cosh t & 0 \\ 0 & 0 & 0 & 1 \end{pmatrix}$

N	something	$\begin{pmatrix} 1-ix & -ix \\ ix & 1+ix \end{pmatrix}$	a 3-dimensional nilpotent group
$\mathfrak{a}^-$	$Z_{\mathfrak{h}}(\mathfrak{a}^+)$	0	$\begin{pmatrix} 0 & & & \\ & c & & \\ & & 0 & \\ & & & -c \end{pmatrix}$, c imaginary

Consider $Sp(2, \mathbf{R})$. Then $\mathfrak{a}^-$ is a compact Cartan subalgebra in $\mathfrak{m}$, and $\Lambda|_{\mathfrak{a}^-} = -ne_2$, $n \geq 2$. Consequently $\Lambda|_{\mathfrak{a}^-}$ gives rise to a holomorphic discrete series representation ω_Λ of $SU(1,1)$. M is generated by $\mathfrak{m}_0$ and -1. Extend ω_Λ to M by putting $\omega_\Lambda(-1) = \xi_\Lambda(-1)I = (-1)^n I$. Then ω_Λ is a square-integrable representation of M.

In the general case, it is possible to proceed similarly. M_s is semisimple and turns out to have holomorphic discrete series. From one of these, we can construct a special square-integrable representation ω_Λ of M.

Now let

$$V_\Lambda = \underset{MA^+N\uparrow G}{\text{induced rep.}} (\omega_\Lambda \otimes 1 \otimes 1).$$

Harish-Chandra has proved that most representations of the form

$$\underset{MA^+N\uparrow G}{\text{induced rep.}} (\omega \otimes 1 \otimes 1),\ \omega \text{ square-integrable, } \chi \text{ unitary character of } A^+,$$

are irreducible. However we have

Theorem 2. Under the above assumptions, U_Λ is unitarily equivalent with a proper subrepresentation of V_Λ, and consequently V_Λ is reducible.

5. Notes

1. The representations $(\mathscr{H}_n, \mathscr{U}_n)$ were constructed by Bargmann [*1*], and the six listed properties were proved by him.

2.1. The bounded symmetric domains were classified by É. Cartan. See [*3*] for a bibliography of the relevant papers, and see [*3*, pp. 339–355], for the classification of noncompact symmetric spaces. Bounded symmetric domains are discussed in detail in the book by L. K. Hua, *Harmonic Analysis of Functions of Several Complex Variables in the Classical Domains.*

2.2. This summary is taken from [*3*, pp. 281–285]. See those pages for more details.

2.3. This material is in [*3*] in a geometric form. For the lemma, see [*3*, pp. 183–186, 194–197]. For the theorem and the proposition, see [*3*, pp. 305, 311].

2.4. Most of the results of this section are contained in [*3*, pp. 312–314]. See also [*2*, pp. 750, 756–761]; our ordering on the roots is such that the totally positive roots in Harish-Chandra's sense are exactly the noncompact positive roots.

2.5. Harish-Chandra's realization of Ω is given in [*2*, pp. 2–5, 589–591] and is reproduced in [*3*, pp. 314–321].

3.1. The idea of defining representation in spaces of holomorphic functions on BG has its origin in the Borel-Weil Theorem, which gives such a realization for all irreducible representations of a compact group; in this case $BG = G^{\mathbf{C}}$, and there are no serious technical problems. Harish-Chandra defined corresponding representations in [*4*, pp. 6, 15], and he developed the basic properties in [*2*, pp. 6–19]. The main lemma appears in Lemmas 6 and 14.

3.2–3. Irreducibility is proved in [*2*, pp. 19–20], the computation of $\|\psi_\Lambda\|^2$ appears in [*2*, pp. 598–609], and the square-integrability is given in [*2*, pp. 30, 612].

4. Bargmann [*1*] observed the identity between the limiting norm of the discrete series for $SU(1,1)$ and the norm of the Hardy class in the disk. The results of this chapter appear in [*4*].

References

[*1*] V. Bargmann, "Irreducible unitary representations of the Lorentz group I," *Ann. of Math.*, **48** (1947), 568–640.

[*2*] Harish-Chandra, "Representations of semisimple Lie groups IV, V, VI," *Amer. J. Math.*, **77** (1955), 743–777; **78** (1956), 1–41, 564–628.

[*3*] S. Helgason, *Differential Geometry and Symmetric Spaces*, Academic Press, New York, 1962.

[*4*] A. W. Knapp and K. Okamoto, "Limits of holomorphic discrete series," *J. of Functional Analysis.*

SCHWARZ LEMMA*

SHOSHICHI KOBAYASHI

Department of Mathematics
University of California
Berkeley, California

*Partially supported by NSF Grant GP-8008.

The lemma of Schwarz is one of the most powerful tools in investigating the behavior of complex analytic functions. It plays an essential role in the proofs of such important theorems as the Riemann mapping theorem, Landau's theorem, Schottky's theorem, and the two theorems of Picard. It is therefore clear that if we are to generalize geometric function theory to functions of several complex variables, we should begin with a Schwarz lemma for several complex variables. The purpose of this article is to give various generalizations of the Schwarz lemma.†

We recall that the classical lemma of Schwarz reads as follows. Let D be the open unit disk in the complex plane $\mathbf{C}$, i.e.,

$$D = \{z \in \mathbf{C} : |z| < 1\}.$$

If $f : D \to D$ is a holomorphic mapping sending the origin into itself, then

$$|f(z)| \leqq |z| \qquad \text{for} \quad z \in D$$

and

$$|f'(0)| \leqq 1,$$

and the equality $|f'(0)| = 1$ or the equality $|f(z)| = z$ at a single point $z \neq 0$ implies $f(z) = \varepsilon z$, where $|\varepsilon| = 1$.

A simple calculation shows that this is equivalent to the following statement. If f is a holomorphic mapping of the disk D into itself, then

$$\frac{|df|}{1 - |f|^2} \leqq \frac{|dz|}{1 - |z|^2} \qquad \text{for} \quad z \in D,$$

and the equality at a single point z of D implies that f is an automorphism of D.

This second formulation of the Schwarz lemma has a nice differential geometric meaning. Consider the Hermitian metric

$$ds_D^2 = \frac{dz\, d\bar{z}}{(1 - |z|^2)^2}$$

on the disk D. A simple calculation shows that its Gaussian curvature is equal to -4. The metric ds_D^2, known as the *Poincaré metric* or the *Poincaré-Bergman metric* of D, defines the so-called non-Euclidean distance in D.

†For more details and an extensive list of references, see Volume 2 of this series, recent monograph by the author: Hyperbolic manifolds and holomorphic mappings.

Then the Schwarz lemma says that every holomorphic mapping $f: D \to D$ is distance-decreasing with respect to ds_D^2, i.e., $f^*(ds_D^2) \leqq ds_D^2$ and that the equality at one point implies that f is an automorphism. If we denote by ρ the non-Euclidean distance on D defined by ds_D^2, then the Schwarz lemma may be rephrased as follows: if f is a holomorphic mapping of D into itself, then $\rho(p, q) \geqq \rho(f(p), f(q))$ for $p, q \in D$, and the equality for a pair of distinct points p, q implies that f is an automorphism. This geometric formulation was obtained by Pick in 1915 [*15*].

It is, however, the paper of Ahlfors [*1*] in 1938 which has really revealed the differential geometric character of the Schwarz lemma. To state Ahlfors' result we denote by D_a the open disk of radius a in **C**, i.e.,

$$D_a = \{z \in \mathbf{C} : |z| < a\}.$$

The metric

$$ds_a^2 = \frac{4a^2\, dz\, d\bar{z}}{A(a^2 - |z|^2)^2} \qquad (A > 0)$$

on D_a has Gaussian curvature $-A$. With these notations, we state the

Theorem of Ahlfors. Let M be a 1-dimensional Kaehler manifold with metric ds_M^2 whose Gaussian curvature is bounded above by a negative constant $-B$. Then every holomorphic map $D_a \to M$ satisfies

$$f^* ds_M^2 \leqq (A/B)\, ds_a^2.$$

The proof can be outlined as follows. Let $0 < r < a$ and denote by $f_r : D_r \to M$ the restriction of f to D_r. We define a nonnegative function u on D_a and a nonnegative function u_r on D_r by

$$f^* ds_M^2 = u \cdot ds_a^2, \qquad f_r^* ds_M^2 = u_r \cdot ds_r^2.$$

Since $ds_r^2 \to ds_a^2$ as $r \to a$, it follows that $\lim_{r \to a} u_r = u$ (pointwise). Hence, it suffices to prove the inequality $u_r \leqq A/B$. Since the coefficient of ds_r^2 goes to infinity at the boundary of D_r and since, being the restriction of $f^* ds_M^2$ to D_r, $f_r^* ds_M^2$ can be continuously extended to the boundary of D_r, the function u_r approaches zero at the boundary of D_r. Hence, u_r attains its maximum in the interior of D_r. At a maximum point, $\partial^2 \log u_r / \partial z\, \partial \bar{z}$ must be nonpositive. By a simple calculation, this implies that $u_r \leqq A/B$.

Actually, Ahlfors proves the result above for a metric ds_M^2 which is not completely smooth. As an application, he obtains, among other things, a sharp estimate of the Bloch constant.

We mention two examples of Riemann surfaces M with metric ds_M^2 satisfying the condition in Ahlfors' theorem: (1) compact Riemann surfaces with genus > 1; (2) the complex plane $\mathbf{C}$ minus two points. We all know that these Riemann surfaces can be uniformized by a unit disk and hence admit metrics of constant negative curvature, but this fact is not easy to prove. On the other hand, as Grauert and Reckziegel [5] have shown, metrics of nonconstant but negatively bounded curvature can be constructed explicitly in an elementary manner. Moreover, their metric is complete even for $\mathbf{C}$ minus two points.

The next natural step is to generalize the theorem of Ahlfors to Hermitian manifolds M of higher dimension. Grauert and Reckziegel [5] considered a certain class of complex manifolds (i.e., manifolds with differential metric of negative curvature) for which Ahlfors theorem can be proved. But a simple differential geometric condition one can impose on M is that M admit a Hermitian metric with negatively bounded holomorphic sectional curvature. More precisely [*10*],

Theorem 1. Let D_a be the open unit disk of radius a with the metric ds_a^2 and let M be a Hermitian manifold whose holomorphic sectional curvature is bounded above by a negative constant $-B$. Then every holomorphic mapping $f: D_a \to M$ satisfies

$$f^*(ds_M^2) \leqq (A/B)\cdot ds_a^2.$$

The main idea of the proof is that the immersed complex submanifold $f(D_a)$ of M has curvature $\leqq -B$ so the theorem of Ahlfors applies to the mapping $f: D_a \to f(D_a)$. Of course, f fails to be an immersion at those points of D_a where f is degenerate, but $f^*(ds_M^2)$ vanishes wherever f is degenerate. Hence, it suffices to consider the points where f is an immersion.

So far, we have been considering holomorphic mappings from a disk. It is a simple matter to extend Theorem 1 to a holomorphic mapping f from a polydisk $D_a^k = D_a \times \cdots \times D_a$ (k times) into M. We note that the holomorphic sectional curvature of D_a^k with respect to the product metric varies between $-A$ and $-A/k$. For a symmetric bounded domain D of rank k, a polydisk of dimension k plays the same role as a maximal torus for a compact Lie group. Hence, Theorem 1 can be extended as follows.

Theorem 2. Let D be a bounded symmetric domain with a canonical invariant metric ds_D^2 whose sectional curvature is $\geqq -A$. Let M be a Hermitian manifold with metric ds_M^2 whose holomorphic sectional curvature is $\leqq -B$. Then every holomorphic mapping $f: D \to M$ satisfies

$$f^*(ds_M^2) \leqq (A/B)ds_D^2.$$

In applications of Theorem 2, the following two extreme cases are important: (1) D is a polydisk; (2) D is a symmetric bounded domain of rank 1, i.e., an open unit ball in $\mathbf{C}^k$. In these two cases, Theorem 2 can be readily derived from Theorem 1 without much fanfare on symmetric spaces.

By specializing Theorem 2 to the case where M is also a symmetric bounded domain, we recover the result of Koranyi [*13*].

For a Riemann surface, a Hermitian metric may be identified with its volume element. Thus, if we regard ds_M^2 and ds_a^2 in Ahlfors theorem as the volume elements for M and D_a, we arrive at the following generalization.

Theorem 3. Let D be a bounded homogeneous domain in $\mathbf{C}^n$ with its Bergman metric ds_D^2 normalized so that $-ds_D^2$ is the Ricci tensor. Let M be an n-dimensional Hermitian manifold with metric ds_M^2 whose Ricci tensor is $\leqq -ds_M^2$. Denote by v_D and v_M the volume elements of D and M. Then every holomorphic mapping $f: D \to M$ satisfies

$$f^*v_M \leqq v_D.$$

Theorems of this type with varying assumptions have been proved by several authors [*3, 4, 6, 7, 9*]. The proof is essentially the same as that of Ahlfors. Setting $f^*v_M = u \cdot v_D$, we look at the complex Hessian of u.

If we apply Theorem 3 to the case where D and M are symmetric bounded domains, we obtain distortion theorems. As a sample, we exhibit the following. Let $D = M$ be the space of symmetric matrices Z of order n such that $I_n - ZZ^* < 0$. Let $W = f(Z)$ be a holomorphic mapping of D into itself and $J(Z)$ be its Jacobian. Then

$$|J(Z)|^2 \leqq \{\det(I_n - WW^*)/\det(I_n - ZZ^*)\}^{n+1} \leqq \{\det(I_n - ZZ^*)\}^{-n-1}.$$

Before we turn in a different direction, we should mention that a theorem of Schwarz-lemma type exists for K-quasiconformal mappings in the 2-dimensional case [*8*], but its higher dimensional analog is still missing. One does not know if the Schwarz lemma could be proved for harmonic mappings.

What we have done so far was to impose suitable geometric conditions on D and M so that every holomorphic mapping $f : D \to M$ is distance-decreasing or volume-decreasing. In his paper [2] published in 1926, Carathéodory approached the Schwarz lemma from the opposite point of view. His idea was to define a distance on every bounded domain in an intrinsic manner so that every holomorphic mapping $f : D \to M$ is automatically distance-decreasing.

Let M be a bounded domain in $\mathbf{C}^n$. Let D denote the open unit disk in $\mathbf{C}$ with non-Euclidean distance ρ. Let F be the family of holomorphic mappings f from M into D. Then the Carathéodory distance c_M on M is defined by

$$c_M(p,q) = \sup_{f \in F} \rho(f(p), f(q)) \qquad \text{for} \quad p,q \in M.$$

The definition above makes sense for all complex manifolds or even complex spaces. Of course, in general, c_M is only a pseudodistance. If M is compact, the family F consists of constant mappings only so that $c_M(p, q) = 0$ identically. The classical Liouville theorem is equivalent to the statement $c_{\mathbf{C}}(p, q) = 0$ for $p, q \in \mathbf{C}$.

In [11], another intrinsic pseudodistance, denoted by d_M for a complex space M, is defined. Given two points p and q in M, we choose a sequence of points $p = p_0, p_1, \ldots, p_{k-1}, p_k = q$ in M and points a_i, b_i $(i = 1, 2, \ldots, k)$ in the disk D together with holomorphic mappings $f_i : D \to M$ such that

$$f_i(a_i) = p_{i-1} \qquad \text{and} \qquad f_i(b_i) = p_i \qquad \text{for} \quad i = 1, 2, \ldots, k.$$

Geometrically speaking, $f_1, \ldots, f_k$ form a chain of analytic disks. We set

$$d_M(p,q) = \inf(\textstyle\sum_{i=1}^k \rho(a_i,b_i)),$$

where the infimum is taken with respect to all choices of p_i, a_i, b_i, f_i.

The pseudodistances c_M and d_M share the following two basic properties:

(1) $c_D = d_D = \rho$;

(2) If f is a holomorphic mapping of one complex space M into another complex space N, then

$$c_M(p,q) \geqq c_N(f(p), f(q)), \qquad d_M(p,q) \geqq d_N(f(p), f(q)) \qquad \text{for} \quad p, q \in M.$$

The property (1) is essentially a restatement of the Schwarz–Pick lemma. If we happen to have any other system of pseudodistances δ_M with properties (1) and (2), then

$$c_M(p,q) \leqq \delta_M(p,q) \leqq d_M(p,q) \qquad \text{for} \quad p,q \in M.$$

As we see from the construction of d_M, the pseudodistance d_M is similar to the distance function defined from a Riemannian metric. This seems to be the underlying reason why a number of results in the geometric theory of functions can be rather nicely explained by means of d_M. If we say that a complex space M is (complete) hyperbolic when d_M is a (complete) distance, then Theorem 1 is essentially equivalent to the statement that a (complete) Hermitian manifold M with holomorphic sectional curvature $\leqq -B$ is (complete) hyperbolic. Since a distance-decreasing map f from a space M with trivial d_M (e.g., $M = \mathbf{C}$) into a hyperbolic space N (e.g., $N = \mathbf{C}$ minus two points) must be a constant map, every holomorphic mapping from **C** into **C** which misses two values must be a constant map, and this is the little Picard theorem. Even the big Picard theorem can be generalized by means of this distance d_M [*14*]. More recently, Royden proved that for a Teichmüller space M the distance d_M coincides with the Teichmüller distance [*16*].

Similarly, two kinds of intrinsic measures can be constructed on each complex space and Theorem 3 can be given an interpretation similar to the one given to Theorem 1.

References

[*1*] L. V. Ahlfors, "An extension of Schwarz's lemma," *Trans. Amer. Math. Soc.*, **43** (1938), 359–364.

[*2*] C. Carathéodory, "Über das Schwarzsche Lemma bei analytischen Funktionen von zwei komplexen Veränderlichen," *Math. Ann.*, **97** (1926), 76–98.

[*3*] S. S. Chern, "On holomorphic mappings of hermitian manifolds of the same dimension," *Proc. Symp. Pure Math., Entire Functions and Related Parts Analysis*, Vol. 11, 1968, pp. 157–170.

[*4*] A. Dinghas, "Ein *n*-dimensionales Analogon des Schwarz–Pickschen Flächensatzes für holomorphe Abbildungen der komplexen Einheitskugel in eine Kähler-Mannigfaltigkeit," *Arbeitsgemeinschaft für Forschung des Landes Norrhein-Westfalen*, Bd. **33** (1965), pp. 477–494.

[*5*] H. Grauert and H. Reckziegel, "Hermitesche Metriken und normale Familien holomorpher Abbildungen," *Math. Z.*, **89** (1965), 108–125.

[*6*] K. T. Hahn and J. Mitchell, "Generalization of Schwarz–Pick lemma to invariant volume in Kähler manifolds," *Trans. Amer. Math. Soc.*, **128** (1967), 221–231.

[*7*] ———, "Generalization of Schwarz–Pick lemma to invariant volume in Kähler manifolds, II," *Canad. J. Math.*, **21** (1969), 669–674.

[*8*] P. Kiernan, "Quasiconformal mappings and Schwarz's lemma," *Trans. Amer. Math. Soc.*, **147** (1970).

[9] S. Kobayashi, "Volume elements, holomorphic mappings and Schwarz lemma," *Proc. Symp. Pure Math., Entire Functions and Related Parts Analysis*," Vol. 11, 1968, pp. 253–260.

[10] ———, "Distance, holomorphic mappings and Schwarz lemma," *J. Math. Soc. Japan*, **19** (1967), 481–485.

[11] ———, "Intrinsic metrics on complex manifolds," *Bull. Amer. Math. Soc.*, **73** (1967), 347–349.

[12] ———, "Invariant distances on complex manifolds and holomorphic mappings," *J. Math. Soc. Japan*, **19** (1967), 460–480.

[13] A. Koranyi, "A Schwarz lemma for bounded symmetric domains," *Proc. Amer. Math. Soc.*, **17** (1966), 210–213.

[14] M. H. Kwack, "Generalization of the big Picard theorem," *Ann. of Math.*, **90** (1969), 9–22.

[15] G. Pick, "Über eine Eigenschaft der konformen Abbildung kreisformiger Bereiche," *Math. Ann.*, (2), **77** (1916).

[16] H. L. Royden, "Report on the Teichmyller metric," *Proz. Nat Acad. Su. USA 65* (1970), 497–499.

Note added in proof: S. Bergman has also obtained a generalized Schwarz lemma in his paper "Zur Theorie von pseudokonformen Abbildungen," *Recueil math. nouv. ser.*, **1** (43) (1936), 79–96.

ON TUBE DOMAINS

YOZO MATSUSHIMA

Department of Mathematics
University of Notre Dame
Notre Dame, Indiana

This paper discusses some results related to holomorphic transformations and the equivalences of tube domains. The expository work here is based on materials from a recent paper [*2*] written with W. Kaup and T. Ochiai. Also included is a section about a recent interesting result of Vey [*5*] which is closely related to the results discussed in the paper. Vey's result is: If a tube domain admits a discontinuous group of holomorphic transformations with compact quotient, then the tube domain is necessarily symmetric. We should remark that in [*2*] and [*5*] the case of a more general class of domains (generalized Siegel domains) is treated, whereas here the case is limited mainly to tube domains for simplicity and because of the importance of tube domains among generalized Siegel domains.*

1. Some Basic Properties of the Group of Holomorphic Transformations of a Bounded Domain and Tube Domains

(A). Let D be a bounded domain in $\mathbf{C}^n$ and $\mathrm{Aut}(D)$ the group of all holomorphic transformations of D onto itself. It is a well-known theorem of H. Cartan that $\mathrm{Aut}(D)$ is a Lie group (with respect to the compact open topology) and that the isotropy subgroup of $\mathrm{Aut}(D)$ at each point of D is compact. We can prove this theorem admitting the fact that the group of all isometries of a Riemannian manifold is a Lie group (with respect to the compact open topology), because $\mathrm{Aut}(D)$ is a closed subgroup of the group of isometries of D with Bergmann metric.

(B). Let X be a (real) vector field on D which is *complete* in the sense that X is the infinitesimal transformation of a one parameter group φ_t ($t \in \mathbf{R}$) of transformations of D. We can write the vector field X in the form

$$X = p^k(\partial/\partial z^k) + \bar{p}^k(\partial/\partial \bar{z}^k) \tag{1.1}$$

where $z^1, \ldots, z^n$ denote the coordinates in $\mathbf{C}^n$, p^k ($k = 1, \cdots, n$) are complex valued functions on D, and $\bar{p}^k$ denotes the conjugate complex of p^k; the summation sign is omitted in (1.1) and will be throughout this paper. Also as a convention, ∂_k and $\bar{\partial}_k$ will be written for $\partial/\partial z^k$ and $\partial/\partial \bar{z}^k$, that is,

$$\partial_k = \partial/\partial z^k, \qquad \bar{\partial}_k = \partial/\partial \bar{z}^k \qquad (k = 1, \cdots, n). \tag{1.2}$$

*I would like to express my gratitude to Professor W. Boothby and Professor J. Hano for their invitation and continued interests.

A one parameter group of transformations φ_t belongs to Aut(D) if and only if the components p^k of X are holomorphic. We call a (real) vector field X holomorphic if the components p^k are holomorphic. We shall denote by $\mathfrak{g}(D)$ the set of all complete holomorphic vector fields on D. Then the $\mathfrak{g}(D)$ form a Lie algebra over $\mathbf{R}$ and can be identified canonically with the Lie algebra of Aut(D).

For each vector field X on D of the form (1.1), JX denotes the vector field with components ip^k for each k, where $i^2 = -1$. It is shown that if X and JX both belong to $\mathfrak{g}(D)$, then $X = 0$. In fact, by the definition of JX, it is easily seen that $[X, JX] = 0$. Let φ_t and ψ_t ($t \in \mathbf{R}$) be the one parameter groups of holomorphic transformations of D corresponding to X and JX, respectively. Then $\varphi_t \cdot \psi_s = \psi_s \cdot \varphi_t$ for any t and s and for each $z = t + is \in \mathbf{C}$, a holomorphic transformation ξ_z by $\xi_z = \varphi_t \cdot \psi_s = \psi_s \cdot \varphi_t$ is defined. Then, for each $u \in D$, the kth coordinate $z^k(\xi_z(u))$ of the point $\xi_z(u) \in D$ is a holomorphic function of the variable $z \in \mathbf{C}$. Since D is bounded, these functions are bounded and hence are constant by the Liouville theorem. Since $\xi_0(u) = u$, then $\xi_z(u) = u$ for any z and in particular, $\xi_t(u) = \varphi_t(u) = u$ for any $t \in \mathbf{R}$. Then $X(u) = 0$ and hence $X = 0$, because u is an arbitrary point of D.

For each real vector field X of the form (1.1) there is a complex vector field $p^k\partial_k$, the correspondence $X \leftrightarrow p^k\partial_k$ is one-to-one, and moreover, if X is holomorphic, this correspondence preserves the bracket. Therefore, throughout this paper, each element $X \in \mathfrak{g}(D)$ is written in the form

$$X = p^k\partial_k.$$

Then the result proved above can be restated as follows: *If X and iX both belong to $\mathfrak{g}(D)$, then $X = 0$.*

(C). A subset Ω in $\mathbf{R}^n$ is said to be a *convex nondegenerate cone* if Ω satisfies the following conditions:

(1) Ω is open and convex and, for each $x \in \Omega$ and $\lambda \in \mathbf{R}^+$, λx belongs also to Ω.

(2) Ω does not contain any full straight line.

Throughout this paper, Ω always denotes a convex nondegenerate cone in $\mathbf{R}^n$. A *tube domain* $T(\Omega)$ over Ω is the open subset of $\mathbf{C}^n$ consisting of all $z = x + iy \in \mathbf{C}^n$ ($x, y \in \mathbf{R}^n$) such that $y \in \Omega$.

A tube domain $T(\Omega)$ is holomorphically equivalent to a bounded domain in $\mathbf{C}^n$ (cf. [4]). In fact, a coordinate system $(x^1, \cdots, x^n)$ of $\mathbf{R}^n$ can be chosen such that the cone Ω is contained in the sector $x^1 > 0$, $x^2 > 0$, $\cdots$, $x^n > 0$. Then $T(\Omega)$ is contained in the product H^n of n copies of the upper halfplane $H = \{z \in \mathbf{C} | \mathrm{Im}\, z > 0\}$ and H^n is equivalent to the product of D^n of the unit disk D. Hence $T(\Omega)$ is equivalent to a domain in D^n which is bounded.

It follows from this in particular, that $\mathrm{Aut}(T(\Omega))$ is a Lie group and the Lie algebra $\mathfrak{g}(T(\Omega))$ of all complete holomorphic vector fields on $T(\Omega)$ is finite dimensional over $\mathbf{R}$ and isomorphic to the Lie algebra of $\mathrm{Aut}(T(\Omega))$.

By the definition of a tube domain $T(\Omega)$, the following one parameter groups of transformations are contained in $\mathrm{Aut}(T(\Omega))$:

(1) $z \to z + ta$, for each $a \in \mathbf{R}^n$, $t \in \mathbf{R}$,

(2) $z \to e^t x$, $t \in \mathbf{R}$.

The infinitesimal transformations of these one parameter groups are $a^k \partial_k$ $(a = (a^k),\ a^k \in \mathbf{R})$, and $z^k \partial_k$, respectively. Therefore *the holomorphic vector fields* ∂_k $(k = 1, \cdots, n)$ *and* $\partial = z^k \partial_k$ *are contained in the Lie algebra* $\mathfrak{g}(T(\Omega))$.

Moreover, for any holomorphic vector field $X = p^j \partial_j$,

$$[\partial_k, X] = \partial_k p^j \cdot \partial_j, \qquad [\partial, X] = (\partial p^j - p^j)\partial_j. \tag{1.3}$$

2. Structure of $\mathfrak{g}(T(\Omega))$

A holomorphic vector field X is said to be a *polynomial* vector field if the components p^k of X are polynomials of $z^1, \cdots, z^n$ and X is *homogeneous* of degree λ, if each p^k is a homogeneous polynomial of degree λ.

Lemma 1. Let D be a domain in $\mathbf{C}^n$ containing the origin O of $\mathbf{C}^n$ and let $\mathfrak{g}$ be a finite dimensional Lie algebra over $\mathbf{C}$ consisting of holomorphic vector fields on D. Assume that $\mathfrak{g}$ contains the vector field $\partial = z^k \partial_k$. Then every vector field in $\mathfrak{g}$ is polynomial.

Proof. Let $X \in \mathfrak{g}$ and $X = p^k \partial_k$. Let $p^k = \sum_{\lambda=0}^{\infty} p^k_\lambda$ be the Taylor expansion of p^k in a neighborhood of the origin O, where p^k_λ denotes a homogeneous polynomial of degree λ. Put $X_\lambda = p^k_\lambda \partial_k$. Then $[\partial, X_\lambda] = (\lambda - 1)X_\lambda$ by (1.3) and $X = \sum_\lambda X_\lambda$; thus $\mathrm{ad}(\partial)X = [\partial, X] = \sum_\lambda (\lambda - 1)X_\lambda$. Let $\varphi(x)$ be the minimal polynomial of the endomorphism $\mathrm{ad}(\partial)$ of $\mathfrak{g}$. Then we have $\varphi(\mathrm{ad}\ \partial)X = \sum_\lambda \varphi(\lambda - 1)X_\lambda = 0$ and hence $\varphi(\lambda - 1)X_\lambda = 0$ for all λ. This implies that $X_\lambda = 0$ except for a finite number of indices λ and hence X is polynomial.

Theorem 1. Let $D = T(\Omega)$ be a tube domain over Ω and let $\mathfrak{g}(D)$ be the Lie algebra of all complete holomorphic vector fields on D. Then every $X \in \mathfrak{g}(D)$ is a polynomial vector field and the coefficients of each component of X are real. Moreover, each homogeneous component of X is also contained in $\mathfrak{g}(D)$. Let $\mathfrak{g}_\lambda$ denote the subspace of $\mathfrak{g}$ consisting of all homogeneous vector fields of degree $\lambda + 1$ ($\lambda = -1, 0, 1, \cdots$). Then

(1) $\mathfrak{g}(D) = \sum_\lambda \mathfrak{g}_\lambda$, $[\mathfrak{g}_\lambda, \mathfrak{g}_\mu] = \mathfrak{g}_{\lambda+\mu}$, that is, $\mathfrak{g}$ is a graded Lie algebra;

(2) an element $X \in \mathfrak{g}(D)$ belongs to $\mathfrak{g}_\lambda$ if and only if $(\mathrm{ad}\, \partial)X = \lambda X$.

*Proof.** Introduce a new system of coordinates $(z'^1, \cdots, z'^n)$ by $z'^k = z^k - ia^k$ ($k = 1, 2, \cdots, n$), where $a = (a^k)$ is a point in Ω. Write $\partial'_k = \partial/\partial z'^k$ and $\partial' = z'^k \partial'_k$. Then $\partial_k = \partial'_k$ and $\partial = z^k \partial_k = \partial' - ia^k \partial'_k$. Since ∂ and $a^k \partial'_k = a^k \partial_k$ are contained in $\mathfrak{g}(D)$, $\partial' = \partial + ia^k \partial'_k$ is contained in $\mathfrak{g}(D)^{\mathbf{C}}$. Moreover, with respect to the coordinates z', D contains the origin O. It follows then from Lemma 1 that every element in $\mathfrak{g}(D)^{\mathbf{C}}$ is a polynomial of z' and so is one of z. In particular, every X in $\mathfrak{g}(D)$ is a polynomial vector field. Let $X \in \mathfrak{g}(D)$ and let $X = \sum_\lambda X_\lambda$ ($\lambda = -1, 0, 1, \cdots$), where X_λ is homogeneous of degree $\lambda + 1$. Then $\mathrm{ad}(\partial) \cdot X = \sum_\lambda \lambda X_\lambda$ by (1.3). Let $X_\lambda \neq 0$. There exists a polynomial $f(x)$ with real coefficients such that $f(\mu) = 0$ for $\mu = -1, 0, \cdots, m$ except $\mu = \lambda$. Then $f(\mathrm{ad}\, \partial)X = f(\lambda)X_\lambda$ and $f(\mathrm{ad}\, \partial)X \in \mathfrak{g}(D)$. Hence, $X_\lambda \in \mathfrak{g}(D)$. This proves that $\mathfrak{g}(D) = \sum_\lambda \mathfrak{g}_\lambda$ and also that an element X belongs to $\mathfrak{g}_\lambda$ if and only if $(\mathrm{ad}\, \partial)X = \lambda X$. Let $X \in \mathfrak{g}_\lambda$ and $Y \in \mathfrak{g}_\mu$. Then $(\mathrm{ad}\, \partial)[X, Y] = [(\mathrm{ad}\, \partial)X, Y] + [X, (\mathrm{ad}\, \partial)Y] = (\lambda + \mu)[X, Y]$ and hence $[X, Y] \in \mathfrak{g}_{\lambda+\mu}$. It remains to show that the coefficients of each component p^k of an element $X \in \mathfrak{g}_\lambda$ are real. We write p^k in the form

$$p^k = \sum p^k_{\lambda_1 \ldots \lambda_n} (z^1)^{\lambda_1}, \cdots, (z^n)^{\lambda_n}.$$

Then by (1.3),

$$(\mathrm{ad}\, \partial_1)^{\lambda_1} \cdots (\mathrm{ad}\, \partial_n)^{\lambda_n} X = p^k_{\lambda_1 \ldots \lambda_n}.$$

Thus $Z = p^k_{\lambda_1 \ldots \lambda_n} \partial_k$ belongs to $\mathfrak{g}(D)$. Moreover, since ∂_k belongs to $\mathfrak{g}(D)$ for every $k = 1, \cdots, n$, $\mathrm{Re}(Z) = \mathrm{Re}(p^k_{\lambda_1 \ldots \lambda_n})\partial_k$ and $\mathrm{Im}(Z) = \mathrm{Im}(p^k_{\lambda_1 \ldots \lambda_n})\partial_k$ also belong to $\mathfrak{g}(D)$. Then $Z - \mathrm{Re}(Z) = i\, \mathrm{Im}(Z)$ is also in $\mathfrak{g}(D)$. From a result proved in Section 1(B), it follows that $\mathrm{Im}(Z) = 0$ and hence, $p^k_{\lambda_1 \ldots \lambda_n}$ is a real number for each k and $\lambda_1, \cdots, \lambda_n$.

It follows from Theorem 1 that $\mathfrak{g}_{-1}$ consists of all vector fields of the form $A = a^k \partial_k$, where $a = (a^k) \in \mathbf{R}^n$, and in particular $\dim_{\mathbf{R}} \mathfrak{g}_{-1} = n$.

*The present proof of Theorem 1 is due to J. Hano. Our original proof was more complicated.

To each $A = a^k\partial_k \in \mathfrak{g}_{-1}$ associate a linear map $\varphi_A : \mathfrak{g}_1 \to \mathfrak{g}_{-1}$ by

$$\varphi_A(X) = \tfrac{1}{2}(\operatorname{ad} A)^2 X, \; X \in \mathfrak{g}_1. \tag{2.1}$$

For each $a \in \Omega$, denote by p_a the point ia of the tube domain $D = T(\Omega)$. By a simple computation it is seen that

$$X(p_a) = -\varphi_A(X)(p_a), \qquad X \in \mathfrak{g}_1, \tag{2.2}$$

where $X(p_a)$ and $\varphi_A(X)(p_a)$ denote the value of the respective vector field at the point p_a and $A = a^k\partial_k$ with $a \in \Omega$.

We shall denote by $\mathfrak{g}_{pa}$ the isotropy subalgebra of $\mathfrak{g}(D)$ consisting of all $Y \in \mathfrak{g}(D)$ such that $Y(p_a) = 0$. Then we have $\varphi_A(X) + X \in \mathfrak{g}_{pa}$ for each $X \in \mathfrak{g}_1$. Since the isotropy subgroup of $\mathrm{Aut}(D)$ at p_a is compact and since $\mathfrak{g}_{pa}$ can be identified with the subalgebra of $\mathfrak{g}(D)$ corresponding to the isotropy subgroup at p_a, it is seen, in particular, that ad Y is a semisimple endomorphism of $\mathfrak{g}(D)$ with purely imaginary eigenvalues.

We now study the structure of the radical of the Lie algebra $\mathfrak{g}(D)$, where $D = T(\Omega)$.

Lemma 2. Any ideal $\mathfrak{i}$ of $\mathfrak{g}(D)$ is a graded ideal, that is,

$$\mathfrak{i} = \sum_\lambda \mathfrak{i}_\lambda, \qquad \mathfrak{i}_\lambda = \mathfrak{i} \cap \mathfrak{g}_\lambda.$$

In particular, the radical $\mathfrak{r}$ of $\mathfrak{g}(D)$ is a graded ideal and we have $\mathfrak{r}_\lambda = \mathfrak{g}_\lambda$ for $\lambda \geqq 2$.

Proof. Let $X \in \mathfrak{i}$ and $X = \sum X_\lambda$, $X_\lambda \in \mathfrak{g}_\lambda$. Then $(\operatorname{ad} \partial)X = \sum \lambda X_\lambda$ belongs also to $\mathfrak{i}$ and by the argument already used in the course of the proof of Theorem 1, it is seen that each X_λ belongs also to $\mathfrak{i}$ and hence $\mathfrak{i}$ is graded. Now let $X \in \mathfrak{g}_\lambda$ with $\lambda \geqq 2$. Then, for any $Y \in \mathfrak{g}_\mu$ $(\mu = -1, 0, \cdots)$, $(\operatorname{ad} X \operatorname{ad} Y)\mathfrak{g}_\nu \subset \mathfrak{g}_{\lambda+\mu+\nu}$ and $\lambda + \mu + \nu > \nu$. It follows from this that ad X ad Y is a nilpotent endomorphism of $\mathfrak{g}(D)$ and hence $B(X, Y) = \mathrm{Tr}(\operatorname{ad} X \operatorname{ad} Y) = 0$, where B denotes the Killing form of $\mathfrak{g}(D)$. We then conclude that $B(X, Y) = 0$ for any $Y \in \mathfrak{g}$ and $X \in \mathfrak{g}_\lambda$, $\lambda \geqq 2$. Each element which is orthogonal to every element of $\mathfrak{g}(D)$ is contained in the radical $\mathfrak{r}$ (see [*1*]), hence $\mathfrak{g}_\lambda = \mathfrak{r}_\lambda$ for $\lambda \geqq 2$.

Lemma 3. Let $\mathfrak{s}$ be a solvable subalgebra of $\mathfrak{g}(D)$. Assume that either $\mathfrak{s}$ is an ideal or $\mathfrak{s}$ contains an element A of the form $A = a^k\partial_k$ with $a =$

$(a^k) \in \Omega$. Then we have

$$\mathfrak{s} \cap \mathfrak{g}_1 = (0).$$

Proof. Let $X \in \mathfrak{s} \cap \mathfrak{g}_1$. Then by assumption, $\varphi_A(X)$ is contained in $\mathfrak{s}$. Since $\mathfrak{s}$ is solvable, one can choose a basis of $\mathfrak{g}(D)^{\mathbf{C}}$ with respect to which the matrix of ad Y is upper triangular for each $Y \in \mathfrak{s}$. Now since $X \in \mathfrak{g}_1$ and $\varphi_A(X) \in \mathfrak{g}_{-1}$, ad X and ad $\varphi_A(X)$ are both nilpotent and hence the matrices of ad X and ad $\varphi_A(X)$ are upper triangular with zero on the main diagonal. It follows from this that ad$(\varphi_A(X) + X)$ is nilpotent. On the other hand, it has already been shown that $\varphi_A(X) + X$ belongs to $\mathfrak{g}_{pa}$ and hence ad$(\varphi_A(X) + X)$ is semisimple. Then ad$(\varphi_A(X) + X) = 0$ and in particular $[\partial, \varphi_A(X) + X] = 0$. Since $\varphi_A(X) \in \mathfrak{g}_{-1}$ and $X \in \mathfrak{g}_1$, we obtain $-\varphi_A(X) + X = 0$ and hence $X = \varphi_A(X) = 0$ which proves that $\mathfrak{s} \cap \mathfrak{g}_1 = (0)$.

The following lemma is a corollary of Lemma 3.

Lemma 4. Let $\mathfrak{r}$ be the radical of $\mathfrak{g}(D)$. Then $\mathfrak{r}_1 = \mathfrak{r} \cap \mathfrak{g}_1 = (0)$.

We show now that $\mathfrak{r}_\lambda = (0)$ for all $\lambda \geqq 1$ by induction on λ. This is true for $\lambda = 1$ by Lemma 4. Assume that it has been shown that $\mathfrak{r}_\mu = (0)$ for all μ such that $1 \leqq \mu < \lambda$ and let $X \in \mathfrak{r}_\lambda$. Then $[\partial_k, X]$ is in $\mathfrak{r}_{\lambda-1}$ and hence we have $[\partial_k, X] = 0$ for all k. It follows then from (1.3) that $X = 0$ and this proves that $\mathfrak{r}_\lambda = (0)$.

On the other hand, it has been shown that $\mathfrak{g}_\lambda = \mathfrak{r}_\lambda$ for $\lambda \geqq 2$ and hence $\mathfrak{g}_\lambda = (0)$ for $\lambda \geqq 2$.

Theorem 2. Let $D = T(\Omega)$ be the tube domain over Ω and let $\mathfrak{g}(D)$ be the Lie algebra of all complete holomorphic vector fields on D. Then we have:

(1) $\mathfrak{g}(D) = \mathfrak{g}_{-1} + \mathfrak{g}_0 + \mathfrak{g}_1$, where $\mathfrak{g}_\lambda$ consists of all vector fields of $\mathfrak{g}(D)$ whose components are homogeneous polynomials of $z^1, \cdots, z^n$ of degree $\lambda + 1$ with real coefficients. Then $[\mathfrak{g}_\lambda, \mathfrak{g}_\mu] \subset \mathfrak{g}_{\lambda+\mu}$ $(\lambda, \mu = -1, 0, 1)$ and $[\partial, X] = \lambda X$ for $X \in \mathfrak{g}_\lambda$.

(2) Let $\mathfrak{r}$ be the radical of $\mathfrak{g}(D)$. Then

$$\mathfrak{r} = \mathfrak{r}_{-1} + \mathfrak{r}_0, \qquad \mathfrak{r}_\lambda = \mathfrak{r} \cap \mathfrak{g}_\lambda \qquad (\lambda = -1, 0).$$

(3) We have

$$\dim_{\mathbf{R}} \mathfrak{g}_{-1} = n, \dim_{\mathbf{R}} \mathfrak{g}_1 = n - \dim_{\mathbf{R}} \mathfrak{r}_{-1}.$$

(4) Let $\mathfrak{a} = \mathfrak{g}_{-1} + \mathfrak{g}_0$. Then $\mathfrak{a}$ is the subalgebra of $\mathfrak{g}(D)$ corresponding to the subgroup $\mathrm{Aff}(D)$ of $\mathrm{Aut}(D)$ consisting of all affine transformations of $\mathbf{C}^n$ leaving D invariant. Moreover, the subalgebra $\mathfrak{g}_0$ is identified with the Lie algebra of the group $\mathrm{Aff}(\Omega)$ consisting of all affine transformations of $\mathbf{R}^n$ leaving Ω invariant.

Proof. Assertions (1) and (2) have already been proved. Assertion (4) follows easily from (1). We prove now assertion (3). Recall that $\dim_{\mathbf{R}}\mathfrak{g}_{-1} = n$, and to show that $\dim_{\mathbf{R}}\mathfrak{g}_1 = n - \dim_{\mathbf{R}}\mathfrak{r}_{-1}$, consider the quotient algebra $\mathfrak{g}' = \mathfrak{g}/\mathfrak{r}$. Then $\mathfrak{g}'$ is a semisimple graded Lie algebra with $\mathfrak{g}' = \sum \mathfrak{g}'_\lambda$, $\mathfrak{g}'_\lambda = \mathfrak{g}_\lambda/\mathfrak{r}_\lambda$ $(\lambda = -1, 0, 1)$. Let B' be the Killing form of $\mathfrak{g}'$ and let $X \in \mathfrak{g}'_{-\lambda}$, $Y \in \mathfrak{g}'_\mu$ $(\lambda, \mu = 0, 1)$. If $\lambda \neq \mu$, then $\operatorname{ad} X \operatorname{ad} Y$ is nilpotent and hence $B'(X, Y) = \operatorname{Tr}(\operatorname{ad} X \operatorname{ad} Y) = 0$. Therefore, if $Y \in \mathfrak{g}'_1$ and if $B'(X, Y) = 0$ for all $X \in \mathfrak{g}_{-1}$, then $B'(Z, Y) = 0$ for all $Z \in \mathfrak{g}'$ and hence $Y = 0$, because B' is nondegenerate. This shows that B' defines a nondegenerate bilinear function on $\mathfrak{g}'_{-1} \times \mathfrak{g}'_1$ and hence $\dim_{\mathbf{R}}\mathfrak{g}'_1 = \dim_{\mathbf{R}}\mathfrak{g}'_{-1} = \dim_{\mathbf{R}}\mathfrak{g}_{-1} - \dim_{\mathbf{R}}\mathfrak{r}_{-1}$. However, $\mathfrak{r}_1 = 0$ by (2) and hence $\dim_{\mathbf{R}}\mathfrak{g}'_1 = \dim_{\mathbf{R}}\mathfrak{g}_1$ and this proves assertion (3).

Remark. The above proof also shows that B' defines an invariant nondegenerate bilinear function on the subalgebra $\mathfrak{g}'_0$ of $\mathfrak{g}'$ and hence $\mathfrak{g}'_0 = \mathfrak{g}_0/\mathfrak{r}_0$ is reductive.

Theorem 3. Let a be a point of Ω and let p_a denote the point ia in the tube domain $D = T(\Omega)$. Let $\mathfrak{g}_{pa}$ be the isotropy subalgebra of $\mathfrak{g}(D)$ at p_a. Then

$$\mathfrak{g}_{pa} = \mathfrak{g}_0 \cap \mathfrak{g}_{pa} + \mathfrak{m} \qquad \text{(direct sum)}$$

where $\mathfrak{m}$ consists of all vector fields of the form $\varphi_A(X) + X$ with $X \in \mathfrak{g}_1$ and

$$\dim_{\mathbf{R}}\mathfrak{m} = \dim_{\mathbf{R}}\mathfrak{g}_1.$$

Proof. It has already been shown that $\varphi_A(X) + X \in \mathfrak{g}_{pa}$ and $\varphi_A(X) + X \in \mathfrak{g}_{-1} + \mathfrak{g}_1$. Moreover the map $X \to \varphi_A(X) + X$ is an injective linear map from $\mathfrak{g}_1$ into $\mathfrak{g}_{-1} + \mathfrak{g}_1$ and hence $\dim_{\mathbf{R}}\mathfrak{m} = \dim_{\mathbf{R}}\mathfrak{g}_1$. Now let $Y \in \mathfrak{g}_{pa}$ and let X be the $\mathfrak{g}_1$-component of Y with respect to the decomposition $\mathfrak{g}(D) = \mathfrak{g}_{-1} + \mathfrak{g}_0 + \mathfrak{g}_1$ and put $Y' = Y - (\varphi_A(X) + X)$. Then $Y' \in \mathfrak{a} \cap \mathfrak{g}_{pa}$, where $\mathfrak{a} = \mathfrak{g}_{-1} + \mathfrak{g}_0$. Then Y' is of the form $Y' = b^k\partial_k + c^k_j z^j \partial_k (b^k,$

$c_j^k \in \mathbf{R}$) and $Y'(p_a) = (b^k + ic_j^k a^j)\partial_k = 0$ and hence $b^k + ic_j^k a^j = 0$ for all k. Since b^k, c_j^k, and a^j are real, $b^k = 0$ and $c_j^k a^j = 0$. Then $Y' = c_j^k z^j \partial_k \in \mathfrak{g}_0 \cap \mathfrak{g}_{pa}$ and $Y = Y' + (\varphi_A(X) + X)$. This shows that $\mathfrak{g}_{pa} = \mathfrak{g}_0 \cap \mathfrak{g}_{pa} + \mathfrak{m}$.

3. The Case where $\mathfrak{g}(D)$ Is Unimodular

The materials of this section are taken from a private communication with J. Vey. The details of his more general results will be found in his paper [5].

Theorem 4. Let $D = T(\Omega)$ be a tube domain over Ω and assume that the Lie algebra $\mathfrak{g}(D)$ is unimodular. Then D is homogeneous, $\mathfrak{g}(D)$ is semisimple, and D is a symmetric domain.

Proof. First prove that $\mathfrak{g}(D)$ is semisimple. Since ad ∂ is the scalar operator $\lambda 1$ on each $\mathfrak{g}_\lambda$, Tr ad $\partial = -\dim_{\mathbf{R}}\mathfrak{g}_{-1} + \dim_{\mathbf{R}}\mathfrak{g}_1 = -\dim_{\mathbf{R}}\mathfrak{r}_{-1}$ by Theorem 2. On the other hand, $\mathfrak{g}(D)$ is unimodular by assumption and Tr ad $\partial = 0$. Thus, $\mathfrak{r}_{-1} = (0)$. Now $[\partial_k, \mathfrak{r}_0] \subset \mathfrak{r}_{-1}$ for $k = 1, \cdots, n$ and hence $[\partial_k, X] = 0$ for each $X \in \mathfrak{r}_0$; this implies $X = 0$ by (1.3). Thus, $\mathfrak{r}_0 = (0)$ and hence $\mathfrak{r} = (0)$ by Theorem 2 and this proves that $\mathfrak{g}(D)$ is semisimple.

Now let $a \in \Omega$ and $A = a^k\partial_k$. Then a simple computation shows that

$$[A, X](p_a) = -2iX(p_a) \tag{3.1}$$

for $X \in \mathfrak{g}_1$. Now every $X \in \mathfrak{g}_1$, $X \neq 0$, does not vanish on D. In fact, if X vanishes at a point of D, then X is contained in the isotropy subalgebra at that point and hence ad X is semisimple. On the other hand, ad X is nilpotent and this implies that ad $X = 0$ and hence $[\partial, X] = X = 0$. It follows from this and from (3.1) that $X \to [A, X](p_a)$, $X \in \mathfrak{g}_1$, is an injective linear map from $\mathfrak{g}_1$ into the tangent space of D at p_a.

We show that the tangent vectors of the form $Y(p_a)$, $Y \in \mathfrak{g}(D)$, span the tangent space of D at p_a. First of all, for each $X \in \mathfrak{g}_1$, $[A, X]$ belongs to $\mathfrak{g}_0$ and hence the one parameter group of linear transformations generated by $[A, X]$ leaves invariant the n-dimensional real submanifold $i\Omega$ of D (cf. Theorem 2). Then $[A, X](p_a)$ is tangent to $i\Omega$. On the other hand, it has already been shown that $\{[A, X](p_a) | X \in \mathfrak{g}_1\}$ is a subspace of the tangent space of D whose dimension is equal to $\dim_{\mathbf{R}}\mathfrak{g}_1$. By Theorem 2, $\dim_{\mathbf{R}}\mathfrak{g}_1 = n - \dim_{\mathbf{R}}\mathfrak{r}_{-1}$ and $\mathfrak{r}_{-1} = (0)$; and hence $\dim_{\mathbf{R}}\mathfrak{g}_1 = n$. These show that $\{[A, X](p_a) | X \in \mathfrak{g}_1\}$ is equal to the tangent space of $i\Omega$ at p_a. On the other hand, it is clear that the vectors $a^k \partial_k$ span the tangent space of the n-dimensional real submanifold $\mathbf{R}^n$ of D. Since $D = \mathbf{R}^n + i\Omega$, it follows from what has been shown that the tangent space of D at p_a is spanned by the vectors of the form $Y(p_a)$, $Y \in \mathfrak{g}(D)$. Then the orbit $\mathrm{Aut}(D)p_a$ is open in D for any $a \in \Omega$. However, for each point, $p = b + ia$, $b \in \mathbf{R}^n$, $a \in \Omega$, of D, the translation $\sigma : z \to z + b$ is contained in $\mathrm{Aut}(D)$ and hence $\mathrm{Aut}(D)p = \mathrm{Aut}(D)\sigma p_a = \mathrm{Aut}(D)p_a$ and this shows that each orbit $\mathrm{Aut}(D)p$ is open in D. It follows immediately from this that $\mathrm{Aut}(D)$ is transitive on D and D is homogeneous. On the other hand, $\mathrm{Aut}(D)$ is semisimple and hence D is symmetric by a well-known theorem of Borel and Koszul.

Theorem 5. Let D be a tube domain and assume that there exists a discontinuous group Γ of holomorphic transformations of D such that the quotient space $\Gamma\backslash D$ is compact. Then D is symmetric.

Proof. It is shown that $\mathrm{Aut}(D)$ is unimodular. Then D is asymmetric by Theorem 4. The group Γ being a discrete subgroup of $\mathrm{Aut}(D)$, by a classical argument (cf. Van Est [6]), it is enough to see that $\Gamma\backslash\mathrm{Aut}(D)$ is compact. By our assumption on Γ, there exists a compact subset K of D such that $\Gamma K = D$. Fix a point p of D and let $B = \{\sigma \in \mathrm{Aut}(D) | \sigma(p) \in K\}$. Since $\mathrm{Aut}(D)$ is a closed subgroup of the group of isometries of D with respect to the Bergmann metric, it is easily seen that B is a compact subset of $\mathrm{Aut}(D)$ and $\mathrm{Aut}(D) = \Gamma \cdot B$ and hence $\Gamma\backslash\mathrm{Aut}(D)$ is compact; this completes the proof.

4. Equivalences of Two Tube Domains

Let $D = T(\Omega)$ and $D' = T(\Omega')$ be tube domains, where Ω and Ω' are convex nondegenerate cones in $\mathbf{R}^n$. If there exists a nonsingular linear transformation τ of $\mathbf{R}^n$ such that $\tau(\Omega) = \Omega'$, then τ can be extended to a nonsingular linear transformation of $\mathbf{C}^n$ and $\tau(D) = D'$. In particular, D and D' are holomorphically equivalent. We shall show that the converse is also true by proving the following theorem.

Theorem 6. Let $D = T(\Omega)$ and $D' = T(\Omega')$ be two tube domains and assume that D and D' are holomorphically equivalent. Then there exists a nonsingular linear transformation of $\mathbf{R}^n$ which maps Ω onto Ω'.

To prove this theorem, we need a little preparation. A subalgebra $\mathfrak{t}$ of a real semisimple Lie algebra $\mathfrak{g}$ is said to be *triangular*, if there exists a basis of $\mathfrak{g}$ with respect to which ad X is represented by an upper triangular matrix for every $X \in \mathfrak{t}$. A triangular subalgebra is always solvable. The following result of Mostow [*3*] is needed.

Lemma 5. Any two maximal triangular subalgebras of a real semisimple Lie algebra are conjugate under an inner automorphism.

Now let $\mathfrak{g}$ be a real Lie algebra. A solvable subalgebra $\mathfrak{t}$ of $\mathfrak{g}$ is said to be maximal triangular, if $\mathfrak{t}$ contains the radical $\mathfrak{r}$ of $\mathfrak{g}$ and if $\mathfrak{t}/\mathfrak{r}$ is a maximal triangular subalgebra of the semisimple Lie algebra $\mathfrak{g}/\mathfrak{r}$.

Lemma 6. Any two maximal triangular subalgebras of a real Lie algebra $\mathfrak{g}$ are conjugate under an inner automorphism.

Proof. Let $\mathfrak{t}$ and $\mathfrak{t}'$ be maximal triangular subalgebras of $\mathfrak{g}$. Then $\mathfrak{t}/\mathfrak{r}$ and $\mathfrak{t}'/\mathfrak{r}$ are maximal triangular subalgebras of $\mathfrak{g}/\mathfrak{r}$. Let $\mathfrak{s}$ be a semisimple part of $\mathfrak{g}$. Then by the canonical isomorphism of $\mathfrak{s}$ onto $\mathfrak{g}/\mathfrak{r}$, there correspond to $\mathfrak{t}/r$ and $\mathfrak{t}'/r$ maximal triangular subalgebras $\mathfrak{m}$ and $\mathfrak{m}'$ of $\mathfrak{s}$, and $\mathfrak{t} = \mathfrak{m} + \mathfrak{r}$, $\mathfrak{t}' = \mathfrak{m}' + \mathfrak{r}$. By Lemma 5, there exists an inner automorphism of $\mathfrak{s}$ which maps $\mathfrak{m}$ onto $\mathfrak{m}'$. This inner automorphism of $\mathfrak{s}$ extends to an inner automorphism of $\mathfrak{g}$ which sends $\mathfrak{t}$ onto $\mathfrak{t}'$.

Lemma 7. There exists a maximal triangular subalgebra $\mathfrak{t}$ of $\mathfrak{g}(D)$ such that

(1) ∂ is contained in $\mathfrak{t}$,

(2) $\mathfrak{t} = \mathfrak{g}_{-1} + \mathfrak{t} \cap \mathfrak{g}_0$.

Proof. Let $\mathfrak{g}' = \mathfrak{g}(D)/\mathfrak{r}$ and let ∂' be the image of ∂ in $\mathfrak{g}'$. Then $\mathfrak{g}'$ is a graded semisimple Lie algebra and ∂' and $\mathfrak{g}_{-1}$ span a triangular subalgebra of $\mathfrak{g}'$. Let $\mathfrak{m}'$ be a maximal triangular subalgebra of $\mathfrak{g}'$ containing $\{\partial', \mathfrak{g}'_{-1}\}$ and $\mathfrak{t}$ be the subalgebra of $\mathfrak{g}(D)$ such that $\mathfrak{t}/\mathfrak{r} = \mathfrak{m}'$. Then $\mathfrak{t}$ is a maximal triangular subalgebra of $\mathfrak{g}(D)$ containing ∂ and $\mathfrak{g}_{-1}$. Since $\mathfrak{t}$ contains ∂, $\mathfrak{t}$ is graded and, moreover, by Lemma 3, $\mathfrak{t} \cap \mathfrak{g}_1 = (0)$ and hence $\mathfrak{t} = \mathfrak{g}_{-1} + \mathfrak{t} \cap \mathfrak{g}_0$.

Now we are going to prove Theorem 6. Let

$$\Phi : D \to D'$$

be a holomorphic diffeomorphism of D onto D'. Denote by (z^k) and (z'^k) respectively, the coordinate systems of the ambiant spaces of D and D' and let

$$\Phi(z) = (\Phi^1(z), \cdots, \Phi^n(z)), \qquad \Phi^k = z'^k \circ \Phi.$$

The map Φ induced a Lie algebra isomorphism $\Phi_* : \mathfrak{g}(D) \to \mathfrak{g}(D')$. Let $\mathfrak{t}$ and $\mathfrak{t}'$ be maximal triangular subalgebras of $\mathfrak{g}(D)$ and $\mathfrak{g}(D')$ respectively, satisfying the conditions in Lemma 7. By Lemma 6, there exists then an element $\sigma \in \mathrm{Aut}(D')$ such that $\sigma_*(\Phi_*\mathfrak{t}) = \mathfrak{t}'$, because $\Phi_*\mathfrak{t}$ is a maximal triangular subalgebra of $\mathfrak{g}(D')$. Then replacing Φ by $\sigma \cdot \Phi$, we may assume that $\Phi_*\mathfrak{t} = \mathfrak{t}'$.

From now on consider a vector field $X = p^k\partial_k$ as column vectors with entries p^k and denote by $(J\Phi)(z)$ the $n \times n$ matrix $(\partial\Phi^j(z)/\partial z^k)$. Then for any $X \in \mathfrak{g}(D)$,

$$(\Phi_*X)(\Phi z) = (J\Phi)(z)X(z), \qquad z \in D. \tag{4.1}$$

Since $\Phi_*\mathfrak{t} = \mathfrak{t}'$, $\mathfrak{g}_{-1} \subset \mathfrak{t}'$, where $\mathfrak{g}(D') = \sum \mathfrak{g}'_\lambda$, there exists an element $Y_k = (q_k^j)$ in $\mathfrak{t}$ for each $k = 1, \cdots, n$ such that $\Phi_* Y_k = \partial'_k$, where $\partial'_k = \partial/\partial z'^k$. Then by (4.1),

$$(J\Phi)(z)\, Y(z) = 1, \qquad z \in D, \tag{4.2}$$

where $Y(z)$ denotes the $n \times n$ matrix $(q_k^j(z))$ and 1 denotes the $n \times n$ unit matrix. Since $q_k^j(z)$ are polynomials (cf. Theorems 1 and 2), it follows from (4.2) that $\partial\Phi^j(z)/\partial z^k$ are rational $(j, k = 1, \cdots, n)$.

Now let $Y = (q^k)$ be the element of $\mathfrak{t}$ such that $\Phi_* Y = \partial' = z'^j\partial'_j$. From (4.1),

$$\Phi^j(z) = \sum_k \frac{\partial\Phi^j(z)}{\partial z^k}\, q^k(z)$$

and hence $\Phi^j(z)$ are also rational.

Let $\Phi_*\partial_k = Y'_k = (q_k^j)$, $k = 1, \cdots, n$. Then, since Y'_k belongs to $\mathfrak{t}'$ and $\mathfrak{t}'$ is contained in $\mathfrak{g}'_{-1} + \mathfrak{g}_0$, the component q_k^j of Y'_k is of the form

$$q_k^j(z) = \sum a_{l,k}^j z'^l + b_k^j, \qquad a_{l,k}^j,\ b_k^j \in \mathbf{R}$$

(cf. Theorem 2). Then we obtain from (4.1) that

$$\partial\Phi(z)/\partial z^k = A_k\Phi(z) + B_k, \tag{4.3}$$

where A_k and B_k denote the $n \times n$ matrix $(a_{l,k}^j)_{j,l=1 \ldots, n}$ and the column vector $(b_k^j)_{j=1,\ldots,n}$. Thus $\Phi(z)$ satisfies the linear system of differential equations (4.3) with constant coefficients. The function $\Phi(z)$ can be rational only if each matrix A_k is nilpotent and then $\Phi(z)$ is a polynomial. We now show that Φ is an affine transformation. To see this, let $\Phi_*\partial = Y' = (q^k)$ with $q^k = a_j^k z'^j + b^k$. Then by (4.1), we get

$$a_j^k \Phi^j(z) + b^k = z^j \, \partial\Phi^k/\partial z^j. \tag{4.4}$$

Let

$$\Phi^k(z) = \sum_\nu \Phi_\nu^k(z) \qquad (k = 1, \ldots, n),$$

where Φ_ν^k is a homogeneous polynomial of degree ν. Then $(\partial\Phi_\nu^k/\partial z^j)z^j = \nu\Phi_\nu^k$ and it follows from (4.4) that

$$\sum a_j^k \Phi_\nu^j(z) = \nu\Phi_\nu^k \qquad (\nu \geqq 1). \tag{4.5}$$

Now let $U' = \sum u^j \partial_j' \; (u^j \in \mathbf{C})$ be an element of $(\mathfrak{g}'_{-1})^{\mathbf{C}}$. Then by a simple computation it is seen that

$$[Y', U'] = -a_j^k u^j \partial_k'. \tag{4.6}$$

Now suppose that $\Phi_\nu(z) = (\Phi_\nu(z)) \neq 0$ for some $\nu \geqq 1$. Then there exists a point $z_0 \in D$ and an index k such that $\Phi_\nu^k(z_0) \neq 0$. Let $u^j = \Phi_\nu^j(z_0)$ and $U' = u^j \partial_j'$. Then $U' \neq 0$ and, from (4.5) and (4.6), $(\operatorname{ad} Y')U' = -{}^\nu U'$. Since $Y' = \Phi_*\partial$, then $[\partial, U] = -{}^\nu U$, where U is the element $(\neq 0)$ of $\mathfrak{t}^{\mathbf{C}}$ such that $\Phi_* U = U'$. However, the eigenvalues of ad ∂ in $\mathfrak{t}^{\mathbf{C}} = \mathfrak{g}_{-1}^{\mathbf{C}} + \mathfrak{t}_0^{\mathbf{C}}$ are -1 and 0. Hence ν must equal 1. This shows that $\Phi_\nu^k = 0$ for all k and $\nu > 1$ and $\Phi^k = \Phi_0^k + \Phi_1^k$ hence Φ is an affine transformation and Φ^k can be written as

$$\Phi^k(z) = A_j^k z^j + B^k.$$

Since the translation $\tau : z'^k \to z'^k - \operatorname{Re} B^k$ is contained in $\operatorname{Aut}(D')$, replacing Φ by $\tau \circ \Phi$, if necessary, it may be assumed that B^k is purely imaginary for each k. Then

$$\Phi_*\partial_j = A_j^k \partial_k',$$

and A_j^k must be real by Theorem 2. Let Y be the element of $\mathfrak{t}$ such that $\Phi_* Y = \partial' = z'^k \partial_k'$. Then

$$A_j^k z^j + B^k = A_l^k c_j^l z^j + A_j^k d^j,$$

where $p^l = c^l_j z^j + d^l$ denote the components of Y. Then $B^k = A^k_j d^j$. Since B^k are purely imaginary and $A^k_j d^j$ are real, then $B^k = 0$. Thus, we have shown that Φ is a nonsingular linear transformation of $\mathbf{C}^n$ with real matrix such that $\Phi(D) = D'$. Then it is clear that Φ maps Ω onto Ω' and the proof of Theorem 6 is completed.

5. Projective Imbedding of a Tube Domain

Denote by r the dimension of the Lie algebra $\mathfrak{g}(D)$, $D = T(\Omega)$, and by $\mathfrak{g}^*$ the dual space of $\mathfrak{g}(D)$. Since $\dim_{\mathbf{R}} \mathfrak{g}_{-1} = n$ and $\mathfrak{g}_0 \neq (0)$, then $r > n$.
Let

$$V = \overset{n}{\Lambda}(\mathfrak{g}^*)^C$$

and define a polynomial map $\tilde{T}$ from $\mathbf{C}^n$ into V as follows: choose a basis $\{X_\nu\}$ of $\mathfrak{g}$ and let $\{f^\nu\}$ be the basis of $\mathfrak{g}^*$ dual to $\{X_\nu\}$. Let $X_\nu = p^k_\nu \partial_k$ and $P^k(z) = p^k_\nu(z) f^\nu$. Then P^k is a polynomial map from $\mathbf{C}^n$ into $(g^*)^{\mathrm{C}}$ and $P^k(z)$ $(k = 1, \cdots, n)$ are linearly independent for each $z \in \mathbf{C}^n$. We put

$$\tilde{T}(z) = \overset{n}{\underset{k=1}{\Lambda}} P^k(z).$$

Then $\tilde{T}$ is a polynomial map from $\mathbf{C}^n$ into V and $\tilde{T}(z) \neq 0$ for each z. Moreover, it is easily seen that the definition of $\tilde{T}$ is independent of the choice of a basis $\{X_\nu\}$ of $\mathfrak{g}$. Since $\tilde{T}(z) \neq 0$ for each $z \in \mathbf{C}^n$, $\tilde{T}$ defines a map T from $\mathbf{C}^n$ into the complex projective space $P(V)$ associated to V.

Choose a basis $\{X_\nu\}$ of $\mathfrak{g}$ such that $X_\nu = \partial_\nu$ $(\nu = 1, \cdots, n)$, $X_{n+1} = \partial = z^k \partial_k$. Then P^k is of the form $P^k(z) = f^k + z^k f^{n+1} + \sum_{\nu > n+1} p^k_\nu f^\nu$. If we denote by $(z_{i_1, \ldots, i_n} | i_1 < \cdots < i_n)$ the homogeneous coordinates of $P(V)$ (with respect to the basis $\{f^{i_1} \wedge \cdots \wedge f^{i_n} | i_1 < \cdots < i_n\}$ of V), then

$$z_{12 \ldots n}(T(z)) = 1,$$

$$z_{1 \ldots \hat{k} \ldots n\, n+1}(T(z)) = z^k \qquad (k = 1, \cdots, n).$$

It follows then that the image $T(\mathbf{C}^n)$ is an algebraic set in the affine subspace $U = \{p \in P(V) | z_{12 \ldots n}(p) \neq 0\}$ and that $T : \mathbf{C}^n \to U \subset P(V)$ is an embedding. Moreover, if an affine map $S : U \to \mathbf{C}^n$ is defined by $S(p) = (z_{1 \ldots \hat{k} \ldots n\, n+1}(p)/z_{1 \ldots n}(p))_{k=1, \ldots, n}$, then

$$ST(z) = z$$

for any $z \in \mathbf{C}^n$. It follows then that $T : \mathbf{C}^n \to T(\mathbf{C}^n)$ is birational. We shall denote by B the closure of $T(\mathbf{C}^n)$ in $P(V)$. Then B is an irreducible algebraic subvariety of $P(V)$ which contains $T(D)$ as an open subset.

Now let Φ be a holomorphic diffeomorphism of $D = T(\Omega)$ onto another tube domain $D' = T(\Omega')$. Then the transpose ${}^t\Phi_*$ of the isomorphism $\Phi_* : \mathfrak{g} \to \mathfrak{g}'$ induces an isomorphism of $V' = \Lambda^n (\mathfrak{g}'^*)^{\mathbf{C}}$ onto V which is denoted by the same letter ${}^t\Phi_*$. Also define the maps $\tilde{T}' : \mathbf{C}^n \to V'$, $T' : \mathbf{C}^n \to P(V')$, and $S' : U' \to \mathbf{C}^n$ starting from D' and $\mathfrak{g}'$. Then, for each $z \in D$,

$${}^t\Phi_* \tilde{T}'(\Phi(z)) = \tilde{T}(z) \cdot \det J\Phi(z),$$

where $J\Phi(z)$ denotes the Jacobian matrix of $\Phi : D \to D'$ at z. If φ denotes the projective transformation from $P(V)$ onto $P(V')$ induced by the non-singular linear map ${}^t\Phi_*^{-1} : V \to V'$, then we obtain the following commutative diagram.

$$\begin{array}{ccc} D & \xrightarrow{\ \Phi\ } & D' \\ T \downarrow & & \downarrow T' \\ T(D) & \xrightarrow[\ \varphi\ ]{} & T(D') \end{array}$$

Since $\Phi = S'^{-1} \cdot \varphi \cdot T$ and T, φ, and S' are rational maps, we conclude that Φ is also rational.

Now consider the case where $D = D'$ and denote by $\mathrm{Aut}(B)$ the group of all projective transformations of $P(V)$ leaving B invariant. Then $\mathrm{Aut}(B)$ is a complex algebraic group and it follows from what has been shown that $\mathrm{Aut}(D)$ is identified with the subgroup of $\mathrm{Aut}(B)$, which leaves the open subset D of B invariant.

Summing up, we obtain the following:

Theorem 7. Let $D = T(\Omega)$ be a tube domain. Then D can be imbedded as an open set in an algebraic subvariety B of a complex projective space in such a way that $\mathrm{Aut}(D)$ is identified with the subgroup of $\mathrm{Aut}(B)$ consisting of all $\sigma \in \mathrm{Aut}(B)$ such that $\sigma(D) = D$, where $\mathrm{Aut}(B)$ denotes the complex algebraic group of all projective transformations leaving B invariant.

Any holomorphic diffeomorphism from D onto another tube domain is a rational map.

References

[*1*] N. Bourbaki, "Algèbre de Lie," *Éléments de Mathématiques*, Hermann, 1961.

[*2*] W. Kaup, Y. Matsushima, and T. Ochiai, "On the automorphisms and equivalences of generalized Siegel domains." *Amer. J. Math.* **92** (1970), 475–498.

[*3*] G. D. Mostow, "On maximal subgroups of real Lie groups," *Ann. of Math.*, **74** (1961), 503–517.

[*4*] I. I. Pjateckii-Sapiro, *Géométrie des domaines classiques et théorie des fonctions automorphes*, French translation, Dunod, Paris, 1966.

[*5*] J. Vey, "Domaines de Siegel divisibles." To appear.

[*6*] Van Est, *Proc. Neder. Akad. Wentensch.*, **61** (1958).

FINE STRUCTURE OF HERMITIAN SYMMETRIC SPACES

*JOSEPH A. WOLF**

Department of Mathematics
Rutgers University
New Brunswick, New Jersey

**Present address:* Department of Mathematics, University of California, Berkeley, California 94720.

1. Introduction

Let X_0 be a Hermitian symmetric space of noncompact type and X its compact dual. We express them as coset spaces of Lie groups

$$X_0 = G_0/K \qquad \text{and} \qquad X = G/P$$

where G_0 and G are the largest connected (compact-open topology) groups of complex analytic automorphisms. Several interesting facts then emerge.

1. G is a complex Lie group, complexification of the real Lie group G_0.
2. X is a projective algebraic variety.
3. Using $G_0 \subset G$, one can arrange $K = G_0 \cap P$, so X_0 has a natural embedding as an open G_0-orbit on X; there, every complex analytic automorphism of X_0 extends to an automorphism of X.
4. There is a natural complex Euclidean space $\mathfrak{m}^+ \subset X$, whose complement is a subvariety of lower dimension in X, such that

$$X_0 \subset \mathfrak{m}^+ \subset X$$

and the inclusion $X_0 \subset \mathfrak{m}^+$ is a canonical realization of X_0 as a bounded symmetric domain.

As a first illustration, consider the case where X_0 is the unit disk in the complex line $\mathbf{C}$,

$$X_0 = \{z \in \mathbf{C} : |z| < 1\}.$$

Then X is the Riemann sphere, $\mathfrak{m}^+ = \mathbf{C} = X - \{\infty\}$ embedded by stereographic projection, and X_0 is the open lower hemisphere of X. The groups are

$$G = \left\{\pm\begin{pmatrix} a & b \\ c & d \end{pmatrix} : ad - bc = 1\right\}, \qquad \text{complex dimension } 3;$$

$$P = \left\{\pm\begin{pmatrix} a & 0 \\ c & d \end{pmatrix} : ad = 1\right\}, \qquad \text{complex dimension } 2;$$

$$G_0 = \left\{\pm\begin{pmatrix} a & b \\ \bar{b} & \bar{a} \end{pmatrix} : |a|^2 - |b|^2 = 1\right\}, \qquad \text{real dimension } 3;$$

$$K = \left\{\pm\begin{pmatrix} a & 0 \\ 0 & \bar{a} \end{pmatrix} : |a|^2 = 1\right\}, \qquad \text{real dimension } 1;$$

they act on X and X_0 by linear fractional transformations

$$\pm\begin{pmatrix} a & b \\ c & d \end{pmatrix}: z \to (az + b)/(cz + d).$$

Under the action of G_0, X decomposes into two open orbits

$X_0 = G_0(0)$ lower hemisphere and $G_0(\infty)$ upper hemisphere

and one closed orbit $G_0(i)$, equator.

As a second illustration, consider the case where $X_0 = \{z \in \mathbf{C}^n: \|z\| < 1\}$ open unit ball in $\mathbf{C}^n$. Then X is the complex projective space $P^n(\mathbf{C})$ with the usual Fubini–Study metric of constant holomorphic sectional curvature, $\mathfrak{m}^+ = \mathbf{C}^n$ complement in X of the polar hyperplane to 0, and X_0 consists of all elements $x \in X$ of distance less than $\frac{1}{2}$(diameter of X) from 0. These facts are more easily seen by viewing X as the space of complex lines through the origin in $\mathbf{C}^{n+1}$ and choosing a basis $\{e_1, \ldots, e_{n+1}\}$, and a Hermitian form $\langle u,v\rangle = -\sum_{k=1}^{n} u^k\bar{v}^k + u^{n+1}\bar{v}^{n+1}$ on $\mathbf{C}^{n+1}$ relative to the basis. Then $\mathbf{C}^n = \mathfrak{m}^+$ injects to X by $z \to [z^1, z^2, \ldots, z^n, 1]$ and X_0 consists of all lines $x \in X$ that are positive definite under $\langle\ ,\ \rangle$. The groups are

G : complex general linear group of $\mathbf{C}^{n+1}$, modulo scalars;
P : subgroup of G preserving the line $[e_{n+1}]$;
G_0 : complex Lorentz group (for $\langle\ ,\ \rangle$), modulo scalars;

$$K : \{\lambda\begin{pmatrix} A & 0 \\ 0 & a \end{pmatrix} : \lambda \neq 0,\ A \in \mathbf{U}(n),\ \det A = a^{-1}\}/\{\lambda \cdot I\}.$$

They act on X through their linear action on $\mathbf{C}^{n+1}$. Again, X decomposes into two open G_0-orbits

$X_0 = G_0[e_{n+1}]$ positive definite lines and $G_0[e_1]$ negative definite lines,

and one closed G_0-orbit

$$G_0[e_1 + e_{n+1}] \text{ isotropic lines.}$$

Here, for $n > 1$, a new feature emerges. The open G_0-orbit $\neq X_0$ is concave and, by Hartogs' Theorem, carries no nonconstant holomorphic function. In fact it has compact subvarieties $gK[e_1]$, $g \in G_0$; for $K[e_1]$ is the polar hyperplane to $[e_{n+1}]$. Further there is a K-equivariant holomorphic fibration

$$\beta : G_0[e_1] \to K[e_1] \quad \text{by} \quad \beta[a^1, \ldots, a^{n+1}] = [a^1, \ldots, a^n, 0],$$

whose fibers are unit disks. That turns out [*17*] to compensate for the lack of holomorphic functions on the orbit.

In general we will see that if X_0 is an irreducible Hermitian symmetric space of noncompact type, and if its symmetric space rank is r, then there are precisely $\frac{1}{2}(r + 1)(r + 2)$ distinct G_0-orbits on X, $r + 1$ of which are open and just one of which is closed. Each orbit turns out to decompose, in a G_0-equivariant manner, into complex submanifolds of X that we call the "holomorphic arc components" of the orbit. The open orbits, being complex manifolds, have just one holomorphic arc component; they turn out to be indefinite metric versions of X_0. The holomorphic arc components of the general orbit turn out to be lower dimensional versions of the open orbits. In a given orbit $G_0(x)$, x properly chosen within the orbit, each holomorphic arc component is fibered holomorphically over its maximal compact subvariety, and those fibrations fit together to form a real analytic bundle $G_0(x) \to K(x)$. These matters are the subject of Part II.

In Part I we develop the now-standard basic material of Hermitian symmetric spaces (Sections 2–4). We then apply that material (Sections 5 and 6) to obtain the G_0-orbit structure of the topological boundary of X_0 in X and the configuration of the holomorphic arc components of the boundary orbits. Those components are called the "boundary components" of X_0 in X. The boundary component theory is somewhat simplified here by systematic use of the restricted root system.

In Part II, we develop the holomorphic arc component theory and orbit configuration in general. Originally [*16*] I needed analytic machinery that went considerably beyond the theory of Hermitian symmetric spaces just in order to reduce these matters to the boundary component theory ([*7*], [*13*]) that Korányi and I had worked out. Here, Part I sheds a little more light on the boundary component theory, allowing development in Part II of the holomorphic arc component theory within the context of Hermitian symmetric spaces. That is a considerable simplification of the corresponding material of [*16*, Chapter III].

In Part III, we apply the general theory of Parts I and II to work out the cases where X_0 belongs to one of the four series of classical domains. Of course this is known [*9*] for the closure of X_0 in X, by *ad hoc* methods. Our method is easier, works for the two spaces X_0 that are not classical domains, and accommodates all of the G_0-orbits.

In this paper, we do not study realizations of X_0 as a Siegel domain of type I, II, or III; that is best done in [*7*], [*13*] and [*16*]. We do not study the function theory of X_0; it is the subject of a good number of papers in this volume. We do not consider arithmetic questions; that seems best done in [*1*]. Somewhat in the spirit of classical algebraic geometry, we just take a close look at a pretty subject.

Part I. Boundary Component Theory

We assume acquaintance with semisimple Lie groups and Riemannian or Hermitian symmetric spaces. The material of [*15*, Chapter *8*] is more than sufficient. Parts of the more basic material are reviewed in Section 2 in order to establish notation and terminology. Section 2 ends with a short proof of the Borel Embedding Theorem.

Sections 2–4 consist of preparatory material, divided so that readers with various degrees of familiarity with symmetric spaces can begin at the appropriate place. In general, however, the reader should at least skim through these sections for notation and terminology.

Section 3 is the first application of Harish-Chandra's maximal set Ψ of strongly orthogonal noncompact positive roots for the Lie algebra $\mathfrak{g}_0$ of a noncompact type Hermitian symmetric space X_0. We construct the polydisk of dimension rank (X_0) that sweeps out X_0 under the isotropy group, and use it to derive the Harish-Chandra Embedding Theorem.

Section 4 is the second application of Ψ. The Cayley transform and partial Cayley transforms are constructed, and the restricted root system is described. We do not prove the characterization of the restricted root system. Properties of the restricted root system are used to prove convexity of X_0.

In Section 5 we work out the G_0-orbit structure of the topological boundary of X_0 and also the boundary components. There is no essential change from Wolf–Korányi [*13*, Section 4].

The G_0-normalizers of the boundary components are worked out in Section 6. Originally [*13*] this was a direct, but complicated, business. Then it was observed [*1*] that those normalizers are just the maximal parabolic subgroups of G_0. Here we use the restricted root system to prove directly that the G_0-normalizers of the boundary components are the maximal parabolic subgroups of G_0, and use that fact to describe the normalizers precisely. The result is an improved version of [*13*, Sections 5 and 6] which allows further improvements in the material of Part II.

2. Borel Embedding

A *Riemannian symmetric space* is a connected Riemannian manifold Y

such that, given $y \in Y$, there is a (globally defined) isometry s_y that preserves y and has differential $-I$ on the tangent space to Y at y. Then y is an isolated fixed point of s_y, and the isometry s_y is called the *symmetry* of Y at y. The symmetry s_y is unique.

Given a complex manifold with Hermitian metric, one obtains a Riemannian manifold consisting of the underlying real differentiable manifold and the real part of the Hermitian metric. If this Riemannian manifold is symmetric, and if its symmetries are Hermitian isometries, then one says that the original complex manifold with Hermitian metric is a *Hermitian symmetric space.*

Let Y be a Hermitian symmetric space, $\tilde{Y} \to Y$ the universal covering space. The complex structure, Hermitian metric, and Riemannian symmetries, all lift from Y to $\tilde{Y}$. Thus $\tilde{Y}$ is a Hermitian symmetric space. Let

$$\tilde{Y} = Y_0 \times Y_1 \times \ldots \times Y_t$$

be the de Rham decomposition of the underlying Riemannian manifold, where Y_0 is a Euclidean space and the other Y_k are irreducible (not Euclidean, not locally isometric to a product of lower dimensional manifolds). Hermitian symmetric spaces are automatically Kaehlerian. Thus $\tilde{Y} = Y_0 \times Y_1 \times \ldots \times Y_t$ product of Hermitian symmetric spaces. We say that Y and $\tilde{Y}$ are

Euclidean if $\tilde{Y} = Y_0$;
irreducible if $\tilde{Y} = Y_1$;
strictly non-Euclidean if $\tilde{Y} = Y_1 \times \ldots \times Y_t$;
compact type if strictly non-Euclidean with each Y_k compact; and
noncompact type if strictly non-Euclidean with each Y_k noncompact.

In general [*14*, Lemma 1],

$Y = Y_0' \times Y_1 \times \cdots \times Y_t$ where Y_0' is the quotient of the complex Euclidean space Y_0 by a discrete group of translations.

In particular,

if Y is strictly non-Euclidean, then it is simply connected,

so

if Y is of compact or noncompact type, it is simply connected.

There is a duality between Hermitian symmetric spaces of compact and of noncompact type, given as follows. Let X be a Hermitian symmetric space of compact type, G_c its largest connected group of (Hermitian) isometries, $x_0 \in X$ and $K = \{g \in G_c : g(x_0) = x_0\}$ the isotropy subgroup there. Then $K \subset G_c$ are compact Lie groups and

$$X \cong G_c/K \quad \text{under} \quad g(x_0) \leftrightarrow gK.$$

Let s be the symmetry to X at x_0 and decompose the Lie algebra of G_c by $\mathfrak{g}_c = \mathfrak{k} + \mathfrak{m}_c$, (± 1)-eigenspaces of $\mathrm{ad}(s)$. That gives another algebra

$$\mathfrak{g}_0 = \mathfrak{k} + \mathfrak{m}_0, \qquad \mathfrak{m}_0 = i\mathfrak{m}_c,$$

with the same complexification as $\mathfrak{g}_c$. Passing to the group level, we get a Hermitian symmetric space $X_0 = G_0/K$ of noncompact type. If that construction is applied to X_0, one obtains X. We say

$$X_0 \quad \text{is the (noncompact) dual of } X,$$

$$X \quad \text{is the (compact) dual of } X_0.$$

This duality will be of constant high importance. Section 2 ends with a special embedding of X_0 in X such that the action of G_0 extends to X and $X_0 = G_0(x_0)$ open orbit.

In the notation above, G_c and G_0 have center reduced to $\{1\}$ and

$$\text{rank } G_c = \text{rank } K = \text{rank } G_0.$$

Further,

$$X \text{ is irreducible} \Leftrightarrow G_c \text{ is simple} \Leftrightarrow X_0 \text{ is irreducible}.$$

In the irreducible case, the fact that X and X_0 are Hermitian (not just Riemannian) symmetric is equivalent to the fact that K has nondiscrete center; then the center of K is a circle group which defines the almost complex structures of X and X_0.

We establish our notation for Hermitian symmetric spaces.

X_0 : Hermitian symmetric space of noncompact type. (2.1)

G_0 : largest connected group of isometries of X_0. (2.2)

x_0 : fixed "base point" in X_0. (2.3)

K : isotropy subgroup of G_0 at x_0. (2.4)

σ : Cartan involution $\mathrm{ad}(s)$ of G_0 where s is the symmetry of X_0 at x_0. (2.5)

$\mathfrak{k} \subset \mathfrak{g}_0$: Lie algebras of $K \subset G_0$. (2.6)

Then G_0 is a connected centerless semisimple Lie group, the center of K is a torus Z_K whose dimension is the number of simple direct factors of G_0, and K is the centralizer of Z_K in G_0. Also we have the Cartan decomposition

$$\mathfrak{g}_0 = \mathfrak{k} + \mathfrak{m}_0, \text{ sum of } +1 \text{ and } -1 \text{ eigenspaces of } \sigma. \tag{2.7}$$

$$\mathfrak{g} = \mathfrak{g}_0^{\mathbf{C}} = \mathfrak{k}^{\mathbf{C}} + \mathfrak{m} \qquad \text{complexification where } \mathfrak{m} = \mathfrak{m}_0^{\mathbf{C}}. \tag{2.8}$$

$$\mathfrak{g}_c = \mathfrak{k} + \mathfrak{m}_c \qquad \text{compact real form of } \mathfrak{g} \text{ where } \mathfrak{m}_c = i\mathfrak{m}_0. \tag{2.9}$$

G : adjoint group of $\mathfrak{g}$, so G_0 is the real analytic subgroup for $\mathfrak{g}_0$. (2.10)

G_c : real analytic subgroup of G for $\mathfrak{g}_c$, compact real form of G. (2.11)

X : G_c/K, Hermitian symmetric space, compact dual of X_0. (2.12)

$\mathfrak{t}$: Cartan subalgebra of $\mathfrak{k}$. (2.13)

As K is the centralizer of the torus Z_K, it has maximal rank in G_0, thus also in G_c. So $\mathfrak{t}$ is a Cartan subalgebra in $\mathfrak{g}_0$ and in $\mathfrak{g}_c$, and $\mathfrak{t}^{\mathbf{C}}$ is a Cartan subalgebra of $\mathfrak{g}$.

Δ : $\mathfrak{t}^{\mathbf{C}}$-root system of $\mathfrak{g}$, so $\mathfrak{g} = \mathfrak{t}^{\mathbf{C}} + \sum_{\Delta} \mathfrak{g}^{\varphi}$. (2.14)

Δ_K : compact roots, $\mathfrak{t}^{\mathbf{C}}$-root system of $\mathfrak{k}^{\mathbf{C}}$. (2.15)

Δ_M : noncompact roots, i.e., roots φ with $\mathfrak{g}^{\varphi} \subset \mathfrak{m}$. (2.16)

z : central element of $\mathfrak{k}$ such that $J = \mathrm{ad}(z)|_{\mathfrak{m}}$ is the almost-complex structure of X and X_0. (2.17)

$\mathfrak{m}^{\pm}$: $(\pm i)$-eigenspace of $J = \mathrm{ad}(z)|_{\mathfrak{m}}$. (2.18)

$\mathfrak{p}$: $\mathfrak{k}^{\mathbf{C}} + \mathfrak{m}^{-}$, parabolic subalgebra of $\mathfrak{g}$ that is the sum of the nonnegative eigenspaces of $\mathrm{ad}(iz) : \mathfrak{g} \to \mathfrak{g}$. (2.19)

P : parabolic subgroup of G that is the complex analytic group for $\mathfrak{p}$. (2.20)

Borel Embedding Theorem. G_c is transitive on the complex coset space G/P with isotropy group $G_c \cap P = K$; thus

$$X = G/P \text{ coset space of complex Lie groups.}$$

Let x_c denote the identity coset $1 \cdot P \in G/P$. Then $G_0 \cap P = K$, so

$gK \to g(x_c)$ embeds X_0 holomorphically as an open G_0-orbit $G_0(x_c) \subset X$.

Proof. Note $\mathfrak{g}_c \cap \mathfrak{p} = \mathfrak{k}$, so $\dim_R G_c(x_c) = \dim_R \mathfrak{g}_c - \dim_R \mathfrak{k} = \dim_R \mathfrak{m}_c = \dim_R \mathfrak{m}^{+} = \dim_R(G/P)$; thus $G_c(x_c)$ is open in G/P. Similarly $G_0(x_c)$ is open in G/P.

As G_c is compact, $G_c(x_c)$ is compact, so the latter is closed (as well as open) in the connected space G/P. Thus $G_c(x_c) = G/P$. Now $gK \to g(x_c)$ defines a complex analytic covering space $X = G_c/K \to G/P$. That gives G/P the structure of Hermitian symmetric space of compact type, so G/P is simply connected. Now $X \to G/P$ is a complex analytic diffeomorphism. Similarly $X_0 \to G_0(x_c)$ is a complex analytic diffeomorphism. Q.E.D.

3. Harish-Chandra Realization

Choose an ordering of the set Δ of $\mathfrak{t}^{\mathbf{C}}$-roots of $\mathfrak{g}$ such that

$$\mathfrak{m}^+ = \sum_{\varphi \in \Delta_M^+} \mathfrak{g}^\varphi \quad \text{and} \quad \mathfrak{m}^- = \sum_{\varphi \in \Delta_M^-} \mathfrak{g}^\varphi. \tag{3.1}$$

For example let $iy \in i\mathfrak{t}$ be in the interior of a Weyl chamber whose closure contains iz, and define the positive root system $\Delta^+ = \{\varphi \in \Delta : \varphi(iy) > 0\}$.

Two roots $\varphi, \psi \in \Delta$ are *strongly orthogonal*, denoted $\varphi \perp\!\!\!\perp \psi$, if neither of $\varphi \pm \psi$ is a root. In that case $\varphi \perp \psi$, ordinary orthogonality. We construct a maximal strongly orthogonal set of noncompact positive roots:

$$\Psi = \{\psi_1, \ldots, \psi_r\} \quad \text{where} \quad \psi_{j+1} \text{ is the lowest element of } \Delta_M^+ \text{ strongly orthogonal to each of } \{\psi_1, \ldots, \psi_j\}. \tag{3.2}$$

If $\varphi \in \Delta$ we have $h_\varphi \in i\mathfrak{t}$ defined by: $2\varphi(h)/\langle\varphi,\varphi\rangle = \langle h_\varphi, h\rangle$ $h \in \mathfrak{t}$. Choose root vectors $e_\varphi \in \mathfrak{g}^\varphi$ normalized by $[e_\varphi, e_{-\varphi}] = h_\varphi$. That choice can be made so that $\mathfrak{m}_0$ has a real basis consisting of all the

$$x_{\varphi,0} = e_\varphi + e_{-\varphi} \quad \text{and} \quad y_{\varphi,0} = i(e_\varphi - e_{-\varphi}), \quad \varphi \in \Delta_M^+, \tag{3.3}$$

related to the almost-complex structure by

$$\begin{aligned} Jx_{\varphi,0} &= [z, x_{\varphi,0}] = y_{\varphi,0}; \\ Jy_{\varphi,0} &= [z, y_{\varphi,0}] = -x_{\varphi,0}; \; [x_{\varphi,0}, y_{\varphi,0}] = -2ih_\varphi. \end{aligned} \tag{3.4}$$

Then $\mathfrak{m}_c$ has a real basis consisting of all

$$x_\varphi = i(e_\varphi + e_{-\varphi}) \quad \text{and} \quad y_\varphi = -(e_\varphi - e_{-\varphi}), \quad \varphi \in \Delta_M^+, \tag{3.5}$$

such that

$$Jx_\varphi = [z, x_\varphi] = y_\varphi; \quad Jy_\varphi = [z, y_\varphi] = -x_\varphi; \quad [x_\varphi, y_\varphi] = 2ih_\varphi. \tag{3.6}$$

From strong orthogonality of Ψ we have abelian subspaces $\mathfrak{a}_0 \subset \mathfrak{m}_0$,

$i\mathfrak{a}_0 = \mathfrak{a}_c \subset \mathfrak{m}_c = i\mathfrak{m}_0$, defined by

$$\mathfrak{a}_0 = \sum_{\Psi} x_{\psi,0}\mathbf{R} \qquad \text{and} \qquad \mathfrak{a}_c = \sum_{\Psi} x_{\psi}\mathbf{R}. \tag{3.7}$$

That provides closed abelian subgroups

$$A_0 = \exp \mathfrak{a}_0 \subset G_0 \qquad \text{and} \qquad A_c = \exp \mathfrak{a}_c \subset G_c. \tag{3.8}$$

If $\varphi \in \Delta$, we have 3-dimensional simple subalgebras

$$\mathfrak{g}[\varphi] = h_{\varphi}\mathbf{C} + \mathfrak{g}^{\varphi} + \mathfrak{g}^{-\varphi} \subset \mathfrak{g} \tag{3.9}$$

and their real forms

$$\mathfrak{g}_0[\varphi] = \mathfrak{g}_0 \cap \mathfrak{g}[\varphi] \qquad \text{and} \qquad \mathfrak{g}_c[\varphi] = \mathfrak{g}_c \cap \mathfrak{g}[\varphi]. \tag{3.10}$$

That defines 3-dimensional simple subgroups

$$G[\varphi] \text{ for } \mathfrak{g}[\varphi], \qquad G_0[\varphi] \text{ for } \mathfrak{g}_0[\varphi], \qquad \text{and } G_c[\varphi] \text{ for } \mathfrak{g}_c[\varphi]. \tag{3.11}$$

Similarly, if $\Gamma \subset \Psi$, then we have direct sums

$$\mathfrak{g}[\Gamma] = \sum_{\Gamma} \mathfrak{g}[\gamma], \qquad \mathfrak{g}_0[\Gamma] = \sum_{\Gamma} \mathfrak{g}_0[\gamma], \qquad \mathfrak{g}_c[\Gamma] = \sum_{\Gamma} \mathfrak{g}_c[\gamma] \tag{3.12}$$

and local direct products as their analytic groups

$$G[\Gamma] = \prod_{\Gamma} G[\gamma], \qquad G_0[\Gamma] = \prod_{\Gamma} G_0[\gamma], \qquad G_c[\Gamma] = \prod_{\Gamma} G_c[\gamma]. \tag{3.13}$$

Polydisk Theorem. If $\Gamma \subset \Psi$, then $G[\Gamma](x_0) = G_c[\Gamma](x_0)$ is a holomorphically embedded, totally geodesic submanifold of X that is a product of $|\Gamma|$ Riemann spheres, and $G_0[\Gamma](x_0)$ is a holomorphically embedded totally geodesic submanifold of X_0 that is the product of the $|\Gamma|$ lower hemispheres of $G_c[\Gamma](x_0)$. K exhausts X with the polysphere $G_c[\Psi](x_0)$ by $K \cdot G_c[\Psi](x_0) = X$. K exhausts X_0 with the polydisk $G_0[\Psi](x_0)$ by $K \cdot G_0[\Psi](x_0) = X_0$.

Proof. $G_c[\Gamma](x_0)$ is totally geodesic in X because $\sigma(\mathfrak{g}_c[\Gamma]) = \mathfrak{g}_c[\Gamma]$, holomorphically embedded in X because $J(\mathfrak{g}_c[\Gamma] \cap \mathfrak{m}_c) = \mathfrak{g}_c[\Gamma] \cap \mathfrak{m}_c$, Riemann polysphere because each $G_c[\gamma](x_0)$ is compact homogeneous, and of complex dimension 1. The same statements on $G_0[\Gamma](x_0)$ now follow.

For the exhaustion, it suffices to prove that

$$\mathfrak{a}_0 \quad \text{is a maximal abelian subspace of} \quad \mathfrak{m}_0, \tag{3.14}$$

$$\mathfrak{a}_c \quad \text{is a maximal abelian subspace of} \quad \mathfrak{m}_c. \tag{3.15}$$

For then standard symmetric space theory says

$$G_0 = KA_0K \qquad \text{and} \qquad G_c = KA_cK \tag{3.16}$$

so that

$$X_0 = G_0(x_0) = KA_0K(x_0) = KA_0(x_0) \subset KG_0[\Psi](x_0),$$
$$X = G_c(x_0) = KA_cK(x_0) = KA_c(x_0) \subset KG_c[\Psi](x_0).$$

Note that (3.14) and (3.15) are equivalent. Now suppose (3.15) false. Then we have $\xi \in \mathfrak{m}_0$, nonzero and orthogonal to $\mathfrak{a}_0$ such that $[\xi, \mathfrak{a}_0] = 0$. Thus there is a set $S \subset \Delta_M^+ - \Psi$ such that

$$0 \neq \xi = \textstyle\sum_s (s^\varphi x_{\varphi,0} + t^\varphi y_{\varphi,0}), \qquad \text{each } (s^\varphi)^2 + (t^\varphi)^2 > 0,$$

and, for every $\psi \in \Psi$,

$$0 = [x_{\psi,0}, \xi] = \textstyle\sum_s (s^\varphi[x_{\psi,0}, x_{\varphi,0}] + t^\varphi[x_{\psi,0}, y_{\varphi,0}])$$

which implies $\varphi \perp\!\!\!\perp \psi$. Thus maximality of Ψ forces (3.15). Q.E.D.

Note that $\mathfrak{m}^\pm$ are commutative subalgebras of $\mathfrak{g}$ consisting of nilpotent elements. That gives us complex analytic unipotent abelian subgroups of G, by

$$M^+ = \exp \mathfrak{m}^+ \qquad \text{and} \qquad M^- = \exp \mathfrak{m}^-. \tag{3.17}$$

Harish Chandra Embedding Theorem. The map $M^+ \times K^C \times M^- \to G$, given by $(m^+, k, m^-) \to m^+km^-$, is a complex analytic diffeomorphism onto a dense open subset of G that contains G_0. In particular,

$$\xi : \mathfrak{m}^+ \to X = G/P \qquad \text{by} \qquad \xi(m) = \exp(m)P$$

is a complex analytic diffeomorphism of $\mathfrak{m}^+$ onto a dense open subset of X that contains X_0. Furthermore, $\xi^{-1}(X_0)$ is a bounded domain in $\mathfrak{m}^+$.

Proof. Let $f : M^+ \times K^C \times M^- \to G$ be the map. It is holomorphic because M^+, K^C, and M^- are complex subgroups of G. Now we check that it is one to one. Suppose that

$$m_1^+ k_1 m_1^- = m_2^+ k_2 m_2^-.$$

Then $(m_1^+)^{-1} m_2^+ \in M^+ \cap K^{\mathbf{C}} M^-$. As it is unipotent, its entire 1-parameter subgroup of M^+ lies in $K^{\mathbf{C}} M^-$. As $\mathfrak{m}^+ \cap (\mathfrak{k}^{\mathbf{C}} + \mathfrak{m}^-) = 0$, we conclude $m_1^+ = m_2^+$. Similarly $m_1^- = m_2^-$. Now $k_1 = k_2$. We have proved f one to one. We see f nonsingular by viewing the Lie algebras as left-invariant vector fields, so that, at (m^+, k, m^-),

$$f_*(\mathfrak{m}^+ \oplus \mathfrak{k}^{\mathbf{C}} \oplus \mathfrak{m}^-) = \mathrm{ad}(k)\mathfrak{m}^+ + \mathfrak{k}^{\mathbf{C}} + \mathfrak{m}^- = \mathfrak{m}^+ + \mathfrak{k}^{\mathbf{C}} + \mathfrak{m}^- = \mathfrak{g}.$$

That also shows the image of f open in G.

Suppose $G = G[\psi]$. Then $G = \mathbf{SL}(2,\mathbf{C})/\{\pm I\}$ and we realize

$$e_\psi = \begin{pmatrix} 0 & 1 \\ 0 & 0 \end{pmatrix}, \qquad e_{-\psi} = \begin{pmatrix} 0 & 0 \\ 1 & 0 \end{pmatrix}, \qquad h_\psi = \begin{pmatrix} 1 & 0 \\ 0 & -1 \end{pmatrix}.$$

In particular,

$$x_{\psi,0} = \begin{pmatrix} 0 & 1 \\ 1 & 0 \end{pmatrix} \quad \text{and} \quad x_\psi = \begin{pmatrix} 0 & i \\ i & 0 \end{pmatrix}.$$

Thus

$$\begin{aligned} \exp(tx_{\psi,0}) &= \begin{pmatrix} \cosh t & \sinh t \\ \sinh t & \cosh t \end{pmatrix} = \begin{pmatrix} 1 & \tanh t \\ 0 & 1 \end{pmatrix} \begin{pmatrix} \frac{1}{\cosh t} & 0 \\ 0 & \cosh t \end{pmatrix} \begin{pmatrix} 1 & 0 \\ \tanh t & 1 \end{pmatrix} \\ &= \exp(\{\tanh t\} e_\psi) \cdot \exp(\{-\log \cosh t\} h_\psi) \cdot \exp(\{\tanh t\} e_{-\psi}) \end{aligned} \tag{3.18}$$

and

$$\begin{aligned} \exp(tx_\psi) &= \begin{pmatrix} \cosh(it) & \sinh(it) \\ \sinh(it) & \cosh(it) \end{pmatrix} = \begin{pmatrix} \cos t & i \sin t \\ i \sin t & \cos t \end{pmatrix} \\ &= \begin{pmatrix} 1 & i \tan t \\ 0 & 1 \end{pmatrix} \begin{pmatrix} \frac{1}{\cos t} & 0 \\ 0 & \cos t \end{pmatrix} \begin{pmatrix} 1 & 0 \\ i \tan t & 1 \end{pmatrix}. \end{aligned} \tag{3.19}$$

Now observe

$$K = \left\{ \begin{pmatrix} e^{i\theta} & 0 \\ 0 & e^{-i\theta} \end{pmatrix} : \theta \text{ real} \right\}.$$

k_θ will denote $\begin{pmatrix} e^{i\theta} & 0 \\ 0 & e^{-i\theta} \end{pmatrix}$,

so

$$\mathrm{ad}(k_\theta)e_\psi = e^{2i\theta}e_\psi.$$

Note also that

$$\xi(\mathfrak{m}^+) = \exp(e_\psi \mathbf{C})\cdot P = \left\{\begin{pmatrix} 1 & z \\ 0 & 1 \end{pmatrix}\cdot P \quad : \quad z \in \mathbf{C}\right\}.$$

As $G_c = KA_cK$, so

$$X = \{k_\theta \exp\,(tx_\psi)\cdot P : \theta,t \text{ real}\},$$

we now have

$$\xi(\mathfrak{m}^+) = \{k_\theta\cdot\exp(tx_\psi)\cdot P : \theta,t \text{ real and } \cos t \neq 0\}.$$

That exhibits $\xi(\mathfrak{m}^+)$ as a dense open subset of X. And

$$X_0 = \left\{\begin{pmatrix} 1 & z \\ 0 & 1 \end{pmatrix}\cdot P : |z| = \tanh t \text{ for some real } t\right\}$$

$$= \left\{\begin{pmatrix} 1 & z \\ 0 & 1 \end{pmatrix}\cdot P : |z| < 1\right\} = \xi(\{ze_\psi : |z| < 1\})$$

exhibits $\xi^{-1}X_0$ as a bounded domain in $\mathfrak{m}^+$.

Our assertions are proved for the case $G = G[\psi]$. They follow for the case $G = G[\Psi]$.

We go to the general case with the Polydisk Theorem. That says

$$X = \{k\cdot(\exp\sum_\Psi t_\psi x_\psi)\cdot P : k \in K \text{ and } t_\psi \text{ real}\}, \tag{3.20}$$

$$\xi(\mathfrak{m}^+) = \{k\cdot(\exp\sum_\Psi t_\psi x_\psi)\cdot P : k \in K,\ t_\psi \text{ real, } \cos t_\psi \neq 0\}, \tag{3.21}$$

$$X_0 = \{k\cdot\xi(\sum_\Psi z_\psi e_\psi) : k \in K \text{ and } |z_\psi| < 1\}. \tag{3.22}$$

Q.E.D.

4. Restricted Root Systems

Recall (3.7), (3.14), (3.15) the maximal abelian subspaces

$$\mathfrak{a}_0 = \sum_\Psi x_{\psi,0}\mathbf{R} \subset \mathfrak{m}_0 \qquad \text{and} \qquad \mathfrak{a}_c = i\mathfrak{a}_0 = \sum_\Psi x_\psi \mathbf{R} \subset i\mathfrak{m}_0 = \mathfrak{m}_c.$$

Strong orthogonality of Ψ, with (3.4) and (3.6), gives

$$[\mathfrak{a}_0, J\mathfrak{a}_0] = \sum_{\Psi} ih_\psi \mathbf{R} = [\mathfrak{a}_c, J\mathfrak{a}_c]. \tag{4.1}$$

Now

$$\mathfrak{t} = \mathfrak{t}^+ + \mathfrak{t}^- \qquad \text{orthogonal direct sum} \quad \text{where} \tag{4.2}$$

$$\mathfrak{t}^- = [\mathfrak{a}_0, J\mathfrak{a}_0] \qquad \text{and} \qquad \mathfrak{t}^+ = \{\tau \in \mathfrak{t} : [\tau, \mathfrak{a}_0] = 0\}. \tag{4.3}$$

If $\Gamma \subset \Psi$, we define the *partial Cayley transform*

$$c_\Gamma = \prod_\Gamma c_\gamma, \qquad c_\gamma = \exp\left(\frac{\pi}{4} y_\gamma\right) \in G_c. \tag{4.4}$$

That gives Cartan subalgebras $\mathfrak{h}_\Gamma \subset \mathfrak{g}_0$ by

$$\mathfrak{h}_\Gamma = \mathfrak{g}_0 \cap \operatorname{ad}(c_\Gamma)\mathfrak{t}^C = \mathfrak{t}^+ + \mathfrak{t}_\Gamma^- + \mathfrak{a}_\Gamma, \tag{4.5}$$

where

$$\mathfrak{t}_\Gamma^- = \sum_{\Psi - \Gamma} ih_\psi \mathbf{R} \qquad \text{and} \qquad \mathfrak{a}_\Gamma = \sum_\Gamma x_{\gamma,0}\mathbf{R}. \tag{4.6}$$

As $\mathfrak{h}_\phi = \mathfrak{t}$ maximally compact Cartan subalgebra of $\mathfrak{g}_0$, and $\mathfrak{h}_\Psi = \mathfrak{t}^+ + \mathfrak{a}_0$ maximally **R**-split Cartan subalgebra of $\mathfrak{g}_0$, it follows that every conjugacy class of Cartan subalgebras of $\mathfrak{g}_0$ is represented by one of the $\mathfrak{h}_\Gamma$.

The full *Cayley transform* is given by

$$c_\Psi = \prod_\Psi c_\psi \qquad \text{so} \qquad \operatorname{ad}(c_\Psi)\mathfrak{t}^C = (\mathfrak{t}^+ + \mathfrak{a}_0)^C. \tag{4.7}$$

Thus its dual map $\operatorname{ad}(c_\Psi)^*$ sends the $(\mathfrak{t}^+ + \mathfrak{a}_0)^C$-root system of $\mathfrak{g}$ to the $\mathfrak{t}^C$-root system Δ of $\mathfrak{g}$. Define

$$\mathfrak{a}_0\text{-root of } \mathfrak{g}_0 : \text{restriction to } \mathfrak{a}_0 \text{ of a } (\mathfrak{t}^+ + \mathfrak{a}_0)^C\text{-root of } \mathfrak{g}. \tag{4.8}$$

As $\mathfrak{a}_0$ is a maximal abelian subspace of $\mathfrak{m}_0$, the $\mathfrak{a}_0$-roots take real values on $\mathfrak{a}_0$, and they are the *real roots* of $\mathfrak{g}_0$. Now

$$\operatorname{ad}(c_\Psi)^* : \mathfrak{a}_0 \text{ roots} \to \{\varphi|_{i\mathfrak{t}^-} : \varphi \in \Delta\}. \tag{4.9}$$

Combinatorial arguments of Harish-Chandra [*5*] and C.C. Moore [*8*], which we omit here, result in

Restricted Root Theorem. Suppose that G is simple, i.e., that the symmetric space X_0 is irreducible. Let ρ denote restriction of roots from $\mathfrak{t}^{\mathbb{C}}$ to $\mathfrak{t}^-$. Identify each element of $\Psi = \{\psi_1, \ldots, \psi_r\}$ with its ρ-image.

1. There are just two cases, as follows.

Case 1. $\rho(\Delta) \cup \{0\} = \{\pm\frac{1}{2}\psi_s \pm \frac{1}{2}\psi_t : 1 \leqq s, \ t \leqq r\}$. In that case, nonzero ρ-images of some subsets of Δ are given by

compact simple roots:	$\{\frac{1}{2}(\psi_{t+1} - \psi_t) : 1 \leqq t \leqq r - 1\}$,
compact positive roots:	$\{\frac{1}{2}(\psi_s - \psi_t) : 1 \leqq t < s \leqq r\}$,
noncompact positive roots:	$\{\frac{1}{2}(\psi_s + \psi_t) : 1 \leqq t \leqq s \leqq r\}$.

Case 2. $\rho(\Delta) \cup \{0\} = \{\pm\frac{1}{2}\psi_s \pm \frac{1}{2}\psi_t, \pm\frac{1}{2}\psi_t : 1 \leqq s, t \leqq r\}$. Then nonzero ρ-images from some subsets of Δ are given by

compact simple roots:	$\{\frac{1}{2}(\psi_{t+1} - \psi_t), 1 \leqq t \leqq r - 1\} \cup \{-\frac{1}{2}\psi_r\}$
compact positive roots:	$\{\frac{1}{2}(\psi_s - \psi_t) : 1 \leqq t < s \leqq r\} \cup \{-\frac{1}{2}\psi_t : 1 \leqq t \leqq r\}$
noncompact positive roots:	$\{\frac{1}{2}(\psi_s + \psi_t) : 1 \leqq t \leqq s \leqq r\} \cup \{\frac{1}{2}\psi_t : 1 \leqq t \leqq r\}$.

2. The ψ_t all have the same length.
3. The subgroup of the Weyl group of G that preserves Ψ induces all signed permutations $\psi_t \to \pm\psi_s$ of Ψ.

Using (4.9), the Restricted Root Theorem gives a description of the $\mathfrak{a}_0$-root system of $\mathfrak{g}_0$ in terms of Ψ.

We use the Restricted Root Theorem to express (3.22) in an invariant manner. Denote

$$\left.\begin{array}{ll} u \to \bar{u} & \text{complex conjugation of } \mathfrak{g} \text{ over } \mathfrak{g}_c, \\ \langle\ ,\ \rangle & \text{Killing form of } \mathfrak{g}, \text{ and} \\ (u,v) = -\langle u, \bar{v}\rangle & \text{positive definite Hermitian form on } \mathfrak{g}. \end{array}\right\} \tag{4.10}$$

If $u \in \mathfrak{g}$, we have the operator norm

$$\|\mathrm{ad}(u)\| = \sup\{|\mathrm{ad}(u)\cdot v| : v \in \mathfrak{g}, |v|^2 = (v,v) = 1\}. \tag{4.11}$$

Also let $\pi_0 : \mathfrak{m}^+ \to \mathfrak{m}_0$ be the projection,

$$\pi_0(u) = \tfrac{1}{2}(u - \bar{u}) \in \mathfrak{m}_0, \qquad \text{all} \quad u \in \mathfrak{m}^+. \tag{4.12}$$

Hermann Convexity Theorem. Let $D = \xi^{-1}(X_0)$, bounded symmetric domain in $\mathfrak{m}^+$ that is the image of X_0 under the Harish-Chandra Embedding. Then

$$D = \{u \in \mathfrak{m}^+ : \|\mathrm{ad}(\pi_0 u)\| < 1\},$$

unit ball in the (generally nondifferentiable) Banach space norm $\|\mathrm{ad}\pi_0 u\|$ of $\mathfrak{m}^+$. In particular D is convex in $\mathfrak{m}^+$.

Proof. Let $\mathfrak{a}^+ = \{\sum_\Psi b_\psi e_\psi : b_\psi \in \mathbf{C}\}$, so $\mathfrak{m}^+ = \mathrm{ad}(K)\mathfrak{a}^+$ and $D = \mathrm{ad}(K)(D \cap \mathfrak{a}^+)$. All norms are invariant under conjugation by elements of K. Thus we need only prove $D \cap \mathfrak{a}^+ = \{u \in \mathfrak{a}^+ : \|\mathrm{ad}(\pi_0 u)\| < 1\}$. Using (3.22), which says $D \cap \mathfrak{a}^+ = \{\sum_\Psi b_\psi e_\psi : |b_\psi| < 1\}$, it suffices to show

$$\text{if} \quad \{t_\psi\} \subset \mathbf{R}, \quad \text{then} \quad \|\mathrm{ad}(\tfrac{1}{2}\textstyle\sum_\Psi t_\psi x_{\psi,0})\| = \sup\{|t_\psi|\}. \tag{4.13}$$

Conjugating by the Cayley transform c_Ψ that says:

$$\text{if} \quad \{t_\psi\} \subset \mathbf{R}, \quad \text{then} \quad \|\mathrm{ad}(\tfrac{1}{2}\textstyle\sum_\Psi t_\psi h_\psi)\| = \sup\{|t_\psi|\}. \tag{4.14}$$

In terms of roots, the latter says, for $\{t_\psi\} \subset \mathbf{R}$,

$$|\mathrm{ad}(\tfrac{1}{2}\textstyle\sum t_\psi h_\psi)e_\varphi| \leqq \sup\{|t_\psi e_\psi|\} \quad \text{for all } \varphi \in \Delta \text{ with equality for at least one root } \varphi \in \Delta. \tag{4.15}$$

The equality of (4.15) comes from $\varphi = \psi_0$ where $|t_{\psi_0}| = \sup\{|t_\psi|\}$. For the inequality it suffices to note that, for every $\varphi \in \Delta$, either Ψ has just one element ψ not $\perp\varphi$ and it has $|\langle\psi,\varphi\rangle| = 2$, or Ψ has just two elements not $\perp\varphi$ and they have $|\langle\psi,\varphi\rangle| = 1$. That comes from the Restricted Root Theorem. Q.E.D.

5. Boundary Components

Suppose that V is a complex analytic space and $S \subset V$ is a subset. By *holomorphic arc* in S, we mean a holomorphic map

$$f : \{z \in \mathbf{C} : |z| < 1\} \to V \quad \text{with image in} \quad S. \tag{5.1}$$

By *chain of holomorphic arcs* in S, we mean a finite sequence $\{f_1, \ldots, f_k\}$ of holomorphic arcs in S such that the image of f_j meets the image of f_{j+1} for $1 \leqq j \leqq k-1$. That provides an equivalence relation on S:

$$v_1 \sim v_2 \text{ iff } \exists \text{ chain } \{f_1, \ldots, f_k\} \text{ of holomorphic arcs in } S, \quad v_1 \in \text{Image } f_1 \quad \text{and} \quad v_2 \in \text{Image } f_k. \tag{5.2}$$

The equivalence classes are the *holomorphic arc components* of S in V.

If S is an open subset of V with topological boundary bd S, then the holomorphic arc components of bd S are the *boundary components* of S.

In this section, we describe the boundary components of the domain $X_0 = G_0(x_0) \subset X$, via convexity of the Harish-Chandra embedding of X_0 as a bounded domain in $\mathfrak{m}^+$. The result is joint work of Wolf and Korányi [*13*].

We must define certain subalgebras $\mathfrak{g}_\Gamma \subset \mathfrak{g}$ corresponding to the subsets Γ of our maximal set Ψ of strongly orthogonal noncompact positive roots. They will be normalized by semisimplicity and

$$\Gamma \text{ is the maximal } \perp\!\!\!\perp \text{ subset of } \Delta_M^+ \text{ for } \mathfrak{g}_\Gamma. \tag{5.3}$$

Then, in particular,

$$\mathfrak{g}_\Psi = \mathfrak{g} \qquad \text{and} \qquad \mathfrak{g}_\phi = 0. \tag{5.4}$$

For that, note that the centralizer of $\mathfrak{t}_\Gamma = \sum_{\Psi-\Gamma} ih_\psi \mathbf{R}$ is $\mathfrak{t}^\mathbf{C} + \sum_{\varphi\perp\Psi-\Gamma}\mathfrak{g}^\varphi$; that the centralizer of $\mathfrak{g}[\Psi - \Gamma] = \sum_{\Psi-\Gamma}(h_\psi\mathbf{C} + \mathfrak{g}^\psi + \mathfrak{g}^{-\psi})$ is $(\mathfrak{t}^+ + \mathfrak{t}_{\Psi-\Gamma})^\mathbf{C} + \sum_{\varphi\perp\Psi-\Gamma}\mathfrak{g}^\varphi$; and that those two centralizers are reductive subalgebras of $\mathfrak{g}$ with the same derived algebra. Now

$$\mathfrak{g}_\Gamma : \text{derived algebra of } \mathfrak{t}^\mathbf{C} + \sum_{\varphi\perp\Psi-\Gamma} \mathfrak{g}^\varphi. \tag{5.5}$$

It has real forms

$$\mathfrak{g}_{\Gamma,0} = \mathfrak{g}_0 \cap \mathfrak{g}_\Gamma \qquad \text{and} \qquad \mathfrak{g}_{\Gamma,c} = \mathfrak{g}_c \cap \mathfrak{g}_\Gamma. \tag{5.6}$$

The analytic groups G_Γ, $G_{\Gamma,0}$, and $G_{\Gamma,c}$ for $\mathfrak{g}_\Gamma$, $\mathfrak{g}_{\Gamma,0}$, and $\mathfrak{g}_{\Gamma,c}$ define symmetric subspaces

$$X_\Gamma = G_\Gamma(x_0) = G_{\Gamma,c}(x_0) \subset X; \qquad X_{\Gamma,0} = G_{\Gamma,0}(x_0) \subset X_0. \tag{5.7}$$

Boundary Orbit Theorem. Retain the notation $\xi : \mathfrak{m}^+ \to X$ of the Harish-Chandra Embedding Theorem. Denote $\mathfrak{m}_\Gamma^+ = \mathfrak{m}^+ \cap \mathfrak{g}_\Gamma$ and $\xi_\Gamma = \xi|_{\mathfrak{m}_\Gamma^+}$.

1. $X_\Gamma \subset X$ and $X_{\Gamma,0} \subset X_0$ are totally geodesic Hermitian symmetric subspaces.
2. $X_{\Gamma,0} \subset X_\Gamma$ is the Borel embedding of $X_{\Gamma,0}$ in its compact dual.
3. $\xi_\Gamma^{-1} : X_{\Gamma,0} \to \mathfrak{m}_\Gamma^+$ is the Harish-Chandra embedding of $X_{\Gamma,0}$ in its holomorphic tangent space at x_0.
4. The G_0-orbits on the topological boundary bd X_0 of X_0 in X are the

sets

$$G_0(c_{\Psi-\Gamma}x_0) = \bigcup_{k\in K} kc_{\Psi-\Gamma}X_{\Gamma,0}, \qquad \Gamma \subsetneqq \Psi.$$

5. Let $\Delta = \Delta_1 \cup \cdots \cup \Delta_q$ where the Δ_t are the root systems of the simple ideals of $\mathfrak{g}$. Let $\Gamma, \Sigma \subsetneqq \Psi$. Then $G_0(c_{\Psi-\Sigma}x_0)$ is in the closure of $G_0(c_{\Psi-\Gamma}x_0)$ if and only if $|\Sigma \cap \Delta_t| \leqq |\Gamma \cap \Delta_t|$ for $1 \leqq t \leqq q$.
6. Boundary orbits $G_0(c_{\Psi-\Gamma}x_0) = G_0(c_{\Psi-\Sigma}x_0)$ if and only if $|\Sigma \cap \Delta_t| = |\Gamma \cap \Delta_t|$ for $1 \leqq t \leqq q$.
7. $G_0(c_\Psi x_0)$ is the unique closed boundary orbit. It is a K-orbit and is the Bergman-Šilov boundary of X_0 in X.

Proof. (1) follows from stability of $\mathfrak{g}_\Gamma$ under both σ and ad(z), whence (2) is immediate. For (3) note that $\mathfrak{a}_\Gamma$ is maximal abelian in $\mathfrak{g}_{\Gamma,0} \cap \mathfrak{m}$.

The polydisk $G_0[\Psi](x_0)$ is the product of the lower hemispheres of the factors of the polysphere $G[\Psi](x_0)$. Thus its boundary is the product where each factor is a lower hemisphere or an equator. Now $G_0[\Psi](x_0)$ has boundary $\bigcup_\Gamma G_0[\Psi](c_{\Psi-\Gamma}x_0)$ where $\Gamma \subsetneqq \Psi$. So (4) follows from the Polydisk Theorem and the observation that $G_0(c_{\Psi-\Gamma}x_0) = K\cdot G_0[\Psi](c_{\Psi-\Gamma}x_0) = K\cdot G_0[\Gamma](c_{\Psi-\Gamma}x_0) = K\cdot G_{\Gamma,0}(c_{\Psi-\Gamma}x_0) = K\cdot c_{\Psi-\Gamma}G_{\Gamma,0}(x_0) = Kc_{\Psi-\Gamma}X_{\Gamma,0}$.

Let $\Sigma,\Gamma \subset \Psi$. If $|\Sigma \cap \Delta_t| \leqq |\Gamma \cap \Delta_t|$ for $1 \leqq t \leqq q$, then part (3) of the Restricted Root Theorem provides $k \in K$ such that ad$(k)^*\Sigma \subset \Gamma$. Thus we may assume $\Sigma \subset \Gamma$. Now $\Sigma \subset \Gamma$ if and only if $G_0[\Psi](c_{\Psi-\Sigma}x_0)$ is in the closure of $G_0[\Psi](c_{\Psi-\Gamma}x_0)$. Applying K, the latter is equivalent to $G_0(c_{\Psi-\Sigma}x_0)$ being in the closure of $G_0(c_{\Psi-\Gamma}x_0)$. That proves (5); now (6) follows, as does the fact that $G_0(c_\Psi x_0)$ is the only closed boundary orbit.

$G_0(c_\Psi x_0) = Kc_\Psi X_{\phi,0} = K(c_\Psi x_0)$ shows it to be a K-orbit.

Let **B** denote the Banach algebra of functions continuous on the closure $X_0 \cup \mathrm{bd}\,X_0$ of X_0 in X and holomorphic on X_0 (point multiplication, sup norm). G acts on **B** by $g(f): x \to f(g^{-1}x)$. Let $f \in \mathbf{B}$ not constant on any irreducible factor of X_0, and x_f a maximum point of $|f(x)|$. Transforming by an element of K, we may assume x_f in the polysphere $G[\Psi](x_0)$. It is a maximum point of the restriction of f to the closure of the polydisk $G_0[\Psi](x_0)$, hence in the boundary $G_0[\Psi](c_\Psi x_0)$ of the polydisk. Thus $x_f \in G_0(c_\Psi x_0)$. By transitivity of G_0 on $G_0(c_\Psi x_0)$ now, the latter is the set of all maximum points of elements of **B**. Q.E.D.

Boundary Component Theorem. The boundary components of X_0 in X are just the sets $kc_{\Psi-\Gamma}X_{\Gamma,0}$ with $k \in K$ and $\Gamma \subsetneqq \Psi$. They are Hermitian symmetric spaces of noncompact type and rank given by

$$\operatorname{rank} kc_{\Psi-\Gamma}X_{\Gamma,0} = |\Gamma|. \tag{5.8}$$

Proof. The $kc_{\Psi-\Gamma}X_{\Gamma,0}$ are complex manifolds, so the boundary components are unions of such sets. Now we need only show that each $c_{\Psi-\Gamma}X_{\Gamma,0}$ is a boundary component.

We use Hermann Convexity of $\xi^{-1}X_0$ in $\mathfrak{m}^+$ to check that each $c_{\Psi-\Gamma}X_{\Gamma,0}$ is a boundary component. Define

$$\mathfrak{o}_\Gamma = \xi^{-1}c_{\Psi-\Gamma}x_0, \qquad \mathfrak{m}_\Gamma^+ = \mathfrak{g}_\Gamma \cap \mathfrak{m}^+, \qquad \mathfrak{e}_\Gamma = \mathfrak{o}_\Gamma + \mathfrak{m}_\Gamma^+. \tag{5.9}$$

Then $\mathfrak{e}_\Gamma$ is a complex affine subspace of $\mathfrak{m}^+$ that contains the boundary point $\mathfrak{o}_\Gamma$ of $\xi^{-1}X_0$. Looking at the polysphere we see that

$$\mathfrak{o}_\Gamma = \xi^{-1}\exp\Big(\frac{\pi}{4}\sum_{\Psi-\Gamma}(e_{-\psi} - e_\psi)\Big)x_0 = -\sum_{\Psi-\Gamma} e_\psi. \tag{5.10}$$

Recall the projection $\pi_0 : \mathfrak{m}^+ \to \mathfrak{m}_0$ from (4.12). Now

$$\pi_0(\mathfrak{o}_\Gamma + \mathfrak{m}_\Gamma^+) = \Big(\tfrac{1}{2}\sum_{\Psi-\Gamma} x_{\psi,0}\Big) + \mathfrak{m}_{\Gamma,0}; \qquad \mathfrak{m}_{\Gamma,0} = \mathfrak{m}_0 \cap \mathfrak{g}_\Gamma. \tag{5.11}$$

Let $\mathfrak{f}_\Gamma$ denote the orthocomplement of $\mathfrak{o}_\Gamma$ in $\mathfrak{m}^+$. The Restricted Root Theorem says that $\{v \in \pi_0(\mathfrak{o}_\Gamma + \mathfrak{f}_\Gamma) : \|\mathrm{ad}_{\mathfrak{g}}(v)\| \leqq 1\} = \frac{1}{2}\sum_{\Psi-\Gamma} x_{\psi,0} + \{w \in \mathfrak{m}_{\Gamma,0} : \|\mathrm{ad}_{\mathfrak{g}_\Gamma}(w)\| \leqq 1\}$. By the Hermann Convexity Theorem now

$$\overline{\xi^{-1}(X_0)} \cap (\mathfrak{o}_\Gamma + \mathfrak{f}_\Gamma) = \overline{\xi^{-1}(X_0)} \cap (\mathfrak{o}_\Gamma + \mathfrak{m}_\Gamma^+) = \overline{\xi^{-1}(c_{\Psi-\Gamma}X_{\Gamma,0})}. \tag{5.12}$$

Now let $U \subset \mathbf{C}$ unit disk and $l : U \to X$ a holomorphic arc in bd X_0 whose image meets $c_{\Psi-\Gamma}X_{\Gamma,0}$. $\lambda : U \to \mathfrak{m}^+$ is given by $\xi \cdot \lambda = l$. Let κ be the linear functional on $\mathfrak{m}^+$ such that the support hyperplane to $\xi^{-1}X_0$ at $\mathfrak{o}_\Gamma$ has equation $\operatorname{Re}\kappa = 1$. Now $\kappa \cdot \lambda$ is a holomorphic function on U, whose real part is $\leqq 1$ and achieves its maximum value. Now $\kappa \cdot \lambda$ is constant, i.e., $\lambda(U) \subset \mathfrak{o}_\Gamma + \mathfrak{f}_\Gamma$. From (5.12) now λ takes values in the closure of $\xi^{-1}c_{\Psi-\Gamma}X_{\Gamma,0}$. Were λ to take a value in the boundary of $\xi^{-1}c_{\Psi-\Gamma}X_{\Gamma,0}$, the same maximum argument would keep all values of λ in that boundary. Thus $\lambda(U) \subset \xi^{-1}c_{\Psi-\Gamma}X_{\Gamma,0}$. Now $l(U) \subset c_{\Psi-\Gamma}X_{\Gamma,0}$. That proves that $c_{\Psi-\Gamma}X_{\Gamma,0}$ is a union of boundary components. As $c_{\Psi-\Gamma}X_{\Gamma,0}$ is a complex submanifold of X, we conclude that it is a boundary component of X_0.

Q.E.D.

Corollary 1. The boundary components of $\xi^{-1}X_0$ in $\mathfrak{m}^+$ are bounded symmetric domains

$$\xi^{-1}kc_{\Psi-\Gamma}X_{\Gamma,0} = \mathrm{ad}(k)\cdot\xi^{-1}c_{\Psi-\Gamma}X_{\Gamma,0} \subset \mathrm{ad}(k)\,(\mathfrak{v}_\Gamma + \mathfrak{m}_\Gamma^+)$$

in Harish-Chandra Embedding.

{Combine (5.12) and the Boundary Component Theorem.}

Corollary 2. Decompose $X_0 = X_0^1 \times \cdots \times X_0^q$ product of irreducible symmetric spaces. Let X^t be the compact dual of X_0^t so $X = X^1 \times \ldots \times X^q$. Then the boundary components of X_0 in X are the products $Y_0^1 \times \cdots \times Y_0^q \neq X_0$ where Y_0^t is X_0^t or a boundary component of X_0^t.

{Everything decomposes as a product.}

Corollary 3. If X_0 is an irreducible Hermitian symmetric space, then its boundary components are irreducible Hermitian symmetric spaces of classical type.

Proof. Let W_Γ denote the subgroup of the Weyl group of G that preserves Ψ and acts trivially on $\Psi - \Gamma$. It induces every permutation of Γ by part 3 of the Restricted Root Theorem. That proves irreducibility of the Hermitian symmetric space $X_{\Gamma,0}$. The fact that it is of clasical type is contained in the classification theorem below. Q.E.D.

Corollary 4. A boundary component of a boundary component of X_0 is a boundary component of X_0.

{The boundary components of $c_{\Psi-\Gamma}X_{\Gamma,0}$ are the $kc_{\Psi-\Gamma}c_{\Gamma-\Sigma}X_{\Sigma,0}$ with $k \in K \cap G_{\Gamma,0}$ and $\Sigma \subsetneqq \Gamma$.}

X_0 is said to be of *tube type* if it is holomorphically equivalent to a tube domain over a self dual cone. That is equivalent to $c_\Psi^4 = 1$. Momentarily taking the latter as definition, we see that X_0 is of tube type if and only if each of its irreducible factors has restricted root system given by Case 1 of the Restricted Root Theorem. That situation persists under passage from $\mathfrak{g}_0$ to $\mathfrak{g}_{\Gamma,0}$. Thus:

Corollary 5. If X_0 is of tube type, then its boundary components are of tube type. If X_0 is irreducible and has a positive-dimensional boundary component of tube type, then X_0 is of tube type.

Suppose $\Sigma \subset \Gamma \subset \Psi$. Then $\mathrm{ad}(c_\Sigma)\mathfrak{g}_\Gamma = \mathfrak{g}_\Gamma$ and we define

$$\mathfrak{g}_\Gamma^\Sigma \quad \text{is the fixed point set of} \quad \mathrm{ad}(c_\Sigma^4) \quad \text{on} \quad \mathfrak{g}_\Gamma. \tag{5.13}$$

As c_Σ has order 1, 4, or 8, c_Σ^4 commutes with complex conjugation of $\mathfrak{g}$ over $\mathfrak{g}_0$, so we have

$$\mathfrak{g}_{\Gamma,0}^\Sigma = \mathfrak{g}_0 \cap \mathfrak{g}_\Gamma^\Sigma, \qquad \text{real form of} \quad \mathfrak{g}_\Gamma^\Sigma. \tag{5.14}$$

Now define

$$G_{\Gamma,0}^\Sigma : \text{analytic subgroup} \quad \text{of} \quad G_0 \quad \text{for} \quad \mathfrak{g}_{\Gamma,0}^\Sigma, \tag{5.15}$$

$$X_{\Gamma,0}^\Sigma = G_{\Gamma,0}^\Sigma(x_0), \text{ Hermitian symmetric subspace of } X_{\Gamma,0}. \tag{5.16}$$

Also observe

$$X_0 \quad \text{is of tube type} \quad \Leftrightarrow X_0 = X_{\Psi,0}^\Psi. \tag{5.17}$$

Lemma. If $\Gamma \subset \Psi$, then $[\mathfrak{g}_\Gamma^\Gamma, \mathfrak{g}_{\Psi-\Gamma}] = 0$, so we have

$$\mathfrak{g}_{\Psi,0}^\Gamma = \mathfrak{g}_{\Psi-\Gamma,0} \oplus \mathfrak{g}_{\Gamma,0}^\Gamma \oplus (\text{compact ideal}), \tag{5.18}$$

$$X_{\Psi,0}^\Gamma = X_{\Psi-\Gamma,0} \times X_{\Gamma,0}^\Gamma. \tag{5.19}$$

Proof. We may assume $\mathfrak{g}$ simple and use the Restricted Root Theorem. Let Δ_Γ^Γ and $\Delta_{\Psi-\Gamma}$ denote the respective root systems of $\mathfrak{g}_\Gamma^\Gamma$ and $\mathfrak{g}_{\Psi-\Gamma}$. Let $\alpha \in \Delta_\Gamma^\Gamma$ noncompact positive and $\beta \in \Delta_{\Psi-\Gamma}$ noncompact negative. Then

$$\alpha|_{\mathfrak{t}^-} = \tfrac{1}{2}(\gamma + \gamma') \qquad \text{for some} \quad \gamma, \gamma' \in \Gamma; \qquad \text{and}$$
$$\beta|_{\mathfrak{t}^-} = 0 \quad \text{or} \quad -\tfrac{1}{2}(\psi + \psi') \quad \text{or} \quad -\tfrac{1}{2}\psi \quad \text{for some} \quad \psi, \psi' \in \Psi - \Gamma.$$

Now $(\alpha + \beta)|_{\mathfrak{t}^-}$ is $\tfrac{1}{2}(\gamma + \gamma')$ or $\tfrac{1}{2}(\gamma + \gamma' - \psi - \psi')$ or $\tfrac{1}{2}(\gamma + \gamma' - \psi)$, so $\alpha + \beta$ is not a root. Similarly, if α is noncompact negative from Δ_Γ^Γ and β is noncompact positive from $\Delta_{\Psi-\Gamma}$, then $\alpha + \beta$ is not a root. This proves

$$[\mathfrak{m} \cap \mathfrak{g}_\Gamma^\Gamma, \mathfrak{m} \cap \mathfrak{g}_{\Psi-\Gamma}] = 0.$$

Now $\mathfrak{m} \cap \mathfrak{g}_{\Psi-\Gamma}$ generates the semisimple algebra $\mathfrak{g}_{\Psi-\Gamma}$, and $\mathfrak{g}_\Gamma^\Gamma = (\mathfrak{g}_\Gamma^\Gamma)' \oplus \mathfrak{l}$ where $\mathfrak{m} \cap \mathfrak{g}_\Gamma^\Gamma$ generates $(\mathfrak{g}_\Gamma^\Gamma)'$ and $\mathfrak{l} \subset \mathfrak{k}^{\mathbf{C}}$. As $[\mathfrak{l}, \mathfrak{a}_{\Psi-\Gamma}] = 0$, we obtain $[\mathfrak{l}, \mathfrak{m} \cap \mathfrak{g}_{\Psi-\Gamma}] = 0$, and thus $[\mathfrak{l}, \mathfrak{g}_{\Psi-\Gamma}] = 0$. Now $[\mathfrak{g}_\Gamma^\Gamma, \mathfrak{g}_{\Psi-\Gamma}] = 0$.

Q.E.D.

We now describe the boundary components explicitly.

Classification of Boundary Components. Let $X_0 \subset X$ be the Borel embedding of an irreducible Hermitian symmetric space of noncompact type and rank r. For each integer m, $0 \leqq m < r$, there is just one G_0-equivalence class $\{X_{m,0}\}$ of boundary component of symmetric space rank m for X_0 in X; $X_{0,0}$ is a single point, and the $X_{m,0}$ are given by one of the following series with $X_0 = X_{r,0}$.

1. $X_{m,0} = \mathbf{SU}(m, m+k) / \mathbf{S}(\mathbf{U}(m) \times \mathbf{U}(m+k))$, $k \geqq 0$ fixed.
2. $X_{m,0} = \mathbf{SO}^*(4m) / \mathbf{U}(2m)$.
3. $X_{m,0} = \mathbf{SO}^*(4m+2) / \mathbf{U}(2m+1)$.
4. $X_{m,0} = \mathbf{Sp}(m\ ;\mathbf{R}) / \mathbf{U}(m)$.
5. $X_0 = \mathbf{SO}(2, n) / \mathbf{SO}(2) \times \mathbf{SO}(n)$, $n > 2$ fixed; here $r = 2$ and $X_{1,0}$ is the unit disk in $\mathbf{C}^1$.
6. $X_0 = \mathbf{E}_6 / \mathbf{SO}(10) \cdot \mathbf{SO}(2)$; here $r = 2$ and $X_{1,0}$ is the open unit ball in $\mathbf{C}^5$.
7. $X_0 = \mathbf{E}_7/\mathbf{E}_6 \cdot \mathbf{SO}(2)$; here $r = 3$, $X_{2,0} = \mathbf{SO}(2,8)/\mathbf{SO}(2) \times \mathbf{SO}(8)$, and $X_{1,0}$ is the unit disk in $\mathbf{C}^1$.

Proof. That there is just one G_0-equivalence class $\{X_{m,0}\}$ of rank m boundary components, follows from the Boundary Component Theorem and the part of the Restricted Root Theorem that ensures that any two subsets of Ψ of the same size are equivalent under the restricted Weyl group. Thus $\{X_{m,0}\}$ is represented by any $c_\Gamma X_{\Psi-\Gamma,0}$ with $|\Psi - \Gamma| = m$.

Let $\Gamma = \{\alpha\}$ so $|\Psi - \Gamma| = r - 1$. Then $\mathrm{ad}(c_\alpha)^4$ is an involutive inner automorphism of $\mathfrak{g}$, so its fixed point set $\mathfrak{g}_\Psi^{\{\alpha\}}$ is a symmetric subalgebra of maximal rank. The lemma says that $\mathfrak{g}_\Psi^{\{\alpha\}}$ has distinct ideals $\mathfrak{g}_{\Psi-\Gamma}$ and $\mathfrak{sl}(2, \mathbf{C}) \subset \mathfrak{g}_\Gamma^\Gamma$. The classification now is immediate from the Borel–de Siebenthal classification of maximal subalgebras of maximal rank in compact simple Lie algebras, and from the corollaries to the Boundary Component Theorem. Q.E.D.

Karpelevič defines boundary components, for noncompact type symmetric spaces, in terms of limiting behavior of geodesic rays. The following result says that his boundary components coincide with ours.

Geodesic Ray Theorem. Let $x \in X_0$ and F a boundary component of X_0 in X. Then there is a unique point $f \in F$ such that some geodesic ray of X_0 from x tends to f.

Proof. Some $g \in G_0$ carries x to x_0, and then some $k \in K$ carries $g(F)$ back to F; thus we may assume $x = x_0$. Applying an element of K, we may also assume $F = c_{\Psi-\Gamma} X_{\Gamma,0}$ for some $\Gamma \subset \Psi$.

Let $\{x_t\}_{0<t<\infty}$ be a geodesic ray in X_0 such that, in X, $\lim_{t\to\infty}\{x_t\} \in F$. We have $b \in \mathfrak{m}_0$ such that $x_t = \exp(tb)\cdot x_0$.

First suppose $b \in \mathfrak{a}_0$, so $b = \sum_\Psi b_\psi x_{\psi,0}$ with b_ψ real. Denote

$$\operatorname{sign}(b_\psi) = 1 \quad \text{if} \quad b_\psi > 0, \qquad 0 \quad \text{if} \quad b_\psi = 0, \qquad -1 \quad \text{if} \quad b_\psi < 0.$$

Now (3.18) says that $\lim_{t\to\infty}\{x_t\} = \xi(\sum \operatorname{sign}(b_\psi)e_\psi)$. As $\lim\{x_t\} \in c_{\Psi-\Gamma}X_{\Gamma,0}$, it follows that this limit is $c_{\Psi-\Gamma}(x_0) = \xi(\sum_{\Psi-\Gamma} e_\psi)$.

Now drop the assumption $b \in \mathfrak{a}_0$. Let $k \in K$ such that $\operatorname{ad}(k)b \in \mathfrak{a}_0$. We then know that $\lim\{k(x_t)\}$ is $c_{\Psi-\Gamma}(x_0) \in F$. As $\lim\{x_t\} \in F$, now $k(F) = F$. However $c_{\Psi-\Gamma}(x_0)$ is the point of $\xi^{-1}(F)$ of least $\mathfrak{m}^+$-distance from the origin, and k does not change that distance. Thus $k(c_{\Psi-\Gamma}(x_0)) = c_{\Psi-\Gamma}(x_0)$. Now $\lim\{x_t\} = k^{-1}\lim\{kx_t\} = c_{\Psi-\Gamma}(x_0)$. Q.E.D.

6. Boundary Groups

We have seen that X_0 has topological boundary

$$\operatorname{bd} X_0 = \bigcup_{\Gamma \subsetneqq \Psi} G_0(c_{\Psi-\Gamma}x_0) \tag{6.1}$$

where the boundary orbits are unions of boundary components

$$G_0(c_{\Psi-\Gamma}x_0) = \bigcup_{k\in K} kc_{\Psi-\Gamma}X_{\Gamma,0} \tag{6.2}$$

which in turn are given by

$$c_{\Psi-\Gamma}X_{\Gamma,0} = c_{\Psi-\Gamma}G_{\Gamma,0}(x_0) = G_{\Gamma,0}(c_{\Psi-\Gamma}x_0). \tag{6.3}$$

Here we will determine the normalizers of the boundary components and the structure of the space of boundary components of a given type. The results are in Wolf and Korányi [*13*], but the proofs here are a considerable simplification of [*13*].

Recall (3.3) $x_{\varphi,0} = e_\varphi + e_{-\varphi} \in \mathfrak{m}_0$ for $\varphi \in \Delta_M^+$. For each subset $\Phi \subset \Psi$, we define

$$x_{\Phi,0} = \sum_{\psi\in\Phi} x_{\psi,0} \in \mathfrak{a}_0. \tag{6.4}$$

We compute

$$\operatorname{ad}(c_\psi)^{-1}x_{\psi,0} = \operatorname{ad}(\exp(\tfrac{\pi}{4}(e_\psi - e_{-\psi})))\cdot(e_\psi + e_{-\psi})$$

in the model $e_\psi = \begin{pmatrix} 0 & 1 \\ 0 & 0 \end{pmatrix}$ and $e_{-\psi} = \begin{pmatrix} 0 & 0 \\ 1 & 0 \end{pmatrix}$. There

$$\exp\frac{\pi}{4}(e_\psi - e_{-\psi}) = \exp\frac{\pi}{4}\begin{pmatrix} 0 & 1 \\ -1 & 0 \end{pmatrix} = \frac{1}{\sqrt{2}}\begin{pmatrix} 1 & 1 \\ -1 & 1 \end{pmatrix},$$

so

$$\mathrm{ad}[\exp \frac{\pi}{4}(e_\psi - e_{-\psi})]\cdot(e_\psi + e_{-\psi}) = \begin{pmatrix} 1 & 0 \\ 0 & -1 \end{pmatrix} = h_\psi.$$

Summing over the subset $\Phi \subset \Psi$, now

$$\mathrm{ad}(c_\Phi)^{-1} x_{\Phi,0} = \sum_{\psi \in \Phi} h_\psi. \tag{6.5}$$

Use the Restricted Root Theorem to compute the eigenspaces of $\mathrm{ad}(\sum_\Phi h_\psi)$ on $\mathfrak{g}$. Let ρ denote the restriction to $\mathfrak{t}^-$ as before. Now the $\mathfrak{t}^C$-root system Δ is the disjoint union,

$$\Delta = E_2(\Phi) \cup E_1(\Phi) \cup E_0(\Phi) \cup E_{-1}(\Phi) \cup E_{-2}(\Phi), \tag{6.6}$$

where

$$E_{\pm 2}(\Phi) = \{\delta \in \Delta : \rho(\delta) = \pm\tfrac{1}{2}(\psi + \psi') \quad \text{with} \quad \psi, \psi' \in \Phi\}; \tag{6.7}$$

$E_{\pm 1}(\Phi)$ is the union of the three sets:

$$\begin{gathered} \{\delta \in \Delta : \rho(\delta) = \pm\tfrac{1}{2}(\psi + \psi') \quad \text{with} \quad \psi \in \Phi, \psi' \in \Psi - \Phi\}, \\ \{\delta \in \Delta : \rho(\delta) = \pm\tfrac{1}{2}(\psi - \psi') \quad \text{with} \quad \psi \in \Phi, \psi' \in \Psi - \Phi\}, \\ \text{and} \quad \{\delta \in \Delta : \rho(\delta) = \pm\tfrac{1}{2}\psi \quad \text{with} \quad \psi \in \Phi\}; \end{gathered} \tag{6.8}$$

$E_0(\Phi)$ is the union of

$$\begin{gathered} \{\delta \in \Delta : \rho(\delta) = \pm\tfrac{1}{2}(\psi \pm \psi') \quad \text{with} \quad \psi, \psi' \in \Psi - \Phi\}, \\ \{\delta \in \Delta : \rho(\delta) = \pm\tfrac{1}{2}\psi \quad \text{with} \quad \psi \in \Psi - \Phi\}, \\ \{\delta \in \Delta : \rho(\delta) = 0\}, \quad \text{and} \\ \{\delta \in \Delta : \rho(\delta) = \tfrac{1}{2}(\psi - \psi') \quad \text{with} \quad \psi, \psi' \in \Phi\}. \end{gathered} \tag{6.9}$$

Applying the Restricted Root Theorem to (6.6), we have

6.10. Lemma. The eigenvalues of $\mathrm{ad}(\sum_{\psi\in\Phi} h_\psi)$ on $\mathfrak{g}$ are ± 2, ± 1, and 0. The eigenspaces are given as follows.

$$\begin{aligned} (\pm 2)\text{-eigenspace:} &\quad \textstyle\sum_{\delta \in E_{\pm 2}(\Phi)} \mathfrak{g}^\delta \\ (\pm 1)\text{-eigenspace:} &\quad \textstyle\sum_{\delta \in E_{\pm 1}(\Phi)} \mathfrak{g}^\delta \\ 0\text{-eigenspace:} &\quad \mathfrak{t}^C + \textstyle\sum_{\delta \in E_0(\Phi)} \mathfrak{g}^\delta. \end{aligned}$$

Combine Lemma 6.10 with (6.5) to obtain

6.11. Lemma. The eigenvalues of $\mathrm{ad}(x_{\Phi,0})$ on $\mathfrak{g}$ are ± 2, ± 1, and 0. The eigenspaces are given as follows.

$$(\pm 2)\text{-eigenspace:} \qquad \mathrm{ad}(c_\Phi)\cdot \textstyle\sum_{\delta\in E_{\pm 2}(\Phi)} \mathfrak{g}^\delta.$$

$$(\pm 1)\text{-eigenspace:} \qquad \mathrm{ad}(c_\Phi)\cdot \textstyle\sum_{\delta\in E_{\pm 1}(\Phi)} \mathfrak{g}^\delta.$$

$$0\text{-eigenspace:} \qquad \mathrm{ad}(c_\Phi)\cdot \{\mathfrak{t}^{\mathbf{C}} + \textstyle\sum_{\delta\in E_0(\Phi)} \mathfrak{g}^\delta\}.$$

With the eigenspace decomposition (6.11) in mind, we define parabolic subalgebras $\mathfrak{n}_\Phi \subset \mathfrak{g}$ and $\mathfrak{n}_{\Phi,0} \subset \mathfrak{g}_0$ by

$$\mathfrak{n}_\Phi : \text{ sum of the nonpositive eigenspaces of } \mathrm{ad}(x_{\Phi,0}) \text{ on } \mathfrak{g}, \tag{6.12a}$$

$$\mathfrak{n}_{\Phi,0} : \text{ sum of the nonpositive eigenspaces of } \mathrm{ad}(x_{\Phi,0}) \text{ on } \mathfrak{g}_0. \tag{6.12b}$$

Then of course $\mathfrak{n}_{\Phi,0}$ is a real form of $\mathfrak{n}_\Phi$. Now define parabolic subgroups $N_\Phi \subset G$ and $N_{\Phi,0} \subset G_0$ by

$$N_\Phi \quad \text{is the complex analytic subgroup of } G \text{ for } \mathfrak{n}_\Phi, \tag{6.13a}$$

$$N_{\Phi,0} = N_\Phi \cap G_0, \quad \text{closed subgroup with Lie algebra } \mathfrak{n}_{\Phi,0}. \tag{6.13b}$$

Boundary Group Theorem. If $\Gamma \subsetneqq \Psi$, then the normalizer of the boundary component $c_{\Psi-\Gamma}X_{\Gamma,0}$ of X_0 in X is given by

$$N_{\Psi-\Gamma,0} = \{g \in G_0 : g c_{\Psi-\Gamma}X_{\Gamma,0} = c_{\Psi-\Gamma}X_{\Gamma,0}\}.$$

Furthermore,

(1) in general, $N_{\Psi-\Gamma,0}$ and $N_{\Psi-\Gamma',0}$ are conjugate in G_0 if, and only if, each $|\Gamma \cap \Delta_t| = |\Gamma' \cap \Delta_t|$ where the Δ_t are the root systems of the simple ideals of $\mathfrak{g}$;

(2) if G is simple, then the $N_{\Psi-\Gamma,0}$, $\Gamma \subsetneqq \Psi$, are maximal parabolic subgroups of G_0, and every maximal parabolic subgroup of G_0 is conjugate to one of them.

Proof. Let $N_0 = \{g \in G_0 : g c_{\Psi-\Gamma}X_{\Gamma,0} = c_{\Psi-\Gamma}X_{\Gamma,0}\}$, $\mathfrak{n}_0$ its Lie algebra, $\mathfrak{n} = \mathfrak{n}_0^{\mathbf{C}}$, and N the complex analytic subgroup of G for $\mathfrak{n}$.

The unipotent radical $\mathfrak{n}^u_{\Psi-\Gamma,0}$ of $\mathfrak{n}_{\Psi-\Gamma,0}$ is the sum of the (-2)-eigenspace and the (-1) eigenspace of $\mathrm{ad}(x_{\Psi-\Gamma,0})$ on $\mathfrak{g}_0$. We check $\mathfrak{n}^u_{\Psi-\Gamma,0} \subset \mathfrak{n}_0$. Let δ be a root. If $\delta \in E_{-2}(\Psi - \Gamma)$, then (6.7) $\rho(\delta) = -\frac{1}{2}(\psi + \psi')$ with $\psi, \psi' \in \Psi - \Gamma$, so δ is a noncompact negative root. If $\delta \in E_{-1}(\Psi - \Gamma)$, then (6.8) there are three possibilities. If $\rho(\delta) = -\frac{1}{2}(\psi + \psi')$ with $\psi \in \Psi - \Gamma$ and

$\psi' \in \Gamma$, then δ is a noncompact negative root. If $\rho(\delta) = -\frac{1}{2}(\psi - \psi')$ with $\psi \in \Psi - \Gamma$ and $\psi' \in \Gamma$, then δ is a compact root. If $\rho(\delta) = -\frac{1}{2}\psi$ with $\psi \in \Psi - \Gamma$, then δ is a noncompact negative root. We have just shown, via Lemma 6.11, that $\mathfrak{n}^u_{\Psi-\Gamma} \subset \mathrm{ad}(c_{\Psi-\Gamma})(\mathfrak{k}^{\mathbf{C}} + \mathfrak{m}^-)$, Lie algebra of the isotropy subgroup of G at $c_{\Psi-\Gamma}x_0$. Thus $\mathfrak{n}^u_{\Psi-\Gamma,0} \subset \mathfrak{g}_0 \cap \mathrm{ad}(c_{\Psi-\Gamma})(\mathfrak{k}^{\mathbf{C}} + \mathfrak{m}^-)$, Lie algebra of the isotropy subgroup $G_0 \cap \mathrm{ad}(c_{\Psi-\Gamma})P$ of G_0 at $c_{\Psi-\Gamma}(x_0)$. Thus the analytic subgroup of G_0 for $\mathfrak{n}^u_{\Psi-\Gamma,0}$ preserves the holomorphic arc component $c_{\Psi-\Gamma}X_{\Gamma,0}$ of $G_0(c_{\Psi-\Gamma}x_0)$ through $c_{\Psi-\Gamma}(x_0)$. That completes our proof of $\mathfrak{n}^u_{\Psi-\Gamma,0} \subset \mathfrak{n}_0$.

The reductive part $\mathfrak{n}^r_{\Psi-\Gamma,0}$ of $\mathfrak{n}_{\Psi-\Gamma,0}$ is the 0-eigenspace of $\mathrm{ad}(x_{\Psi-\Gamma,0})$ on $\mathfrak{g}_0$. We check $\mathfrak{n}^r_{\Psi-\Gamma,0} \subset \mathfrak{n}_0$. Looking at (6.9), we see that

$$E_0(\Psi - \Gamma) = E_0'(\Psi - \Gamma) \cup E_0''(\Psi - \Gamma) \qquad \text{disjoint} \tag{6.14a}$$

where

$$E_0'(\Psi - \Gamma) = \{\delta \in E_0(\Psi - \Gamma) : \rho(\delta) = \tfrac{1}{2}(\psi - \psi');\ \psi, \psi' \in \Psi - \Gamma\}. \tag{6.14b}$$

That decomposes, via Lemma 6.11,

$$\mathfrak{n}^r_{\Psi-\Gamma} = \mathfrak{n}'_{\Psi-\Gamma} + \mathfrak{n}''_{\Psi-\Gamma} \qquad \text{vector space direct sum} \tag{6.15a}$$

where

$$\mathfrak{n}'_{\Psi-\Gamma} = \mathrm{ad}(c_{\Psi-\Gamma})\{\mathfrak{t}^{\mathbf{C}} + \sum_{\delta \in E_0'(\Psi-\Gamma)} \mathfrak{g}^{\delta}\} \tag{6.15b}$$

and

$$\mathfrak{n}''_{\Psi-\Gamma} = \mathrm{ad}(c_{\Psi-\Gamma}) \sum_{\delta \in E_0''(\Psi-\Gamma)} \mathfrak{g}^{\delta}. \tag{6.15c}$$

Every root $\delta \in E_0'(\Psi - \Gamma)$ is compact; thus $\mathfrak{n}'_{\Psi-\Gamma} \subset \mathrm{ad}(c_{\Psi-\Gamma})(\mathfrak{k}^{\mathbf{C}} + \mathfrak{m}^-)$, and it follows as above that $\mathfrak{n}_0$ contains the real form $\mathfrak{g}_0 \cap \mathfrak{n}'_{\Psi-\Gamma}$ of $\mathfrak{n}'_{\Psi-\Gamma}$. If $\delta \in E_0''(\Psi - \Gamma)$, then $-\delta \in E_0''(\Psi - \Gamma)$ and $\mathfrak{g}^{\delta} + \mathfrak{g}^{-\delta}$ centralizes $\mathfrak{g}[\Psi - \Gamma]$; thus $\mathfrak{n}''_{\Psi-\Gamma} \subset \mathfrak{g}_{\Gamma}$, so the real form $\mathfrak{g}_0 \cap \mathfrak{n}''_{\Psi-\Gamma}$ of $\mathfrak{n}''_{\Psi-\Gamma}$ is contained in $\mathfrak{g}_{\Gamma,0} \subset \mathfrak{n}_0$. That completes our proof that $\mathfrak{n}^r_{\Psi-\Gamma,0} \subset \mathfrak{n}_0$.

We have proved $\mathfrak{n}_{\Psi-\Gamma,0} \subset \mathfrak{n}_0$.

We digress to prove assertion (2). Suppose that G is simple and $\Psi = \{\psi_1, \ldots, \psi_r\}$. Let $\Gamma_k = \{\psi_1, \psi_2, \ldots, \psi_k\}$ for $0 \leqq k \leqq r - 1$. Then the orthocomplement of $E_0(\Psi - \Gamma_k)$ in the real dual of it^- is the span of the element $\psi_{k+1} + \cdots + \psi_r$, by (6.9) and the Restricted Root Theorem. Thus the reductive part $N^r_{\Psi-\Gamma_k,0}$ has identity component whose connected center is the product of a torus and a 1-dimensional vector group. It follows that the $N_{\Psi-\Gamma_k,0}$ are maximal parabolic subgroups of G_0. Again by the Restricted Root Theorem, the $N_{\Psi-\Gamma_k,0}$ are mutually nonconjugate, so, there being r of them, they represent every conjugacy class of maximal parabolic subgroup of G_0. Now (2) is proved, and incidentally (1) follows without further argument.

We had proved $\mathfrak{n}_{\Psi-\Gamma,0} \subset \mathfrak{n}_0$. To prove them equal, it suffices to consider the case where G is simple. There $\mathfrak{n}_{\Psi-\Gamma,0}$ is a maximal subalgebra of $\mathfrak{g}_0$ by the fact (2) just proved. If $\mathfrak{n}_0 = \mathfrak{g}_0$, then G_0 preserves $c_{\Psi-\Gamma}X_{\Gamma,0}$, hence acts on it by analytic automorphisms. Rendering that action effective, we obtain $G_{\Gamma,0}$ as a homomorphic image of the simple Lie group G_0, so either $X_0 = X_{\Gamma,0}$ or $X_{\Gamma,0}$ is reduced to a point. As $X_{\Gamma,0}$ has rank $|\Gamma| < |\Psi|$ that is impossible. Thus $\mathfrak{n}_{\Psi-\Gamma,0} = \mathfrak{n}_0$.

We have just proved $\mathfrak{n}_0 = \mathfrak{n}_{\Psi-\Gamma,0}$. Thus N_0 and $N_{\Psi-\Gamma,0}$ have the same identity component. The parabolic subgroup $N_{\Psi-\Gamma,0}$ of G_0 is the normalizer in G_0 of that identity component. Now $N_0 \subset N_{\Psi-\Gamma,0}$ open subgroup, and to prove equality there we need only show that every topological component of $N_{\Psi-\Gamma,0}$ has an element that normalizes $c_{\Psi-\Gamma}X_{\Gamma,0}$.

Recall the Cartan involution σ of G_0 with fixed point set K. As $\sigma(x_{\Psi-\Gamma,0}) = -x_{\Psi-\Gamma,0}$ now

$$\sigma(\mathfrak{n}^r_{\Psi-\Gamma,0}) = \mathfrak{n}^r_{\Psi-\Gamma,0} \qquad \text{and} \qquad \mathfrak{g}_0 = \mathfrak{n}_{\Psi-\Gamma,0} + \sigma(\mathfrak{n}^u_{\Psi-\Gamma,0}). \tag{6.16a}$$

Let $N^u_{\Psi-\Gamma,0}$ denote the unipotent radical $\exp(\mathfrak{n}^u_{\Psi-\Gamma,0})$ of $N_{\Psi-\Gamma,0}$, so $N_{\Psi-\Gamma,0}$ is its G_0-normalizer. Now

$$N^r_{\Psi-\Gamma,0} = \{g \in G_0 : \mathrm{ad}(g)x_{\Psi-\Gamma,0} = x_{\Psi-\Gamma,0}\} \tag{6.16b}$$

is a σ-stable subgroup of G_0, with Lie algebra $\mathfrak{n}^r_{\Psi-\Gamma,0}$, such that

$$N_{\Psi-\Gamma,0} = N^r_{\Psi-\Gamma,0} \cdot N^u_{\Psi-\Gamma,0} \qquad \text{semidirect product.} \tag{6.16c}$$

In particular,

$$K \cap N_{\Psi-\Gamma,0} = K \cap N^r_{\Psi-\Gamma,0}$$
$$\text{and meets every component of } N_{\Psi-\Gamma,0}. \tag{6.17a}$$

Thus the remaining assertion $N_0 = N_{\Psi-\Gamma,0}$ of our theorem is reduced to the assertion that

$$\text{every element of } K \cap N^r_{\Psi-\Gamma,0} \text{ leaves } c_{\Psi-\Gamma}(x_0) \text{ fixed.} \tag{6.17b}$$

Let $k \in K \cap N^r_{\Psi-\Gamma,0}$. Recall (2.17) that $\mathfrak{k}$ has a central element z such that

$J = \mathrm{ad}(z)|_{\mathfrak{m}}$ is the almost complex structure of X_0. Now apply $\mathrm{ad}(\exp(\pi/2)z)$ to the identity $\mathrm{ad}(k)x_{\Psi-\Gamma,0} = x_{\Psi-\Gamma,0}$. According to (3.4), the result is

$$\mathrm{ad}(k)y_{\Psi-\Gamma,0} = y_{\Psi-\Gamma,0}.$$

Multiply by i and compare (3.3) and (3.5):

$$\mathrm{ad}(k)y_{\Psi-\Gamma} = y_{\Psi-\Gamma}.$$

Exponentiate $\pi/4$ times each side, using (4.4):

$$kc_{\Psi-\Gamma}k^{-1} = c_{\Psi-\Gamma}.$$

Thus

$$k(c_{\Psi-\Gamma}x_0) = c_{\Psi-\Gamma}(kx_0) = c_{\Psi-\Gamma}x_0.$$

That completes our proof of $N_0 = N_{\Psi-\Gamma,0}$. Q.E.D.

Corollary 1. Let F be a boundary component of X_0 in X. Then the G_0-normalizer of F is a parabolic subgroup of G_0.

{For $F = k \cdot c_{\Psi-\Gamma}X_{\Gamma,0}$ with $k \in K$ and $\Gamma \subsetneqq \Psi$, so its G_0-normalizer is $\mathrm{ad}(k)N_{\Psi-\Gamma,0}$.}

Corollary 2. Let F be a boundary component of X_0 in X. Then the G_0-normalizer of F is transitive on the space X_0.

{For if N_0 is a parabolic subgroup of G_0, then $KN_0 = G_0$.}

Corollary 3. The space of all boundary components of X_0 in X that are contained in a given G_0-orbit on the boundary, is the set of real points in a G-homogeneous projective variety defined over the rational number field.

{For the boundary orbit is $G_0(c_{\Psi-\Gamma}(x_0))$, $\Gamma \subsetneqq \Psi$, and its space of holomorphic arc components is $G_0/N_{\Psi-\Gamma,0} = (G/N_{\Psi-\Gamma})_{\mathbf{R}}$.}

Corollary 3 says that the space of boundary components $kc_{\Psi-\Gamma}X_{\Gamma,0}$, $k \in K$, of a given type Γ, is a "real flag manifold" $G_0/N_{\Psi-\Gamma,0}$. We take a careful look at the structure of that real flag manifold.

Boundary Flag Theorem. Let $\Gamma \subsetneqq \Psi$ and consider the fibration

$$\pi : G_0(c_{\Psi-\Gamma}x_0) \to G_0/N_{\Psi-\Gamma,0} \qquad \text{by} \qquad \pi(gc_{\Psi-\Gamma}x_0) = gN_{\Psi-\Gamma,0} \quad (6.18a)$$

of the corresponding boundary orbit over the space of boundary components of type Γ.

1. $K(c_{\Psi-\Gamma}x_0)$ meets each π-fiber (i.e., each boundary component contained in $G_0(c_{\Psi-\Gamma}x_0)$) in just one point, so it can be identified with the base of the fibration under $kN_{\Psi-\Gamma,0} \leftrightarrow k(c_{\Psi-\Gamma}x_0)$. Thus π is a fibration

$$\pi : G_0(c_{\Psi-\Gamma}x_0) \to K(c_{\Psi-\Gamma}x_0) \qquad K\text{-equivariant} \qquad (6.18b)$$

such that

$$\pi^{-1}(kc_{\Psi-\Gamma}x_0) = kc_{\Psi-\Gamma}X_{\Gamma,0} \quad \text{boundary component through} \quad kc_{\Psi-\Gamma}x_0. \qquad (6.18c)$$

2. The space $K(c_{\Psi-\Gamma}x_0) = G_0/N_{\Psi-\Gamma,0}$ of boundary components contained in the orbit $G_0(c_{\Psi-\Gamma}x_0)$, is fibered

$$\eta : K(c_{\Psi-\Gamma}x_0) \to K(c^2_{\Psi-\Gamma}x_0) \qquad \text{by} \qquad \eta(kc_{\Psi-\Gamma}x_0) = kc^2_{\Psi-\Gamma}x_0 \quad (6.19)$$

over the simply connected complex totally geodesic submanifold $K(c^2_{\Psi-\Gamma}x_0)$ of X, with fibers connected and totally geodesic in X. Thus the base of η is a compact Hermitian symmetric space and the η-fibers are compact Riemannian symmetric spaces.

3. The following conditions are equivalent:

(3a) The base $K(c^2_{\Psi-\Gamma}x_0)$ of η is a single point.

(3b) The total space $K(c_{\Psi-\Gamma}x_0)$ of η is totally geodesic in X.

(3c) The partial Cayley transform $c_{\Psi-\Gamma}$ has $c^4_{\Psi-\Gamma} = 1$.

(3d) Decompose the boundary component $c_{\Psi-\Gamma}X_{\Gamma,0} = c_{\Psi_1-\Gamma_1}X_{\Gamma_1,0} \times \cdots \times c_{\Psi_q-\Gamma_q}X_{\Gamma_q,0}$ where the factors are the corresponding domains or boundary components of the irreducible factors of X_0. Then for $1 \leqq t \leqq q$, either $c_{\Psi_t-\Gamma_t}X_{\Gamma_t,0}$ is the t-factor of X_0, or that factor is of tube type and $c_{\Psi_t-\Gamma_t}X_{\Gamma_t,0}$ is a point on its Bergman–Šilov boundary.

Proof. Assertion (1) follows from (6.17) and the fact that K is transitive on the set of boundary components of X_0 that are in the orbit $G_0(c_{\Psi-\Gamma}x_0)$.

Recall the symmetry s of X_0 and X at x_0. Now $\sigma = \mathrm{ad}(s)$ sends $c_{\Psi-\Gamma}$ to $c^{-1}_{\Psi-\Gamma}$ because each $y_\psi \in \mathfrak{m}$. Similarly the conjugation τ of G over G_0 sends $c_{\Psi-\Gamma}$ to $c^{-1}_{\Psi-\Gamma}$ because $y_\psi \in i\mathfrak{g}_0$. As each $c^8_\psi = 1$, so $c^8_{\Psi-\Gamma} = 1$, now

$$\sigma(c^4_{\Psi-\Gamma}) = c^4_{\Psi-\Gamma} \qquad \text{and} \qquad \tau(c^4_{\Psi-\Gamma}) = c^4_{\Psi-\Gamma}. \qquad (6.20)$$

Suppose that $k \in K$ leaves $c_{\Psi-\Gamma}(x_0)$ fixed. Then k commutes with the symmetry of X at $c_{\Psi-\Gamma}x_0$. That symmetry is $\mathrm{ad}(c_{\Psi-\Gamma})s = c_{\Psi-\Gamma}^2 s$. Thus k commutes with $c_{\Psi-\Gamma}^2$. Now $\mathrm{ad}(k)$ leaves $\mathrm{ad}(c_{\Psi-\Gamma}^2)z$ fixed. Thus $k(c_{\Psi-\Gamma}^2 x_0) = c_{\Psi-\Gamma}^2 x_0$. This proves $\eta : K(c_{\Psi-\Gamma}x_0) \to K(c_{\Psi-\Gamma}^2 x_0)$ well defined.

The symmetry s' of X at $c_{\Psi-\Gamma}^2 x_0$ is $\mathrm{ad}(c_{\Psi-\Gamma}^2)s = c_{\Psi-\Gamma}^2 s c_{\Psi-\Gamma}^{-2} = c_{\Psi-\Gamma}^2 \cdot \sigma(c_{\Psi-\Gamma}^{-2}) \cdot s = c_{\Psi-\Gamma}^4 s$. Now, using (6.20),

$$\mathrm{ad}(s')|_K : K \to K \qquad \text{by} \qquad k \to \mathrm{ad}(c_{\Psi-\Gamma}^4)k. \tag{6.21}$$

First, this shows that $K(c_{\Psi-\Gamma}^2 x_0)$ is a (compact) totally geodesic submanifold of X. Second, it shows that the isotropy subgroup of K at $c_{\Psi-\Gamma}^2 x_0$ contains the center of K. Thus $K(c_{\Psi-\Gamma}^2 x_0)$ is a symmetric space of compact type.

The almost complex structure on X at $c_{\Psi-\Gamma}^2 x_0$ is the transformation

$$J' = \mathrm{ad}(c_{\Psi-\Gamma}^2)z : \mathrm{ad}(c_{\Psi-\Gamma}^2)\mathfrak{m} \to \mathrm{ad}(c_{\Psi-\Gamma}^2)\mathfrak{m}.$$

Let $v \in \mathfrak{k} \cap \mathrm{ad}(c_{\Psi-\Gamma}^2)\mathfrak{m}$, i.e., $v \in \mathfrak{k}$ such that (by (6.21)) $\mathrm{ad}(c_{\Psi-\Gamma}^4)v = -v$.

$$\begin{aligned} \text{Now } \sigma[\mathrm{ad}(c_{\Psi-\Gamma}^2)z,v] &= [\mathrm{ad}(\sigma(c_{\Psi-\Gamma})^2)\cdot\sigma z, \sigma v] \\ &= [\mathrm{ad}(c_{\Psi-\Gamma}^{-2})z, v] = [\mathrm{ad}(c_{\Psi-\Gamma}^2)z, v], \text{ so } [\mathrm{ad}(c_{\Psi-\Gamma}^2)z, v] \in \mathfrak{k}. \end{aligned}$$

In particular $K(c_{\Psi-\Gamma}^2 x_0)$ is a complex submanifold of X. We have proved that $K(c_{\Psi-\Gamma}^2 x_0)$ is a Hermitian symmetric subspace of compact type in X. It follows that $K(c_{\Psi-\Gamma}^2 x_0)$ is simply connected. This completes the proof of the assertions of (2) concerning the base space of the fibration (6.19).

Let L be the isotropy subgroup of K at $c_{\Psi-\Gamma}^2 x_0$. Thus $L(c_{\Psi-\Gamma}x_0) = \eta^{-1}(c_{\Psi-\Gamma}^2 x_0)$, η-fiber. L is connected because $K(c_{\Psi-\Gamma}^2 x_0) \cong K/L$ is simply connected. Now η has connected fibers, and the Lie algebra of L is given (6.21) by

$$\mathfrak{l} = \{v \in \mathfrak{k} : \mathrm{ad}(c_{\Psi-\Gamma}^4)\, v = v\}.$$

Now observe

$$\tau\cdot\mathrm{ad}(c_{\Psi-\Gamma}^2)\mathfrak{l} = \mathrm{ad}(c_{\Psi-\Gamma}^{-2})\,(\tau\mathfrak{l}) = \mathrm{ad}(c_{\Psi-\Gamma}^{-2})\mathfrak{l} = \mathrm{ad}(c_{\Psi-\Gamma}^2)\mathfrak{l}$$

and

$$\sigma\cdot\mathrm{ad}(c_{\Psi-\Gamma}^2)\mathfrak{l} = \mathrm{ad}(c_{\Psi-\Gamma}^{-2})\,(\sigma\mathfrak{l}) = \mathrm{ad}(c_{\Psi-\Gamma}^{-2})\mathfrak{l} = \mathrm{ad}(c_{\Psi-\Gamma}^2)\mathfrak{l}.$$

Thus

$$\mathrm{ad}(c_{\Psi-\Gamma}^2)\mathfrak{l} \subset \mathfrak{g}_0 \cap \mathfrak{k}^{\mathbb{C}} = \mathfrak{k}.$$

It follows that

$$\mathrm{ad}(\mathrm{ad}(c_{\Psi-\Gamma})s) : L \to L \qquad \text{by} \qquad l \to \mathrm{ad}(c^2_{\Psi-\Gamma})l.$$

Now the space $L(c_{\Psi-\Gamma}x_0)$ is totally geodesic in X. That completes the proof of the assertions of (2) concerning the fibers of the fibration (6.19), thus completing the proof of (2).

In proving (3) we may, and do, assume G simple.

(3a) implies (3b) because the η-fibers (6.19) are totally geodesic in X.

Assume (3b). Then K is stable under conjugation by $\mathrm{ad}(c_{\Psi-\Gamma})s = c^2_{\Psi-\Gamma}s$, hence under $\mathrm{ad}(c^2_{\Psi-\Gamma})$. Now $\mathrm{ad}(c^2_{\Psi-\Gamma})s$ is a central element of order 2 in K, hence necessarily equal to s. Thus $c^2_{\Psi-\Gamma}$ commutes with s, so $c^4_{\Psi-\Gamma} = c^2_{\Psi-\Gamma} \cdot \mathrm{ad}(s)c^2_{\Psi-\Gamma} = c^2_{\Psi-\Gamma}c^{-2}_{\Psi-\Gamma} = 1$. Now (3b) implies (3c).

(3c) implies (3a) by a glance at the symmetry to X at $c_{\Psi-\Gamma}(x_0)$.

Now (3a), (3b), and (3c) are equivalent.

(3d) implies (3c) rather trivially. Now assume (3c). Then (5.19), with roles of Γ and $\Psi - \Gamma$ interchanged, says $X_0 = X_{\Psi-\Gamma,0} \times X_{\Gamma,0}$. However we have assumed G simple, thus X_0 irreducible; so either Γ or $\Psi - \Gamma$ is empty. If $\Psi - \Gamma$ is empty, then $c_{\Psi-\Gamma}X_{\Gamma,0} = X_0$. If Γ is empty, then $c_{\Psi-\Gamma}x_0 = c_{\Psi-\Gamma}X_{\Gamma,0}$ point on the Šilov boundary, and X_0 is tube type by (3c). Now (3c) implies (3d), so (3c) and (3d) are equivalent. This completes the proof of (3). Q.E.D.

Part II. Holomorphic Arc Component Theory

Suppose that X_0 is irreducible and of rank r. In Section 5, we saw that the topological boundary bd X_0 of X_0 in X is the union of r distinct G_0-orbits, say $G_0(x_k)$ for $1 \leqq k \leqq r$ where $x_k = c_\Gamma x_0$ and $|\Gamma| = k$, and that the holomorphic arc components of $G_0(x_k)$ are noncompact type irreducible Hermitian symmetric spaces of rank $r - k$ which are transitively permuted by K. In Section 6 we saw that the G_0-normalizers of these boundary components are just the maximal proper parabolic subgroups of G_0 and that $K(x_k)$ represented the space of boundary components contained in $G_0(x_k)$. In Part II, we extend those results to arbitrary G_0-orbits on X.

The G_0-orbit structure of X is worked out in Section 7, simplifying the methods of Takeuchi [*10*] and Wolf [*16*]. The idea is to use the whole polysphere $G[\Psi](x_0)$ rather than the polydisk $G_0[\Psi](x_0)$. The problem comes down to that of distinguishing open orbits, and we do that by using results of Section 6.

The holomorphic arc components of a G_0-orbit are worked out in Section 8. They turn out to be the K-translates of the open $G_{\Psi-\Gamma,0}$-orbits on a certain translate of X_Γ. The method here eliminates a lot of machinery from my original method [*16*], relying on the results of Sections 5–7.

The general open G_0-orbits are indefinite metric versions of X_0 but may have compact subvarieties. See the example of the ball in $\mathbf{C}^n$ in the Introduction. Takeuchi [*10*] showed that an open G_0-orbit is holomorphically fibered over a certain maximal compact subvariety, and I showed [*16*] that the corresponding fibrations of the holomorphic arc components of a G_0-orbit fit together. That material is worked out in Section 9 with a few simplifications. For the full story, however, one should see Section 11 of [*16*].

7. Orbit Structure

The Boundary Orbit Theorem of Section 5 gives the G_0-orbit structure of the topological closure of X_0 in X. There the main trick was the use of the polydisk $G_0[\Psi](x_0)$. Here we use the whole polysphere $G[\Psi](x_0)$ to obtain the G_0-orbit structure of X. The result is in Takeuchi [*10*] and Wolf [*16*], but we avoid most of the heavy machinery used in the arguments of [*10*] and [*16*].

To see the idea, let X be a complex Grassmannian, consisting of all k-dimensional linear subspaces of $\mathbf{C}^n$. The group $\mathbf{GL}(n,\mathbf{C})$ of all invertible linear transformations of $\mathbf{C}^n$ acts on X. The kernel of the action consists of the scalar transformations. Dividing out that kernel, we obtain

$G = \mathbf{GL}(n,\mathbf{C})$ / (scalars) : complex group of automorphisms of X.

Consider the nondegenerate Hermitian form on $\mathbf{C}^n$ given by

$$b(u,v) = -\sum_{j=1}^{k} u^j\bar{v}^j + \sum_{j=k+1}^{n} u^j\bar{v}^j.$$

The subgroup of $\mathbf{GL}(n,\mathbf{C})$ preserving b is the indefinite unitary group usually denoted $\mathbf{U}^k(n)$ or $\mathbf{U}(k, n-k)$. Dividing out the kernel of its action on X, we obtain

$$G_0 = \mathbf{U}^k(n) / (\text{scalars}) : \text{real group of automorphisms of } X.$$

Witt's Theorem says: two k-dimensional subspaces of $\mathbf{C}^n$ are $\mathbf{U}^k(n)$-equivalent if, and only if, the restrictions of b to those subspaces have the same rank and same signature. That gives the G_0-orbit structure of X. The noncompact Hermitian symmetric space X_0 consists of the (using b) positive definite subspaces of $\mathbf{C}^n$; its boundary orbits are the sets consisting of all positive semidefinite subspaces on which b has rank t, $n-k-r \leqq t <$

$n - k$ where $r = \min(k, n - k)$ is the symmetric space rank of X_0 and X. The open G_0 orbits are the sets of k-planes on which b has a fixed nondegenerate signature; there are $r + 1$ of them. There are $\frac{1}{2}(r + 1)(r + 2)$ orbits in all.

We now return to the general case.

To minimize notation, we denote

$$\Delta_t,\ 1 \leqq t \leqq q : \text{root systems of the simple ideals of } \mathfrak{g}. \tag{7.1}$$

$$x_{\Gamma,\Sigma} = c_\Gamma c_\Sigma^2 x_0 \qquad \text{whenever} \quad \Gamma, \Sigma \subset \Psi. \tag{7.2}$$

Orbit Structure Theorem. The G_0-orbits on X are just the $G_0(x_{\Gamma,\Sigma})$ where Γ and Σ are disjoint subsets of Ψ. Furthermore, $G_0(x_{\Gamma',\Sigma'})$ is in the closure of $G_0(x_{\Gamma,\Sigma})$ if, and only if, for $1 \leqq t \leqq q$,

$$|(\Sigma' - \Gamma') \cap \Delta_t| \leqq |(\Sigma - \Gamma) \cap \Delta_t| \quad \text{and} \quad |(\Sigma \cup \Gamma) \cap \Delta_t| \leqq |(\Sigma' \cup \Gamma') \cap \Delta_t|.$$

In particular,

(i) $G_0(x_{\Gamma',\Sigma'}) = G_0(x_{\Gamma,\Sigma})$ if, and only if, for $1 \leqq t \leqq q$, $|(\Sigma' - \Gamma') \cap \Delta_t| = |(\Sigma - \Gamma) \cap \Delta_t|$ and $|\Gamma' \cap \Delta_t| = |\Gamma \cap \Delta_t|$;

(ii) the number of G_0-orbits on X is $\frac{1}{2}\sum_{t=1}^{q}(r_t + 1)(r_t + 2)$ where $r_t = |\Psi \cap \Delta_t|$;

(iii) $G_0(x_{\Gamma,\Sigma})$ is open in X if, and only if, Γ is empty; so there are $\sum_{t=1}^{q}(r_t + 1) = |\Psi| + q$ open orbits;

(iv) the boundary of a typical open orbit $G_0(x_{\phi,\Sigma})$ is the union of the $G_0(x_{\Gamma',\Sigma'})$, $\Sigma' - \Gamma' \subset \Sigma \subset \Sigma' \cup \Gamma'$;

(v) the Bergman–Šilov boundary $G_0(x_{\Psi,\phi})$ of X_0 in X is in the closure of every orbit and is the unique closed G_0-orbit on X.

Proof. Let φ be a noncompact positive root, $\mathfrak{g}[\varphi] = h_\varphi \mathbf{C} + \mathfrak{g}^\varphi + \mathfrak{g}^{-\varphi}$ three-dimensional complex simple algebra, $\mathfrak{g}_0[\varphi] = \mathfrak{g}_0 \cap \mathfrak{g}[\varphi]$ noncompact real form; $G[\varphi] \subset G$ and $G_0[\varphi] \subset G_0$ analytic subgroups for $\mathfrak{g}[\varphi]$ and $\mathfrak{g}_0[\varphi]$; $S[\varphi] = G[\varphi](x_0)$ Riemann sphere and $S_0[\varphi] = G_0[\varphi](x_0)$ lower hemisphere. Then the $G_0[\varphi]$-orbits on $S[\varphi]$ are

lower hemisphere:	$G_0[\varphi]\,(\exp((\pi/4)y_\varphi))^n x_0,$	$n \equiv 0 \bmod 4,$
equator	$G_0[\varphi]\,(\exp((\pi/4)y_\varphi))^n x_0,$	$n \equiv 1 \bmod 2,$
upper hemisphere:	$G_0[\varphi]\,(\exp((\pi/4)y_\varphi))^n x_0,$	$n \equiv 2 \bmod 4.$

Let $S[\Psi] = G[\Psi](x_0)$, product of the $|\Psi|$ Riemann spheres $G[\psi](x_0) = S[\psi]$, $\psi \in \Psi$. The $G_0[\Psi]$-orbits on $S[\Psi]$ are the products of the form

$\prod_{\psi\in\Psi} D_\psi$ where D_ψ is a $G_0[\psi]$-orbit on $S[\psi]$. Given such an orbit, define

$$\Gamma = \{\psi \in \Psi : D_\psi \text{ is the equator of } S[\psi]\}, \qquad \text{and}$$
$$\Sigma = \{\psi \in \Psi : D_\psi \text{ is the upper hemisphere of } S[\psi]\}.$$

Then $\Gamma, \Sigma \subset \Psi$ disjoint, and the orbit is $G_0[\Psi](x_{\Gamma,\Sigma})$.

Recall (3.8) and the corresponding decomposition $G_c = KA_cK$ of the compact real form of G. Now

$$X = G_c(x_0) = KA_cK(x_0) = KA_c(x_0) = K\cdot S[\Psi].$$

If $x \in X$, now $x = k(x')$ where $k \in K$ and $x' \in S[\Psi]$. We have disjoint Γ, $\Sigma \subset \Psi$ such that $G_0[\Psi](x') = G_0[\Psi](x_{\Gamma,\Sigma})$. Now $G_0(x) = G_0(x') = G_0\cdot G_0[\Psi](x') = G_0\cdot G_0[\Psi](x_{\Gamma,\Sigma}) = G_0(x_{\Gamma,\Sigma})$. This proves the first assertion of the Orbit Structure Theorem.

We now prove that an orbit $G_0(x_{\Gamma,\Sigma})$ is open in X if, and only if, $\Gamma = \phi$ empty set. First suppose $G_0(x_{\Gamma,\Sigma})$ open in X. Then $G_0(x_{\Gamma,\Sigma}) \cap S[\Psi]$ is an open $G_0[\Psi]$-invariant subset of $S[\Psi]$, hence a union of sets $G_0[\Psi](x_{\Gamma',\Sigma'})$ that are open in $S[\Psi]$. In particular $G_0[\Psi](x_{\Gamma,\Sigma})$ is open in $S[\Psi]$; thus each of its factors along the $S[\psi]$, $\psi \in \Psi$, is a hemisphere; that proves $\Gamma = \phi$ empty. Conversely consider an orbit $G_0(x_{\phi,\Sigma})$. The isotropy subgroup of G at $x_{\phi,\Sigma}$ is $\mathrm{ad}(c_\Sigma^2)P$, so G_0 has isotropy $G_0 \cap \mathrm{ad}(c_\Sigma^2)P$ there. That isotropy subgroup of G_0 has Lie algebra $\mathfrak{g}_0 \cap \mathrm{ad}(c_\Sigma^2)\mathfrak{p}$ which contains $\mathfrak{t} = \mathrm{ad}(c_\Sigma^2)\mathfrak{t}$ and is stable under the complex conjugation τ of $\mathfrak{g}$ over $\mathfrak{g}_0$. From the latter, $\mathfrak{g}_0 \cap \mathrm{ad}(c_\Sigma^2)\mathfrak{p} = \mathfrak{g}_0 \cap [\mathrm{ad}(c_\Sigma^2)\mathfrak{p} \cap \tau\, \mathrm{ad}(c_\Sigma^2)\mathfrak{p}]$, real form of

$$\mathrm{ad}(c_\Sigma^2)\mathfrak{p} \cap \tau\,\mathrm{ad}(c_\Sigma^2)\mathfrak{p} = \mathfrak{t}^C + \sum_{E\cap\tau^*E} \mathfrak{g}^\varphi \tag{7.3a}$$

where E is the set of all $\mathfrak{t}^C$-roots φ such that $\mathfrak{g}^\varphi \subset \mathrm{ad}(c_\Sigma^2)\mathfrak{p}$. As $\tau^*\varphi = -\varphi$ for every root φ, this says

$$\mathrm{ad}(c_\Sigma^2)\mathfrak{p} \cap \tau\,\mathrm{ad}(c_\Sigma^2)\mathfrak{p} = (\mathrm{ad}(c_\Sigma^2)\mathfrak{p})^r = \mathrm{ad}(c_\Sigma^2)\mathfrak{p}^r = \mathrm{ad}(c_\Sigma^2)\mathfrak{k}^C. \tag{7.3b}$$

Thus $\dim_R G_0(x_{\phi,\Sigma}) = \dim_R \mathfrak{g}_0 - \dim_R \mathfrak{g}_0 \cap \mathrm{ad}(c_\Sigma^2)\mathfrak{p}$
$= \dim_R \mathfrak{g}_0 - \dim_C \mathrm{ad}(c_\Sigma^2)\mathfrak{k}^C = \dim_R \mathfrak{g}_0 - \dim_R \mathfrak{k} = \dim_R X$, proving $G_0(x_{\phi,\Sigma})$ open in X.

Next, we prove that open orbits $G_0(x_{\phi,\Sigma'}) = G_0(x_{\phi,\Sigma})$ if, and only if, $|\Sigma' \cap \Delta_t| = |\Sigma \cap \Delta_t|$ for $1 \leqq t \leqq q$. For that we may, and do, assume G simple. Then if $|\Sigma'| = |\Sigma|$, the Restricted Root Theorem provides a Weyl group element $\mathrm{ad}(k)|_{\mathfrak{t}^c}$, $k \in K$, that carries Σ' to Σ; so $G_0(x_{\phi,\Sigma}) = G_0(x_{\phi,ad(k)^*\Sigma'}) = G_0k(x_{\phi,\Sigma'}) = G_0(x_{\phi,\Sigma'})$. Conversely suppose that $G_0(x_{\phi,\Sigma})$

$= G_0(x_{\phi,\Sigma'})$. Then some $g \in G_0$ sends $x_{\phi,\Sigma}$ to $x_{\phi,\Sigma'}$. Replacing g by an element of $g \cdot [G_0 \cap \mathrm{ad}(c_{\Sigma'}^2)P]$ if necessary, we may also assume that $\mathrm{ad}(g)$ preserves $\mathrm{ad}(c_{\Sigma'}^2)\mathfrak{t} = \mathfrak{t}$. Now the computation (7.3) says

$$\mathrm{ad}(g)\ \mathrm{ad}(c_{\Sigma}^2)\mathfrak{k}^{\mathbf{C}} = \mathrm{ad}(c_{\Sigma'}^2)\mathfrak{k}^{\mathbf{C}}, \qquad \mathrm{ad}(g)\mathfrak{t} = \mathfrak{t}.$$

In other words, we have Weyl group elements $w_1, w_2 \in W_K$ such that

$$\mathrm{ad}(c_{\Sigma'}^2)\big|_{\mathfrak{t}} = w_1 \cdot \mathrm{ad}(c_{\Sigma}^2)\big|_{\mathfrak{t}} \cdot w_2.$$

Let z denote the central element (2.17) of $\mathfrak{k}$ that gives the almost complex structure of X_0 and X. If $\Phi \subset \Psi$, then $\mathrm{ad}(c_\Phi^2)z - z = \sum_\Phi i h_\varphi$, so $w_1(z) = z = w_2(z)$ says

$$\begin{aligned} -\sum_{\psi \in \Sigma'} h_\psi &= i\{\mathrm{ad}(c_{\Sigma'}^2)z - z\} = i\{w_1 \cdot \mathrm{ad}(c_\Sigma^2) \cdot w_2 z - z\} \\ &= i\{w_1 \cdot \mathrm{ad}(c_\Sigma^2)z - w_1 z\} = w_1\Big(-\sum_{\psi \in \Sigma} h_\psi\Big). \end{aligned}$$

Now (6.12) and Lemma 6.10 say that the parabolic subgroups $N_{\Sigma',0}$ and $N_{\Sigma,0}$ of G_0 are conjugate, and then the Boundary Group Theorem ensures $|\Sigma'| = |\Sigma|$.

Note that we have proved assertion (iii) on open orbits in the Orbit Structure Theorem.

We return for a moment to the $G_0[\Psi]$-orbit structure of $S[\Psi]$. The equator of $S[\Psi]$ had descriptions $G_0[\Psi](c_\psi x_0) = G_0[\Psi]\,(c_\psi^3 x_0)$. If $\Gamma, \Sigma \subset \Psi$, it follows that

$$x_{\Gamma,\Sigma} = c_{\Gamma-\Sigma} \cdot c_{\Gamma\cap\Sigma}^3 \cdot c_{\Sigma-\Gamma}^2 x_0 \in G_0[\Psi]\,(x_{\Gamma,\Sigma-\Gamma}). \tag{7.4a}$$

In other words,

$$G_0[\Psi](x_{\Gamma,\Sigma}) = G_0[\Psi](x_{\Gamma,\Sigma-\Gamma}). \tag{7.4b}$$

If $\Gamma, \Sigma, \Gamma', \Sigma' \subset \Psi$, it follows that $G_0[\Psi](x_{\Gamma',\Sigma'})$ is in the closure of $G_0[\Psi](x_{\Gamma,\Sigma})$ if, and only if,

(i) $\Gamma \subset \Gamma'$

(ii) $(\Sigma - \Gamma) \subset (\Sigma' \cup \Gamma')$. and

(iii) $\{\Psi - (\Sigma \cup \Gamma)\} \subset \{[\Psi - (\Sigma' \cup \Gamma')] \cup \Gamma'\}$.

In other words,

$$G_0[\Psi]\,(x_{\Gamma',\Sigma'}) \text{ is in the closure of } G_0[\Psi]\,(x_{\Gamma,\Sigma}) \text{ if and only if}$$
$$(\Sigma' - \Gamma') \subset (\Sigma - \Gamma) \quad \text{and} \quad (\Sigma \cup \Gamma) \subset (\Sigma' \cup \Gamma'). \tag{7.5}$$

Now consider an open orbit $G_0(x_{\phi,\Sigma})$. Its intersection with $S[\Psi]$ is $G_0[\Psi]$ invariant and open in $S[\Psi]$, hence is the union of all sets $G_0[\Psi](x_{\phi,B})$ such that $G_0(x_{\phi,B}) = G_0(x_{\phi,\Sigma})$. Thus

$$G_0(x_{\phi,\Sigma}) \cap S[\Psi] = \bigcup_{\mathscr{B}} G_0[\Psi](x_{\phi,B}) \tag{7.6a}$$

where

$$\mathscr{B} = \{B \subset \Psi : |B \cap \Delta_t| = |\Sigma \cap \Delta_t| \text{ for } 1 \leqq t \leqq q\}. \tag{7.6b}$$

Now an orbit $G_0(x_{\Gamma',\Sigma'})$ is in the closure of $G_0(x_{\phi,\Sigma})$ if and only if $x_{\Gamma',\Sigma'}$ is in that closure. Thus (7.5) with $\Gamma = \phi$ gives us

$$G_0(x_{\Gamma',\Sigma'}) \quad \text{is in the closure of} \quad G_0(x_{\phi,\Sigma}) \quad \text{iff some}$$
$$B \in \mathscr{B} \quad \text{has} \quad (\Sigma' - \Gamma') \subset B \subset (\Sigma' \cup \Gamma'). \tag{7.6c}$$

Now we reformulate (7.6) as

$$G_0(x_{\Gamma',\Sigma'}) \quad \text{is in the closure of} \quad G_0(x_{\phi,\Sigma}) \quad \text{iff, for} \quad 1 \leqq t \leqq q,$$
$$|(\Sigma' - \Gamma') \cap \Delta_t| \leqq |\Sigma \cap \Delta_t| \leqq |(\Sigma' \cup \Gamma') \cap \Delta_t|. \tag{7.7}$$

Now consider arbitrary orbits $G_0(x_{\Gamma,\Sigma})$ and $G_0(x_{\Gamma',\Sigma'})$. From (7.7), $G_0\ (x_{\Gamma,\Sigma})$ is in the closure of an open orbit $G_0(x_{\phi,\Phi})$ if and only if

$$|(\Sigma - \Gamma) \cap \Delta_t| \leqq |\Phi \cap \Delta_t| \leqq |(\Sigma \cup \Gamma) \cap \Delta_t|.$$

In particular

$$G_0(x_{\Gamma,\Sigma}) \quad \text{is in the closures of} \quad G_0(x_{\phi,\Sigma-\Gamma}) \quad \text{and} \quad G_0(x_{\phi,\Sigma\cup\Gamma}). \tag{7.8a}$$

Again using (7.7),

$$\text{if} \quad G_0(x_{\Gamma',\Sigma'}) \text{ is in the closure of } G_0(x_{\phi,\Sigma-\Gamma}),$$
$$\text{then} \quad |(\Sigma' - \Gamma') \cap \Delta_t| \leqq |(\Sigma - \Gamma) \cap \Delta_t| \quad \text{for} \quad 1 \leqq t \leqq q, \tag{7.8b}$$

and

$$\text{if} \quad G_0(x_{\Gamma',\Sigma'}) \quad \text{is in the closure of} \quad G_0(x_{\phi,\Sigma\cup\Gamma})$$
$$\text{then} \quad |(\Sigma \cup \Gamma) \cap \Delta_t| \leqq |(\Sigma' \cup \Gamma') \cap \Delta_t| \quad \text{for} \quad 1 \leqq t \leqq q. \tag{7.8c}$$

If $G_0(x_{\Gamma',\Sigma'})$ is in the closure of $G_0(x_{\Gamma,\Sigma})$ then (7.8a) says that it is in the closures of $G_0(x_{\phi,\Sigma-\Gamma})$ and $G_0(x_{\phi,\Sigma\cup\Gamma})$, whence (7.8b) and (7.8c) say that it satisfies

$$|(\Sigma' - \Gamma') \cap \Delta_t| \leqq |(\Sigma - \Gamma) \cap \Delta_t| \quad \text{and}$$
$$|(\Sigma \cup \Gamma) \cap \Delta_t| \leqq |(\Sigma' \cup \Gamma') \cap \Delta_t| \quad \text{for} \quad 1 \leqq t \leqq q.$$

Conversely, if $G_0(x_{\Gamma',\Sigma'})$ satisfies those numerical conditions, then some $\mathrm{ad}(k)|_{\mathfrak{t}} \in W_K$ sends $\Sigma' - \Gamma'$ into $\Sigma - \Gamma$ and $\Sigma \cup \Gamma$ into $\Sigma' \cup \Gamma'$, so we may assume $(\Sigma' - \Gamma') \subset (\Sigma - \Gamma)$ and $(\Sigma \cup \Gamma) \subset (\Sigma' \cup \Gamma')$, whence (7.5) shows that $G_0(x_{\Gamma',\Sigma'})$ is in the closure of $G_0(x_{\Gamma,\Sigma})$. This proves the second assertion of the Orbit Structure Theorem.

Orbits $G_0(x_{\Gamma',\Sigma'}) = G_0(x_{\Gamma,\Sigma})$ if and only if each is in the closure of the other. Now that is the same as

$$|(\Sigma' - \Gamma') \cap \Delta_t| = |(\Sigma - \Gamma) \cap \Delta_t| \qquad \text{and}$$
$$|(\Sigma' \cup \Gamma') \cap \Delta_t| = |(\Sigma \cup \Gamma) \cap \Delta_t|$$

for $1 \leqq t \leqq q$, which is equivalent to the condition of the particular assertion (i). Now enumerate

$$\Psi \cap \Delta_t = \{\psi_{t,1}, \ldots, \psi_{t,r_t}\}.$$

Then the choices of (Γ,Σ) leading to distinct G_0-orbits are given by

$$\Gamma \cap \Delta_t = \{\psi_{t,1}, \ldots, \psi_{t,j}\}, \qquad 0 \leqq j \leqq r_t$$
$$\Sigma \cap \Delta_t = \{\psi_{t,j+1}, \ldots, \psi_{t,k}\}, \qquad j \leqq k \leqq r_t.$$

The number of such partitions is

$$\sum_{t=1}^{q} \sum_{j=0}^{r_t} (r_t + 1 - j) = \sum_{t=1}^{q} \{(r_t + 1)^2 - \sum_{j=0}^{r_t} j\} = \tfrac{1}{2} \sum_{t=1}^{q} (r_t + 1)(r_t + 2).$$

That proves the particular assertion (ii). Recall that we have already proved (iii). (iv) and (v) follow easily. Q.E.D.

8. Holomorphic Arc Components

The Orbit Structure Theorem of Section 7 says that the G_0-orbits on X are the $G_0(x_{\Gamma,\Sigma})$, where $x_{\Gamma,\Sigma} = c_\Gamma c_\Sigma^2 x_0$ and $\Gamma, \Sigma \subset \Psi$ are disjoint. It also gives the relative position of those orbits. Here we work out the decomposition of an arbitrary orbit $G_0(x_{\Gamma,\Sigma})$ into holomorphic arc components, due to Wolf [*16*]. This is done in Section 5 for the boundary orbits $G_0(x_{\Gamma,\phi})$, and the method of [*16*] is a rather complicated reduction to the boundary orbit case. Here we simplify the reduction, substituting some specific information about Hermitian symmetric spaces for the general theory [*16*, Chapter II] of holomorphic arc components.

Recall that the boundary components of $X_0 = G_0(x_0)$ in X are the holomorphic arc components of the topological boundary bd X_0. They are the

$$kG_{\Psi-\Gamma,0}(c_\Gamma x_0) = kG_{\Psi-\Gamma,0}(x_{\Gamma,\phi}); \qquad \Gamma \subsetneqq \Psi, \quad k \in K$$

as described in Section 5. Thus the boundary components of X_0 in X are the holomorphic arc components of G_0-orbits on bd X_0. The general idea of the reduction is to show that the holomorphic arc components of an arbitrary orbit $G_0(x_{\Gamma,\Sigma})$, Γ and Σ disjoint in Ψ, are the

$$kG_{\Psi-\Gamma,0}(x_{\Gamma,\Sigma}), \qquad k \in K,$$

by comparing $G_0(x_{\Gamma,\Sigma}) = G_0(c_\Sigma^2 x_{\Gamma,\phi})$ with $G_0(x_{\Gamma,\phi})$.

Holomorphic Arc Component Theorem. Let Γ, $\Sigma \subset \Psi$ disjoint. Then the holomorphic arc components of $G_0(x_{\Gamma,\Sigma})$ are the $k \cdot G_{\Psi-\Gamma,0}(x_{\Gamma,\Sigma})$, $k \in K$. Furthermore, they have the following structure.

(1) $G_{\Psi-\Gamma}(x_{\Gamma,\Sigma}) = c_\Gamma X_{\Psi-\Gamma}$ complex totally geodesic submanifold of X that is a Hermitian symmetric space of compact type and rank $|\Psi - \Gamma|$.

(2) $G_{\Psi-\Gamma,0}(x_{\Gamma,\Sigma})$ is an open $G_{\Psi-\Gamma,0}$-orbit on $G_{\Psi-\Gamma}(x_{\Gamma,\Sigma})$ and is an indefinite-Kaehler symmetric space of $G_{\Psi-\Gamma,0}$; the isotropy subgroups of $G_{\Psi-\Gamma,0}$ at $x_{\Gamma,\Sigma}$ and $x_{\Gamma,\phi}$ have complexifications isomorphic under $\mathrm{ad}(c_\Sigma^2)$.

(3) The G_0-normalizer of the holomorphic arc component $G_{\Psi-\Gamma,0}(x_{\Gamma,\Sigma})$ is $N_{\Gamma,0}$, independent of Σ; the action of $N_{\Gamma,0}$ on $G_{\Psi-\Gamma,0}(x_{\Gamma,\Sigma})$ factors through the action of $G_{\Psi-\Gamma,0}$ there.

Proof. Using the definition of $x_{\Gamma,\Sigma}$, then $c_\Sigma \in G_{\Psi-\Gamma}$ (because Γ and Σ are disjoint), then the fact that c_Γ centralizes $G_{\Psi-\Gamma}$, and finally the definition of $X_{\Psi-\Gamma}$, we compute $G_{\Psi-\Gamma}(x_{\Gamma,\Sigma}) = G_{\Psi-\Gamma}(c_\Sigma^2 c_\Gamma x_0) = G_{\Psi-\Gamma}(c_\Gamma x_0) = c_\Gamma G_{\Psi-\Gamma}(x_0)$ $= c_\Gamma X_{\Psi-\Gamma}$. That proves (1).

Apply the Orbit Structure Theorem of Section 7 to $c_\Gamma X_{\Psi-\Gamma}$. It shows that $G_{\Psi-\Gamma,0}(x_{\Gamma,\Sigma})$ is an open $G_{\Psi-\Gamma,0}$ orbit there. Now (7.3) says that the complexified isotropy groups of $G_{\Psi-\Gamma,0}$ at $x_{\Gamma,\Sigma}$ and $x_{\Gamma,\phi}$ are $\mathrm{ad}(c_\Sigma^2)$-conjugate. As the boundary component $G_{\Psi-\Gamma,0}(x_{\Gamma,\phi})$ of X_0 in X is a Hermitian symmetric space of $G_{\Psi-\Gamma,0}$, the assertions of (2) on $G_{\Psi-\Gamma,0}(x_{\Gamma,\Sigma})$ follow.

We may, and do, assume G simple. Thus the normalizer $N_{\Gamma,0}$ of the boundary component $G_{\Psi-\Gamma,0}(x_{\Gamma,\phi})$ is a maximal parabolic subgroup of G_0. From its action on the boundary component,

$$N_{\Gamma,0} = G_{\Psi-\Gamma,0} \cdot Z_{\Gamma,0} \qquad \text{semidirect product,} \tag{8.1a}$$

$$Z_{\Gamma,0} \quad \text{leaving fixed every point of} \quad G_{\Psi-\Gamma,0}(x_{\Gamma,\phi}). \tag{8.1b}$$

Now by complex analyticity,

$$Z_{\Gamma,0} \quad \text{acts trivially on} \quad G_{\Psi-\Gamma}(x_{\Gamma,\phi}) = c_\Gamma X_{\Psi-\Gamma}. \tag{8.1c}$$

In particular $Z_{\Gamma,0}$ preserves $G_{\Psi-\Gamma,0}(x_{\Gamma,\Sigma})$. Thus

$$N_{\Gamma,0} \subset \{g \in G_0 : gG_{\Psi-\Gamma,0}(x_{\Gamma,\Sigma}) = G_{\Psi-\Gamma,0}(x_{\Gamma,\Sigma})\}.$$

If that inclusion is strict, then maximality of $N_{\Gamma,0}$ in G_0 implies G_0-stability of $G_{\Psi-\Gamma,0}(x_{\Gamma,\Sigma})$, hence of its Zarisky closure $c_\Gamma X_{\Psi-\Gamma}$, and thus of the interior of bd $X_0 \cap c_\Gamma X_{\Psi-\Gamma}$. That interior being $G_{\Psi-\Gamma,0}(x_{\Gamma,\phi})$, its G_0-stability implies $\Gamma = \phi$, so then $G_{\Psi-\Gamma,0} = G_0 = N_{\Gamma,0}$. Now we have proved that

$$N_{\Gamma,0} = \{g \in G_0 : gG_{\Psi-\Gamma,0}(x_{\Gamma,\Sigma}) = G_{\Psi-\Gamma,0}(x_{\Gamma,\Sigma})\}. \tag{8.2}$$

Thus (3) is proved, subject to completion of the proof that $G_{\Psi-\Gamma,0}(x_{\Gamma,\Sigma})$ is a holomorphic arc component of $G_0(x_{\Gamma,\Sigma})$.

Let D be the unit disk in $\mathbf{C}$ and $f : D \to X$ a holomorphic arc in $G_0(x_{\Gamma,\Sigma})$. We will prove $f(D) \subset kG_{\Psi-\Gamma,0}(x_{\Gamma,\Sigma})$ for some $k \in K$. For that it suffices to show that every $z \in D$ has a neighborhood U_z such that $f(U_z) \subset k_z G_{\Psi-\Gamma,0}(x_{\Gamma,\Sigma})$ where $k_z \in K$; for then every compact connected subset of D has image in some $kG_{\Psi-\Gamma,0}(x_{\Gamma,\Sigma})$, and we exhaust D by an increasing sequence of compact connected subsets. Preceding f by a linear fractional transformation of D, we may assume the given $z \in D$ to be 0. Observe (8.2) that the G_0-normalizers of the $gG_{\Psi-\Gamma,0}(x_{\Gamma,\Sigma})$ are real parabolic subgroups, hence self-normalizing in G_0, so the action of G_0 permutes them; thus we may follow f by an element of G_0 and assume $f(0) = x_{\Gamma,\Sigma}$. Now the proof that $f(D)$ is in some $kG_{\Psi-\Gamma}(x_{\Gamma,\Sigma})$ is reduced to: if $f(0) = x_{\Gamma,\Sigma}$ then $f(U) \subset G_{\Psi-\Gamma,0}(x_{\Gamma,\Sigma})$ for some neighborhood U of 0 in D.

$f : D \to X$ is a holomorphic arc in $G_0(x_{\Gamma,\Sigma})$ such that $f(0) = x_{\Gamma,\Sigma}$. If $0 < r \leqq 1$, then

$$D_r = \{z \in \mathbf{C} : |z| < r\}, \qquad \text{so} \quad D = D_1. \tag{8.3}$$

We will check that there exist $r > 0$ and holomorphic maps

$$f_1 : D_r \to K^{\mathbf{C}} \qquad \text{and} \qquad f_2 : D_r \to N_\Gamma \tag{8.4a}$$

such that

$$f(z) = f_1(z) \cdot f_2(z) \cdot x_{\Gamma,\Sigma} \qquad \text{for} \quad z \in D_r, \tag{8.4b}$$

$$f_1(0) = 1 = f_2(0) \quad \text{and} \quad f_1(D) \subset K \cdot (K^{\mathbf{C}} \cap N_\Gamma). \tag{8.4c}$$

First recall that $g \to g(x_{\Gamma,\Sigma})$ defines a holomorphic fiber bundle $G \to X$. For r_1 sufficiently small, $f(D_{r_1})$ lies in a locally trivializing open set of X for the bundle. Thus the restriction of f lifts to a holomorphic map,

$$f' : D_{r_1} \to G \quad \text{with} \quad f'(0) = 1 \quad \text{and} \quad f(z) = f'(z)\cdot x_{\Gamma,\Sigma}.$$

Next observe $G_0 = KN_{\Gamma,0}$ as $N_{\Gamma,0}$ is parabolic in G_0. Thus $\mathfrak{g}_0 = \mathfrak{k} + \mathfrak{n}_{\Gamma,0}$, so $\mathfrak{g} = \mathfrak{k}^{\mathbf{C}} + \mathfrak{n}_\Gamma$. Using canonical coordinates on G, that gives us neighborhoods U of 1 in $K^{\mathbf{C}}$ and V of 1 in N_Γ, and a nonsingular subvariety W through 1 in U, such that $W \times V$ is holomorphically equivalent to a neighborhood of 1 in G under $(w, v) \to wv$. For $r \leqq r_1$ and sufficiently small, $f'(D_r)$ lies in that neighborhood of 1 in G. Thus $z \in D_r$ gives unique factorization $f'(z) = f_1(z)f_2(z)$ with $f_1 : D_r \to W$ and $f_2 : D_r \to V$ holomorphic. Now f_1 and f_2 satisfy (8.4a), (8.4b), and $f_1(0) = 1 = f_2(0)$. But $f(D) \subset G_0(x_{\Gamma,\Sigma}) = KN_{\Gamma,0}(x_{\Gamma,\Sigma}) \subset KN_\Gamma(x_{\Gamma,\Sigma})$ implies $f'(D_r) \subset KN_\Gamma$. Thus $f_2(D_r) \subset N_\Gamma$ implies $f_1(D_r) \subset K^{\mathbf{C}} \cap KN_\Gamma = K\cdot(K^{\mathbf{C}} \cap N_\Gamma)$, completing the proof of (8.4c). Now (8.4) is proved.

Using (8.4) we will find a number s, $0 < s \leqq r$, such that $f(D_s) \subset G_{\Psi-\Gamma,0}(x_{\Gamma,\Sigma})$. Define

$$f^* : D_r \to X \quad \text{holomorphic by} \quad f^*(z) = f_1(z)f_2(z)x_{\Gamma,\phi}. \tag{8.5a}$$

From (8.4), we have

$$f^*(D_r) \subset KN_\Gamma(x_{\Gamma,\phi}) = KG_{\Psi-\Gamma}(x_{\Gamma,\phi}). \tag{8.5b}$$

As $G_{\Psi-\Gamma,0}(x_{\Gamma,\phi})$ is open in $G_{\Psi-\Gamma}(x_{\Gamma,\phi})$, and as $f^*(0) = x_{\Gamma,\phi}$, (8.5b) says that for s_1 sufficiently small $f^*(D_{s_1}) \subset KG_{\Psi-\Gamma,0}(x_{\Gamma,\phi}) = G_0(x_{\Gamma,\phi})$. Thus we have $0 < s_1 \leqq r$ such that

$$f^* : D_{s_1} \to X \quad \text{is a holomorphic arc in} \quad G_0(x_{\Gamma,\phi}). \tag{8.6a}$$

As $G_{\Psi-\Gamma,0}(x_{\Gamma,\phi})$ is a holomorphic arc component of $G_0(x_{\Gamma,\phi})$ now (8.6a) says that

$$f^*(D_{s_1}) \subset G_{\Psi-\Gamma,0}(x_{\Gamma,\phi}). \tag{8.6b}$$

Now, from (8.4c) and (8.5a), we see that

$$f_1(D_{s_1}) \subset K^{\mathbf{C}} \cap N_\Gamma. \tag{8.6c}$$

Combine (8.4) with (8.6c); the result is that

$$f(D_{s_1}) \subset N_\Gamma(x_{\Gamma,\Sigma}) = G_{\Psi-\Gamma}(x_{\Gamma,\Sigma}). \tag{8.7a}$$

As $G_{\Psi-\Gamma,0}(x_{\Gamma,\Sigma})$ is open in $G_{\Psi-\Gamma}(x_{\Gamma,\Sigma})$, now for $s \leqq s_1$ small but positive, we have

$$f(D_s) \subset G_{\Psi-\Gamma,0}(x_{\Gamma,\Sigma}). \tag{8.7b}$$

We have just proved that every holomorphic arc $f: D \to X$ in $G_0(x_{\Gamma,\Sigma})$ has image contained in $kG_{\Psi-\Gamma,0}(x_{\Gamma,\Sigma})$ for some $k \in K$. As the $k \cdot G_{\Psi-\Gamma,0}(x_{\Gamma,\Sigma})$ are connected complex submanifolds of X by (1) and (2), it follows that they are the holomorphic arc components of $G_0(x_{\Gamma,\Sigma})$. That completes the proof of the Holomorphic Arc Component Theorem. Q.E.D.

Corollary. Let $\Gamma, \Sigma \subset \Psi$ disjoint. Then there is a K-equivariant fibration

$$\pi : G_0(x_{\Gamma,\Sigma}) \to K(x_{\Gamma,\phi})$$

whose fibers are the holomorphic arc components of $G_0(x_{\Gamma,\Sigma})$.

Proof. We combine the Holomorphic Arc Component Theorem for $G_0(x_{\Gamma,\Sigma})$ with the Boundary Flag Theorem for $G_0(x_{\Gamma,\phi})$. First we have

$$\pi : G_0(x_{\Gamma,\Sigma}) \to G_0/N_{\Gamma,0} \quad \text{by} \quad \pi(gx_{\Gamma,\Sigma}) = gN_{\Gamma,0},$$

fibration whose fibers are the holomorphic arc components of $G_0(x_{\Gamma,\Sigma})$. Then we identify the base of π with $K(x_{\Gamma,\phi})$, under $kN_{\Gamma,0} \leftrightarrow k(x_{\Gamma,\phi})$. Now π is given by

$$\pi(gx_{\Gamma,\Sigma}) = k(x_{\Gamma,\phi}) \qquad \text{where} \quad g \in kN_{\Gamma,0}, \quad k \in K.$$

Q.E.D.

9. Compact Subvarieties and Structure of an Orbit

The Holomorphic Arc Component Theorem of Section 8 describes the decomposition of an arbitrary orbit $G_0(x_{\Gamma,\Sigma}) \subset X$ as a fiber bundle over a real flag manifold $G_0/N_{\Gamma,0} = K(x_{\Gamma,\phi})$. The fibers, which are the holomorphic arc components of the orbit, are indefinite-Kaehler symmetric spaces isomorphic to the open orbit $G_{\Psi-\Gamma,0}(x_{\phi,\Sigma}) \subset X_{\Psi-\Gamma}$. Here we factor that fibration π as

$$\begin{array}{ccc} G_0(x_{\Gamma,\Sigma}) & \xrightarrow{\ \beta\ } & K(x_{\Gamma,\Sigma}) \\ & \searrow^{\pi} & \downarrow{\scriptstyle \pi\cdot\beta^{-1}} \\ & & G_0/N_{\Gamma,0} = K(x_{\Gamma,\phi}) \end{array}$$

such that the restriction of β to a holomorphic arc component of $G_0(x_{\Gamma,\Sigma})$ is a holomorphic fiber bundle over a maximal compact subvariety of that arc component. The result is due to Takeuchi [*10*] for the case $\Gamma = \phi$ of open G_0-orbit, where the notion of a holomorphic arc component is irrelevant, without identification of the fiber. The general result is due to Wolf [*16*]; his identification of the β-fibers as certain bounded symmetric domains is important for applications [*17*] to the theory of unitary representations of groups locally isomorphic to G_0.

To formulate the description of $\beta : G_0(x_{\Gamma,\Sigma}) \to K(x_{\Gamma,\Sigma})$, Γ and Σ disjoint subsets of Ψ, we need some notation, part of which was established in Section 5. First, we have

$$\mathfrak{g}_{\Psi-\Gamma,0} = \mathfrak{g}_{\Psi-\Gamma} \cap \mathfrak{g}_0 \qquad \text{real form of} \quad \mathfrak{g}_{\Psi-\Gamma} \tag{9.1a}$$

stable under σ, so that

$$\mathfrak{g}_{\Psi-\Gamma,0} = \mathfrak{k}_{\Psi-\Gamma} + \mathfrak{m}_{\Psi-\Gamma,0} \qquad \text{vector space direct sum} \tag{9.1b}$$

where

$$\mathfrak{k}_{\Psi-\Gamma} = \mathfrak{k} \cap \mathfrak{g}_{\Psi-\Gamma} \qquad \text{and} \qquad \mathfrak{m}_{\Psi-\Gamma,0} = \mathfrak{m}_0 \cap \mathfrak{g}_{\Psi-\Gamma}. \tag{9.1c}$$

As $\mathrm{ad}(c_\Sigma^4)$ commutes with complex conjugation τ over $\mathfrak{g}_0$, commutes with σ, and preserves $\mathfrak{g}_{\Psi-\Gamma}$, now we define

$$\mathfrak{g}_{\Psi-\Gamma}^\Sigma = \{v \in \mathfrak{g}_{\Psi-\Gamma} : \mathrm{ad}(c_\Sigma^4)v = v\}; \tag{9.2a}$$

then

$$\mathfrak{g}_{\Psi-\Gamma,0}^\Sigma = \mathfrak{g}_0 \cap \mathfrak{g}_{\Psi-\Gamma}^\Sigma \qquad \text{is a real form of} \quad \mathfrak{g}_{\Psi-\Gamma}^\Sigma, \tag{9.2b}$$

and

$$\mathfrak{g}_{\Psi-\Gamma,0}^\Sigma = \mathfrak{k}_{\Psi-\Gamma}^\Sigma + \mathfrak{m}_{\Psi-\Gamma,0}^\Sigma \qquad \text{vector space direct sum} \tag{9.2c}$$

where

$$\mathfrak{k}_{\Psi-\Gamma}^\Sigma = \mathfrak{k} \cap \mathfrak{g}_{\Psi-\Gamma}^\Sigma \qquad \text{and} \qquad \mathfrak{m}_{\Psi-\Gamma,0}^\Sigma = \mathfrak{m}_0 \cap \mathfrak{g}_{\Psi-\Gamma}^\Sigma. \tag{9.2d}$$

Now denote the missing pieces of $\mathfrak{g}_{\Psi-\Gamma,0}$ by

$$\mathfrak{q}_{\Psi-\Gamma}^\Sigma = \{v \in \mathfrak{k}_{\Psi-\Gamma} : \mathrm{ad}(c_\Sigma^4)v = -v\}, \tag{9.3a}$$

$$\mathfrak{r}_{\Psi-\Gamma,0}^\Sigma = \{v \in \mathfrak{m}_{\Psi-\Gamma,0} : \mathrm{ad}(c_\Sigma^4)v = -v\}. \tag{9.3b}$$

That gives us symmetric decompositions under $\mathrm{ad}(c_\Sigma^4)$:

$$\mathfrak{k}_{\Psi-\Gamma} = \mathfrak{k}_{\Psi-\Gamma}^\Sigma + \mathfrak{q}_{\Psi-\Gamma}^\Sigma \qquad \text{and} \qquad \mathfrak{g}_{\Psi-\Gamma,0} = \mathfrak{g}_{\Psi-\Gamma,0}^\Sigma + (\mathfrak{q}_{\Psi-\Gamma}^\Sigma + \mathfrak{r}_{\Psi-\Gamma,0}^\Sigma). \tag{9.3c}$$

Also, as $\mathrm{ad}(c_\Sigma^2)$ preserves $\mathfrak{g}_{\Psi-\Gamma}$ and satisfies

$$\sigma\cdot\mathrm{ad}(c_\Sigma^2)\cdot\sigma^{-1} = \mathrm{ad}(c_\Sigma^{-2}) = \tau\cdot\mathrm{ad}(c_\Sigma^2)\cdot\tau^{-1},$$

we have

$$\mathrm{ad}(c_\Sigma^2)\mathfrak{q}_{\Psi-\Gamma}^\Sigma = i\mathfrak{r}_{\Psi-\Gamma,0}^\Sigma \qquad \text{and} \qquad \mathrm{ad}(c_\Sigma^2)\mathfrak{r}_{\Psi-\Gamma,0}^\Sigma = i\mathfrak{q}_{\Psi-\Gamma}^\Sigma. \tag{9.3d}$$

Finally, we denote

$$K_{\Psi-\Gamma}: \quad \text{analytic subgroup of } K \text{ for } \mathfrak{k}_{\Psi-\Gamma}, \tag{9.4a}$$

$$K_{\Psi-\Gamma}^\Sigma: \quad \text{analytic subgroup of } K \text{ for } \mathfrak{k}_{\Psi-\Gamma}^\Sigma, \tag{9.4b}$$

$$G_{\Psi-\Gamma,0}: \quad \text{analytic subgroup of } G_0 \text{ for } \mathfrak{g}_{\Psi-\Gamma,0}, \tag{9.4c}$$

$$G_{\Psi-\Gamma,0}^\Sigma: \quad \text{analytic subgroup of } G_0 \text{ for } \mathfrak{g}_{\Psi-\Gamma,0}^\Sigma. \tag{9.4d}$$

We collect some information on the groups (9.4).

9.5. Lemma. $K_{\Psi-\Gamma} = K \cap G_{\Psi-\Gamma,0}$ and is a maximal compact subgroup of $G_{\Psi-\Gamma,0}$.

Proof. $G_{\Psi-\Gamma,0}(x_0) \cong G_{\Psi-\Gamma,0} / (K \cap G_{\Psi-\Gamma,0})$ is the Hermitian symmetric space $X_{\Psi-\Gamma,0}$ of noncompact type. Thus $K \cap G_{\Psi-\Gamma,0}$ is connected and is a maximal compact subgroup of $G_{\Psi-\Gamma,0}$. Now $K_{\Psi-\Gamma} = K \cap G_{\Psi-\Gamma,0}$ because they are connected and they have the same Lie algebra. Q.E.D.

9.6. Lemma. $K_{\Psi-\Gamma}^\Sigma = K \cap G_{\Psi-\Gamma,0}^\Sigma$ and is a maximal compact subgroup of $G_{\Psi-\Gamma,0}^\Sigma$. Both $K_{\Psi-\Gamma}^\Sigma$ and $G_{\Psi-\Gamma,0}^\Sigma$ are stable under $\mathrm{ad}(c_\Gamma c_\Sigma^2)$. $K_{\Psi-\Gamma}^\Sigma$ is the isotropy group of $G_{\Psi-\Gamma,0}^\Sigma$ at x_0, and also at $x_{\Gamma,\Sigma}$. In particular

$$G_{\Psi-\Gamma,0}^\Sigma(x_{\Gamma,\Sigma}) = c_\Gamma c_\Sigma^2 G_{\Psi-\Gamma,0}^\Sigma(x_0) \cong G_{\Psi-\Gamma,0}^\Sigma / K_{\Psi-\Gamma}^\Sigma. \tag{9.7a}$$

Further

$$G_{\Psi-\Gamma,0}^\Sigma(x_0) = G_{\Psi-(\Gamma\cup\Sigma),0}(x_0) \times G_{\Sigma,0}^\Sigma(x_0), \tag{9.7b}$$

totally geodesic Hermitian symmetric subspace of noncompact type and rank $|\Psi - \Gamma|$ in X_0. Thus

$$G^{\Sigma}_{\Psi-\Gamma,0}(x_{\Gamma,\Sigma}) = G_{\Psi-(\Gamma\cup\Sigma),0}(x_{\Gamma,\Sigma}) \times G^{\Sigma}_{\Sigma,0}(x_{\Gamma,\Sigma}), \tag{9.7c}$$

totally geodesic Hermitian symmetric subspace of noncompact type and rank $|\Psi - \Gamma|$ in $c_\Gamma c_\Sigma^2 X_0$.

Proof. In the Lemma following (5.17), we replace Ψ by $\Psi - \Gamma$ and Γ by Σ, thus obtaining (9.7b). That proves the isotropy subgroup $K \cap G^{\Sigma}_{\Psi-\Gamma,0}$ of $G^{\Sigma}_{\Psi-\Gamma,0}$ at x_0 to be connected and to be a maximal compact subgroup; now $K^{\Sigma}_{\Psi-\Gamma} = K \cap G^{\Sigma}_{\Psi-\Gamma,0}$ and is a maximal compact subgroup of $G^{\Sigma}_{\Psi-\Gamma,0}$.

As c_Γ centralizes $G_{\Psi-\Gamma}$, both $K^{\Sigma}_{\Psi-\Gamma}$ and $G^{\Sigma}_{\Psi-\Gamma,0}$ are trivially stable under $\mathrm{ad}(c_\Gamma)$.

As $c_\Sigma \in G_{\Psi-\Gamma}$, the latter is stable under $\mathrm{ad}(c_\Sigma)$. Thus $\mathfrak{g}_{\Psi-\Gamma}$ is stable under $\mathrm{ad}(c_\Sigma)$, hence also under $\mathrm{ad}(c_\Sigma^2)$, so $\mathfrak{g}^{\Sigma}_{\Psi-\Gamma}$ is $\mathrm{ad}(c_\Sigma^2)$-stable. On $\mathfrak{g}^{\Sigma}_{\Psi-\Gamma}$, $\mathrm{ad}(c_\Sigma^2)$ is equal to its own inverse, so

$$\mathrm{ad}(c_\Sigma^2) \text{ commutes with } \sigma \text{ and with } \tau \text{ on } \mathfrak{g}^{\Sigma}_{\Psi-\Gamma}.$$

That proves $\mathrm{ad}(c_\Sigma^2)$-stability of both $\mathfrak{k}^{\Sigma}_{\Psi-\Gamma}$ and $\mathfrak{m}^{\Sigma}_{\Psi-\Gamma,0}$, hence of $K^{\Sigma}_{\Psi-\Gamma}$ and $G^{\Sigma}_{\Psi-\Gamma,0}$.

Now $K^{\Sigma}_{\Psi-\Gamma}$ and $G^{\Sigma}_{\Psi-\Gamma,0}$ are stable under $\mathrm{ad}(c_\Gamma)$ and $\mathrm{ad}(c_\Sigma^2)$, hence under $\mathrm{ad}(c_\Gamma c_\Sigma^2)$. As $K^{\Sigma}_{\Psi-\Gamma}$ was seen to be the isotropy subgroup of $G^{\Sigma}_{\Psi-\Gamma,0}$ at x_0, it also is the isotropy subgroup at $c_\Gamma c_\Sigma^2 x_0 = x_{\Gamma,\Sigma}$. (9.7a) follows. Now (9.7c) follows from (9.7a) and (9.7b). Q.E.D.

9.8. Lemma. $K^{\Sigma}_{\Psi-\Gamma}$ is the isotropy subgroup of $K_{\Psi-\Gamma}$ at $x_{\Gamma,\Sigma}$, and the orbit $K_{\Psi-\Gamma}(x_{\Gamma,\Sigma}) \cong K_{\Psi-\Gamma}/K^{\Sigma}_{\Psi-\Gamma}$ is a totally geodesic Hermitian symmetric subspace of compact type in X.

Proof. In the part of the Boundary Flag Theorem of Section 6, concerning the base space of the fibration (6.19), we replace Ψ by $\Psi - \Gamma$ and Γ by Σ, and then apply c_Γ; that shows $K_{\Psi-\Gamma}(x_{\Gamma,\Sigma})$ to be a totally geodesic Hermitian symmetric subspace of compact type in X. Thus the isotropy subgroup of $K_{\Psi-\Gamma}$ at $x_{\Gamma,\Sigma}$ is connected and has Lie algebra equal to

$$\begin{aligned}\{v \in \mathfrak{k}_{\Psi-\Gamma} &: \mathrm{ad}(c_\Gamma c_\Sigma^2)\cdot\sigma\cdot\mathrm{ad}(c_\Gamma c_\Sigma^2)^{-1}v = v\}\\ = \{v \in \mathfrak{k}_{\Psi-\Gamma} &: \mathrm{ad}(c_\Gamma^2)\mathrm{ad}(c_\Sigma^4)\sigma v = v\}\\ = \{v \in \mathfrak{k}_{\Psi-\Gamma} &: \mathrm{ad}(c_\Sigma^4)v = v\} = \mathfrak{k}^{\Sigma}_{\Psi-\Gamma}.\end{aligned}$$

Thus $K^{\Sigma}_{\Psi-\Gamma}$ is the isotropy subgroup of $K_{\Psi-\Gamma}$ at $x_{\Gamma,\Sigma}$. Q.E.D.

9.9. Lemma. The isotropy subgroup of $G_{\Psi-\Gamma,0}$ at $x_{\Gamma,\Sigma}$ is the analytic subgroup with Lie algebra $\mathfrak{k}^{\Sigma}_{\Psi-\Gamma} + \mathfrak{r}^{\Sigma}_{\Psi-\Gamma,0} = \mathrm{ad}(c_\Gamma c_\Sigma^2)\mathfrak{k}^{\mathbf{C}}_{\Psi-\Gamma} \cap \mathfrak{g}_{\Psi-\Gamma,0}$.

Proof. $G_{\Psi-\Gamma,0}$ has isotropy subgroup $K_{\Psi-\Gamma}$ at $c_\Gamma x_0 = x_{\Gamma,\phi}$. Statement (2) of the Holomorphic Arc Component Theorem now says that the isotropy subgroup of $G_{\Psi-\Gamma,0}$ at $x_{\Gamma,\Sigma}$ has Lie algebra

$$\begin{aligned}&\mathrm{ad}(c_\Sigma^2)\mathfrak{k}^{\mathbf{C}}_{\Psi-\Gamma} \cap \mathfrak{g}_{\Psi-\Gamma,0}\\ &\quad= \{\mathrm{ad}(c_\Sigma^2)\mathfrak{k}^{\Sigma\mathbf{C}}_{\Psi-\Gamma} + \mathrm{ad}(c_\Sigma^2)\mathfrak{q}^{\Sigma\mathbf{C}}_{\Psi-\Gamma}\} \cap \mathfrak{g}_{\Psi-\Gamma,0}\\ &\quad= \{\mathfrak{k}^{\Sigma\mathbf{C}}_{\Psi-\Gamma} + \mathfrak{r}^{\Sigma\mathbf{C}}_{\Psi-\Gamma,0}\} \cap \mathfrak{g}_{\Psi-\Gamma,0} = \mathfrak{k}^{\Sigma}_{\Psi-\Gamma} + \mathfrak{r}^{\Sigma}_{\Psi-\Gamma,0}.\end{aligned}$$

That algebra being σ-stable, there is a deformation retraction of $G_{\Psi-\Gamma,0}(x_{\Gamma,\Sigma})$ onto $K_{\Psi-\Gamma}(x_{\Gamma,\Sigma})$, which is simply connected by Lemma 9.8. Thus the isotropy subgroup of $G_{\Psi-\Gamma,0}$ at $x_{\Gamma,\Sigma}$ is connected; so it must be the analytic subgroup for $\mathfrak{k}^{\Sigma}_{\Psi-\Gamma} + \mathfrak{r}^{\Sigma}_{\Psi-\Gamma,0}$. Q.E.D.

9.10. Lemma. Define $f: K_{\Psi-\Gamma} \times \mathfrak{m}^{\Sigma}_{\Psi-\Gamma,0} \times \mathfrak{r}^{\Sigma}_{\Psi-\Gamma,0} \to G_{\Psi-\Gamma,0}$ by $f(k, v_1, v_2) = k\cdot\exp(v_1)\cdot\exp(v_2)$. Then f is a diffeomorphism onto $G_{\Psi-\Gamma,0}$.

Proof. $X_{\Psi-\Gamma,0} = G_{\Psi-\Gamma,0}(x_0)$ is a Riemannian symmetric space of noncompact type, and hence is a complete simply connected Riemannian manifold of nonpositive sectional curvature. $X^{\Sigma}_{\Psi-\Gamma,0} = G^{\Sigma}_{\Psi-\Gamma,0}(x_0)$ is a complete totally geodesic submanifold. Let $\mathfrak{N} \to X^{\Sigma}_{\Psi-\Gamma,0}$ be the normal bundle in $X_{\Psi-\Gamma,0}$; the fiber $\mathfrak{N}_x$ over $x \in X^{\Sigma}_{\Psi-\Gamma,0}$ consists of all real tangent vectors to $X_{\Psi-\Gamma,0}$ at x that are orthogonal to the tangent space of $X^{\Sigma}_{\Psi-\Gamma,0}$ there. Then as in the classical Cartan–Hadamard Theorem,

$$\mathrm{Exp}: \mathfrak{N} \to X_{\Psi-\Gamma,0} \text{ is a diffeomorphism} \tag{9.11a}$$

where Exp is defined from the Riemannian exponential map:

$$\text{if } \xi \in \mathfrak{N}_x, \qquad \text{then } \mathrm{Exp}(\xi) = \exp_x(\xi) \in X_{\Psi-\Gamma,0}. \tag{9.11b}$$

We will translate (9.11) into the statement of the Lemma.

Identifying the infinitesimal transvections of $X_{\Psi-\Gamma,0}$ at x_0 (i.e., the elements of $\mathfrak{m}_{\Psi-\Gamma,0}$) with the real tangent vectors which are their values at x_0, we have $\mathfrak{m}^{\Sigma}_{\Psi-\Gamma,0} \perp \mathfrak{r}^{\Sigma}_{\Psi-\Gamma,0}$ because they are orthogonal under the Killing form of $\mathfrak{g}_0$. Thus

$$\mathfrak{N}_{x_0} = \mathfrak{r}^{\Sigma}_{\Psi-\Gamma,0}. \tag{9.12a}$$

It follows that

$$\operatorname{Exp} \mathfrak{N}_{x_0} = \{\exp(v)\cdot x_0 : v \in \mathfrak{r}^{\Sigma}_{\Psi-\Gamma,0}\}. \tag{9.12b}$$

As $G^{\Sigma}_{\Psi-\Gamma,0}$ acts by isometries on $X_{\Psi-\Gamma,0}$ and that action preserves $X^{\Sigma}_{\Psi-\Gamma,0}$, now

$$\mathfrak{N}_{gx_0} = g_*\mathfrak{N}_{x_0} \quad \text{and} \quad \operatorname{Exp} g_*\mathfrak{N}_{x_0} = g\cdot \operatorname{Exp} \mathfrak{N}_{x_0} \quad \text{for} \quad g \in G^{\Sigma}_{\Psi-\Gamma,0}.$$

In other words,

$$\operatorname{Exp} \mathfrak{N}_{gx_0} = g\cdot\operatorname{Exp} \mathfrak{N}_{x_0} \qquad \text{for} \quad g \in G^{\Sigma}_{\Psi-\Gamma,0}. \tag{9.12c}$$

Every $x \in X^{\Sigma}_{\Psi-\Gamma,0}$ has unique expression $x = gx_0$ where $g \in \exp(\mathfrak{m}^{\Sigma}_{\Psi-\Gamma,0}) \subset G^{\Sigma}_{\Psi-\Gamma,0}$. Combining this fact with (9.11) and (9.12), we see that

$$F'(u,v) = \exp(u)\cdot\exp(v)\cdot x_0$$

defines a diffeomorphism F' of $\mathfrak{m}^{\Sigma}_{\Psi-\Gamma,0} \times \mathfrak{r}^{\Sigma}_{\Psi-\Gamma,0}$ onto $X_{\Psi-\Gamma,0}$. Replacing the isometries $\exp(u)\cdot\exp(v)$ by their inverses, we obtain another map

$$F : \mathfrak{r}^{\Sigma}_{\Psi-\Gamma,0} \times \mathfrak{m}^{\Sigma}_{\Psi-\Gamma,0} \to X_{\Psi-\Gamma,0} \qquad \text{diffeomorphism} \tag{9.13a}$$

defined by

$$F(v,u) = \exp(-v)\cdot\exp(-u)\cdot x_0. \tag{9.13b}$$

Note that $\exp(-v)\cdot\exp(-u)\cdot K_{\Psi-\Gamma}$ is the set of all elements of $G_{\Psi-\Gamma,0}$ that send x_0 to $F(v,u)$. Thus F "lifts" to a map

$$f' : \mathfrak{r}^{\Sigma}_{\Psi-\Gamma,0} \times \mathfrak{m}^{\Sigma}_{\Psi-\Gamma,0} \times K_{\Psi-\Gamma} \to G_{\Psi-\Gamma,0} \quad \text{diffeomorphism} \tag{9.14a}$$

defined by

$$f'(v,u,k) = \exp(-v)\cdot\exp(-u)\cdot k^{-1}. \tag{9.14b}$$

The Lemma follows on observing that

$$f(k,v_1v_2) = f'(v_2,v_1,k)^{-1}.$$

Q.E.D.

We remark that our proof of Lemma 9.10 is based on the same general idea as the corresponding proofs in [*10*] and [*16*].

Orbit Fibration Theorem. Let Γ and Σ be disjoint subsets of Ψ. If

$g \in G_{\Psi-\Gamma,0}$, use Lemma 9.10 to factor

$$g = g_K \cdot g_M \cdot g_R \tag{9.15a}$$

where

$$g_K \in K_{\Psi-\Gamma}, \qquad g_M \in \exp(\mathfrak{m}^{\Sigma}_{\Psi-\Gamma,0}), \qquad g_R \in \exp(\mathfrak{r}^{\Sigma}_{\Psi-\Gamma,0}). \tag{9.15b}$$

If $k \in K$, define

$$\beta_k : k \cdot G_{\Psi-\Gamma,0}(x_{\Gamma,\Sigma}) \to k \cdot K_{\Psi-\Gamma}(x_{\Gamma,\Sigma}) \quad \text{by} \quad \beta_k(kgx_{\Gamma,\Sigma}) = kg_K x_{\Gamma,\Sigma}. \tag{9.16a}$$

Further define

$$\beta : G_0(x_{\Gamma,\Sigma}) \to K(x_{\Gamma,\Sigma}) \qquad \text{by} \quad \beta|_{k \cdot G_{\Psi-\Gamma,0}(x_{\Gamma,\Sigma})} = \beta_k. \tag{9.16b}$$

1. $\beta_k : k \cdot G_{\Psi-\Gamma,0}(x_{\Gamma,\Sigma}) \to k \cdot K_{\Psi-\Gamma}(x_{\Gamma,\Sigma})$ is a well defined holomorphic fiber bundle with

(1a) structure group: the connected reductive complex Lie group $K^{\Sigma C}_{\Psi-\Gamma}$;

(1b) total space: the holomorphic arc component $k \cdot G_{\Psi-\Gamma,0}(x_{\Gamma,\Sigma})$ of $G_0(x_{\Gamma,\Sigma})$ that contains $k(x_{\Gamma,\Sigma})$;

(1c) base space: the maximal compact subvariety $k \cdot K_{\Psi-\Gamma}(x_{\Gamma,\Sigma})$ of the total space, which is a totally geodesic Hermitian symmetric subspace of compact type in X; and

(1d) fiber over $kk'(x_{\Gamma,\Sigma})$, $k' \in K_{\Psi-\Gamma}$: the totally geodesic Hermitian symmetric subspace

$$\begin{aligned} kk' \cdot G^{\Sigma}_{\Psi-\Gamma,0}(x_{\Gamma,\Sigma}) &= kk' c_\Gamma c_\Sigma^2 \cdot G^{\Sigma}_{\Psi-\Gamma,0}(x_0) \\ &= kk' c_\Gamma c_\Sigma^2 \{G_{\Psi-(\Gamma\cup\Sigma),0}(x_0) \times G^{\Sigma}_{\Sigma,0}(x_0)\} \end{aligned}$$

of noncompact type and rank $|\Psi - \Gamma|$ in $kk' c_\Gamma c_\Sigma^2 X_0$.

2. Let $\nu_k : k_* \mathfrak{M}^{\Sigma}_{\Psi-\Gamma} \to k \cdot K_{\Psi-\Gamma}(x_{\Gamma,\Sigma})$ denote the holomorphic normal bundle in $k \cdot G_{\Psi-\Gamma,0}(x_{\Gamma,\Sigma})$. Then ν_k is the homogeneous holomorphic vector bundle over the compact Hermitian symmetric coset space $K_{\Psi-\Gamma}/K^{\Sigma}_{\Psi-\Gamma} \cong k \cdot K_{\Psi-\Gamma}(x_{\Gamma,\Sigma})$ defined by the representation of $K^{\Sigma}_{\Psi-\Gamma}$ on the holomorphic tangent space of $G^{\Sigma}_{\Psi-\Gamma,0}/K^{\Sigma}_{\Psi-\Gamma} \cong k \cdot G^{\Sigma}_{\Psi-\Gamma,0}(x_{\Gamma,\Sigma})$. The bundle β_k is holomorphically fiber-equivalent to the relatively compact tubular neighborhood of the zero-section of ν_k, whose intersection with the fiber over $kk'(x_{\Gamma,\Sigma})$, $k' \in K_{\Psi-\Gamma}$, is the image of the Harish-Chandra embedding of $G^{\Sigma}_{\Psi-\Gamma,0}/K^{\Sigma}_{\Psi-\Gamma}$ as a bounded symmetric domain in $\mathfrak{m}^{\Sigma C}_{\Psi-\Gamma} \cap \mathrm{ad}(c_\Gamma c_\Sigma^2)\mathfrak{m}^+$.

3. $\beta : G_0(x_{\Gamma,\Sigma}) \to K(x_{\Gamma,\Sigma})$ is a well-defined real analytic fiber bundle whose fibers and structure group coincide with those of its restrictions β_k to the holomorphic arc components $k \cdot G_{\Psi-\Gamma,0}(x_{\Gamma,\Sigma})$ of $G_0(x_{\Gamma,\Sigma})$. The bundle β is real analytically fiber-equivalent to the homorgeneous complex vector bundle over $K(x_{\Gamma,\Sigma})$, whose fiber is the holomorphic tangent space $\mathfrak{m}^{\Sigma C}_{\Psi-\Gamma} \cap \mathrm{ad}(c_\Gamma c^2_\Sigma)\mathfrak{m}^+$ to $G^\Sigma_{\Psi-\Gamma,0}(x_{\Gamma,\Sigma})$.

Proof. The factorization (9.15) of $G_{\Psi-\Gamma,0}$ is smooth and well defined by Lemma 9.10. Fix $k \in K$. Now the map β_k of (9.16a) is smooth and well defined. Its total space $k \cdot G_{\Psi-\Gamma,0}(x_{\Gamma,\Sigma})$ is the holomorphic arc component of $G_0(x_{\Gamma,\Sigma})$ through $k(x_{\Gamma,\Sigma})$, by the Holomorphic Arc Component Theorem of Section 8, and the base space $k \cdot K_{\Psi-\Gamma}(x_{\Gamma,\Sigma})$ is a totally geodesic Hermitian symmetric subspace of compact type in X by Lemma 9.8. As $k \cdot K_{\Psi-\Gamma}(x_{\Gamma,\Sigma}) \cong K_{\Psi-\Gamma}/K^\Sigma_{\Psi-\Gamma}$ by Lemma 9.8, now β_k is a real analytic fiber bundle with structure group $K^\Sigma_{\Psi-\Gamma}$. We extend the structure group to $K^{\Sigma C}_{\Psi-\Gamma}$.

Let $k' \in K_{\Psi-\Gamma}$. Then $\beta_k^{-1}(kk'x_{\Gamma,\Sigma})$ consists of all $kgx_{\Gamma,\Sigma}$ such that $g \in G_{\Psi-\Gamma,0}$ and $kg_K x_{\Gamma,\Sigma} = kk'x_{\Gamma,\Sigma}$. The latter condition is $g_K(x_{\Gamma,\Sigma}) = k'(x_{\Gamma,\Sigma})$, which just says $g_K \in k'K^\Sigma_{\Psi-\Gamma}$ by Lemma 9.8. Thus the fiber over $kk'(x_{\Gamma,\Sigma})$ consists of all $kk'k_1 g_M g_R x_{\Gamma,\Sigma}$ such that $k_1 \in K^\Sigma_{\Psi-\Gamma}$, $g_M \in \exp(\mathfrak{m}^\Sigma_{\Psi-\Gamma,0})$, and $g_R \in \exp(\mathfrak{r}^\Sigma_{\Psi-\Gamma,0})$. As $g_R(x_{\Gamma,\Sigma}) = x_{\Gamma,\Sigma}$ by Lemma 9.9, and as $k_1 g_M x_{\Gamma,\Sigma} = \mathrm{ad}(k_1) g_M \cdot x_{\Gamma,\Sigma} \in \exp(\mathfrak{m}^\Sigma_{\Psi-\Gamma,0}) \cdot x_{\Gamma,\Sigma}$ by Lemma 9.9, now

$$\beta_k^{-1}(kk'x_{\Gamma,\Sigma}) = kk'\exp(\mathfrak{m}^\Sigma_{\Psi-\Gamma,0}) \cdot x_{\Gamma,\Sigma}.$$

Lemma 9.6 now says that the β_k-fiber over $kk'(x_{\Gamma,\Sigma})$ is $kk' \cdot G^\Sigma_{\Psi-\Gamma,0}(x_{\Gamma,\Sigma}) = kk'c_\Gamma c^2_\Sigma G^\Sigma_{\Psi-\Gamma,0}(x_0) = kk'c_\Gamma c^2_\Sigma \{G_{\Psi-(\Gamma\cup\Sigma),0}(x_0) \times G^\Sigma_{\Sigma,0}(x_0)\}$, totally geodesic Hermitian symmetric subspace of noncompact type and rank $|\Psi - \Gamma|$ in $kk'c_\Gamma c^2_\Sigma X_0$.

To complete the proof of (1), it now suffices to show that β_k has holomorphic $K^{\Sigma C}_{\Psi-\Gamma}$-valued transition functions. For that, it suffices to prove (2).

In order to discuss holomorphic tangent spaces, we introduce the notation

$$\mathfrak{m}^\pm_{\Psi-\Gamma} = \mathfrak{m}^C_{\Psi-\Gamma,0} \cap \mathfrak{m}^\pm, \tag{9.17a}$$

$$\mathfrak{m}^{\Sigma\pm}_{\Psi-\Gamma} = \mathfrak{m}^{\Sigma C}_{\Psi-\Gamma,0} \cap \mathfrak{m}^\pm, \tag{9.17b}$$

$$\mathfrak{r}^{\Sigma\pm}_{\Psi-\Gamma} = \mathfrak{r}^{\Sigma C}_{\Psi-\Gamma} \cap \mathfrak{m}^\pm, \tag{9.17c}$$

$$\mathfrak{q}^{\Sigma\pm}_{\Psi-\Gamma} = \mathrm{ad}(c^2_\Sigma)\mathfrak{r}^{\Sigma\pm}_{\Psi-\Gamma}. \tag{9.17d}$$

Thus we have

$$\mathfrak{m}^{\pm}_{\Psi-\Gamma} = \mathfrak{m}^{\Sigma\pm}_{\Psi-\Gamma} + \mathfrak{r}^{\Sigma\pm}_{\Psi-\Gamma} \tag{9.18a}$$

and

$$\mathrm{ad}(c_\Gamma c_\Sigma^2)\mathfrak{m}^{\pm}_{\Psi-\Gamma} = \mathrm{ad}(c_\Sigma^2)\mathfrak{m}^{\Sigma\pm}_{\Psi-\Gamma} + \mathfrak{q}^{\Sigma\pm}_{\Psi-\Gamma}. \tag{9.18b}$$

In the notation (9.17), the holomorphic tangent space to $G_{\Psi-\Gamma,0}(x_0)$ at x_0 is $\mathfrak{m}^+_{\Psi-\Gamma}$, and $\mathfrak{m}^{\Sigma+}_{\Psi-\Gamma}$ is the subspace that is holomorphic tangent space to $G^{\Sigma}_{\Psi-\Gamma,0}(x_0)$. Now Lemma 9.6, Lemma 9.8, and (9.18) say that, at $x_{\Gamma,\Sigma}$,

$G_{\Psi-\Gamma,0}(x_{\Gamma,\Sigma})$ has holomorphic tangent space $\mathrm{ad}(c_\Sigma^2)\mathfrak{m}^{\Sigma+}_{\Psi-\Gamma} + \mathfrak{q}^{\Sigma+}_{\Psi-\Gamma}$, (9.19a)

$G^{\Sigma}_{\Psi-\Gamma,0}(x_{\Gamma,\Sigma})$ has holomorphic tangent space $\mathrm{ad}(c_\Sigma^2)\mathfrak{m}^{\Sigma+}_{\Psi-\Gamma}$, (9.19b)

$K_{\Psi-\Gamma}(x_{\Gamma,\Sigma})$ has holomorphic tangent space $\mathfrak{q}^{\Sigma+}_{\Psi-\Gamma}$. (9.19c)

Now the holomorphic normal space to $K_{\Psi-\Gamma}(x_{\Gamma,\Sigma})$ in $G_{\Psi-\Gamma,0}(x_{\Gamma,\Sigma})$ at $x_{\Gamma,\Sigma}$ is $\mathrm{ad}(c_\Sigma^2)\mathfrak{m}^{\Sigma+}_{\Psi-\Gamma}$. That space is $K^{\Sigma C}_{\Psi-\Gamma}$-stable, the representation being that of $K^{\Sigma C}_{\Psi-\Gamma}$ on the holomorphic tangent space to $G^{\Sigma}_{\Psi-\Gamma,0}(x_{\Gamma,\Sigma})$ at $x_{\Gamma,\Sigma}$. Now $v_k : k_*\mathfrak{M}^{\Sigma}_{\Psi-\Gamma} \to k\cdot K_{\Psi-\Gamma}(x_{\Gamma,\Sigma})$ is identified as stated in (2), and the total space of β_k is injected into $k_*\mathfrak{M}^{\Sigma}_{\Psi-\Gamma}$, as stated, by the Harish-Chandra embeddings of the $kk'G^{\Sigma}_{\Psi-\Gamma,0}(x_{\Gamma,\Sigma})$ into the $\mathrm{ad}(kk')$-images of the $\mathrm{ad}(c_\Sigma^2)\mathfrak{m}^{\Sigma+}_{\Psi-\Gamma} = \mathfrak{m}^{\Sigma C}_{\Psi-\Gamma,0} \cap \mathrm{ad}(c_\Gamma c_\Sigma^2)\mathfrak{m}^+$. Now (1) and (2) are proved.

We check that the map $\beta : G_0(x_{\Gamma,\Sigma}) \to K(x_{\Gamma,\Sigma})$ of (9.16b) is well-defined. In other words, given $k, k' \in K$ with $k\cdot G_{\Psi-\Gamma,0}(x_{\Gamma,\Sigma}) = k'\cdot G_{\Psi-\Gamma,0}(x_{\Gamma,\Sigma})$, we show that $\beta_k = \beta_{k'}$. For that, suppose

$$kg(x_{\Gamma,\Sigma}) = x = k'g'(x_{\Gamma,\Sigma}); \qquad k, k' \in K \qquad \text{and} \qquad g, g' \in G_{\Psi-\Gamma,0}.$$

Then $k^{-1}k'$ preserves the holomorphic arc component $G_{\Psi-\Gamma,0}(x_{\Gamma,\Sigma})$ of $G_0(x_{\Gamma,\Sigma})$, i.e., $k^{-1}k' \in K \cap N_{\Gamma,0}$, so

$$k^{-1}k' = k_1k_2, \qquad k_1 \in K_{\Psi-\Gamma}, \qquad k_2 \ \text{ trivial on } \ G_{\Psi-\Gamma,0}(x_{\Gamma,\Sigma}).$$

Thus we may ignore k_2 and assume that

$$k' = kk_1 \qquad \text{with} \quad k_1 \in K_{\Psi-\Gamma}.$$

As in (9.15), we factor

$$g = g_K g_M g_R \qquad \text{and} \qquad g' = g'_K g'_M g'_R.$$

Using $g_R(x_{\Gamma,\Sigma}) = x_{\Gamma,\Sigma} = g'_R(x_{\Gamma,\Sigma})$, we now have

$$k g_K g_M x_{\Gamma,\Sigma} = x = k k_1 g'_K g'_M x_{\Gamma,\Sigma}.$$

Thus $k_1 g'_K g'_M = g_K g_M h$ where $h \in G_{\Psi-\Gamma,0}$ leaves $x_{\Gamma,\Sigma}$ fixed. Lemma 9.9 says $h = h_K h_R$ with $h_K \in K^{\Sigma}_{\Psi-\Gamma}$ and $h_R \in \exp(\mathfrak{r}^{\Sigma}_{\Psi-\Gamma,0})$. Now

$$(k_1 g'_K) g'_M = g_K g_M h_K h_R = (g_K h_K)(h_K^{-1} g_M h_K) h_R.$$

Uniqueness of the factorization (9.15) thus says

$$k_1 g'_K = g_K h_K, \qquad g'_M = h_K^{-1} g_M h_K, \qquad 1 = h_R.$$

Thus we compute

$$\begin{aligned} \beta_k(x) &= k g_K(x_{\Gamma,\Sigma}) = k g_K h_K(x_{\Gamma,\Sigma}) = k k_1 g'_K(x_{\Gamma,\Sigma}) \\ &= k' g'_K(x_{\Gamma,\Sigma}) = \beta_{k'}(x). \end{aligned}$$

Now β is proved well defined.

Now that β is well defined, it also is real analytic. If $x \in G_0(x_{\Gamma,\Sigma})$, then x and $\beta(x)$ are in the same holomorphic arc component of $G_0(x_{\Gamma,\Sigma})$. This proves that $\beta : G_0(x_{\Gamma,\Sigma}) \to K(x_{\Gamma,\Sigma})$ is a real analytic bundle whose fibers coincide with those of its restrictions β_k.

The structure group of β may be taken to be the isotropy subgroup of K at $x_{\Gamma,\Sigma}$. That isotropy group preserves the holomorphic arc component $G_{\Psi-\Gamma,0}(x_{\Gamma,\Sigma})$, hence is in $K \cap N_{\Gamma,0}$, so its every element has factorization

$$k = k_1 k_2, \qquad k_1 \in K^{\Sigma}_{\Psi-\Gamma}, \qquad k_2 \text{ trivial on } G_{\Psi-\Gamma,0}(x_{\Gamma,\Sigma}).$$

Now we may reduce the structure group of β to $K^{\Sigma}_{\Psi-\Gamma}$, then increase it to the structure group $K^{\Sigma C}_{\Psi-\Gamma}$ of the restrictions β_k. Q.E.D.

Mostly for reference later, we write down the case $\Gamma = \phi$ of an open orbit for the Orbit Fibration Theorem.

Corollary. Every open G_0-orbit on X has a holomorphic fibration

$$\beta : G_0(x_{\phi,\Sigma}) \to K(x_{\phi,\Sigma}) \quad \text{by} \quad \beta(g x_{\phi,\Sigma}) = g_K x_{\phi,\Sigma} \tag{9.20a}$$

where

$$g = g_K g_M g_R; \qquad g_K \in K, \quad g_M \in \exp(\mathfrak{m}^{\Sigma}_{\Psi,0}), \quad g_R \in \exp(\mathfrak{r}^{\Sigma}_{\Psi,0}). \tag{9.20b}$$

This fibration has structure group, the connected reductive complex Lie group $K^{\Sigma C}_{\Psi}$. The total space is the open orbit $G_0(x_{\phi,\Sigma})$; the base space is the maximal compact subvariety $K(x_{\phi,\Sigma})$ of $G_0(x_{\phi,\Sigma})$, which is a totally geodesic Hermitian symmetric subspace of compact type in X; and the fiber over $k(x_{\phi,\Sigma})$, $k \in K$, is the totally geodesic Hermitian symmetric subspace

$$k \cdot G^{\Sigma}_{\Psi,0}(x_{\phi,\Sigma}) = kc^2_{\Sigma} \cdot G^{\Sigma}_{\Psi,0}(x_0) = kc^2_{\Sigma}\{G_{\Psi-\Sigma,0}(x_0) \times G^{\Sigma}_{\Sigma,0}(x_0)\}$$

of noncompact type and equal rank $|\Psi|$ in $kc^2_{\Sigma}X_0$.

For further details on the bundles β_k and β see [*16*, pp. 1213–1223].

Part III. Examples: The Classical Domains

Hermitian symmetric spaces were first studied by É. Cartan [*4*], who classified them by means of his classification [*3*] of Riemannian symmetric spaces. See [*2*], [*12*], [*11*], and [*14*] for increasingly direct proofs of Cartan's classification. The result is that the irreducible Hermitian symmetric spaces $X = G_c/K$ of compact type are

(i) the Grassmannian $\mathbf{SU}(m+n) / [\mathbf{SU}(m+n) \cap \{\mathbf{U}(m) \times \mathbf{U}(n)\}]$ of n-dimensional linear subspaces of $\mathbf{C}^{m+n}$, $1 \leqq m \leqq n$;

(ii) the subvariety $\mathbf{Sp}(n)/\mathbf{U}(n)$ of $\mathbf{SU}(2n)/[\mathbf{SU}(2n) \cap \{\mathbf{U}(n) \times \mathbf{U}(n)\}]$ consisting of the n-dimensional linear subspaces of $\mathbf{C}^{2n}$ annihilated by a nondegenerate antisymmetric bilinear form, $n \geqq 1$;

(iii) the subvariety $\mathbf{SO}(2n)/\mathbf{U}(n)$ of that same Grassmannian consisting of the n-dimensional linear subspaces of $\mathbf{C}^{2n}$ annihilated by a nondegenerate symmetric bilinear form, $n \geqq 3$;

(iv) the nonsingular quadric hypersurface $\mathbf{SO}(n+2)/\mathbf{SO}(n) \times \mathbf{SO}(2)$ in complex projective $(n+1)$-space, $n \geqq 3$;

(v) an "exceptional" space $\mathbf{E}_7/\mathbf{E}_6 \cdot \mathbf{SO}(2)$; and

(vi) an "exceptional" space $\mathbf{E}_6/\mathbf{SO}(10) \cdot \mathbf{SO}(2)$.

In the next four sections we work out the material of Parts I and II for the four series just described. The method is: use the general theory, first to guess the matrix and then to check the guess. It's quite efficient. Without the general theory, however, most of the facts would be extremely difficult to see, much less to prove; thus the books of Hua L.-K. [*6*] and Pjateckĭi-Šapiro [*9*] remain all the more impressive.

10. Complex Grassmann Manifolds

We work out the G_0-orbit structure, the holomorphic arc component structure of an arbitrary G_0-orbit, and the orbit fibration, for the case where X is a complex Grassmannian. The reader should notice that our method is that of constant application of the general theory, and he might enjoy working out some *ad hoc* proofs of those applications. The less energetic reader should at least glance at Pjateckĭi-Šapiro's *ad hoc* determination of the boundary components of X_0 in X for the case of the complex Grassmannian [9, Section 6].

Let X be the complex Grassmann manifold consisting of n-planes through the origin in complex number space $\mathbf{C}^{m+n}$ ($m \geqq 1$, $n \geqq 1$). If $\{v_1, \ldots, v_n\} \subset \mathbf{C}^{m+n}$ is linearly independent, then

$$v_1 \wedge \cdots \wedge v_n \in X \text{ is the linear span of } \{v_1, \ldots, v_n\}.$$

The complex general linear group

$$\mathbf{GL}(m+n, \mathbf{C}) = \{g : \mathbf{C}^{m+n} \to \mathbf{C}^{m+n} : g \text{ is linear and invertible}\}$$

acts on X by

$$g(v_1 \wedge \cdots \wedge v_n) = g(v_1) \wedge \cdots \wedge g(v_n).$$

An element $g \in \mathbf{GL}(m+n, \mathbf{C})$ acts trivially on X if, and only if, it is a scalar multiplication. Thus $\mathbf{GL}(m+n, \mathbf{C})$ induces the complex Lie transformation group

$$G = \mathbf{GL}(m+n, \mathbf{C}) \,/\, \{aI : 0 \neq a \in \mathbf{C}\}$$

of complex dimension $(m+n)^2 - 1$ on X.

We can almost normalize representing matrices for elements of G by choosing them in the complex special linear group

$$\mathbf{SL}(m+n, \mathbf{C}) = \{g \in \mathbf{GL}(m+n, \mathbf{C}) : \det g = 1\}.$$

That realizes G as the quotient,

$$G = \mathbf{SL}(m+n, \mathbf{C})/\{e^{2\pi i l/(m+n)} \cdot I : 0 \leqq l < m+n\}$$

of $\mathbf{SL}(m+n, \mathbf{C})$ by a cyclic group of order $m+n$. In particular, now, the Lie algebra $\mathfrak{g}$ of G consists of all complex $(m+n) \times (m+n)$ matrices of trace zero.

Choose a basis of $\mathbf{C}^{m+n}$ and denote it by $\{e_1, \ldots, e_{m+n}\}$. That gives us

$$x_0 = e_{m+1} \wedge \cdots \wedge e_{m+n} \in X \qquad \text{base point.}$$

The linear transformations of $\mathbf{C}^{m+n}$ will be written in matrix form relative to the basis $\{e_j\}$. Thus every linear transformation of $\mathbf{C}^{m+n}$ is represented by a complex matrix $\begin{pmatrix} A & B \\ C & D \end{pmatrix}$ where the blocks have size

$$A \text{ is } m \times m, \quad B \text{ is } m \times n, \quad C \text{ is } n \times m, \quad D \text{ is } n \times n.$$

In particular,

$$P = \{g \in G : g(x_0) = x_0\}$$

is represented by

$$\left\{\begin{pmatrix} A & B \\ C & D \end{pmatrix} : B = 0 \text{ and } (\det A)(\det D) = 1\right\}.$$

Thus P is a connected complex Lie subgroup of dimension $m^2 + n^2 + mn - 1$ in G, and

$$X = G/P \qquad \text{compact complex } (mn)\text{-manifold.}$$

G has maximal compact subgroup G_c which is its compact real form, given by

$$G_c \text{ is the image of the unitary group } \mathbf{U}(m + n) \text{ in } G.$$

The isotropy subgroup of G_c at x_0 is the group

$$K = G_c \cap P$$

which consists of all elements of G represented by matrices

$$\begin{pmatrix} A & 0 \\ 0 & D \end{pmatrix} \qquad \text{with} \quad A \in \mathbf{U}(m), \quad D \in \mathbf{U}(n), \quad (\det A)(\det D) = 1.$$

Now $G_c(x_0)$ is open in X because it has the same real dimension, closed in X because it is compact, so

$$X = G_c/K \qquad \text{compact presentation.}$$

This exhibits X as a Hermitian symmetric space of compact type; the symmetry at x_0 is the element of G_c represented by any

$$s = a\begin{pmatrix} -I_m & 0 \\ 0 & I_n \end{pmatrix}, \qquad (-1)^m a^{m+n} = 1.$$

Consider the Hermitian form on $\mathbf{C}^{m+n}$ given by

$$\langle u,v \rangle = -\sum_{j=1}^{m} u^j \bar{v}^j + \sum_{k=1}^{n} u^{m+k} \bar{v}^{m+k}.$$

That defines an indefinite unitary group

$$\mathbf{U}^m(m+n) = \mathbf{U}(m,n)$$
$$= \{g \in \mathbf{GL}(m+n, \mathbf{C}) : \langle g(u), g(v)\rangle = \langle u,v\rangle, \text{ all } u, v \in \mathbf{C}^{m+n}\}$$

and an indefinite special unitary group

$$\mathbf{SU}^m(m+n) = \mathbf{SU}(m,n) = \mathbf{U}(m,n) \cap \mathbf{SL}(m+n, \mathbf{C}).$$

Those two groups have the same image

$$\mathbf{U}(m,n)/\{aI : |a| = 1\} = G_0 = \mathbf{SU}(m,n)/\{aI : a^{m+n} = 1\}$$

in G. That group G_0 is a real form of G, so

$$G_0^{\mathbf{C}} = G = G_c^{\mathbf{C}},$$

and it is readily checked that

$$G_0 \cap P = K \qquad \text{so} \qquad G_0/K \cong G_0(x_0) \qquad \text{open in } X.$$

This exhibits

$$X_0 = G_0(x_0) \subset X \qquad \text{open } G_0\text{-orbit}$$

as a Hermitian symmetric space of noncompact type, with the same symmetry s at x_0, inside its compact dual under Borel Embedding.

We work out the Harish-Chandra embedding of X_0 as a bounded domain in the holomorphic tangent space at x_0. In our block form matrix representations, e_j is the column vector, i.e., $(m+n) \times 1$ matrix, with all entries zero except for a 1 in the jth place. $\mathbf{SL}(m+n, \mathbf{C}) \to G$ has finite kernel, so we identify Lie algebras there; now G has Lie algebra

$$\mathfrak{g} = \left\{\begin{pmatrix} A & B \\ C & D \end{pmatrix} : \text{trace}(A) + \text{trace}(D) = 0\right\}.$$

Now

$$\mathfrak{k}^{\mathbf{C}} = \left\{\begin{pmatrix} A & 0 \\ 0 & D \end{pmatrix} : \text{trace}(A) + \text{trace}(D) = 0\right\}$$

and

$$\mathfrak{p} = \left\{\begin{pmatrix} A & 0 \\ C & D \end{pmatrix} : \text{trace}(A) + \text{trace}(D) = 0\right\}.$$

Thus

$$\mathfrak{m}^- = \left\{\begin{pmatrix} 0 & 0 \\ C & 0 \end{pmatrix}\right\} \quad \text{and} \quad \mathfrak{m}^+ = \left\{\begin{pmatrix} 0 & B \\ 0 & 0 \end{pmatrix}\right\}.$$

In particular,

$$Z \to \tilde{Z} = \begin{pmatrix} 0 & Z \\ 0 & 0 \end{pmatrix} \qquad \text{maps } (m \times n \text{ matrices}) \cong \mathfrak{m}^+.$$

Harish-Chandra's map $\xi : \mathfrak{m}^+ \to X$, given by $\xi(\tilde{Z}) = (\exp \tilde{Z})\,(x_0)$, now may be viewed as a map from the complex vector space of all $m \times n$ matrices. Compute

$$\exp \tilde{Z} = \sum_{k=0}^{\infty} \frac{1}{k!} \begin{pmatrix} 0 & Z \\ 0 & 0 \end{pmatrix}^k = \begin{pmatrix} I & Z \\ 0 & I \end{pmatrix}.$$

Thus

$$\xi(Z) = (\exp \tilde{Z})\,(x_0) = v_1 \wedge \cdots \wedge v_n \quad \text{where} \quad (v_1, \ldots, v_n) = \begin{pmatrix} Z \\ I \end{pmatrix}.$$

The condition $\xi(Z) \in X_0$ is, by Witt's theorem, that $\langle\ ,\ \rangle$ be positive definite on $v_1 \wedge \cdots \wedge v_n$, i.e., that the matrix

$$(\langle v_j, v_l \rangle) = I_n - {}^t Z \cdot \bar{Z}, \qquad {}^t Z = \text{transpose } (Z),$$

be positive definite, i.e., such that its complex conjugate $I_n - Z^* \cdot Z, Z^* = {}^t\bar{Z}$, be positive definite. Thus

$$\xi^{-1}(X_0) = \{m \times n \text{ matrices } Z : I_n - Z^* \cdot Z \gg 0\}.$$

For comparison with more classical work on that domain, we work out the action of G_0 on $\xi^{-1}(X_0)$. Let $g \in G_0$ be represented by $\begin{pmatrix} A & B \\ C & D \end{pmatrix} \in \mathbf{SL}(m+n, \mathbf{C})$.

Then $g \cdot \exp \tilde{Z}$ is represented by $\begin{pmatrix} A & AZ + B \\ C & CZ + D \end{pmatrix}$, so

$$(g \cdot \exp \tilde{Z})\,(x_0) = w_1 \wedge \cdots \wedge w_n \qquad \text{where} \quad (w_1, \ldots, w_n) = \begin{pmatrix} AZ + B \\ CZ + D \end{pmatrix}.$$

The columns of $\begin{pmatrix} AZ + B \\ CZ + D \end{pmatrix}$ and $\begin{pmatrix} (AZ + B)\,(CZ + D)^{-1} \\ I \end{pmatrix}$ have the same span, the inverse there existing because $(g \cdot \exp \tilde{Z})(x_0) \in X_0$. Thus

$$g = \begin{pmatrix} A & B \\ C & D \end{pmatrix} \quad \text{implies} \quad g(Z) = (AZ + B)(CZ + D)^{-1}.$$

Finally, we note that the operator norm of the Hermann Convexity Theorem, i.e., the Banach space norm on the $m \times n$ matrices for which $\xi^{-1}(X_0)$ is the open unit ball, is given by

$$\|Z\|^2 = \max \{\text{eigenvalues of } Z^* \cdot Z\}.$$

The common rank of the symmetric spaces X and X_0 is

$$r = \min(m,n).$$

The partial Cayley transforms $c_j \in G_c$, $1 \leqq j \leqq r$, are given by

$$c_j(e_j) = \frac{1}{\sqrt{-2}}(ie_j + e_{m+j}), \qquad c_j(e_{m+j}) = \frac{1}{\sqrt{-2}}(e_j + ie_{m+j}),$$

$$c_j(e_k) = e_k \qquad \text{for} \quad j \neq k \neq m + j.$$

Now define

$$x_{s,t} = c_1 c_2 \cdots c_s c_{s+1}^2 c_{s+2}^2 \cdots c_t^2 x_0, \qquad 0 \leqq s \leqq t \leqq r.$$

These points will play the role of the $x_{\Gamma,\Sigma}$.

To $x \in X$, we associate the triple (a, b, c) where $\langle\ ,\ \rangle$ has rank $a + b$ on x with a positive squares and b negative squares, and $c = n - (a + b)$ is the nullity of $\langle\ ,\ \rangle$ on X. The only restrictions are that a, b, c be integers and

$$a + b + c = n, \qquad 0 \leqq a \leqq n, \quad 0 \leqq b \leqq n, \quad 0 \leqq c \leqq r.$$

For example,

to $x_{s,t}$, we associate $(n - t, t - s, s)$.

Thus

each admissible triple is associated to just one of the $x_{s,t}$.

If $y \in G_0(x)$, one associates the same triple to x and y. If x and y give the same triple, then Witt's Theorem says $y \in G_0(x_{s,t})$. Now

the G_0-orbits on X are the $G_0(x_{s,t})$, $\quad 0 \leqq s \leqq t \leqq r$.

In particular,

there are precisely $\frac{1}{2}(r + 1)(r + 2)$ G_0-orbits on X;
the $r + 1$ orbits $G_0(x_{0,t})$, $0 \leqq t \leqq r$, are the open orbits;
the orbit $G_0(x_{r,r})$ is the unique closed orbit.

Note that $\xi(\mathfrak{m}^+)$ consists of all $x \in X$ whose orthogonal projection to x_0 is surjective, i.e.,

$$\xi(\mathfrak{m}^+) = \{x \in X : e_1 \wedge \cdots \wedge e_m \wedge x \neq 0\}.$$

In other words, the complement

$$X - \xi(\mathfrak{m}^+) = \{x \in X : e_1 \wedge \cdots \wedge e_m \wedge x = 0\},$$

which exhibits it as a proper subvariety of X. Now consider the sets

$$D_{s,t} = \xi^{-1} G_0(x_{s,t}) \subset \mathfrak{m}^+.$$

Continuing to view $\mathfrak{m}^+$ as the space of all $m \times n$ complex matrices, we checked

$$D_{0,0} = \xi^{-1}(X_0) = \{Z \in \mathfrak{m}^+ : I_n - Z^* \cdot Z \gg 0\}.$$

Let us agree that an $n \times n$ Hermitian matrix H has "signature" (a, b, c) if it has a positive eigenvalues, b negative eigenvalues, and c zero eigenvalues. Then H and $\overline{H}$ have the same signature. Now, as in the proof of the characterization of $\xi^{-1}(X_0)$, we see that

$$D_{s,t} = \{Z \in \mathfrak{m}^+ : I_n - Z^* \cdot Z \text{ has signature } (n - t, t - s, s)\}.$$

Recall that $(n - t, t - s, s)$ is the triple associated to $x_{s,t}$. Thus

if $x \in \xi(\mathfrak{m}^+)$, say $x = \xi(Z)$, then the triple

associated to x is the signature of $I_n - Z^*Z$.

The topological boundary of X_0 in X is given by

$$\text{bd } X_0 = G_0(x_{1,1}) \cup G_0(x_{2,2}) \cup \cdots \cup G_0(x_{r,r}).$$

The boundary orbit $G_0(x_{s,s})$ is the union of those boundary components that are symmetric spaces of rank $r - s$. To obtain one boundary component of rank $r - s$, define, for $0 \leqq s \leqq r$,

$X_{(m-s,n-s)}$ is the set of all elements $x \in X$ such that

(i) x is contained in $e_{s+1} \wedge e_{s+2} \wedge \cdots \wedge e_{m+n}$

(ii) x contains $e_{m+1}, e_{m+2}, \ldots, e_{m+s}$.

To $x \in X_{(m-s,n-s)}$, we assign its interior product with $e_{m+1} \wedge \cdots \wedge e_{m+s}$. That gives an isomorphism of $X_{(m-s,n-s)}$ onto the complex Grassmannian of $(n - s)$-planes through the origin in the $\mathbf{C}^{(m-s)+(n-s)}$ with basis $\{e_{s+1}, \ldots, e_m; e_{m+s+1}, \ldots, e_{m+n}\}$. Let

$$X_{(m-s,n-s),0} = X_0 \cap X_{(m-s,n-s)}.$$

Then for $1 \leqq s \leqq r$,

$$c_1 c_2 \cdots c_s X_{(m-s,n-s),0} \text{ is a boundary component of rank } r - s$$

so

$$G_0(x_{s,s}) = \bigcup_{k \in K} k c_1 c_2 \cdots c_s X_{(m-s,n-s),0}.$$

By Witt's Theorem, two totally $\langle\, , \rangle$-isotropic subspaces of $\mathbf{C}^{m+n}$ are $\mathbf{SU}(m,n)$-equivalent if, and only if, they have the same dimension. The possible dimensions are $\{0, 1, \ldots, r\}$. The span

$$V_s = (e_1 + ie_{m+1}) \wedge \cdots \wedge (e_s + ie_{m+s})$$

is an s-dimensional totally $\langle\, , \rangle$-isotropic subspace of $\mathbf{C}^{m+n}$, and for $1 \leqq s \leqq r$,

$$N_{s,0} = \{g \in G_0 : g(V_s) = V_s\}$$

is a maximal parabolic subgroup of G_0. Further, for $1 \leqq s \leqq r$

$$N_{s,0} = \{g \in G_0 : gc_1 \cdots c_s X_{(m-s,n-s),0} = c_1 \cdots c_s X_{(m-s,n-s),0}\}.$$

We pull the boundary components of X_0 in X back, by ξ^{-1}, to the boundary components of

$$\xi^{-1}(X_0) = \{m \times n \text{ matrices } Z : I_n - Z^* \cdot Z \gg 0\}.$$

The result is that

$$\xi^{-1}(c_1 c_2 \cdots c_s X_{(m-s,n-s),0}) \text{ consists of all } Z = \begin{pmatrix} I_s & 0 \\ 0 & 'Z \end{pmatrix} \text{ where}$$

$$'Z \text{ is an } (m - s) \times (n - s) \text{ matrix such that } I_{n-s} - 'Z^* \cdot 'Z \gg 0.$$

Thus, as indeed we see from our discussion of $D_{s,s}$,

$$K \cdot \xi^{-1}(c_1 c_2 \cdots c_s X_{(m-s,n-s),0}) = \xi^{-1} G_0(x_{s,s}) \text{ consists of all } m \times n$$
$$\text{matrices } Z \text{ such that } I_n - Z^* \cdot Z \text{ has signature } (n - s, 0, s).$$

Here $K = \{\mathbf{U}(m) \times \mathbf{U}(n)\}/\{aI : |a| = 1\}$ acts by $(k_1,k_2) : Z \to k_1 Z k_2^{-1}$.

From the general theory, we now see that the holomorphic arc components of an arbitrary orbit $G_0(x_{s,t})$ are the

$$k\cdot\{c_1c_2\cdots c_sX_{(m-s,n-s)}\cap G_0(x_{s,t})\},\qquad k\in K,$$

and that

$$N_{s,0}\text{ is the }G_0\text{-normalizer of }c_1c_2\cdots c_sX_{(m-s,n-s)}\cap G_0(x_{s,t}).$$

We describe the K-equivariant holomorphic fibration

$$\beta: G_0(x_{0,t})\to K(x_{0,t}),\qquad 0\leqq t\leqq r,$$

of an open orbit over its maximal compact subvariety. Decompose $\mathbf{C}^{m+n}=U\oplus V$ where $U=e_1\wedge\cdots\wedge e_m$ and $V=e_{m+1}\wedge\cdots\wedge e_{m+n}$. K consists of all transformations of X represented by block matrices of shape

$$\begin{pmatrix} A & 0\\ 0 & D\end{pmatrix}\qquad\text{with}\quad A\in\mathbf{U}(m)\quad\text{and}\quad D\in\mathbf{U}(n).$$

Thus $x_{0,t}=e_1\wedge\cdots\wedge e_t\wedge e_{m+t+1}\wedge\cdots\wedge e_{m+n}$ implies

$$K(x_{0,t})=\{y\wedge z : y\text{ is a }t\text{-plane in }U,\ z\text{ is an }(n-t)\text{-plane in }V\}.$$

To accomodate $c_1c_2\cdots c_t$, we break $(m+n)\times(m+n)$ matrices into refined block form, four blocks by four, with diagonal blocks of indicated size:

$$\left(\begin{array}{c|c|c|c} t\times t & & & \\ \hline & (m-t)\times(m-t) & & \\ \hline & & t\times t & \\ \hline & & & (n-t)\times(n-t)\end{array}\right)$$

Let

$$\begin{aligned} &\mathfrak{k}_1\text{ correspond to }\mathfrak{k}_{\Psi-\Sigma}, &&\mathfrak{m}_1\text{ to }\mathfrak{m}_{\Psi-\Sigma};\\ &\mathfrak{k}_2\text{ correspond to }\mathfrak{k}_{\Sigma}^{\Sigma}, &&\mathfrak{m}_2\text{ to }\mathfrak{m}_{\Sigma}^{\Sigma}.\end{aligned}$$

Then the support of representing matrices for the algebra corresponding to $\mathfrak{g}_{\Psi}^{\Sigma}\sim\mathfrak{g}_{\Psi-\Sigma}\oplus\mathfrak{g}_{\Sigma}^{\Sigma}$ is indicated by

$$\left(\begin{array}{c|c|c|c} \mathfrak{k}_2 & & \mathfrak{m}_2 & \\ \hline & \mathfrak{k}_1 & & \mathfrak{m}_1\\ \hline \mathfrak{m}_2 & & \mathfrak{k}_2 & \\ \hline & \mathfrak{m}_1 & & \mathfrak{k}_1\end{array}\right)$$

From this and the general theory we see that $\beta^{-1}(x_{0,t})$ is the product of the manifold

$$\{e_1 \wedge \cdots \wedge e_t \wedge z : z \text{ is a positive definite } (n-t)\text{-plane in } e_{t+1} \wedge \cdots \wedge e_m \wedge e_{m+t+1} \wedge \cdots \wedge e_{m+n}\}$$

corresponding to $G_{\Psi-\Sigma,0}(x_{\phi,\Sigma})$, with the manifold

$$\{y \wedge e_{m+t+1} \wedge \cdots \wedge e_{m+n} : y \text{ is a negative definite } t\text{-plane in } e_1 \wedge \cdots \wedge e_t \wedge e_{m+1} \wedge \cdots \wedge e_{m+t}\}$$

corresponding to $G^{\Sigma}_{\Sigma,0}(x_{\phi,\Sigma})$. By K-equivariance all β-fibers are described.

Now we can specify the map β as follows. Define unit quadrics

$$Q = \{u \in \mathbf{C}^{m+n} : \langle u,u\rangle = -1\} \quad \text{and} \quad S = \{v \in \mathbf{C}^{m+n} : \langle v,v\rangle = +1\}$$

and orthogonal projections

$$\pi_U : \mathbf{C}^{m+n} \to U \qquad \text{and} \qquad \pi_V : \mathbf{C}^{m+n} \to V.$$

Let $x \in G_0(x_{0,t})$. Thus x is an n-dimensional linear subspace of $\mathbf{C}^{m+n}$ on which $\langle\ ,\ \rangle$ has t negative eigenvalues and $n-t$ positive eigenvalues. Define $\{u_1, \ldots, u_t\} \subset x$ by: $u_0 = 0$ and, for $1 \leqq j \leqq t$, $u_j \in Q \cap x$ is a maximum point for the negative function

$$f_j : Q \cap x \cap \{u_0, \ldots, u_{j-1}\}^{\perp} \to \mathbf{R} \qquad \text{by} \qquad f_j(u) = \langle \pi_U(u), \pi_U(u)\rangle.$$

Then $u_1 \wedge \cdots \wedge u_t$ is the t-dimensional negative definite subspace of x that is "nearest" to U. Define $\{v_1, \ldots, v_{n-t}\} \subset x$ by: $v_0 = 0$ and, for $1 \leqq k \leqq n-t$, $v_k \in S \cap x$ is a minimum point for the positive function

$$h_k : S \cap x \cap \{v_0, \ldots, v_{k-1}\}^{\perp} \to \mathbf{R} \qquad \text{by} \quad h_k(v) = \langle \pi_V(v), \pi_V(v)\rangle.$$

Then $v_1 \wedge \cdots \wedge v_{n-t}$ is the $(n-t)$-dimensional positive definite subspace of x that is "nearest" to V. Now

$$x = (u_1 \wedge \cdots \wedge u_t) \wedge (v_1 \wedge \cdots \wedge v_{n-t}) \in G_0(x_{0,t})$$

and we have

$$\beta(x) = \pi_U(u_1 \wedge \cdots \wedge u_t) \wedge \pi_V(v_1 \wedge \cdots \wedge v_{n-t}) \in K(x_{0,t}).$$

Note that this construction simply imitates the proof of Lemma 9.10.

The K-equivariant partially holomorphic fibration

$$\beta : G_0(x_{s,t}) \to K(x_{s,t}), \qquad 0 \leqq s \leqq t \leqq r,$$

of an arbitrary orbit, is specified by its restrictions to the holomorphic arc components of the orbit. We have described the holomorphic arc components, and the restrictions β_k there iterate the open orbit case.

11. Manifolds Corresponding to the Siegel Half Planes

We work out the G_0-orbit structure, holomorphic arc components, and orbit fibration, for the case where $X_0 \subset X$ is the Borel Embedding of the "Siegel upper half plane" in its compact dual. That is the case where Harish-Chandra's map $\xi : \mathfrak{m}^+ \to X$ satisfies

$$\xi^{-1} : c_\Psi X_0 \cong \{n \times n \text{ matrices } Z : Z = {}^tZ \text{ and } Im\, Z \gg 0\}.$$

X is a subvariety in the complex Grassmann manifold of n-planes in $\mathbf{C}^{2n}$, given as follows. Let us fix

$$J : \text{nondegenerate antisymmetric bilinear form on } \mathbf{C}^{2n}.$$

Then

$$X = \{n\text{-planes } x \subset \mathbf{C}^{2n} : J(x, x) = 0\}.$$

A different choice of J would alter X only by an automorphism of the Grassmannian, for any two nondegenerate antisymmetric bilinear forms on $\mathbf{C}^{2n}$ are $\mathbf{GL}(2n, \mathbf{C})$-equivalent.

The complex symplectic group is

$$\mathbf{Sp}(n, \mathbf{C}) = \{g : \mathbf{C}^{2n} \to \mathbf{C}^{2n} \text{ linear} : g \text{ preserves } J\}.$$

The latter condition is

$$J(u,v) = J(gu,gv) \qquad \text{for all} \quad u, v \in \mathbf{C}^{2n}.$$

It implies $\det g = \pm 1$, so g is nonsingular and g^{-1} also preserves J. $\mathbf{Sp}(n,\mathbf{C})$ visibly preserves X and acts there by holomorphic diffeomorphisms. The center of $\mathbf{Sp}(n,\mathbf{C})$ is $\{\pm I_{2n}\}$, which is the kernel of its action on X, so

$$G = \mathbf{Sp}(n,\mathbf{C})/\{\pm I_{2n}\}$$

is a complex Lie transformation group on X.

Choose a basis $\{e_1, \ldots, e_{2n}\}$ of $\mathbf{C}^{2n}$ in which

$$J = \begin{pmatrix} 0 & I \\ -I & 0 \end{pmatrix}, \qquad \text{i.e,} \qquad J(u,v) = {}^tuJv \qquad \text{for} \quad u,v \in \mathbf{C}^{2n}.$$

In that basis, $\mathbf{Sp}(n,\mathbf{C})$ consists of all transformations with $(n \times n)$-block form matrix $\begin{pmatrix} A & B \\ C & D \end{pmatrix}$ such that

$$ {}^t\begin{pmatrix} A & B \\ C & D \end{pmatrix} \begin{pmatrix} 0 & I \\ -I & 0 \end{pmatrix} \begin{pmatrix} A & B \\ C & D \end{pmatrix} = \begin{pmatrix} 0 & I \\ -I & 0 \end{pmatrix}, $$

i.e.,

$$ {}^tA \cdot C = {}^tC \cdot A, \qquad {}^tB \cdot D = {}^tD \cdot B, \qquad \text{and} \qquad {}^tA \cdot D - {}^tC \cdot B = I. $$

The matrix form of J shows $e_{n+1} \wedge \cdots \wedge e_{2n} \in X$. Define

$$ x_0 = e_{n+1} \wedge \cdots \wedge e_{2n} \in X \qquad \text{base point.} $$

The isotropy subgroup

$$ P = \{g \in G : g(x_0) = x_0\} $$

now consists of all linear transformations of $\mathbf{C}^{2n}$ with matrix

$$ \begin{pmatrix} A & 0 \\ C & D \end{pmatrix} \qquad \text{where} \quad {}^tA \cdot C = {}^tC \cdot A \quad \text{and} \quad {}^tA \cdot D = I. $$

Witt's Theorem proves G transitive on X. Thus

$$ X = G/P \qquad \text{compact complex } \tfrac{1}{2}n(n+1)\text{-manifold.} $$

The maximal compact subgroups of G are the conjugates of its compact real form

$$ G_c = \mathbf{Sp}(n)/\{\pm I_{2n}\}, $$

image in G of the unitary symplectic group

$$ \mathbf{Sp}(n) = \mathbf{Sp}(n,\mathbf{C}) \cap \mathbf{U}(2n) \subset \mathbf{SU}(2n). $$

The isotropy subgroup of G_c at x_0 is

$$ K = G_c \cap P; $$

it consists of all

$$ \pm \begin{pmatrix} A & 0 \\ 0 & \bar{A} \end{pmatrix} \qquad \text{such that} \qquad A \in \mathbf{U}(n). $$

Now $G_c(x_0)$ is open in X by dimension, closed in X by compactness, so

$$ X = G_c/K \qquad \text{compact presentation.} $$

This exhibits X as the Hermitian symmetric space $\mathbf{Sp}(n)/\mathbf{U}(n)$ of compact type. The symmetry at x_0 is

$$s = \pm\begin{pmatrix} iI & 0 \\ 0 & -iI \end{pmatrix}.$$

Consider the Hermitian form on $\mathbf{C}^{2n}$ given by

$$\langle u,v\rangle = -\sum_{j=1}^{n} u^j\bar{v}^j + \sum_{k=1}^{n} u^{n+k}\bar{v}^{n+k}.$$

It specifies the indefinite unitary group $\mathbf{U}(n,n)$, hence the group $\mathbf{Sp}(n,\mathbf{C}) \cap \mathbf{U}(n,n)$. Define

$$G_0 : \text{transformation group on } X \text{ from } \mathbf{Sp}(n,\mathbf{C}) \cap \mathbf{U}(n,n).$$

As $\mathbf{U}(n,n)$ is a real form of $\mathbf{GL}(2n,\mathbf{C})$, we can see that

$$G_0 \text{ is a noncompact real form of } G.$$

Furthermore

$$G_0 \cap P = K \qquad \text{so} \qquad G_0/K \cong G_0(x_0) \qquad \text{open in } X.$$

Thus

$$X_0 = G_0(x_0) \subset X \qquad \text{open } G_0\text{-orbit}$$

is the Hermitian symmetric space of noncompact type, dual to $G_c/K = X$ with the same symmetry s at x_0, inside X under Borel Embedding. Note now $X_0 \cong \mathbf{Sp}(n,\mathbf{R})/\mathbf{U}(n)$ and $G_0 \cong \mathbf{Sp}(n,\mathbf{R})/\{\pm I_{2n}\}$.

Now we come to the Harish-Chandra embedding of X_0 in its holomorphic tangent space. Identify the Lie algebra $\mathfrak{g}$ of G with the Lie algebra of $\mathbf{Sp}(n,\mathbf{C})$ under $G = \mathbf{Sp}(n,\mathbf{C})/\{\pm I\}$. Then

$$\mathfrak{g} = \left\{\begin{pmatrix} A & B \\ C & D \end{pmatrix} : {}^t\begin{pmatrix} A & B \\ C & D \end{pmatrix}\cdot\begin{pmatrix} 0 & I \\ -I & 0 \end{pmatrix} + \begin{pmatrix} 0 & I \\ -I & 0 \end{pmatrix}\cdot\begin{pmatrix} A & B \\ C & D \end{pmatrix} = 0\right\}.$$

In other words,

$$\mathfrak{g} = \left\{\begin{pmatrix} A & B \\ C & D \end{pmatrix} : {}^tA = -D,\ {}^tB = B, \text{ and } {}^tC = C\right\}.$$

Thus

$$\mathfrak{p} = \left\{\begin{pmatrix} A & 0 \\ C & D \end{pmatrix} : {}^tA = -D \text{ and } {}^tC = C\right\}$$

and

$$\mathfrak{m}^+ = \left\{\begin{pmatrix} 0 & B \\ 0 & 0 \end{pmatrix} : {}^tB = B\right\}.$$

In particular,

$$Z \to \tilde{Z} = \begin{pmatrix} 0 & Z \\ 0 & 0 \end{pmatrix} \qquad \text{maps} \quad (n \times n \text{ symmetric matrices}) \cong \mathfrak{m}^+.$$

We now view Harish-Chandra's map $\xi : \mathfrak{m}^+ \to X$, given by $\xi(Z) = (\exp \tilde{Z})(x_0)$, as a map from the complex vector space of all $n \times n$ symmetric matrices. As

$$\exp\begin{pmatrix} 0 & Z \\ 0 & 0 \end{pmatrix} = \begin{pmatrix} I & Z \\ 0 & I \end{pmatrix}$$

now

$$\xi(Z) = (\exp \tilde{Z})(x_0) = v_1 \wedge \cdots \wedge v_n \qquad \text{where} \quad (v_1, \ldots, v_n) = \begin{pmatrix} Z \\ I \end{pmatrix}.$$

By Witt's Theorem, $\xi(Z) \in X_0$ if and only if $\langle\ ,\ \rangle$ is positive definite on $v_1 \wedge \cdots \wedge v_n$. That condition is that $(\langle v_k, v_l \rangle) = I - {}^tZ \cdot \bar{Z}$ be positive definite. Thus

$$\xi^{-1}(X_0) = \{n \times n \text{ symmetric } Z : I - Z^* \cdot Z \gg 0\}.$$

If

$$g = \pm\begin{pmatrix} A & B \\ C & D \end{pmatrix} \in G_0,$$

then

$$g \cdot \exp \tilde{Z} = \pm\begin{pmatrix} A & AZ + B \\ C & CZ + D \end{pmatrix},$$

which sends x_0 to the span of the columns of $\begin{pmatrix} AZ + B \\ CZ + D \end{pmatrix}$. As $(CZ + D)^{-1}$ exists, that is the same as the span of the columns of $\begin{pmatrix} (AZ+B)(CZ+D)^{-1} \\ I \end{pmatrix}$.

Thus G_0 acts on $\xi^{-1}(X_0)$ by

$$\pm\begin{pmatrix} A & B \\ C & D \end{pmatrix}: Z \to (AZ + B)\cdot(CZ + D)^{-1},$$

as expected. Finally, note that the Banach norm on $\mathfrak{m}^+$, for which $\xi^{-1}(X_0)$ is the unit ball, is

$$\|Z\|^2 = \max\{\text{eigenvalues of } Z^*\cdot Z\}.$$

The symmetric spaces X_0 and X have rank n. The partial Cayley transforms $c_j \in G_c$, $1 \leqq j \leqq n$, are given by

$$c_j(e_j) = \frac{1}{\sqrt{-2}}(ie_j + e_{n+j}), \qquad c_j(e_{n+j}) = \frac{1}{\sqrt{-2}}(e_j + ie_{n+j}),$$
$$c_j(e_k) = e_k \quad \text{for} \quad j \neq k \neq n + j,$$

because X has equal rank and is totally geodesic in the ambient Grassmann manifold. As before, define

$$x_{s,t} = c_1c_2 \cdots c_s c_{s+1}^2 c_{s+2}^2 \cdots c_t^2 x_0, \qquad 0 \leqq s \leqq t \leqq n,$$

same centers as for the ambient Grassmannian of n-planes in $\mathbf{C}^{2n}$. Again,

the G_0-orbits on X are the $G_0(x_{s,t})$, $\quad 0 \leqq s \leqq t \leqq n$.

In particular,

there are precisely $\frac{1}{2}(n + 1)(n + 2)$ G_0-orbits on X,

the $n + 1$ orbits $G_0(x_{0,t})$, $0 \leqq t \leqq n$, are the open orbits, and

the orbit $G_0(x_{n,n})$ is the unique closed orbit.

The orbit $G_0(x_{s,t})$ consists of all elements of X on which $\langle\ ,\ \rangle$ has $n - t$ positive eigenvalues, $t - s$ negative eigenvalues, and s eigenvalues zero. Again as in the ambient Grassmannian,

$$X - \xi(\mathfrak{m}^+) = \{x \in X : e_1 \wedge \cdots \wedge e_n \wedge x = 0\},$$

and

$$D_{s,t} = \xi^{-1}G_0(x_{s,t}) \subset \mathfrak{m}^+$$

consists of all $n \times n$ symmetric matrices Z such that the Hermitian matrix $I - Z^*\cdot Z$ has $n - t$ positive eigenvalues, $t - s$ negative eigenvalues, and s eigenvalues zero.

The topological boundary of X_0 in X is bd $X_0 = G_0(x_{1,1}) \cup G_0(x_{2,2}) \cup \cdots \cup G_0(x_{n,n})$. Here $G_0(x_{s,s})$ is the union of those boundary components that are symmetric spaces of rank $n - s$. For $0 \leqq s \leqq n$, define

$$X_{(n-s)} \text{ is the set of all elements } x \in X \text{ such that}$$
$$e_{n+1} \wedge \cdots \wedge e_{n+s} \subset x \subset e_{s+1} \wedge e_{s+2} \wedge \cdots \wedge e_{2n}.$$

The interior product with $e_{n+1} \wedge \cdots \wedge e_{n+s}$ maps $X_{(n-s)}$ isomorphically onto the subvariety of the complex Grassmannian of $(n - s)$-planes in $\mathbf{C}^{2(n+s)} = (e_{s+1} \wedge \cdots \wedge e_n) \wedge (e_{n+s+1} \wedge \cdots \wedge e_{2n})$ defined by $J(y,y) = 0$. The Borel-Embedded noncompact dual of $X_{(n-s)}$ is

$$X_{(n-s),0} = X_0 \cap X_{(n-s)}.$$

For $1 \leqq s \leqq n$,

$$c_1 c_2 \cdots c_s X_{(n-s),0} \text{ is a boundary component of rank } n - s,$$

so

$$G_0(x_{s,s}) = \bigcup_{k \in K} k c_1 c_2 \ldots c_s X_{(n-s),0}.$$

As in the case of the Grassmannian, the maximal parabolic subgroups of G_0 are the

$$N_{s,0} = \{g \in G_0 : g(V_s) = V_s\}$$

where the

$$V_s = (e_1 + ie_{n+1}) \wedge \cdots \wedge (e_s + ie_{n+s}), \qquad 0 \leqq s \leqq n,$$

represent the G_0-equivalence classes of totally $\langle\ ,\ \rangle$-isotropic J-isotropic subspaces of $\mathbf{C}^{2n}$; and

$$N_{s,0} = \{g \in G_0 : gc_1 \cdots c_s X_{(n-s),0} = c_1 \cdots c_s X_{(n-s),0}\}.$$

We locate the ξ^{-1}-images of the boundary components of X_0 in X; they are the boundary components of

$$\{n \times n \text{ symmetric matrices } Z : I - Z^*Z \gg 0\}$$

in the space of $n \times n$ symmetric matrices. The result is that

$$\xi^{-1}(c_1 c_2 \cdots c_s X_{(n-s),0}) \text{ consists of all } n \times n \text{ symmetric matrices}$$
$$Z = \begin{pmatrix} I_s & 0 \\ 0 & 'Z \end{pmatrix} \qquad \text{such that} \qquad I_{n-s} - 'Z^* \cdot 'Z \gg 0.$$

$$K = \{\pm \begin{pmatrix} A & 0 \\ 0 & \bar{A} \end{pmatrix} : A \in \mathbf{U}(n)\} \text{ acts on } \mathfrak{m}^+ \text{ by } \pm \begin{pmatrix} A & 0 \\ 0 & \bar{A} \end{pmatrix} : Z \to AZ\bar{A}^{-1} = A \cdot Z \cdot {}^tA.$$

Thus

$$K\cdot\xi^{-1}(c_1c_2\cdots c_sX_{(n-s),0}) = \xi^{-1}G_0(x_{s,s}) \text{ consists}$$
$$\text{of all } n\times n \text{ symmetric matrices } Z \text{ such that}$$
$$I_n - Z^*\cdot Z \text{ is positive semidefinite and of rank } n-s.$$

We know from general theory that the holomorphic arc components of an arbitrary orbit $G_0(x_{s,t})$ are the

$$k\cdot\{c_1c_2\cdots c_sX_{(n-s)}\cap G_0(x_{s,t})\},\qquad k\in K,$$

and that

$$N_{s,0} \text{ is the } G_0\text{-normalizer of } c_1c_2\cdots c_sX_{(n-s)}\cap G_0(x_{s,t}).$$

We describe the K-equivariant holomorphic fibration

$$\beta: G_0(x_{0,t})\to K(x_{0,t}),\qquad 0\leqq t\leqq n,$$

of an open orbit over its maximal compact subvariety. As before,

$\mathbf{C}^{2n} = U\oplus V$ where $U = e_1\wedge\cdots\wedge e_n$ and $V = e_{n+1}\wedge\cdots\wedge e_{2n}$.

As

$$x_{0,t} = (e_1\wedge\cdots\wedge e_t)\wedge(e_{n+t+1}\wedge\cdots\wedge e_{2n}),$$

now

$K(x_{0,t}) = \{x = y\wedge z : x\in X, y \text{ is } t\text{-plane in } U, z \text{ is } (n-t)\text{-plane in } V\}$.

Our refined matrix block form is

$$\left(\begin{array}{c|c|c|c}
t\times t & & & \\ \hline
 & (n-t)\times(n-t) & & \\ \hline
 & & t\times t & \\ \hline
 & & & (n-t)\times(n-t)
\end{array}\right)$$

Let

$$\mathfrak{k}_1 \text{ correspond to } \mathfrak{k}_{\Psi-\Sigma},\qquad \mathfrak{m}_1 \text{ to } \mathfrak{m}_{\Psi-\Sigma},$$
$$\mathfrak{k}_2 \text{ correspond to } \mathfrak{k}_{\Sigma}^{\Sigma},\qquad \mathfrak{m}_2 \text{ to } \mathfrak{m}_{\Sigma}^{\Sigma}.$$

Then representing matrices for the algebra corresponding to $\mathfrak{g}_{\Psi}^{\Sigma}\sim\mathfrak{g}_{\Psi-\Sigma}\oplus\mathfrak{g}_{\Sigma}^{\Sigma}$ have support

$$\left(\begin{array}{c|c|c|c}
\mathfrak{k}_2 & & \mathfrak{m}_2 & \\ \hline
 & \mathfrak{k}_1 & & \mathfrak{m}_1 \\ \hline
\mathfrak{m}_2 & & \mathfrak{k}_2 & \\ \hline
 & \mathfrak{m}_1 & & \mathfrak{k}_1
\end{array}\right)$$

Thus $\beta^{-1}(x_{0,t})$ is the product of the manifold

$$\{e_1 \wedge \cdots \wedge e_t \wedge z : z \subset (e_{t+1} \wedge \cdots \wedge e_n) \wedge (e_{n+t+1} \wedge \cdots \wedge e_{2n}) \text{ is a positive definite } (n-t)\text{-plane with } J(z,z) = 0\}$$

corresponding to $G_{\Psi-\Sigma,0}(x_{\phi,\Sigma})$, with the manifold

$$\{y \wedge e_{n+t+1} \wedge \cdots \wedge e_{2n} : y \subset (e_1 \wedge \cdots \wedge e_t) \wedge (e_{n+1} \wedge \cdots \wedge e_{n+t}) \text{ is a negative definite } t\text{-plane with } J(y,y) = 0\}$$

corresponding to $G^{\Sigma}_{\Sigma,0}(x_{\phi,\Sigma})$. Via K-equivariance, this describes all β-fibers.

We can now describe the map β, as before, by a minimax construction from the unit quadrics

$$Q = \{u \in \mathbf{C}^{2n} : \langle u,u \rangle = -1\} \qquad \text{and} \qquad S = \{v \in \mathbf{C}^{2n} : \langle v,v \rangle = +1\}$$

and the orthogonal projections

$$\pi_U : \mathbf{C}^{2n} \to U \qquad \text{and} \qquad \pi_V : \mathbf{C}^{2n} \to V.$$

Let $x \in G_0(x_{0,t})$. In other words, x is an n-dimensional subspace of $\mathbf{C}^{2n}$, on which $\langle\ ,\ \rangle$ has t negative eigenvalues and $n-t$ positive eigenvalues, such that $J(x,x) = 0$. Within the ambient Grassmannian of X, we decompose

$$x = (u_1 \wedge \cdots \wedge u_t) \wedge (v_1 \wedge \cdots \wedge v_{n-t})$$

where $\{u_1, \ldots, u_t\} \subset Q \cap x$ are mutually orthogonal, spanning the t-dimensional negative definite subspace of x closest to U, and $\{v_1, \ldots, v_{n-t}\} \subset S \cap x$ are mutually orthogonal, spanning to the $(n-t)$-dimensional positive definite subspace of x closest to V. As $G_0(x_{0,t})$ is totally geodesic in the open subset $\mathbf{SU}(n,n)\,(x_{0,t})$ of the ambient Grassmannian, it follows that

$$\beta(x) = \pi_U(u_1 \wedge \cdots \wedge u_t) \wedge \pi_V(v_1 \wedge \cdots \wedge v_{n-t}) \in K(x_{0,t}).$$

As before, the K-equivariant partially holomorphic fibration

$$\beta : G_0(x_{s,t}) \to K(x_{s,t}), \qquad 0 \leqq s \leqq t \leqq n,$$

of an arbitrary orbit, is now specified by our knowledge of the holomorphic arc components and the open orbit case.

12. Subvarieties of Grassmannians Defined by Symmetric Forms

We work out the G_0-orbit structure, the holomorphic arc components, and the orbit fibrations, for the symmetric spaces corresponding to $X = \mathbf{SO}(2n)/\mathbf{U}(n)$. They differ from the Siegel Half Plane case in that J is replaced by a symmetric bilinear form. That change does not destroy the simplifying feature that X be totally geodesic in an ambient Grassmannian.

Fix a nondegenerate symmetric bilinear form S on $\mathbf{C}^{2n}$ and define

$$\tilde{X} = \{n\text{-planes } x \subset \mathbf{C}^{2n} : S(x,x) = 0\}.$$

Then $\tilde{X}$ is homogeneous under the complex orthogonal group

$$\mathbf{O}(2n,\mathbf{C}) = \{g : \mathbf{C}^{2n} \to \mathbf{C}^{2n} \text{ linear} : g \text{ preserves } S\}.$$

That orthogonal group has two topological components; the identity component is

$$\mathbf{SO}(2n,\mathbf{C}) = \{g \in \mathbf{O}(2n,\mathbf{C}) : \det g = +1\}$$

and the other component is

$$\{g \in \mathbf{O}(2n,\mathbf{C}) : \det g = -1\}.$$

From now on assume $n > 1$. Then the centers of $\mathbf{SO}(2n,\mathbf{C})$ and $\mathbf{O}(2n,\mathbf{C})$ are $\{\pm I_{2n}\}$, which is the common kernel of their actions on $\tilde{X}$, so

$$G = \mathbf{SO}(2n,\mathbf{C})/\{\pm I\} \qquad \text{and} \qquad \tilde{G} = \mathbf{O}(2n,\mathbf{C})/\{\pm I\}$$

are complex Lie transformation groups on $\tilde{X}$, of complex dimension $2n^2 - n$, with G connected and of index 2 in $\tilde{G}$.

Choose a basis $\{e_1, \ldots, e_{2n}\}$ of $\mathbf{C}^{2n}$ in which

$$S = \begin{pmatrix} 0 & I \\ I & 0 \end{pmatrix}, \qquad \text{i.e.,} \qquad S(u,v) = \sum_{k=1}^{n} (u^k v^{n+k} + u^{n+k} v^k).$$

In that basis, using $n \times n$ matrix blocks,

$$\tilde{G} = \{\pm \begin{pmatrix} A & B \\ C & D \end{pmatrix} : {}^t A \cdot C + {}^t C \cdot A = 0 = {}^t B \cdot D + {}^t D \cdot B,\ {}^t A \cdot D + {}^t C \cdot B = I\}$$

and

$$G = \{\pm \begin{pmatrix} A & B \\ C & D \end{pmatrix} \in \tilde{G} : \det \begin{pmatrix} A & B \\ C & D \end{pmatrix} = 1\}.$$

Now define

$$x_0 = e_{n+1} \wedge \cdots \wedge e_{2n} \in \tilde{X} \qquad \text{base point.}$$

The isotropy subgroup $P = \{g \in \tilde{G} : g(x_0) = x_0\}$ is

$$P = \{\pm\begin{pmatrix} A & 0 \\ C & D \end{pmatrix} : {}^tA\cdot C + {}^tC\cdot A = 0,\ {}^tA\cdot D = I\}.$$

If

$$\pm\begin{pmatrix} A & 0 \\ C & D \end{pmatrix} \in P$$

we compute

$$\det\begin{pmatrix} A & 0 \\ C & D \end{pmatrix} = (\det A)(\det D) = \det({}^tA\cdot D) = \det I = 1.$$

Now

$$P \subset G \qquad \text{so} \qquad \tilde{X} = \tilde{G}/P \qquad \text{has two components.}$$

Now define

$$X = G(x_0), \qquad \text{component of } \tilde{X} \text{ that contains } x_0.$$

Thus

$$X = G/P \qquad \text{compact complex } \tfrac{1}{2}n(n-1)\text{-manifold}$$

and

$$x_0 = e_{n+1} \wedge \cdots \wedge e_{2n} \in X \qquad \text{base point.}$$

The maximal compact subgroups of G are the conjugates of its compact real form

$$G_c = \{\mathbf{SO}(2n,\mathbf{C}) \cap \mathbf{U}(2n)\}/\{\pm I_{2n}\}.$$

That group obviously is compact. To check that it is a real form observe that

$$T = \frac{1}{\sqrt{2i}}\begin{pmatrix} I & iI \\ iI & I \end{pmatrix} \in \mathbf{U}(2n) \qquad \text{and} \qquad {}^tT\cdot T = \begin{pmatrix} 0 & I \\ I & 0 \end{pmatrix},$$

so that

$$T^{-1}\cdot\mathbf{SO}(2n)\cdot T = \mathbf{SO}(2n,\mathbf{C}) \cap \mathbf{U}(2n)$$

where $\mathbf{SO}(2n)$ denotes the group of all real orthogonal matrices of determinant 1 and degree $2n$. The isotropy subgroup of G_c at x_0 is

$$K = G_c \cap P = \{\pm \begin{pmatrix} A & 0 \\ 0 & \bar{A} \end{pmatrix} : A \in \mathbf{U}(n)\}.$$

Now $G_c(x_0)$ is open in X by dimension and closed in X by compactness, so

$$X = G_c/K \qquad \text{compact presentation.}$$

This exhibits X as the Hermitian symmetric space $\mathbf{SO}(2n)/\mathbf{U}(n)$ of compact type; the symmetry at x_0 is

$$s = \pm \begin{pmatrix} iI & 0 \\ 0 & -iI \end{pmatrix}.$$

We again use the Hermitian form on $\mathbf{C}^{2n}$ given by

$$\langle u,v \rangle = -\sum_{j=1}^{n} u^j \bar{v}^j + \sum_{k=1}^{n} u^{n+k} \bar{v}^{n+k}.$$

It specifies the indefinite unitary group $\mathbf{U}(n,n)$ and thus specifies $\mathbf{SO}(2n,\mathbf{C}) \cap \mathbf{U}(n,n)$. Define

$$G_0 : \text{transformation group on } X \text{ from } \mathbf{SO}(2n,\mathbf{C}) \cap \mathbf{U}(n,n).$$

Then

$$G_0 \text{ is a noncompact real form of } G$$

and

$$G_0 \cap P = K \qquad \text{so} \qquad G_0/K \cong G_0(x_0) \qquad \text{open in } X.$$

Thus

$$X_0 = G_0(x_0) \subset X \qquad \text{open } G_0\text{-orbit}$$

is the Hermitian symmetric space of noncompact type dual to $X = G_c/K$, with the same symmetry s at x_0, in X under Borel Embedding. In particular, $X_0 \cong \mathbf{SO}^*(2n)/\mathbf{U}(n)$ and $G_0 \cong \mathbf{SO}^*(2n)/\{\pm I_{2n}\}$.

Now we work out the Harish-Chandra embedding. Identify the Lie algebra $\mathfrak{g}$ of $G = \mathbf{SO}(2n,\mathbf{C})/\{\pm I\}$ with that of $\mathbf{SO}(2n,\mathbf{C})$. Thus

$$\mathfrak{g} = \left\{\begin{pmatrix} A & B \\ C & D \end{pmatrix} : {}^t\begin{pmatrix} A & B \\ C & D \end{pmatrix}\begin{pmatrix} 0 & I \\ I & 0 \end{pmatrix} + \begin{pmatrix} 0 & I \\ I & 0 \end{pmatrix}\begin{pmatrix} A & B \\ C & D \end{pmatrix} = 0\right\}.$$

That says

$$\mathfrak{g} = \left\{ \begin{pmatrix} A & B \\ C & D \end{pmatrix} : B + {}^tB = 0,\ C + {}^tC = 0, \text{ and } D = -{}^tA \right\}.$$

It follows that

$$\mathfrak{p} = \left\{ \begin{pmatrix} A & 0 \\ C & D \end{pmatrix} : C + {}^tC = 0, \text{ and } D = -{}^tA \right\}.$$

and

$$\mathfrak{m}^+ = \left\{ \begin{pmatrix} 0 & B \\ 0 & 0 \end{pmatrix} : B + {}^tB = 0 \right\}.$$

In particular,

$$Z \to \tilde{Z} = \begin{pmatrix} 0 & Z \\ 0 & 0 \end{pmatrix} \text{ maps } \quad (\text{antisymmetric } n \times n \text{ matrices}) \cong \mathfrak{m}^+.$$

We view Harish-Chandra's map $\xi : \mathfrak{m}^+ \to X$ as the map $\xi(Z) = (\exp \tilde{Z})(x_0)$ from the space of antisymmetric complex $n \times n$ matrices. Then

$$\xi(Z) = (\exp \tilde{Z})(x_0) = v_1 \wedge \ldots \wedge v_n \text{ where } (v_1, \ldots, v_n) = \begin{pmatrix} Z \\ I \end{pmatrix}$$

and we use Witt's Theorem to see that

$$\xi^{-1}(X_0) = \{n \times n \text{ antisymmetric } Z : I - Z^* \cdot Z \gg 0\}.$$

The same calculations as in Sections 10 and 11 show that G_0 acts on $\xi^{-1}(X_0)$ by

$$\pm \begin{pmatrix} A & B \\ C & D \end{pmatrix} : Z \to (AZ + B)(CZ + D)^{-1}$$

and that the Banach norm on $\mathfrak{m}^+$, for which $\xi^{-1}(X_0)$ is the unit ball, is

$$\|Z\|^2 = \max\{\text{eigenvalues of } Z^* \cdot Z\}.$$

The symmetric spaces X_0 and X have the same rank r given by

$$2r = n \quad \text{for} \quad n \text{ even}, \qquad 2r + 1 = n \quad \text{for} \quad n \text{ odd}.$$

Observe that the matrix

$$c = \frac{1}{\sqrt{2}}\begin{pmatrix} 1 & 0 & 0 & 1 \\ 0 & 1 & -1 & 0 \\ 0 & 1 & 1 & 0 \\ -1 & 0 & 0 & 1 \end{pmatrix} \text{ satisfies: } c^2 = \begin{pmatrix} 0 & 0 & 0 & 1 \\ 0 & 0 & -1 & 0 \\ 0 & 1 & 0 & 0 \\ -1 & 0 & 0 & 0 \end{pmatrix},$$

so we can check that

$$c \in \mathbf{SO}(4,\mathbf{C}) \cap \mathbf{U}(4) \qquad \text{and} \qquad c^4 = -I_4.$$

Now, the partial Cayley transforms $c_i \in G_c$, $1 \leqq j \leqq r = [n/2]$, are given by

$$c_j(e_{2j-1}) = \frac{1}{\sqrt{2}}(e_{2j-1} - e_{n+2j}), \qquad c_j(e_{2j}) = \frac{1}{\sqrt{2}}(e_{2j} + e_{n+2j-1});$$

$$c_j(e_{n+2j-1}) = \frac{1}{\sqrt{2}}(-e_{2j} + e_{n+2j-1}), \qquad c_j(e_{n+2j}) = \frac{1}{\sqrt{2}}(e_{2j-1} + e_{n+2j});$$

$$c_j(e_k) = e_k \qquad \text{for} \quad k \notin \{2j-1, 2j, n+2j-1, n+2j\}.$$

These do not coincide with the partial Cayley transforms on the ambient Grassmannian of n-planes in $\mathbf{C}^{2n}$; for although X is totally geodesic there, it has lower rank $r < n$. Now our centers for G_0-orbits are the

$$x_{s,t} = c_1 c_2 \cdots c_s c_{s+1}^2 c_{s+2}^2 \cdots c_t^2 x_0, \qquad 0 \leqq s \leqq t \leqq r = [n/2],$$

and

$$\text{the } G_0\text{-orbits on } X \text{ are the } G_0(x_{s,t}), \qquad 0 \leqq s \leqq t \leqq r.$$

In particular,

there are precisely $\frac{1}{2}(r+1)(r+2)$ G_0-orbits on X,

the $r+1$ orbits $G_0(x_{0,t})$, $0 \leqq t \leqq r$, are the open ones, and

the orbit $G_0(x_{r,r})$ is the unique closed orbit.

The orbit $G_0(x_{s,t})$ consists of all elements $x \in X$ on which $\langle\ ,\ \rangle$ has $n - 2t$ positive eigenvalues, $2(t-s)$ negative eigenvalues, and $2s$ eigenvalues zero. As before,

$$X - \xi(\mathfrak{m}^+) = \{x \in X : e_1 \wedge \cdots \wedge e_n \wedge x = 0\},$$

and

$$D_{s,t} = \xi^{-1} G_0(x_{s,t}) \subset \mathfrak{m}^+$$

consists of all $n \times n$ antisymmetric matrices Z such that the Hermitian matrix $I - Z^* \cdot Z$ has $n - 2t$ positive eigenvalues, $2(t - s)$ negative eigenvalues, and $2s$ eigenvalues zero.

The topological boundary of X_0 in X is $G_0(x_{1,1}) \cup G_0(x_{2,2}) \cup \cdots \cup G_0(x_{r,r})$ where $G_0(x_{s,s})$ is the union of those boundary components that are symmetric spaces of rank $r - s$. For $0 \leqq s \leqq r$, define

$X_{(r-s)}$ is the set of all elements $x \in X$ such that

$$e_{n+1} \wedge \cdots \wedge e_{n+2s} \subset x \subset e_{2s+1} \wedge e_{2s+2} \wedge \cdots \wedge e_{2n}.$$

Interior product with $e_{n+1} \wedge \cdots \wedge e_{n+2s}$ maps $X_{(r-s)}$ isomorphically onto the component of $e_{n+2s+1} \wedge \cdots \wedge e_{2n}$ in the two-component subvariety of the complex Grassmannian of $(n - 2s)$-planes in $\mathbf{C}^{2n-4s} = (e_{2s+1} \wedge \cdots \wedge e_n) \wedge (e_{n+2s+1} \wedge \cdots \wedge e_{2n})$ defined by $S(y,y) = 0$. The Borel-Embedded noncompact dual of $X_{(r-s)}$ is

$$X_{(r-s),0} = X_0 \cap X_{(r-s)}.$$

For $1 \leqq s \leqq r$,

$c_1 c_2 \cdots c_s X_{(r-s),0}$ is a boundary component of rank $r - s$,

so

$$G_0(x_{s,s}) = \bigcup_{k \in K} k c_1 c_2 \cdots c_s X_{(r-s),0}.$$

The even dimensional G_0-equivalence classes of totally S-isotropic $\langle\ ,\ \rangle$-isotropic subspaces of $\mathbf{C}^{2n}$ are represented by the $2s$-dimensional spaces V_s, $0 \leqq s \leqq r$, defined by

$$V_s = \{(-e_2 + e_{n+1}) \wedge (e_1 + e_{n+2})\} \wedge \cdots \wedge \{(-e_{2s} + e_{n+2s-1}) \wedge (e_{2s-1} + e_{n+2s})\}.$$

The maximal parabolic subgroups of G_0 are the

$$N_{s,0} = \{g \in G_0 : g(V_s) = V_s\}$$

and

$$N_{s,0} = \{g \in G_0 : g c_1 c_2 \cdots c_s X_{(r-s),0} = c_1 c_2 \cdots c_s X_{(r-s),0}\}.$$

We locate the ξ^{-1}-images of the boundary components of X_0 in X; they are the boundary components of

$$\{n \times n \text{ antisymmetric matrices } Z\colon I - Z^*\cdot Z \gg 0\}$$

in the space of $n \times n$ antisymmetric matrices:

$\xi^{-1}(c_1c_2 \cdots c_sX_{(r-s),0})$ consists of all $n \times n$ antisymmetric matrices $Z = \begin{pmatrix} J_{2s} & 0 \\ 0 & 'Z \end{pmatrix}$, where

$$J_{2s} = \begin{pmatrix} 0 & 1 & & & \\ -1 & 0 & & & \\ & & \ddots & & \\ & & & 0 & 1 \\ & & & -1 & 0 \end{pmatrix} \quad \text{and} \quad I_{n-2s} - 'Z^* \cdot 'Z \gg 0.$$

$K = \left\{ \pm\begin{pmatrix} A & 0 \\ 0 & \bar{A} \end{pmatrix} : A \in \mathbf{U}(n) \right\}$ acts on $\mathfrak{m}^+$ by

$$\pm\begin{pmatrix} A & 0 \\ 0 & \bar{A} \end{pmatrix} : Z \to AZ\bar{A}^{-1} = A\cdot Z\cdot {}^t A.$$

Thus $K\cdot\xi^{-1}(c_1c_2 \cdots c_sX_{(r-s),0}) = \xi^{-1}G_0(x_{s,s})$ consists of all $n \times n$ antisymmetric matrices Z such that $I_n - Z^*\cdot Z$ is positive semidefinite and of rank $n - 2s$.

The general theory ensures that the holomorphic arc components of an arbitrary orbit $G_0(x_{s,t})$ are the

$$k\{c_1c_2 \cdots c_sX_{(r-s)} \cap G_0(x_{s,t})\}, \qquad k \in K,$$

and that their normalizers are:

$$N_{s,0} \text{ is the } G_0\text{-normalizer of } c_1c_2 \cdots c_sX_{(r-s)} \cap G_0(x_{s,t}).$$

We describe the K-equivariant holomorphic fibration

$$\beta : G_0(x_{0,t}) \to K(x_{0,t}), \qquad 0 \leqq t \leqq r = [n/2],$$

of an open orbit over its maximal compact subvariety. As before, $\mathbf{C}^{2n} = U \oplus V$ where $U = e_1 \wedge \cdots \wedge e_n$ and $V = e_{n+1} \wedge \cdots \wedge e_{2n}$. Now

$$x_{0,t} = (e_1 \wedge \cdots \wedge e_{2t}) \wedge (e_{n+2t+1} \wedge \cdots \wedge e_{2n})$$

gives us

$$K(x_{0,t}) = \{x = y \wedge z : x \in X,\ y \text{ is } 2t\text{-plane in } U,\ z \text{ is } (n-2t)\text{-plane in } V\}.$$

We use refined block form $2n \times 2n$ matrices divided as follows.

$$\left(\begin{array}{c|c|c|c}
2t \times 2t & & & \\ \hline
 & (n-2t)\times(n-2t) & & \\ \hline
 & & 2t \times 2t & \\ \hline
 & & & (n-2t)\times(n-2t)
\end{array}\right)$$

As before, let

$$\mathfrak{k}_1 \text{ correspond to } \mathfrak{k}_{\Psi-\Sigma}, \qquad \mathfrak{m}_1 \text{ to } \mathfrak{m}_{\Psi-\Sigma};$$
$$\mathfrak{k}_2 \text{ correspond to } \mathfrak{k}_{\Sigma}^{\Sigma}, \qquad \mathfrak{m}_2 \text{ to } \mathfrak{m}_{\Sigma}^{\Sigma}.$$

Then matrices representing the algebra corresponding to $\mathfrak{g}_{\Psi}^{\Sigma} \sim \mathfrak{g}_{\Psi-\Sigma} \oplus \mathfrak{g}_{\Sigma}^{\Sigma}$ have support

$$\left(\begin{array}{c|c|c|c}
\mathfrak{k}_2 & & \mathfrak{m}_2 & \\ \hline
 & \mathfrak{k}_1 & & \mathfrak{m}_1 \\ \hline
\mathfrak{m}_2 & & \mathfrak{k}_2 & \\ \hline
 & \mathfrak{m}_1 & & \mathfrak{k}_1
\end{array}\right)$$

Thus $\beta^{-1}(x_{0,t})$ is the product of the manifold

$$\{e_1 \wedge \cdots \wedge e_{2t} \wedge z : z \subset (e_{2t+1} \wedge \cdots \wedge e_n) \wedge (e_{n+2t+1} \wedge \cdots \wedge e_{2n})$$
$$\text{is a positive definite } (n-2t)\text{-plane with } S(z,z) = 0\}$$

corresponding to $G_{\Psi-\Sigma,0}(x_{\phi,\Sigma})$, with the manifold

$$\{y \wedge e_{n+2t+1} \wedge \cdots \wedge e_{2n} : y \subset (e_1 \wedge \cdots \wedge e_{2t}) \wedge (e_{n+1} \wedge \cdots \wedge e_{n+2t})$$
$$\text{is a negative definite } 2t\text{-plane with } S(y,y) = 0\}$$

corresponding to $G_{\Sigma,0}^{\Sigma}(x_{\phi,\Sigma})$. By means of K-equivariance, this describes all β-fibers.

As X is totally geodesic in the ambient Grassmannian of n-planes in $\mathbf{C}^{2n}$, and as the center $x_{0,t}$ of an open G_0-orbit on X is the center of an open $\mathbf{SU}(n,n)$-orbit on the Grassmannian, we now can use the minimax construction of Section 10 to describe the map β. Denote the unit quadrics

$$Q = \{u \in \mathbf{C}^{2n} : \langle u,u \rangle = -1\} \qquad \text{and} \qquad L = \{v \in \mathbf{C}^{2n} : \langle v,v \rangle = +1\}$$

and the orthogonal projections

$$\pi_U : \mathbf{C}^{2n} \to U \qquad \text{and} \qquad \pi_V : \mathbf{C}^{2n} \to V.$$

If $x \in G_0(x_{0,t})$, we construct, as in Section 10,

$$\{u_1, \ldots, u_{2t}\} \subset Q \cap x \quad \text{mutually orthogonal,}$$
$$\{v_1, \ldots, v_{n-2t}\} \subset L \cap x \quad \text{mutually orthogonal,}$$

with

$$x = (u_1 \wedge \cdots \wedge u_{2t}) \wedge (v_1 \wedge \cdots \wedge v_{n-2t}) \in G_0(x_{0,t}).$$

That construction is such that $u_1 \wedge \cdots \wedge u_{2t}$ is the negative definite $2t$-dimensional subspace of x nearest to U and $v_1 \wedge \cdots \wedge v_{n-2t}$ is the positive definite $(n - 2t)$-dimensional subspace of x nearest to V. Thus, as described in the first sentence of this paragraph,

$$\beta(x) = \pi_U(u_1 \wedge \cdots \wedge u_{2t}) \wedge \pi_V(v_1 \wedge \cdots \wedge v_{n-2t}) \in K(x_{0,t}).$$

Now, in the case of an arbitrary orbit $G_0(x_{s,t})$, the K-equivariant partially holomorphic fibration

$$\beta : G_0(x_{s,t}) \to K(x_{s,t}), \qquad 0 \leqq s \leqq t \leqq r,$$

is specified because we know the holomorphic arc components and the open orbit case.

13. The Complex Quadrics

In this section we work out the G_0-orbit structure, holomorphic arc components, and orbit fibration for the remaining classical series of irreducible Hermitian symmetric spaces. That is the case where X is a quadric in complex projective space $P^{n+1}(\mathbf{C})$, $n > 2$, which is not contained in a hyperplane. Here matters are rather more complicated than in Sections 11 and 12 because X is not totally geodesic in its ambient Grassmannian $P^{n+1}(\mathbf{C})$, and because G_0 is properly contained in its real algebraic hull.

Given an integer $n \geqq 0$, the complex projective space of complex dimension $n + 1$ is

$$P^{n+1}(\mathbf{C}) : \text{complex Grassmannian of lines through 0 in } \mathbf{C}^{n+2}.$$

Fix a nondegenerate symmetric bilinear form S on $\mathbf{C}^{n+2}$ and define

$$X = \{x \in P^{n+1}(\mathbf{C}) : S(x,x) = 0\}.$$

That is the complex quadric. It is homogeneous under the complex orthogonal group $\mathbf{O}(n + 2, \mathbf{C})$ associated to S. Thus X carries complex Lie transformation groups

$$\tilde{G} = \mathbf{O}(n + 2, \mathbf{C})/\{\pm I\} \qquad \text{and} \qquad G = \mathbf{SO}(n + 2, \mathbf{C})\cdot\{\pm I\}/\{\pm I\}$$

of complex dimension $\frac{1}{2}(n + 1)(n + 2)$, with G connected. If n is odd, so $\det(-I) = -1$, then $G = \tilde{G}$. If n is even, then G has index 2 in $\tilde{G}$.

As we do not use exterior products in expressing elements of X by their bases, we adopt the convention

$$\text{if } 0 \neq v \in \mathbf{C}^{n+2} \qquad \text{then } [v] \in P^{n+1}(\mathbf{C}) \text{ is the span } v\mathbf{C}.$$

Now choose a basis $\{e_1, \ldots, e_{n+2}\}$ of $\mathbf{C}^{n+2}$ in which $S(u,v) = \sum_{k=1}^{n+2} u^k v^k$. Then

$$\tilde{G} = \{\pm M : M \text{ is a complex } (n+2) \times (n+2) \text{ matrix and } {}^t M \cdot M = I\}.$$

Now choose base point

$$x_0 = [e_{n+1} + ie_{n+2}] \in X.$$

The isotropy subgroup $\tilde{P} = \{g \in \tilde{G} : g(x_0) = x_0\}$ is

$$\tilde{P} = \left\{\pm\begin{pmatrix} A & B \\ C & D \end{pmatrix} \in \tilde{G} : B = (B', iB'),\ D = \begin{pmatrix} a & b \\ c & d \end{pmatrix},\ ia - b = c + id\right\}$$

where the matrix blocks are of size given by

$$A \text{ is } n \times n, \qquad B \text{ is } n \times 2, \qquad B' \text{ is } n \times 1, \qquad C \text{ is } 2 \times n, \qquad D \text{ is } 2 \times 2.$$

As $\tilde{P}$ contains the transformation $\pm\begin{pmatrix} -1 & 0 \\ 0 & I_{n+1} \end{pmatrix}$ represented by a matrix of determinant -1, it meets every component of $\tilde{G}$. Thus

$$X = G(x_0) \cong G/P \qquad \text{connected where} \quad P = G \cap \tilde{P}.$$

The maximal compact subgroups of G are the conjugates of its compact real form

$$G_c = \{\mathbf{SO}(n+2, \mathbf{C}) \cap \mathbf{U}(n+2)\} \cdot \{\pm I\}/\{\pm I\}.$$

Here note that

$$\mathbf{SO}(n+2, \mathbf{C}) \cap \mathbf{U}(n+2) = \mathbf{SO}(n+2)$$

where the real special orthogonal group

$$\mathbf{SO}(n+2) = \{g \in \mathbf{SO}(n+2, \mathbf{C}) : g \text{ preserves real span of } \{e_1, \ldots, e_{n+2}\}\}.$$

The isotropy subgroup of G_c at x_0 is

$$K = G_c \cap P = \{\pm\begin{pmatrix} A & 0 \\ 0 & D \end{pmatrix} : A \in \mathbf{SO}(n) \text{ and } D \in \mathbf{SO}(2)\}.$$

Now dimension and compactness of $G_c(x_0)$ gives us

$$X = G_c/K \qquad \text{compact presentation.}$$

That exhibits X as the Hermitian symmetric space $\mathbf{SO}(n+2)/\mathbf{SO}(n) \times \mathbf{SO}(2)$ of compact type. The symmetry at x_0 is

$$s = \pm\begin{pmatrix} I_n & 0 \\ 0 & -I_2 \end{pmatrix}$$

We use the indefinite Hermitian form on $\mathbf{C}^{n+2}$ given by $\langle u,v \rangle = -\sum_{k=1}^{n} u^k \bar{v}^k + (u^{n+1}\bar{v}^{n+1} + u^{n+2}\bar{v}^{n+2})$. That defines the indefinite unitary group $\mathbf{U}(n,2)$ and thus the intersections

$$\mathbf{O}(n,2) = \mathbf{O}(n+2,\mathbf{C}) \cap \mathbf{U}(n,2) \qquad \text{and}$$
$$\mathbf{O}^+(n,2) = \mathbf{SO}(n+2,\mathbf{C}) \cap \mathbf{U}(n,2).$$

The indefinite orthogonal group $\mathbf{O}(n,2)$ has 4 topological components and $\mathbf{O}^+(n,2)$ has 2 components. Define

$\mathbf{SO}(n,2)$: common identity component of $\mathbf{O}(n,2)$ and $\mathbf{O}^+(n,2)$;

that is our indefinite special orthogonal group. Now define

G_0 : connected transformation group on X from $\mathbf{SO}(n,2)$.

Then

$$G_0 \text{ is a noncompact real form of } G$$

and

$$G_0 \cap P = K \qquad \text{so} \qquad G_0/K \cong G_0(x_0) \qquad \text{open in } X.$$

Now

$$X_0 = G_0(x_0) \subset X \qquad \text{open } G_0\text{-orbit}$$

is the Hermitian symmetric space of noncompact type dual to X, inside X by Borel Embedding. Note $X_0 \cong \mathbf{SO}(n,2) / \mathbf{SO}(n) \times \mathbf{SO}(2)$.

We come to the Harish-Chandra embedding. The Lie algebra $\mathfrak{g}$ of G is identified with the Lie algebra of $\mathbf{SO}(n + 2, \mathbf{C})$. Thus, in block form,

$$\mathfrak{g} = \left\{\begin{pmatrix} A & B \\ C & D \end{pmatrix} : {}^tA = -A,\ {}^tD = -D,\ {}^tB = -C\right\}.$$

Now

$$\mathfrak{p} = \left\{\begin{pmatrix} A & B \\ C & D \end{pmatrix} : {}^tA = -A,\ {}^tB = -C,\ B = (B', iB'),\ D = \begin{pmatrix} 0 & b \\ -b & 0 \end{pmatrix}\right\}$$

and

$$\mathfrak{m}^+ = \left\{\begin{pmatrix} 0 & B \\ -{}^tB & 0 \end{pmatrix} : B = (iB'', B'') \text{ where } B'' \text{ is } n \times 1\right\}.$$

Now, viewing the elements of $\mathbf{C}^n$ as column vectors ($n \times 1$ matrices),

$$Z \to \tilde{Z} = \begin{pmatrix} 0 & Z' \\ -{}^tZ' & 0 \end{pmatrix}, \qquad Z' = (iZ, Z), \qquad \text{maps } \mathbf{C}^n \cong \mathfrak{m}^+.$$

View Harish-Chandra's map $\xi : \mathfrak{m}^+ \to X$ as a map from $\mathbf{C}^n$. Now compute

$$Z' \cdot {}^tZ' = (iZ, Z)\begin{pmatrix} i{}^tZ \\ {}^tZ \end{pmatrix} = -Z \cdot {}^tZ + Z \cdot {}^tZ = 0$$

and

$${}^tZ' \cdot Z' = \begin{pmatrix} i{}^tZ \\ {}^tZ \end{pmatrix}(iZ, Z) = \begin{pmatrix} -{}^tZ \cdot Z & i{}^tZ \cdot Z \\ i{}^tZ \cdot Z & {}^tZ \cdot Z \end{pmatrix};$$

thus also

$$Z' \cdot ({}^tZ' \cdot Z') = (iZ, Z)\begin{pmatrix} -{}^tZ \cdot Z & i{}^tZ \cdot Z \\ i{}^tZ \cdot Z & {}^tZ \cdot Z \end{pmatrix} = 0.$$

That gives us

$$\tilde{Z} \cdot \tilde{Z} = \begin{pmatrix} 0 & 0 \\ 0 & Z'' \end{pmatrix} \text{ where } \quad Z'' = \begin{pmatrix} {}^tZ \cdot Z & -i{}^tZ \cdot Z \\ -i{}^tZ \cdot Z & -{}^tZ \cdot Z \end{pmatrix}$$

and

$$(\tilde{Z})^3 = 0, \qquad \text{so} \quad (\tilde{Z})^k = 0 \quad \text{for} \quad k \geqq 3.$$

Now

$$\exp \tilde{Z} = \sum_{k=0}^{\infty} \frac{1}{k!}(\tilde{Z})^k = \begin{pmatrix} I & Z' \\ -{}^tZ' & I + \frac{1}{2}Z'' \end{pmatrix}.$$

As $\xi(Z) = (\exp \tilde{Z})(x_0)$ and $x_0 = [e_{n+1} + ie_{n+2}]$ now

$$\xi(Z) = [v] \text{ where } v = \begin{pmatrix} 2iZ \\ 1 + {}^tZ\cdot Z \\ i(1 - {}^tZ\cdot Z) \end{pmatrix}.$$

In order to locate $\xi^{-1}(X_0) \subset \mathfrak{m}^+$, we consider the open set

$$\Omega = \{Z \in \mathbf{C}^n : \xi(Z) = [v] \text{ with } \langle v,v\rangle > 0\}.$$

The formula for $\xi(Z)$ says

$$\Omega = \{Z \in \mathbf{C}^n : 1 + |{}^tZ\cdot Z|^2 - 2Z^*\cdot Z > 0\}.$$

Thus

$$\Omega = \{Z \in \mathbf{C}^n : (1 - Z^*\cdot Z)^2 > (Z^*\cdot Z)^2 - |{}^tZ\cdot Z|^2\}.$$

As $Z^*\cdot Z \geqq |{}^tZ\cdot Z| \geqq 0$, we take positive square roots, thus expressing

$$\Omega = \Omega_1 \cup \Omega_2 \qquad \text{disjoint union}$$

where Ω_1 is the nonempty bounded domain star-shaped from 0 given by

$$\Omega_1 = \{Z \in \mathbf{C}^n : 1 - Z^*\cdot Z > [(Z^*\cdot Z)^2 - |{}^tZ\cdot Z|^2]^{1/2}\}$$

and Ω_2 is the nonempty unbounded domain star shaped from ∞ given by

$$\Omega_2 = \{Z \in \mathbf{C}^n : Z^*\cdot Z - 1 > [(Z^*\cdot Z)^2 - |{}^tZ\cdot Z|^2]^{1/2}\}.$$

From Witt's Theorem, $\xi^{-1}(X_0)$ is the topological component of Ω containing 0, thus is Ω_1:

$$\begin{aligned}\xi^{-1}(X_0) &= \{Z \in \mathbf{C}^n : 1 - Z^*\cdot Z > [(Z^*\cdot Z)^2 - |{}^tZ\cdot Z|^2]^{1/2}\} \\ &= \{Z \in \mathbf{C}^n : 1 + |{}^tZ\cdot Z|^2 - 2Z^*\cdot Z > 0 \text{ and } 1 - Z^*\cdot Z > 0\}.\end{aligned}$$

Recall that G_0 is the image in G of the identity component of $\mathbf{O}^+(n,2) = \mathbf{SO}(n+2, \mathbf{C}) \cap \mathbf{U}(n,2)$. Thus we have a subgroup of G in which G_0 has index 2, given by

$$G_{\mathbf{R}} = G_0 \cup c^2 G_0, \qquad c^2 = \pm\begin{pmatrix} -1 & & & \\ & I_{n-1} & & \\ & & -1 & \\ & & & 1 \end{pmatrix}.$$

The notation c^2 comes from $c = c_\Psi$ in the Cayley transform theory. Here the point is that $\xi^{-1}(c^2 X_0) = \Omega_2$ and that $c^2 G_0(x_0) = G_0(c^2 x_0)$ open G_0-orbit on X. This explains the occurrence of Ω_2. Compare with the special case $n = 1$.

The action of G_0 on $\xi^{-1}(X_0)$ is also a bit of a mess. Denote

$$q : \mathbf{C}^n \to \mathbf{C} \quad \text{by} \quad q(Z) = {}^t Z \cdot Z; \qquad \text{so } |q(Z)| < 1 \qquad \text{on} \quad \xi^{-1}(X_0).$$

Now the formula for $\xi(Z)$ says that

$$\text{if} \quad q = q(Z),\ |q| \neq 1, \qquad \text{then} \quad \xi(Z) = [z] \quad \text{where} \quad z = \begin{pmatrix} \dfrac{2}{1-q} Z \\ -i\dfrac{1+q}{1-q} \\ 1 \end{pmatrix}.$$

Now we invert $\xi : \xi^{-1}(X_0) \to X_0$. First note from our new formula for ξ that

$$\text{if } [v] \in X_0, \text{ say } v = \begin{pmatrix} V \\ v_1 \\ v_2 \end{pmatrix}, \text{ then } v_2 \neq 0 \neq (iv_1/v_2) + 1.$$

Using that notation and the identity $p = -i(1+q)/(1-q) \Rightarrow q = (ip-1)/(ip+1)$, the new formula for ξ says $\xi^{-1}[v] = \{(1-q)/2v_2\} V$ where $q = \dfrac{i(v_1/v_2) - 1}{i(v_1/v_2) + 1}$. Substituting q into $\xi^{-1}[v]$ now

$$\text{if} \quad [v] \in X_0, \qquad v = \begin{pmatrix} V \\ v_1 \\ v_2 \end{pmatrix}, \qquad \text{then } \xi^{-1}[v] = \frac{1}{v_2 + iv_1} V.$$

The action of G_0 on $\xi^{-1}(X_0)$ is, of course,

$$g(Z) = \xi^{-1} g \xi(Z).$$

If we write

$$g = \pm \begin{pmatrix} A & B \\ C & D \end{pmatrix} \qquad \text{and} \qquad \begin{pmatrix} 2iZ \\ 1 + q(Z) \\ i - iq(Z) \end{pmatrix} = \begin{pmatrix} Z_1 \\ Z_2 \end{pmatrix},$$

the inversion formula for ξ says

$$g(Z) = \frac{1}{(1,i)\,(CZ_1 + DZ_2)}\,(AZ_1 + BZ_2).$$

The symmetric spaces X_0 and X are of rank $r = 2$ provided $n \geqq 2$, $r = 1$ in the degenerate case $n = 1$ where X is the Riemann sphere $P^1(\mathbf{C})$. If $n = 2$, then $X = P^1(\mathbf{C}) \times P^1(\mathbf{C})$; if $n > 2$, then X is irreducible.

X_0 is of tube type. Except for the reducible case ($n = 2$), now $G_0(c_\Psi^2 x_0) = c_\Psi^2 X_0$ is the only orbit $\neq X_0$ but $\cong X_0$. As $[e_{n+1} + ie_{n+2}] = x_0$, that says

$$c_\Psi^2 X_0 = G_0([e_{n+1} - ie_{n+2}]) \supset \xi(\Omega_2).$$

That consideration suggests the condition, which defines the partial Cayley transform in the case $n = 1$ (where $r = 1$ and $c_\Psi = c_1$), that

$$c_\Psi : e_1 \to -e_{n+1}, \qquad e_{n+1} \to e_1, \qquad e_k \to e_k \qquad \text{for} \quad 1 \neq k \neq n+1.$$

Note that this definition implies that

$$c_\Psi^2 \text{ represents the nonidentity component of } G_{\boldsymbol{R}},$$

i.e., that

$$\Omega_2 = \xi^{-1} G_0(c_\Psi^2 x_0) = \xi^{-1} c_\Psi^2 X_0.$$

If $n > 1$, so $r = 2$, we define the partial Cayley transforms $c_1, c_2 \in G_c$ such that $c_\Psi = c_1 c_2$ by

$$c_1 = \exp \frac{\pi}{4}\begin{pmatrix} 0 & B_1 \\ -{}^tB_1 & 0 \end{pmatrix}, \qquad {}^tB_1 = \begin{pmatrix} 1 & 0 & 0 & \cdots & 0 \\ 0 & 1 & 0 & \cdots & 0 \end{pmatrix}.$$

and

$$c_2 = \exp \frac{\pi}{4}\begin{pmatrix} 0 & B_2 \\ -{}^tB_2 & 0 \end{pmatrix}, \qquad {}^tB_2 = \begin{pmatrix} 1 & 0 & 0 & \cdots & 0 \\ 0 & -1 & 0 & \cdots & 0 \end{pmatrix}.$$

Thus

$$c_1(e_1) = \frac{1}{\sqrt{2}}(e_1 - e_{n+1}), \qquad c_1(e_2) = \frac{1}{\sqrt{2}}(e_2 - e_{n+2}),$$
$$c_1(e_{n+1}) = \frac{1}{\sqrt{2}}(e_1 + e_{n+1}), \qquad c_1(e_{n+2}) = \frac{1}{\sqrt{2}}(e_2 + e_{n+2}),$$
$$\text{and} \qquad c_1(e_k) = e_k \qquad \text{for} \quad 2 < k \leqq n;$$

and

$$c_2(e_1) = \frac{1}{\sqrt{2}}(e_1 - e_{n+1}), \qquad c_2(e_2) = \frac{1}{\sqrt{2}}(e_2 + e_{n+2}),$$
$$c_2(e_{n+1}) = \frac{1}{\sqrt{2}}(e_1 + e_{n+1}), \qquad c_2(e_{n+2}) = \frac{1}{\sqrt{2}}(-e_2 + e_{n+2}),$$
$$\text{and} \qquad c_2(e_k) = e_k \qquad \text{for} \quad 2 < k \leqq n.$$

Note that the c_j have order 4 if $n = 2$, order 8 if $n > 2$.

We now assume $n > 2$. Then X is irreducible and of rank $r = 2$, so there are six orbits $G_0(x_{s,t})$, $0 \leqq s \leqq t \leqq 2$ with centers as follows.

$$\begin{array}{llll}
x_{0,0} = & x_0 & = [e_{n+1} + ie_{n+2}] & \text{positive} \\
x_{0,1} = & c_1^2 x_0 & = [e_1 + ie_2] & \text{negative} \\
x_{0,2} = & c_1^2 c_2^2 x_0 & = [e_{n+1} - ie_{n+2}] & \text{positive} \\
x_{1,1} = & c_1 x_0 & = [e_1 + ie_2 + e_{n+1} + ie_{n+2}] & \text{isotropic} \\
x_{1,2} = & c_1 c_2^2 x_0 & = [e_1 - ie_2 - e_{n+1} + ie_{n+2}] & \text{isotropic} \\
x_{2,2} = & c_1 c_2 x_0 & = [e_1 + ie_{n+2}] & \text{isotropic}
\end{array}$$

The six orbits are the

3 open orbits $\quad G_0(x_{0,0}),\quad G_0(x_{0,1}),\quad$ and $\quad G_0(x_{0,2})$,
2 intermediate orbits $\quad G_0(x_{1,1})\quad$ and $\quad G_0(x_{1,2})$,
1 closed orbit $\quad G_0(x_{2,2})$.

Now

$$\{Z \in \mathbf{C}^n : 1 + |{}^t Z \cdot Z| - 2Z^* \cdot Z > 0\} = \xi^{-1}(G_0(x_{0,0}) \cup G_0(x_{0,2})),$$
$$\{Z \in \mathbf{C}^n : 1 + |{}^t Z \cdot Z| - 2Z^* \cdot Z < 0\} = \xi^{-1}(G_0(x_{0,1})), \qquad \text{and}$$
$$\{Z \in \mathbf{C}^n : 1 + |{}^t Z \cdot Z| - 2Z^* \cdot Z = 0\} = \xi^{-1}(G_0(x_{1,1}) \cup G_0(x_{1,2}) \cup G_0(x_{2,2})).$$

The topological boundary of X_0 in X is $G_0(x_{1,1}) \cup G_0(x_{2,2})$. There $G_0(x_{1,1})$ is the union of those boundary components that are symmetric spaces of rank 1 (in fact, unit disks in $\mathbf{C}^1$) and $G_0(x_{2,2})$ is the Bergman–Šilov boundary. Define compact totally geodesic submanifolds of X by $X_{(0)} = \{x_0\}$ point, $X_{(2)} = X$, and

$$\begin{aligned} X_{(1)} &= \{[a(e_1 + ie_2) + b(e_{n+1} + ie_{n+2})] : (a,b) \neq (0,0)\} \\ &= \{x \in X : x \subset (e_1 + ie_2) \wedge (e_{n+1} + ie_{n+2})\}. \end{aligned}$$

Note that $X_{(1)}$ is a Riemann sphere. The Borel Embedded noncompact dual of $X_{(1)}$ is

$$X_{(1),0} = X_0 \cap X_{(1)} = \{[z(e_1 + ie_2) + (e_{n+1} + ie_{n+2})] : |z| < 1\}.$$

Now

$\{c_1 c_2 x_0\}$ and $c_1 X_{(1),0}$ are boundary components of X_0

and

$$G_0(x_{2,2}) = K\cdot c_1c_2x_0 \qquad \text{and} \qquad G_0(x_{1,1}) = K\cdot c_1 X_{(1),0}.$$

The maximal parabolic subgroups of G_0 are $N_{0,0} = G_0$,

$$N_{1,0} = \{g \in G_0 : g \text{ preserves the 2-plane } (e_1 + ie_2) \wedge (e_{n+1} + ie_{n+2})\},$$
$$N_{2,0} = \{g \in G_0 : g \text{ preserves the line } [e_1 + ie_{n+2}]\}.$$

The boundary components of X_0 have normalizers

$$N_{1,0} = \{g \in G_0 : gc_1X_{(1),0} = c_1X_{(1),0}\},$$
$$N_{2,0} = \{g \in G_0 : gc_1c_2x_0 = c_1c_2x_0\}.$$

We locate the ξ^{-1}-images of the boundary components of X_0 in X; they are the boundary components of

$$\{Z \in \mathbf{C}^n : 1 - Z^*\cdot Z > 0 \quad \text{and} \quad 1 + |{}^t Z\cdot Z|^2 - 2Z^*\cdot Z > 0\}$$

in $\mathbf{C}^n$:

$$\xi^{-1}(c_1X_{(1),0}) = \{Z \in \mathbf{C}^n : {}^tZ = (z, iz, 0, \ldots, 0) \text{ with } \operatorname{Im} z < 0\}$$

and

$$\xi^{-1}(c_1c_2x_0) = {}^t(-i, 0, \ldots, 0).$$

$K = \{\pm\begin{pmatrix} A & 0 \\ 0 & D\end{pmatrix} : A \in \mathbf{SO}(n) \text{ and } D \in \mathbf{SO}(2)\}$ acts on $\mathfrak{m}^+$ by

$$\pm\begin{pmatrix} A & 0 \\ 0 & D\end{pmatrix} : Z \to W \qquad \text{where} \quad (iW, W) = A\cdot(iZ, Z)\cdot D^{-1};$$

thus

$$\text{if} \quad D = \begin{pmatrix} \cos\theta & \sin\theta \\ -\sin\theta & \cos\theta\end{pmatrix}, \qquad \text{then} \quad \pm\begin{pmatrix} A & 0 \\ 0 & D\end{pmatrix} : Z \to e^{i\theta}AZ.$$

Now

$$\begin{aligned} K\cdot\xi^{-1}(c_1c_2x_0) &= \xi^{-1}G_0(x_{2,2}) \\ &= \{Z \in \mathbf{C}^n : Z^*Z = 1 \text{ and } e^{i\theta}Z \in \mathbf{R}^n \text{ for some } \theta\} \end{aligned}$$

and

$$\begin{aligned} K\cdot\xi^{-1}(c_1X_{(1),0}) &= \xi^{-1}G_0(x_{1,1}) \\ &= \{Z \in \mathbf{C}^n : Z = U + iV \text{ with } U, V \in \mathbf{R}^n, U^*U = V^*V \neq 0 = U^*V\}. \end{aligned}$$

The general theory ensures that the holomorphic arc components of $G_0(x_{s,t})$ are the $k \cdot \{c_1 \cdots c_s X_{(2-s)} \cap G_0(x_{s,t})\}$, $k \in K$, and that $N_{s,0}$ is the normalizer of $c_1 \cdots c_s X_{(2-s)} \cap G_0(x_{s,t})$.

We describe the K-equivariant holomorphic fibrations

$$\beta : G_0(x_{0,t}) \to K(x_{0,t}), \qquad t = 0, 1, 2,$$

of an open orbit over its maximal compact subvariety. If $t = 0$ or $t = 2$, then $K(x_{0,t})$ is reduced to a point; thus

$$\text{if} \quad t = 0 \quad \text{or} \quad t = 2, \qquad \text{then} \quad \beta(gx_{0,t}) = x_{0,t}.$$

Now suppose $t = 1$. We note

$$x_{0,1} = [e_1 + ie_2]$$

Denoting real span by Re,

$$K(x_{0,1}) = \{[u+iv] : u,v \in \mathrm{Re}(e_1, \ldots, e_n), \langle u,u \rangle = \langle v,v \rangle = -1, \langle u,v \rangle = 0\}.$$

Note also that

$$G_0(x_{0,1}) = \{[z_1(u + iv) + z_2(e_{n+1} + ie_{n+2})] : u,v \in \mathrm{Re}(e_1, \ldots, e_n),$$
$$\langle u,u \rangle = \langle v,v \rangle = -1, \langle u,v \rangle = 0, |z_1| < 1, |z_2| < 1\}.$$

Now

$$\beta[z_1(u + iv) + z_2(e_{n+1} + ie_{n+2})] = [u + iv]$$

and the β-fibers are products of two unit disks.

The holomorphic arc components of $G_0(x_{1,1})$ and $G_0(x_{1,2})$ are unit disks; their maximal compact subvarieties are reduced to points. Thus the K-equivariant partially holomorphic fibrations

$$\beta : G_0(x_{1,t}) \to K(x_{1,t}), \qquad t = 1, 2,$$

are given by

$$\beta(k \cdot \{c_1 X_{(1)} \cap G_0(x_{1,t})\}) = k(x_{1,t}).$$

Finally, of course, $\beta : G_0(x_{2,2}) \to K(x_{2,2})$ has fiber reduced to a point.

References

[1] W. L. Baily and A. Borel, Compactification of arithmetic quotients of bounded symmetric domains, *Ann. of Math.* **84** (1966), 442–528.

[2] A. Borel and J. de Siebenthal, Les sous-groupes fermés de rang maximum des groupes de Lie clos, *Comment. Math. Helv.* **23** (1949), 200–221.

[3] É. Cartan, Sur une classe remarquable d'espaces de Riemann, *Bull. Soc. Math. France*, **54** (1926), 214–264; ibid, **55** (1927), 114–134.

[4] É. Cartan, Sur les domaines bornes homogènes de l'espace de *n* variables complexes, *Abh. Math. Sem. Univ. Hamburg*, **11** (1935), 116–162.

[5] Harish-Chandra, Representations of semisimple Lie groups, VI, *Amer. J. Math.* **78** (1956), 564–628.

[6] Hua L.-K., "*Harmonic Analysis of Functions of Several Complex Variables in Classical Domains*," Science Press, Peking, 1958 (in Chinese). Russian translation from the Chinese: Izdat. Inostr. Lit., Moscow, 1959. English translation from the Russian: Amer. Math. Soc., Providence, 1963.

[7] A. Korányi and J. A. Wolf, Realization of hermitian symmetric spaces as generalized half planes, *Ann. of Math.*, **81** (1965), 265–288.

[8] C. C. Moore, Compactifications of symmetric spaces. II (the Cartan domains), *Amer. J. Math.*, **86** (1964), 358–378.

[9] I. I. Pjateckĭi-Šapiro, "*Geometry of Classical Domains and Theory of Automorphic Functions.*" Moscow, 1961 (in Russian). French translation from the Russian: Dunod, Paris, 1966.

[10] M. Takeuchi, On orbits in a compact hermitian symmetric space, *Amer. J. Math.*, **90** (1968), 657–680.

[11] J. Tits, Espaces homogènes complexes compacts, *Comment. Math. Helv.*, **37** (1962), 111–120.

[12] H.-C. Wang, Closed manifolds with homogeneous complex structure, *Amer. J. Math.*, **76** (1954), 1–32.

[13] J. A. Wolf and A. Korányi, Generalized Cayley transformations of bounded symmetric domains, *Amer. J. Math.*, **87** (1965), 899–939.

[14] J. A. Wolf, On the classification of hermitian symmetric spaces, *J. Math. Mech.*, **13** (1964), 489–496.

[15] J. A. Wolf, "*Spaces of Constant Curvature*," McGraw-Hill, New York, 1967.

[16] J. A. Wolf, The action of a real semisimple group on a complex flag manifold, I: Orbit structure and holomorphic arc components, *Bull. Amer. Math. Soc.*, **75** (1969), 1121–1237.

[17] J. A. Wolf, The action of a real semisimple group on a complex flag manifold, III: Induced representations based on hermitian symmetric spaces. In preparation.

BOUNDARIES OF RIEMANNIAN SYMMETRIC SPACES

HARRY FURSTENBERG

Herbrew University of Jerusalem
Jerusalem, Israel

In this paper, we shall discuss a certain compact space that may be associated with a Riemannian symmetric space of noncompact type. If X denotes the symmetric space, $B(X)$ will denote the corresponding compact space. The space $B(X)$ plays a major role in the theory of harmonic functions on X. For example, if X is the interior of the unit disk with the Poincaré metric, $B(X)$ will be the bounding circle. The connection with harmonic functions is that that there exists an essentially one-to-one correspondence between (say) bounded measurable functions on $B(X)$ and bounded harmonic functions on X. This correspondence is established in the classical case by the Poisson representation formula. Usually one thinks of the function on the circle as constituting the boundary values of the function in the disk. For an abstract symmetric space X, however, there is no obvious boundary to play the role of the circle, and we must find some other characterization of the space $B(X)$. We shall discuss several such characterizations. It will develop that the main role is played not by the symmetric space X, but by its group G of isometries, and that the space $B(X)$, which we shall also denote $B(G)$, is associated with G in a natural way. We shall end with an application of the space $B(G)$ to resolving some questions regarding discrete subgroups of G.

1. Group Representations of Compact Convex Sets

To begin our discussion, we let G be an arbitrary topological group and Q a compact convex set. We may think of Q as a subset of a linear topological space; however, the structure of a compact convex set can be defined without reference to an ambient space. We shall say that G has a representation ρ on Q if to each $g \in G$ there is assigned an affine transformation $\rho(g)$ of Q into itself [i.e., $\rho(g)(\lambda x + (1 - \lambda)y) = \lambda\, \rho(g)x + (1 - \lambda)\rho(g)y$] such that $\rho(g)x$ is a continuous function on $G \times Q$ and $\rho(g_1 g_2) = \rho(g_1)\rho(g_2)$. Such a representation is *irreducible* if no proper closed convex subset of Q is invariant under the action of G.

There are groups G with the property that any representation of the group on a compact convex set leaves some point fixed. We call these FPP (fixed-point-property) groups. It is well known that Abelian and compact groups have this property. If G is a connected Lie group, it may be shown that G is FPP if and only if G has a normal solvable subgroup S with G/S compact [*2*]. Clearly an *FPP* group has no interesting irreducible representations on compact convex sets.

Let ρ, ρ' be two representations of G on Q, Q', respectively. An *equivariant map taking ρ onto ρ'* is a continuous affine map φ of Q onto Q' satisfying $\varphi \circ \rho(g) = \rho'(g) \circ \varphi$. We then say that (ρ', Q') is an equivariant image of (ρ, Q). An equivariant image of an irreducible representation is again irreducible. Unlike the situation for linear representations, it is possible for a nontrivial irreducible representation to be an equivariant image of an *inequivalent irreducible* representation. Indeed we have the following result.

Theorem 1. For any group G, there exists a compact convex set $Q_0(G)$ and a representation ρ_0 of G on $Q_0(G)$ such that ρ_0 is irreducible and every irreducible representation of G is an equivariant image of $(\rho_0, Q_0(G))$.

Proof. Let $\{\rho, Q_\rho\}$ denote the set of inequivalent irreducible representations of G on compact convex sets Q_ρ. One can show that the cardinality of Q_ρ is restricted by the irreducibility condition, so that the above family does indeed form a set. Now the product set $\prod_\rho Q_\rho$ is compact and convex and G acts on it in a natural way. Let Q_0 be a minimal G-invariant closed convex subset of $\prod_\rho Q_\rho$ and let ρ_0 denote the restriction to Q_0 of the action of G. (ρ_0, Q_0) is then irreducible and the projection map π_ρ : $Q_0 \to Q_\rho$ is an equivariant map of (ρ_0, Q_0) onto (ρ, Q_ρ).

Of course, if G is an FPP group, then $Q_0(G)$ reduces to a point.

Let M be a compact topological space. A *measure* on M is, for our purposes, a bounded linear functional on the Banach space $\mathscr{C}(M)$ of continuous functions on M. A measure μ is positive if $\mu(f) \geqq 0$ whenever $f \geqq 0$. μ is a *probability measure* if $\mu(M) = 1$. We denote the set of these by $\mathscr{M}(M)$. We endow $\mathscr{M}(M)$ with the weak* topology it inherits as a subset of the dual to $\mathscr{C}(M)$. As a weakly closed bounded subset of the latter, $\mathscr{M}(M)$ is compact. It is, of course, also convex.

To each $x \in M$, we can assign a probability measure on M by setting $\delta_x(f) = f(x)$. These "point measures" constitute the set of extremals of the compact convex set $\mathscr{M}(M)$.

We shall also have occasion to speak of probability measures on a locally compact space M. A probability measure on a locally compact M is a probability measure on the one-point compactification of M, $M \cup \{\infty\}$, which has no component at ∞; i.e., which does not have the form $\lambda\delta_\infty + (1 - \lambda)\nu$, $\lambda > 0$.

If $\varphi : M \to M'$ is a continuous map, then $f \to f \circ \varphi$ takes $\mathscr{C}(M') \to \mathscr{C}(M)$ and the adjoint map takes $\mathscr{M}(M) \to \mathscr{M}(M')$. We denote this map by φ as well. Note that $\varphi(\delta_x) = \delta_{\phi(x)}$.

If μ is a measure and f is a function, the expression $\mu(f)$ is commonly expressed as an integral $\int_M f(x)\, d\mu(x)$. The following important formula follows immediately from the foregoing definition:

$$\int_M f(\varphi x)\, d\mu(x) = \int_{M'} f(x)\, d\varphi\mu(x).$$

We say that M is a G-space if there is defined an action of G on M. That is, we assume there is defined a map $G \times M \to M$, which we denote $(g, x) \to gx$, which is continuous and which satisfies $(g_1 g_2)x = g_1(g_2 x)$. M is a *homogeneous* G-space if G is transitive on M, i.e., if for each x, $y \in M$ there exists $g \in G$ with $gx = y$. When G is a connected Lie group, every homogeneous G-space is topologically as well as set-theoretically of the form G/H, where H is a closed subgroup of G. If M' is another G-space and φ is a map of M onto M' satisfying $\varphi(gx) = g(\varphi x)$, then we say that φ is an *equivariant* map of M onto M', and M' is called an *equivariant image* of M.

Now let M be a G-space. There is a natural representation of G on $\mathcal{M}(M)$ defined by $\rho_M(g)\mu = g\mu$. Our next result deals with the problem of determining when ρ_M will be irreducible.

Definition 1: Let M be a homogeneous G-space. M is a *boundary* of G if for every probability measure ν on M, the set $G\nu \subset \mathcal{M}(M)$ has in its closure measures of the form δ_x, $x \in M$.

Theorem 2: Let G be a connected Lie group and M a homogeneous G-pace. Then ρ_M is irreducible if and only if M is a boundary.

Proof. Suppose M is a boundary, and let Q be a closed convex subset of $\mathcal{M}(M)$ invariant under G. Then Q contains some point measure, and so by transitivity it contains every δ_x, $x \in M$. But the convex span of $\{\delta_x, x \in M\}$ is dense in $\mathcal{M}(M)$ so that $Q = \mathcal{M}(M)$ and ρ_M is irreducible.

Suppose next that ρ_M is irreducible. If $\nu \in \mathcal{M}(M)$, then the convex span of $G\nu$ has each δ_x as a limit point. We shall show that this already implies that $G\nu$ has δ_x as a limit point. If G is a connected Lie group, then each homogeneous G-space is a manifold and hence a metrizable space. For each $x \in M$, there exists a continuous function $f \geqq 0$ with $f(y) = 0$ only for $y = x$. It is easy to see that a basis of neighborhoods of δ_x in $\mathcal{M}(M)$ is given by $\{\mu | \mu(f) < \varepsilon\}$. But if $\sum \alpha_i g_i \nu(f) < \varepsilon$ where $\alpha_i \geqq 0$, $\sum \alpha_i = 1$, then at least one g_i satisfies $g_i \nu(f) < \varepsilon$. From this our assertion follows.

2. $Q_0(G)$ for Connected Lie Groups

We shall see that when G is a connected Lie group, the universal irreducible representation $(\rho_0, Q_0(G))$ may be determined explicitly. We shall find that $\rho_0 = \rho_M$ where M is a homogeneous G-space which, according to Theorem 2, must be a boundary of G. This boundary will be uniquely determined and will be denoted $B(G)$.

In the course of our discussion, we shall need the notion of the "center of mass" of a measure on a compact convex set. If ν is a probability measure on the compact convex set Q, we can form the integral

$$\bar{\nu} = \int_Q x \, d\nu(x).$$

As to the definition of this integral, we merely point out that Q may always be regarded as a set of linear functionals on some topological linear space with the corresponding weak topology. In this case, we may write $\bar{\nu}(y) = \int_Q x(y) \, d\nu(x)$. We note that $\nu \to \bar{\nu}$ is a continuous map and that $\bar{\nu}$ will be an extremal point $x \in Q$ if and only if $\nu = \delta_x$.

Now let ρ be a representation of the group G on the compact convex set Q and let u be a point of Q. We obtain a continuous map of G into Q by setting $r(g) = \rho(g)u$. Clearly $r(gg') = \rho(g)r(g')$. If μ is a probability measure on G, then $r(\mu)$ is a probability measure on Q and we can form its center of mass $\overline{r(\mu)}$.

We turn now to the determination of $Q_0(G)$ for the case that G is a semisimple Lie group with finite center. In this case, G admits a decomposition $G = K \cdot A \cdot N$ where K is a maximal compact subgroup of G, A is a vector subgroup, N is nilpotent, and the product $A \cdot N$ is solvable. We denote $A \cdot N$ by S so that $G = K \cdot S$. Note that S is an FPP group. Let ρ be a representation of G on the compact convex Q. The restriction of ρ to S has at least one fixed point; let u be one such point. We then define the function r as above.

Lemma. If μ is a measure on G, then there exists a measure μ' on K with $r(\mu) = r(\mu')$.

Proof. Let the map $k : G \to K$ be defined by $g \in k(g)S$, and set $\mu' = k(\mu)$. Then $r(g) = r(k(g)s) = \rho(k(g))r(s) = \rho(k(g))u = r(k(g))$. Hence $r(\mu) = r(\mu')$.

Lemma. $r(\mathcal{M}(K))$ is a compact convex subset of Q invariant under G.

Proof. The compactness and convexity are clear by the continuity and linearity of r. The last property follows from the fact that $r(\mathcal{M}(K)) = r(\mathcal{M}(G))$.

Corollary. If ρ is irreducible, then $r(\mathcal{M}(G)) = r(\mathcal{M}(K)) = Q$. In this case the extremals of Q are the points of Q of the form $r(g)$, $g \in G$, or, equivalently, $r(k)$, $k \in K$.

Proof. Only the last assertion requires comment. Suppose $r(\mu)$ is an extremal of Q. Then for any μ' with $\mu' \leqq c\mu$, we have $r(\mu') = r(\mu)$. From the integral expression $r(\mu) = \int r(g)\, d\mu(g)$, we conclude that $r(g)$ is constant almost everywhere with respect to μ. Hence the extremal in question has the form $r(g)$. Since some extremal exists, and since the group G preserves the convex structure of Q, it follows that all the points $r(g)$ are indeed extremals.

Lemma. When ρ is irreducible, $r(G) \subset Q$ is a boundary of G.

Proof. By Theorem 2, it suffices to show that the representation of G on $\mathcal{M}(r(G))$ is irreducible. Let Q' be an invariant closed convex subset of $\mathcal{M}(r(G))$. We have a map $v \to \bar{v}$ of Q' into Q, and on account of the irreducibility of ρ, $\bar{Q}' = Q$. In particular, there exist $v \in Q'$ with $\bar{v} = x$, x an extremal of Q. But then $v = \delta_x$ It follows that for each $r(g)$, the corresponding point measure is in Q', and hence $Q' = \mathcal{M}(r(G))$.

Lemma. If M, M' are boundaries of G and α is an equivariant map of $\mathcal{M}(M)$ into $\mathcal{M}(M')$, then α takes point measures into point measures.

Proof. By homogeneity it suffices to show that there is one point $x \in M$ with $\alpha(\delta_x)$ a point measure on M'. Let $y \in M$ and consider the measure $\alpha(\delta_y)$ on M'. Since M' is a boundary, there is some sequence g_n in G with $g_n\alpha(\delta_y) \to \delta_{x'}$ for some x'. Passing to a subsequence such that $g_n y \to x$, say, we then have $\alpha(\delta_x) = \delta_{x'}$.

Corollary 1. If M, M' are boundaries of G, there exists at most one equivariant map of $\mathcal{M}(M)$ into $\mathcal{M}(M')$.

Proof. If α and β are two such maps so is $\frac{1}{2}(\alpha + \beta)$. But this map cannot take point measures into point measures unless $\alpha(\delta_x) = \beta(\delta_x)$. But then $\alpha = \beta$.

Corollary 2. If M, M' are boundaries of G, there exists at most one equivariant map of M into M'.

We can now prove

Theorem 3. Let G be a semisimple Lie group with finite center. There exists a universal boundary $B(G)$ such that every boundary of G is an equivariant image of $B(G)$, and we have $Q_0(G) = \mathcal{M}(B(G))$. If we write $B(G) = G/H$, then the subgroup H is determined up to an inner automorphism of G. If we write $G = K \cdot A \cdot N$, then we can take H to be the group $M \cdot A \cdot N$ where M is the centralizer of A in K.

Proof. We consider the universal irreducible representation ρ_0 of G on $Q_0 = Q_0(G)$. With the map r defined as above, we can identify Q_0 with $r(\mathcal{M}(G))$. We set $B(G) = r(G)$. This set forms a boundary of G and $\mathcal{M}(B(G))$ is the space of an irreducible representation of G. Now the map $v \to \bar{v}$ is an affine map of $\mathcal{M}(B(G))$ into Q_0 and by the irreducibility of ρ_0 it must be onto. On the other hand, since ρ_0 is universal, we must have a map β of Q_0 onto $\mathcal{M}(B(G))$. Let the map $v \to \bar{v}$ be denoted by α. Then $\beta\alpha$ is an equivariant map of $\mathcal{M}(B(G))$ into itself and so by Corollary 1 above, $\beta\alpha$ is the identity. Since α is onto, it follows that $\alpha\beta$ is also the identity. Hence Q_0 may be identified with $\mathcal{M}(B(G))$ and ρ_0 with $\rho_{B(G)}$.

Let us now show that every boundary of G is an equivariant image of $B(G)$. Let M be any boundary of G. There exists an equivariant map of $\mathcal{M}(B(G))$ into $\mathcal{M}(M)$. Since such a map must take point measures to point measures, we find that M is an equivariant image of $B(G)$.

Suppose next that G/H_1 and G/H_2 are both universal boundaries of G. Then there exists an equivariant map of G/H_1 onto G/H_2 and vice versa. Their composition in both directions must be identity maps by Corollary 2 above. From this it follows that the two subgroups are conjugate. For the proof of the identification of H with $M \cdot A \cdot N$, we refer the reader to [5].

We give one example of the space $B(G)$. If $G = SL(n, R)$, the subgroup H is seen to be a conjugate of the group of all upper triangular matrices. The quotient G/H will then be the "flag space" of all sequences $V_1 \subset V_2 \subset \cdots \subset V_{n-1}$, where V_i is a subspace of R^n of dimension i. In particular, the projective space P^{n-1} is an equivariant image of $B(G)$, and hence is a boundary of G.

We remark for later reference that the maximal compact subgroup K, and therefore any maximal compact subgroup of G, is always transitive on $B(G)$, for $B(G) = r(G) = r(K)$ in the notation of the foregoing proof.

Theorem 4. If G is any connected Lie group, there exists a universal boundary $B(G)$, and $Q_0(G) = \mathcal{M}(B(G))$.

Proof. If R is the radical of G, then G/R is semisimple. Let R' be the counterimage in G of the center of G/R; then R' is solvable and G/R' is semisimple with a trivial center. One easily checks that since R' is an FPP group, $Q_0(G) = Q_0(G/R')$. From this, one sees that the foregoing theorem applies to the non-semisimple case as well.

3. Embedding Symmetric Spaces into Compact Convex Sets

In what follows, X will denote a Riemannian symmetric space of non-compact type, and G its group of isometries. We fix a point $p_0 \in X$, and let K denote its isotropy group. G is a semisimple Lie group, K a maximal compact subgroup, and X can be identified with G/K.

We now set $B(X) = B(G)$, $Q_0(X) = Q_0(G)$ so that $Q_0(X) = \mathcal{M}(B(X))$. We have remarked that K is transitive on $B(G) = B(X)$. Hence there exists a unique K-invariant measure m in $\mathcal{M}(B(X))$. The map $g \to gm$ takes G/K into $\mathcal{M}(B(X)) = Q_0(X)$. We thus obtain a continuous map $\theta : X \to Q_0(X)$, where $\theta(gp_0) = gm$. It can be shown that this map is one-to-one [2, Theorem 2.1], and that θ is a homeomorphism of X into the compact convex set $Q_0(X)$.

Note that since $B(X)$ is a boundary of G, there is a sequence $\{g_n\}$ in G with $g_n m \to$ a point measure. If we identify the points of $B(X)$ with the point measures on that space, then we find that $B(X)$ can be viewed as a component of the boundary of $\theta(X) \subset Q_0(X)$. It may be shown that the closure $\theta(X) \subset Q_0(X)$ consists of finitely many G-orbits of which only $B(X)$ is compact. We remark that the "compactification" of X induced by our embedding X into $Q_0(X)$ coincides with one of the Satake compactifications [5].

4. Harmonic Functions on X

In this section, we shall identify the Riemannian symmetric space X with its image in $Q_0(X)$. The space $Q_0(X)$, as a compact convex space, possesses a distinguished family of real-valued functions, namely those that preserve its convex structure. We call these *affine* functions; f is affine if $f(\lambda x + (1 - \lambda)y) = \lambda f(x) + (1 - \lambda)f(y)$. In this section, we shall see that the restriction to X of a continuous, affine function on $Q_0(X)$ is a harmonic function on X. Moreover, every harmonic function with "continuous boundary values" arises in this way.

Definition 2. A function $h(p)$ on X is called an *H-function* if it is the restriction to X of a continuous affine function on $Q_0(X)$.

We now verify certain functional equations for H-functions. Let $h(p)$ be the restriction to X of the continuous affine function f. The function f on $Q_0(X)$ is determined by its restriction to the extremal points of $Q_0(X) = \mathcal{M}(B(X))$; i.e., by the values of $f(\delta_x)$. In other words, there exists a continuous function $\varphi(x)$ on $B(X)$ with $f(\delta_x) = \varphi(x)$ and so

$$h(gp_0) = h(gm) = \int_{B(X)} f(\delta_x)\, dgm(x) = \int_{B(X)} \varphi(x)\, dgm(x)$$
$$= \int_{B(X)} \varphi(gx) dm(x).$$

We thus obtain the representation

$$h(p) = \int_{B(X)} \varphi(gx)\, dm(x) = \int_{B(X)} \varphi(x)\, dgm(x) = \int_{B(X)} \varphi(x) \frac{dgm}{dm}(x)\, dm(x)$$
$$= \int_{B(X)} \varphi(x)\, P(p, x)\, dm(x), \qquad (1)$$

where g satisfies $p = gp_0$. We regard (1) as a version of the Poisson representation formula; the H-function h being represented by a "boundary" function φ. In this formula dgm/dm denotes the Radon–Nikodym derivative, and since $gm = gkm$ for $k \in K$, it depends only on gK, and therefore on p. $P(p, x)$ is the analog of the Poisson kernel.

In general, if f is a function on a space on which G operates, we shall denote by f^g the function $f^g(x) = f(gx)$. It is obvious that if h is an H-function on X, then so is h^g. From the Poisson formula, we deduce that if h is an H-function

$$\int_K h(kp)\,dk = \int_K \int_{B(X)} \varphi(kgx)\,dm(x)\,dk.$$

Now for any $y \in B(X)$, $L(\varphi) = \int_K \varphi(ky)\,dk$ defines a K-invariant functional on $\mathscr{C}(B(X))$. Hence $L(\varphi) = m(\varphi)$. Hence

$$\int_K h(kp)\,dk = m(\varphi) = \int_{B(X)} \varphi(x)\,dm(x) = h(p_0),$$

whence

$$\int_K h(kp)\,dk = h(p_0).$$

Replacing h by h^g, we find

$$\int_K h(gkp)\,dk = h(gp_0), \tag{2}$$

which is a mean-value property for H-functions analogous to the mean-value property for ordinary harmonic functions. Indeed, fixing g, the points gkp, $k \in K$, are at a fixed distance from gp, since $d(gkp, gp_0) = d(p, k^{-1}p_0) = d(p, p_0)$.

Another consequence of the Poisson formula is that H-functions on X are C^∞. We now show that they are annihilated by the Laplace–Beltrami operator on X. In fact, let Δ be any G-invariant operator on X which annihilates constants. By G-invariance we mean that

$$(\Delta h)^g = \Delta(h^g), \qquad g \in G.$$

Let h be an H-function. It suffices to show that $\Delta h(p_0) = 0$. Now

$$\Delta h(p_0) = (\Delta h)^k(p_0) = \Delta(h^k)(p_0) = \int_K \Delta(h^k)(p_0)\,dk$$

$$= \Delta\left\{\int_K h^k(p)\,dk\right\}_{p_0} = \Delta\left\{\int_K h(kp)\,dk\right\}_{p_0}.$$

But $\int_K h(kp)\,dk = h(p_0)$, so the function in parentheses is a constant, and applying Δ we obtain 0.

Definition 3. A function $h(p)$ on X is called LUC (left uniformly continuous) if $h(g_n p) \to h(p)$ uniformly for $p \in X$ as $g_n \to$ identity in G.

One sees from (1) that an H-function is necessarily LUC. This shows that not every bounded harmonic function on X is an H-function, since it can be shown that there exist non-LUC harmonic functions. For example, when X is the disk, a bounded harmonic function is LUC if and only if its boundary function is continuous. At any rate, one may show that a LUC harmonic function is an H-function. This will follow from the analysis of the next section.

5. Characterization of Harmonic Functions

In the last section, we saw that considerations revolving about the boundary $B(X)$ lead naturally to a special class of harmonic functions, the H-functions. In this section, we will be concerned with the theory of all bounded harmonic functions, and here too the space $B(X)$ will play a fundamental role. First some preliminary considerations.

If f is a function on $X = G/K$, we denote by $\tilde{f}$ the lift of f to the group G, $\tilde{f}(g) = f(gp_0)$.

We shall be concerned with $L^\infty(G)$, the space of bounded measurable functions on the group G. We shall regard $L^\infty(G)$ as the dual space to $L^1(G)$ and endow it with the corresponding weak* topology. It is well known that in this topology the unit ball of $L^\infty(G)$ is compact. Now let σ denote the regular representation of G on $L(G)$; $\sigma(\gamma)f(g) = f(g\gamma)$. If V is a compact convex invariant subset of $L^\infty(G)$, then the restriction of σ to V will be denoted by the same letter. Now every $f \in L^\infty(G)$ belongs to some invariant, compact convex set in $L^\infty(G)$ (e.g., $\{f' \mid \|f'\|^\infty \leqq \|f\|^\infty\}$). Hence there is a minimal such set for any function in $L^\infty(G)$.

Definition 4. If $f \in L^\infty(G)$, $V(f)$ will denote the smallest G-invariant, compact convex set containing f.

Clearly $V(f)$ is the closure of the convex hull of the right translates of f. We also need the following definition:

Definition 5. A *Laplace operator* Δ on X is any elliptic second-order differential operator invariant under G and annihilating constants.

Theorem 5. The following conditions for a bounded, continuous function $f(p)$ on X are equivalent:

(a) $\Delta f = 0$ for some Laplace operator Δ.

(b) $\tilde{f}$ satisfies an integral equation

$$\tilde{f}(g) = \int_G \tilde{f}(gg')\,\psi(g')\,dg' \tag{3}$$

for some integrable function ψ satisfying $\psi > 0$, $\int_G \psi(g)\,dg = 1$, $\psi(kg) = \psi(gk) = \psi(g)$ for $k \in K$.

(c) f satisfies (3) for every L^1-function ψ satisfying $\psi \geqq 0$, $\int_G \psi(g)dg = 1$, $\psi(kg) = \psi(gk) = \psi(g)$ for $k \in K$.

(d) $\tilde{f}(g) = \int_K \tilde{f}(gkg')\,dk$ for every $g, g' \in G$.

(e) The regular representation of G on $V(f)$ is irreducible.

(f) There exists a bounded measurable function φ on $B(X)$ such that

$$\begin{aligned} f(g) &= \int_{B(X)} \varphi(gx)\,dm(x) = \int_{B(X)} \varphi(x)\,dgm(x) = \int_{B(X)} \varphi(x)\frac{dgm}{dm}(x)\,dm(x) \\ &= \int_{B(X)} \varphi(x)\;P(gK, x)\;dm(x). \end{aligned} \tag{4}$$

We call a function $f(p)$ satisfying these conditions harmonic. We shall also speak of the lifted function $\tilde{f}(g)$ as being harmonic. From the foregoing theorem, we see that harmonic functions can be defined using differential equations or integral equations, or by a Poisson representation formula, or by an irreducibility property. This latter property, which is not as well known even in the classical case of harmonic functions on the disk, says in effect that $\tilde{f}$ may be retrieved by convex combinations of translates of any function that may be obtained from $\tilde{f}$ in this manner.

We shall sketch the proof of Theorem 5. Many of the details omitted may be found in Ref [2]. The sequence of implications we follow is (a) ⇒ (b) ⇒ (f) ⇒ (c) ⇒ (d) ⇒ (e) ⇒ (f), (d) ⇒ (a).

(a) ⇒ (b). One shows first of all that under condition (a), there exist invariant elliptic differential operators on G with $\Delta\tilde{f} = 0$. The operator Δ may be seen to generate a semigroup of convolution operators on G,

$$T_t f(g) = \int_G f(gg')\psi_t(g')\,dg'.$$

Since $dT_t/dt = T_t\Delta$ and $\Delta f = 0$, we have $T_t\tilde{f} = T_0\tilde{f} = \tilde{f}$.

(b) $\Rightarrow$ (f). This is the deepest part of the theorem. We shall discuss some of the relevant theory in the next section.

(f) $\Rightarrow$ (c). Straightforward verification.

(c) $\Rightarrow$ (d). Set $F(g') = \int_K \tilde{f}(gkg')\,dk$. F is continuous since $\tilde{f}$ is, so, in order to prove (d), it will suffice to show that

$$\int_G F(g')\varphi(g')\,dg' = \tilde{f}(g) \quad \text{for all} \quad \varphi \in L'(G)$$

with

$$\varphi \geqq 0, \qquad \int_G \varphi(g)\,dg = 1.$$

But

$$\begin{aligned}\int_G F(g')\varphi(g')\,dg' &= \int_G\int_K \tilde{f}(gkg')\varphi(g')\,dk\,dg'\\ &= \int_G\int_K\int_K \tilde{f}(gkg'k_1)\varphi(g')dk\,dk_1\,dy'\end{aligned}$$

since $\tilde{f}(g) = \tilde{f}(gk_1)$. So

$$\begin{aligned}\int_G F(g')\varphi(g')\,dg' &= \int_G\int_K\int_K \tilde{f}(gg')\varphi(k^{-1}g'k^{-1})\,dk\,dk_1\,dg'\\ &= \int_G \tilde{f}(gg')\psi(g')dg'\end{aligned}$$

where $\psi(g') = \int_K\int_K \varphi(k_1^{-1}g'k^{-1})\,dk_1\,dk$ so that ψ satisfies the conditions of (c). Hence (c) $\Rightarrow$ (d).

(d) $\Rightarrow$ (e). Let

$$W = \{F \in L^\infty(G),\ \|F\|_\infty \leqq \|\tilde{f}\|_\infty \text{ and } \tilde{f}(g) = \int_K F(gkg')dk\}$$

where the above equality for $\tilde{f}$ is to be valid almost everywhere in g, g'. W is a weak*-closed and bounded subset of $L^\infty(G)$ and it is clearly invariant under right translation. Since $\tilde{f} \in W$ according to (d), we must have $V(\tilde{f}) \subset W$. Hence for any $F \in V(\tilde{f})$, $\tilde{f}(g) = \int_K F(gkg')\,dk$ for almost every

g, g'. Let ψ be an L function with $\psi \geqq 0$, $\int_G \psi(g)\, dg = 1$; then

$$\tilde{f}(g) = \int_G \int_K F(gkg')\psi(g')\, dk\, dg'$$

for almost every g, i.e., the right-hand left-hand sides represent the same member of $L^\infty(G)$. Now the right-hand member is in $V(F)$ so that we find that whenever $F \in V(\tilde{f})$, $\tilde{f} \in V(F)$. This implies that $V(\tilde{f})$ defines an irreducible representation.

(e) $\Rightarrow$ (f). We now invoke the theory of Sections 2 and 3. Since $V(\tilde{f})$ defines an irreducible representation of G, there must exist an equivariant map τ from $\mathscr{M}(B(X))$ to $V(\tilde{f})$. Let $x_0 \in B(X)$ and let $\tau(\delta_{x_0}) = \Phi$. Let H_0 be the isotropy group of x_0 in $G : H_0 = \{h | hx_0 = x_0\}$. Then $\sigma(h)\Phi = \tau(h\delta_{x_0}) = \tau(\delta_{x_0}) = \Phi$ so that $\Phi(gh) = \Phi(g)$ a.e. in g for each $h \in H$. It is not hard to show from this that Φ may be taken to be independent of h; i.e., that for almost all $g, \Phi(gh) = \Phi(g)$ for *all* h. Hence we can define a function $\varphi(x)$ on $B(X)$ by $\varphi(gx_0) = \Phi(g)$, and φ will be a measurable function on $B(X)$.

Next we determine $\tau(\delta_x)$ for any $x \in B(X)$. If $x = \gamma x_0$, then $\delta_x = \gamma\delta_{x_0}$ and $\tau(\delta_x) = \sigma(\gamma)\tau(\delta_x) = \sigma(\gamma)\Phi$. So if $\tau(\delta_x) = \Phi_x$, we have $\Phi_x(g) = \Phi_{x_0}(g\gamma) = \varphi(g\gamma x_0) = \varphi(gx)$ and $\Phi_x(g) = \varphi(gx)$.

Note that the convex set $V(\tilde{f})$ has at least one K-invariant element; namely $\tilde{f}$ ($\tilde{f}(g) = \tilde{f}(gk)$). The convex set $\mathscr{M}(B(X))$ has exactly one K-invariant element, namely m. We use this to show that $\tau(m) = \tilde{f}$. For $\tau^{-1}(\tilde{f})$ is a compact convex set invariant under K. Hence it possesses a K-invariant element. This must be m, and hence $\tau(m) = \tilde{f}$. Using this, we may now write

$$\tilde{f} = \tau(m) = \int_{B(x)} \tau(\delta_x)\, dm(x) = \int_{B(x)} \Phi_x\, dm(x).$$

This gives

$$\tilde{f}(g) = \int_{B(X)} \Phi_x(g)\, dm(x) = \int_{B(X)} \varphi(gx)\, dm(x)$$

which is the Poisson representation (f).

(d) $\rightarrow$ (a). This was proved in Section 4.

6. Poisson Boundaries and μ-Harmonic Functions

It is significant that the Poisson representation of the harmonic function

$\tilde{f}$ can be established by knowing merely that $\tilde{f}$ satisfies a single integral equation. This may be seen to be part of a more general phenomenon whereby the solutions to a wide class of convolution equations are expressed in terms of "boundary" functions on an appropriate space.

Definition 6. Let μ be a probability measure on the group G. A bounded Borel measurable function $f(g)$ is called *μ-harmonic* if it satisfies

$$f(g) = \int_G f(gg')\, d\mu(g') \tag{5}$$

for all $g \in G$.

There need not always exist nonconstant μ-harmonic functions. It may be shown, for example, that if the group G is Abelian, then a μ-harmonic function is always constant. In what follows, we describe a procedure for generating (usually nontrivial) μ-harmonic functions.

We make use of the notion of convolution of two measures. Let M be a locally compact G-space (for example, G itself), μ a probability measure on G, and ν a probability measure on M. There is then defined a product measure $\mu \times \nu$ on $G \times M$, and the map of $G \times M$ into M associated with the action of G on M carries this measure onto a measure $\mu * \nu$ on M. In particular the convolution of two probability measures on G is defined, and if μ_1, μ_2 are two such measures on G, ν a probability measure on a G-space M, one shows that $(\mu_1 * \mu_2) * \nu = \mu_1 * (\mu_2 * \nu)$. Note that by the definition of convolution

$$\int_M f(x)\, d\mu * \nu(x) = \int_M \int_G f(gy)\, d\mu(g)\, d\nu(y).$$

Now let M be a *compact* G-space so that the space $\mathscr{M}(M)$ of probability measures on M is a compact convex set. If μ is a probability measure on G, then the operation $\nu \to \mu * \nu$ defines a continuous affine map of $\mathscr{M}(M)$ into itself. As such there necessarily exists a fixed point in $\mathscr{M}(M)$ for this map, and thus there exists a solution on M to the equation $\mu * \nu = \nu$. Now let φ be a bounded Borel function on M and form the integral

$$f(g) = \int_M \varphi(gy)\, d\nu(y). \tag{6}$$

Then

$$\int_G f(gg')\, d\mu(g') = \int_G \int_M \varphi(gg'y)\, d\nu(y)\, d\mu(g')$$
$$= \int_M \varphi(gx)\, d\mu * \nu(x) = \int_M \varphi(gx)\, d\nu(x) = f(g).$$

So each function f formed in this way is μ-harmonic.

We cannot be certain that a function f formed in this manner will not be constant. However one sees that for a given ν, all the μ-harmonic functions formed in this way will be constant if and only if the group G preserves the measure ν. In many cases, one knows that *no* probability measure is preserved under the action of the entire group G. This is the case, for example, if $G = SL(n, R)$ and $M = P^{n-1}$. Thus for any μ on $SL(n, R)$, there exist nontrivial μ-harmonic functions.

Formula (6) plays the role of the Poisson representation for μ-harmonic functions. Note that if the measure μ has the property that $\delta_k * \mu = \mu$ for k is an element in a compact subgroup $K \subset G$, then any solution to $\mu * \nu = \nu$ satisfies $k\nu = \nu$. If furthermore K is transitive on M, then ν must be the unique K-invariant measure on M. This is exactly what happens in the case of ordinary harmonic functions where the measure μ comes from the L^1-function $\psi(g)$ which satisfies $\psi(kg) = \psi(g)$, and where, furthermore, $B(X)$ is homogeneous for K so that the solution to $\mu * \nu = \nu$ is $\nu = m$. We see that in this case, (6) reduces to (4).

Definition 7. If μ is a probability measure on the group G, M a G-space, and ν a probability measure on M satisfying $\mu * \nu = \nu$, we call ν a *stationary* measure for μ, and (M, ν) is called an *admissible pair* for (G, μ). We denote the set of μ-harmonic functions on G given by (6) by $\mathscr{H}(M, \nu)$.

In view of the analogy with ordinary harmonic functions, it is natural to inquire whether there exists an admissible pair (M, ν) which gives rise to *all* μ-harmonic functions. We notice incidentally that if α is an equivariant map of M onto M' and $\alpha(\nu) = \nu'$, then if (M, ν) is admissible, so is (M', ν'). Moreover $\mathscr{H}(M', \nu') \subset \mathscr{H}(M, \nu)$. In particular, if an admissible pair exists that gives rise to all μ-harmonic functions, it would not be unique, since an admissible pair is always an equivariant image of other admissible pairs.

At any rate, the following general result is valid:

Theorem 6. [*4*]: For any locally compact group G and probability measure μ on G, there exists an admissible pair (M, ν) such that if $f(g)$ is a μ-harmonic function on G, there exists a Borel function $\varphi(x)$ on M with

$$f(g) = \int_M \varphi(gy)\, d\nu(y). \qquad \text{(6,bis)}$$

In order to achieve uniqueness, some additional condition must be imposed on the pair (M, ν). The condition we impose will make sense if M is a metriable space. In this case, the convex set $\mathscr{M}(M)$ is also metrizable, and suppose it endowed with some metric. For every $\nu \in \mathscr{M}(M)$, we let $D(\nu)$ denote the distance of ν to the set of point measures on M. Now let μ be a probability measure on G. We denote by $\mu^{(n)}$ the n-fold convolution $\mu * \mu * \cdots * \mu$ on G.

The following definition is a measure-theoretic analog of the notion of boundary introduced earlier.

Definition 8. An admissible pair (M, ν) is called a *boundary pair* of (G, μ) if for each $\varepsilon > 0$,

$$\mu^{(n)}\{g \,|\, D(g\nu) > \varepsilon\} \to 0$$

as $n \to \infty$.

We now have

Theorem 7. If G is a separable locally compact group, μ an absolutely continuous probability measure on G, then there exists a boundary pair (M, ν) such that every μ-harmonic function belongs to $\mathscr{H}(M, \nu)$. If (M', ν') is any other boundary pair of (G, μ), there exists a *measurable* equivariant map $\alpha : M \to M'$ with $\alpha(\nu) = \nu'$. In particular, the pair (M, ν) is unique up to a measure-theoretic isomorphism. We write $(M, \nu) = P(G, \mu)$ and we call it the Poisson boundary of (G, μ).

We refer the reader to Ref. [*1*] for an exposition of Poisson boundaries. In case G is a semisimple Lie group, the situation is particularly pleasant inasmuch as the G-space M for the Poisson boundary will always be one of finitely many spaces. Moreover, if μ satisfies the condition that its support (or that of some $\mu^{(n)}$) contains a neighborhood of the identity, then M can be taken as $B(G)$. This is what happens for Eq. (3) in Theorem 5 where, in addition, the measure μ satisfies $\delta_k * \mu = \mu$ from which it follows that the Poisson boundary must be $(B(G), m)$. This provides the missing link (b) → (f) in the proof of Theorem 5.

7. An Application of the Poisson Boundary to Discrete Groups

We end by sketching an application of the foregoing theory. The reader is referred to Ref. [*3*] for more details.

A discrete subgroup Γ of a connected Lie group G is called a *lattice subgroup* (l.s.) if the space G/Γ has finite invariant measure. This is the case, for example, if $\Gamma = SL(n, Z)$ and $G = SL(n, R)$. We ask the following question. How different can two connected Lie groups be if they contain isomorphic lattice subgroups? For example, if $G = SL(3, R)$ and $G' = SL(2, R)$ and if Γ, Γ' are lattice subgroups of G, G', respectively, can $\Gamma \cong \Gamma'$? We shall see how the theory of Poisson boundaries may be applied to prove that the answer to this last question is negative.

We have seen that when G is a semisimple Lie group, there are measures μ on G for which the Poisson boundary of (G, μ) may be explicitly determined: $P(G, \mu) = (B(G), m)$. This is the case when μ-harmonic functions coincide with harmonic functions on G. In general, however, very little is known as to the identity of the Poisson boundary. Thus if θ is a measure on the lattice Γ, one is unable, in general, to identify $P(\Gamma, \theta)$. However one can show that if Γ is a lattice subgroup of the semisimple group G, there does exist *some* measure θ on Γ with $P(\Gamma, \theta) = P(G, \mu)$. This identity means, in effect, that

(a) The restriction of a harmonic function on G to Γ is θ-harmonic.

(b) A bounded harmonic function on G is determined by its restriction to Γ.

(c) Every bounded θ-harmonic function on Γ extends to a harmonic function on G.

Assume now that Γ and Γ' are isomorphic l.s. of $G = SL(3, R)$ and $G' = SL(2, R)$, respectively. If we could infer from this that $B(G) = B(G')$, we would surely have a contradiction, since we know what these spaces are. We cannot arrive at such a strong conclusion, but we do obtain a result which is almost as good.

We find θ such that $P(\Gamma, \theta) = (B(G), m)$. Since Γ is isomorphic to a subgroup of G' and $B(G')$ is a G'-space, we find that Γ operates on $B(G')$. So there exists a stationary measure for θ on $B(G')$. If ν' is this measure, then $(B(G'), \nu')$ is an admissible pair for (Γ, θ). It is not hard to show that it is in fact a boundary pair. But then by Theorem 7, there exists a measure preserving map of $(B(G), m)$ to $(B(G'), v')$ equivariant with respect to the action of Γ. It turns out that the spaces $B(G)$, $B(G')$ are sufficiently different that this leads to a contradiction.

References

[1] R. Azencott, Espaces de Poisson des groups localement compact, *Lecture Notes in Mathematics* 145, Springer, New York, 1970.

[2] H. Furstenberg, A Poisson formula for semi-simple Lie groups, *Ann. of Math.*, **77** (1963), 335–386.

[3] H. Furstenberg, Poisson boundaries and envelopes of discrete groups, *Bull. Amer. Math. Soc.*, **73** (1967), 350–356.

[4] H. Furstenberg, Random walks and discrete subgroups of Lie groups, to appear in *Advances in Probability and Related Topics.*

[5] C. C. Moore, Compactifications of symmetric spaces, *Amer. J. Math.*, **86** (1964), 201–218.

mean value property, however, is not the usual one; it refers to non-Euclidean circles with respect to their non-Euclidean centers.

In these notes we will be mainly concerned with a generalization of this case, namely with symmetric spaces G/K of noncompact type. Here the algebra of G-invariant operators is commutative and has l algebraically independent generators, where $l = \text{rank } G/K$ [9, pp. 395, 421]. These latter facts, however, will never be used; all will follow from the characterization of harmonic functions by the Godement theorem.

Remarks. 1. If $X_1 = G_1/K_1$ and $X_2 = G_2/K_2$, then $X = X_1 \times X_2$ is identified with $G_1 \times G_2/K_1 \times K_2$. It is clear (either by the definition or by the mean value property) that F is harmonic on X if and only if it is harmonic on both X_1 and X_2 separately.

2. If G is compact, every harmonic function on G/K is constant, since the mean value property obviously implies the maximum principle.

3. If there is given a G-invariant Riemannian metric on $X = G/K$ (clearly there always exists at least one), then the corresponding Laplace–Beltrami operator Δ is G-invariant. If $\Delta F = 0$, F is said to be *weakly harmonic*. In some cases (e.g., $SO(n) \times_s \mathbf{R}^n/SO(n)$, or G/K symmetric of rank one), weakly harmonic implies harmonic. In other cases, F weakly harmonic, everywhere defined and *bounded* on X, implies that F is harmonic; this happens if G is compact (by the maximum principle for the Laplace–Beltrami operator), if $X = \mathbf{R}^n/\{0\}$ (by Liouville's theorem), and if X is a symmetric space of noncompact type (by a theorem of Furstenberg [5]; cf. the discussion after Theorem 7.2).

Definition 1.2. A *weak boundary* of $X = G/K$ is a homogenous space U of G on which K acts transitively. Let m denote the normalized K-invariant measure on U; for $g \in G$ let gm be defined by $\int f\, dgm = \int f \circ g\, dm$. For any integrable f on U, the function F defined on X by $F(g \cdot o) = \int f \circ g\, dm$ is called the *Poisson integral* of f. The Poisson kernel for U is defined on $X \times U$ by

$$P(g \cdot o, u) = \left.\frac{d(gm)}{dm}\right|_u = \frac{dm(g^{-1}u)}{dm\,(u)}.$$

Since m is given on U by a smooth differential form and since G acts smoothly on U, it is clear that the Poisson kernel is a smooth function on $X \times U$. The Poisson integral of f can now also be written as

$$F(g\cdot o) = \int_U P(g\cdot o, u)\, f(u)\, dm(u)$$

and it makes sense to talk about Poisson integrals of measures or distributions on U. We note that we obviously have $\int_U P(g\cdot o, u)\, dm(u) = 1$ for all $g \in G$.

Theorem 1.2. The Poisson integral of any integrable function is harmonic.

Proof. Let F be the Poisson integral of f. By the Godement theorem, it suffices to show that F has the mean value property. Let $x = g_1 \cdot o$, then

$$\int_k F(gk\cdot x)\, dk = \int_K \int_U f\, d(gkg_1 m)\, dk = \int_K \int_U f(gkg_1 u)\, dm(u)\, dk.$$

We fix a point $u_0 \in U$ and write $g_1 u = k' u_0$ ($k' \in K$). Using Fubini's theorem and the right invariance of dk, our last integral is seen to be equal to

$$\int_U \int_K f(gku_0)\, dk\, du = \int_K f(gku_0)\, dk$$
$$= \int_U f(gu)\, dm(u) = F(g\cdot o)$$

and the proof is finished.

Corollary. For every fixed $u \in U$, $P(x,u)$ is a harmonic function on X.

Remark. It can happen that even though there are many harmonic functions on X and there exist several weak boundaries, only the constant functions can be written as Poisson integrals. If $X = K \times_s S/K$ with K compact and S solvable (as in Example (b) above), then by the Markov–Kakutani theorem $K \times_s S$ leaves fixed some probability measure (i.e., a positive Borel measure of total mass 1) on U. This measure must be equal to m, and it follows that every Poisson integral is constant. Still, every homogenous space of K (with S acting on it trivially) is a weak boundary of X. It is in order to avoid these rather useless weak boundaries that the following definition is introduced (cf. also the Remark after Theorem 1.3).

Definition 1.3. A *boundary* for $X = G/K$ is a weak boundary U such that, for some sequence $\{g_n\}$ in G, $g_n m$ converges to a delta measure (i.e., there exists $u_0 \in U$ such that $\lim_{n\to\infty} \int f\, d(g_n m) = f(u_0)$ for all continuous f on U).

Remarks. 1. If $g_n m$ tends to the delta measure at u_0, then $kg_n m$ tends to the delta measure at ku_0. So all values of a *continuous* f on U can be recovered from knowledge of the Poisson integral of f.

2. If μ is any probability measure on a boundary U, then there exists a sequence $\{g_n'\}$ in G such that $g_n'\mu$ tends to a delta measure. (It is this property that is taken as definition of a boundary by Furstenberg [5].) In fact, let σ be a smooth probability measure on G; then $\sigma * \mu$ is smooth, and hence $g_n(\sigma * \mu)$ tends to a delta measure at, say, μ. Let $\{V_i\}$ be a neighborhood base at μ. Then $g_n(\sigma*\mu)(V_i) = \int g_n g \mu(V_i)\, d\sigma(g) > 1-1/i$ for large n. Hence $g_n g \mu(V_i) > 1-1/i$ for some g. Let $g_i' = g_n g$.

Theorem 1.3. If U is a boundary, then the Poisson integral is an injective map from $L^\infty(U)$ to the bounded harmonic functions on X.

Proof. It suffices to show that if F is the Poisson integral of f and $F = 0$, then $f = 0$. In case f is continuous, this is clear since for every $u_0 \in U$, $f(u_0) = \lim F(g_n \cdot o)$ with some sequence $\{g_n\}$ by Remark 1 after Definition 1.3. In the case of arbitrary f we take $\psi \in L^1(G)$, write $f_1 = \psi * f$, and let F_1 be the Poisson integral of f_1. Now

$$F_1(g\cdot o) = \int_U f_1\, d(gm) = \int_U \int_G \psi(h) f(h^{-1}gu)\, dh\, dm(u)$$
$$= \int_G \psi(h)\, F(h^{-1}g\cdot o)\, dh$$

is identically zero. f_1 is continuous, so $f_1 = 0$ by the first part of the proof. This shows $\psi * f = 0$ for any choice of ψ; it follows that $f = 0$.

Remark. It would seem reasonable to say that a weak boundary having the property of this theorem is *effective*. One might try to develop the theory taking this property as a definition and avoiding Definition 1.3.

The following notion is meaningful even in the generality of weak boundaries.

Definition 1.4. A weak boundary U of G/K is said to be *faithful* if $\{g \in G \mid gm = m\} = K$. ($m$ still denotes the normalized K-invariant measure on U.)

Remarks. 1. U is faithful if and only if the map of X into the space of regular measures on U defined by $g \cdot o \to gm$ is injective.

2. In general, writing $H = \{g \in G | \; gm = m\}$, it is clear that $K \subset H$ and that every Poisson integral is a function on G/H (more exactly, constant on fibers if we regard G/K as a fiber space over G/H).

3. If U consists of a single point, then it cannot be faithful unless $K = G$.

4. $U_1 \times U_2$ is a faithful weak boundary for $X_1 \times X_2$ if and only if U_1 and U_2 are both faithful for X_1 and X_2, respectively.

2. The Boundaries of a Symmetric Space

From now on X will be a Riemannian symmetric space of noncompact type. We will write $X = G/K$ with G connected semisimple and K maximal compact in G. G is not uniquely determined by these conditions; however, G divided by the kernel of its action on X is unique: It is centerless and is exactly the connected component of the isometry group of G [*9*, p. 214]. It is clear therefore that every possible choice of G leads to the same notion of harmonic function on X. Furthermore, if $g \in G$ acts trivially on X, then by Definition 1.3 it also acts trivially on every boundary. It follows that for our study of harmonic functions and boundaries it makes no difference which of the possible G's we use. (If we were to study weak boundaries, this would not be true any longer.)

Let $\mathfrak{g} = \mathfrak{k} + \mathfrak{p}$ be the decomposition of the Lie algebra of G under the Cartan involution θ such that $\mathfrak{k}$ is the Lie algebra of K. Let $\mathfrak{a}$ be a maximal Abelian subspace of $\mathfrak{p}$. $\mathrm{ad}_{\mathfrak{g}}(\mathfrak{a})$ is a commutative algebra of linear transformations symmetric with respect to the bilinear form $B(\theta X, Y)$ (where B denotes the Killing form); hence there exist finitely many linear functions λ on $\mathfrak{a}$ and subspaces $\mathfrak{g}_\lambda$ of $\mathfrak{g}$ such that $\mathrm{ad}(H)X = \lambda(H)X$ for all $H \in \mathfrak{a}$ and $X \in \mathfrak{g}_\lambda$. Furthermore $\mathfrak{g}$ is the direct sum of all the $\mathfrak{g}_\lambda$. If $\lambda \neq 0$, λ is called a *restricted root*. We have $[\mathfrak{g}_\lambda, \mathfrak{g}_\mu] \subset \mathfrak{g}_{\lambda+\mu}$.

$\mathfrak{a}$ can be extended to a Cartan subalgebra $\mathfrak{h}$ of $\mathfrak{g}$ and one can consider the root structure of the complexification $\mathfrak{g}^C$ with respect to $\mathfrak{h}^C$. Restricting these roots to the subspace $\mathfrak{a}$, one gets exactly the above defined restricted roots (independent of the choice of $\mathfrak{h}$).

The restricted roots form a (possibly nonreduced) root system in the sense of [*3*]; this seems to have been first proved in Ref. [*24*, p. 106], there is a simpler proof in [*1*, p. 6], and another approach is furnished by the theory of algebraic groups. We choose an ordering once and for all, and denote the corresponding system of simple roots by F.

For any subset $E \subset F$, we define $\mathfrak{a}(E) = \{H \in \mathfrak{a} | \lambda(H) = 0, \forall \lambda \in E\}$ and $\mathfrak{n}(E) = \sum \mathfrak{g}_\lambda$, the sum over those $\lambda > 0$ which do not vanish on $\mathfrak{a}(E)$. We define $\bar{\mathfrak{n}}(E) = \theta\mathfrak{n}(E)$; if $E = \phi$, we have $\mathfrak{a}(E) = \mathfrak{a}$ and we write $\mathfrak{n}$, $\bar{\mathfrak{n}}$ instead of $\mathfrak{n}(E)$, $\bar{\mathfrak{n}}(E)$.

The centralizer of $\mathfrak{a}(E)$ is invariant under θ, hence reductive (it is the sum of all $\mathfrak{g}_\lambda$ such that λ vanishes on $\mathfrak{a}(E)$). Its derived algebra, which is semisimple, will be denoted by $\mathfrak{g}^E$; it is still invariant under θ.

Let $\mathfrak{k}^E = \mathfrak{g}^E \cap \mathfrak{k}$, $\mathfrak{p}^E = \mathfrak{g}^E \cap \mathfrak{p}$. Then $\mathfrak{g}^E = \mathfrak{k}^E + \mathfrak{p}^E$ is a Cartan decomposition. Let $\mathfrak{a}^E = \mathfrak{g}^E \cap \mathfrak{a}$. The restricted roots of $\mathfrak{g}^E$ with respect to $\mathfrak{a}^E$ are then identical with the restricted roots of $\mathfrak{g}$ which vanish on $\mathfrak{a}(E)$, i.e., which are linear combinations of elements of E. It is now easy to see that $\mathfrak{a} = \mathfrak{a}^E \oplus \mathfrak{a}(E)$.

We denote by $\mathfrak{n}^E$ (resp. $\bar{\mathfrak{n}}^E$) the sum of the $\mathfrak{g}_\lambda$ such that λ is a positive (resp. negative) combination of elements of E. We have $\mathfrak{n} = \mathfrak{n}^E + \mathfrak{n}(E)$ as a semidirect sum of Lie algebras; in fact, $\mathfrak{n}(E)$ is an ideal since it is the sum of all positive $\mathfrak{a}(E)$-eigenspaces while $\mathfrak{n}^E$ is in the zero $\mathfrak{a}(E)$-eigenspace.

Let capitals denote the analytic subgroups of G corresponding to all these subalgebras. Then A is a vector group, N, $\bar{N}$ are simply connected nilpotent groups normalized by A, and we have the Iwasawa decomposition $G = KAN$ (or, equivalently, $KA\bar{N}$, $K\bar{N}A$, etc.) [9, p. 234]. $N = N^E \cdot N(E)$ is a semidirect product, $A = A^E \cdot A(E)$ a direct product. The orbit of the identity coset o of X under G^E is a totally geodesic submanifold which we denote by X^E. This is a symmetric space of noncompact type; therefore its isotropy group $G^E \cap K$ is connected. So $G^E \cap K = K^E$ and $X^E = G^E/K^E$.

We have the following generalization of the Iwasawa decomposition.

Lemma 2.1. Every $x \in X$ can be uniquely written in the form $x = na \cdot x^E$ ($n \in N(E)$, $a \in A(E)$, $x^E \in X^E$ with n, a, x^E depending continuously on x).

Proof. By the ordinary Iwasawa decomposition and by the preceding remarks we have $x = n'a' \cdot o$ ($n' \in N$, $a' \in A$); $n' = nn^E$, $a' = aa^E$ ($n \in N(E)$, $n^E \in N^E$, $a \in A(E)$, $a^E \in A^E$). All these decompositions are continuous.

Also $an^E = n^E a$ since g^E centralizes $a(E)$. Now $x = na \cdot x^E$ with $x^E = n^E a^E \cdot o$. Uniqueness follows from the uniqueness of the Iwasawa decomposition.

Corollary. Every $x \in X$ can be written in the form $x = nak^E a^E \cdot o$ ($n \in N(E)$, $a \in A(E)$, $k^E \in K^E$, $a^E \in A^E$) with n and a unique and depending continuously on x.

It is known [*9*, p. 244] that the centralizer of $\mathfrak{a}$ in $\mathfrak{g}$ is of the form $\mathfrak{g}_0 = \mathfrak{a} + \mathfrak{m}$, where $\mathfrak{m}$ is the centralizer of $\mathfrak{a}$ in $\mathfrak{k}$. If M, M' are the centralizer (resp. normalizer) of $\mathfrak{a}$ in K, then both have Lie algebra $\mathfrak{m}$, and the Weyl group W of the system of restricted roots is exactly M'/M [*9*, pp. 246, 249]. M commutes with A and normalizes N since $\mathrm{Ad}_{\mathfrak{g}}(M)$ preserves each of the spaces $\mathfrak{g}_\lambda$. Defining $B = MAN$, this is therefore a closed subgroup, semidirect product of a solvable and a compact group.

The Lemma of Bruhat and Harish-Chandra (for a fairly simple and self-contained proof see [*8*]) states that the map $kM \to BkB$ of $M'/M = W$ to the set of all double cosets of B in G is bijective.

For arbitrary $E \subset F$, let $B(E)$ be the normalizer of $\mathfrak{n}(E)$ in G and let $M(E)$ be the centralizer of $\mathfrak{a}(E)$ in K. Clearly $B \subset B(E)$, and even $M(E)AN \subset B(E)$, since $\mathrm{Ad}_{\mathfrak{g}}(M(E))$ preserves $\mathfrak{n}(E)$, the sum of positive $\mathfrak{a}(E)$-weight spaces. It can be shown directly [*23*, pp. 207–208] and it also follows from our next lemma that $B(E) \cap K = M(E)$; by the Iwasawa decomposition, this implies that $B(E) = M(E)AN$. (In particular, we have $B(\phi) = B$; so all the way through, whenever $E = \phi$, it can simply be omitted from all parentheses.)

Lemma 2.2. Every closed subgroup of G containing B is a $B(E)$ with some $E \subset F$.

Proof. It is immediate that the Lie algebra of $B(E)$ is $\mathfrak{b}(E) = \sum \mathfrak{g}_\lambda$, the sum over $\lambda \geq 0$ and those $\lambda < 0$ that vanish on $\mathfrak{a}(E)$. We show that given any subalgebra $\mathfrak{c}$ of $\mathfrak{g}$ containing $\mathfrak{b}$, it must be equal to a $\mathfrak{b}(E)$.

For this, let $\mathfrak{c}^C$ be the complexification of $\mathfrak{c}$, and let $\mathfrak{h}^C$ be a Cartan subalgebra of $\mathfrak{g}^C$ containing $\mathfrak{a}^C$. Since $\mathfrak{c}^C \supset \mathfrak{b}^C \supset \mathfrak{h}^C$, we have $\mathfrak{c}^C = \mathfrak{h}^C + \sum_{\alpha \in S} \mathbf{C} X_\alpha$ where S is some set of $\mathfrak{h}^C$-roots. We order the $\mathfrak{h}^C$ roots so that a root is nonnegative if and only if its restriction to $\mathfrak{a}$ is nonnegative. Now S contains all positive roots (since $\mathfrak{b} \subset \mathfrak{c}$); let E be the set of elements $\lambda \in F$ such that $\alpha|_{\mathfrak{a}} = \lambda$ and $X_{-\alpha} \in \mathfrak{c}^C$ for some $\mathfrak{h}^C$ root α. We will show that the negative elements in S are exactly those $\alpha < 0$ for which $\alpha|_{\mathfrak{a}}$ is a linear combination of elements in E. From this it will follow at once that $\mathfrak{c}^C = \mathfrak{b}(E)^C$, and hence $\mathfrak{c} = \mathfrak{b}(E)$.

So let $\alpha \in S$, $\alpha < 0$. Denoting the simple $\mathfrak{h}^C$-roots by $\{\alpha_k\}$, we have $\alpha = \sum n_i \alpha_i$ with each $n_i \leq 0$. If $n_i \neq 0$, $X_{-\alpha_i}$ is in the algebra generated by X_α and the elements X_{α_k}. Hence $X_{-\alpha_i} \in c^C$, so $\alpha_i|_{\mathfrak{a}} \in E \cup \{0\}$.

Conversely assume that $\alpha < 0$ and $\alpha|_{\mathfrak{a}}$ is a linear combination of elements in E. Then $\alpha = \sum n_i \alpha_i$ with $n_i \neq 0$ only if $\alpha_i|_{\mathfrak{a}} \in E \cup \{0\}$. If $\alpha_i|_{\mathfrak{a}} = 0$, then

$X_{-\alpha_i} \in \mathfrak{b}^C \subset \mathfrak{c}^C$. If $\alpha_i|_\mathfrak{a} \neq 0$, there exists β such that $X_\beta \in \mathfrak{c}^C$ and $\beta|_\mathfrak{a} = -\alpha_i|_\mathfrak{a}$. It is easy to see ([*24*, p. 80] or [*1*, p. 6]) that either $\beta = -\alpha_i - \sum n_j \alpha_j$ or $\beta = -\sigma\alpha_i - \sum n_j \alpha_j$ where the sum in both cases is over the α_j such that $\alpha_j|_\mathfrak{a} = 0$ and σ denotes the conjugation of $\mathfrak{g}^C$ with respect to $\mathfrak{g}$ which acts on roots in a natural way. It follows that $X_{-\alpha_i}$ or $X_{-\sigma\alpha_i}$ is in the algebra generated by X_β and the X_{α_j}, hence in $\mathfrak{c}^C$. But $\mathfrak{c}^C$ is invariant under σ, so $X_{-\alpha_i} \in \mathfrak{c}^C$, finishing the proof of the statement about subalgebras.

To prove the Lemma we note that by the Bruhat Lemma every subgroup C of G containing B is of the form $\bigcup_{s \in S} BsB$ where S is some subset of W. If $s \in S$, clearly $C \supset sBs^{-1} = MA(sNs^{-1})$. Now $sNs^{-1} = \exp \sum_{\lambda > 0} \mathfrak{g}_{s\lambda}$. Clearly $\mathfrak{b}(E)$ contains $\sum_{\lambda > 0} \mathfrak{g}_{s\lambda}$ if and only if s is contained in the subgroup of W generated by reflections in the elements of E. It follows that to every $\mathfrak{b}(E)$ there corresponds exactly one Lie subgroup of G containing B; this group must be equal to $B(E)$.

Remarks. 1. There are ways of arriving at this result different from the one sketched here. One can assume that G is the adjoint group of $\mathfrak{g}$ (which for our purposes is no restriction of the generality by the remarks at the beginning of this section); then G is the identity component of a real algebraic group, and everything follows from the general theory of such groups. Another way would be to show directly that (G, B, M', F) is a Tits system in the sense of Ref. [*3*]; then the Bruhat Lemma and the results above would follow from the general theory in [*3*].

2. It is clear from the proof above that $B(E)$, which was defined as the normalizer of $\mathfrak{n}(E)$, is also the normalizer of $\mathfrak{b}(E)$ and of itself.

Lemma 2.3. For every $E \subset F$, the orbit of the identity coset of $G/B(E)$ under $\overline{N}(E)$ is open and its complement has measure 0.

Proof. First let $E = \phi$. The orbit under B of any point in G/B is the same as its orbit under N; in fact, by the Bruhat Lemma every double coset of B is of the form BkB $(k \in M')$, and $BkB = NAMkB = Nk\,AMB = NkB$ since $kM = Mk$ (M being normal in M') and $kA = Ak$ (since k normalizes A).

To find the dimension of an orbit of N (or B), we note that

$$BkB/B = k((k^{-1}Bk)B/B) \cong B/(k^{-1}Bk) \cap B$$
$$= MAN/MA\,(\mathrm{ad}(k^{-1})N \cap N).$$

Since $\exp : \mathfrak{n} \to N$ is a diffeomorphism, it is clear that the dimension of the orbit of $\dot{k} = kB$ is $\dim \mathfrak{n} - \dim(\mathrm{Ad}(k^{-1})\mathfrak{n} \cap \mathfrak{n})$. This is equal to $\dim \mathfrak{n}$ if and only if k represents that element of W which maps the positive Weyl chamber onto the negative one.

We see now that N has finitely many orbits on G/B and exactly one of these, $N(\dot{k})$, has full dimension. We have $\overline{N}(\dot{e}) = (\mathrm{Ad}(k^{-1})N)(\dot{e}) = kN(\dot{k})$; this shows that $\overline{N}(\dot{e})$ has full dimension while the other $\overline{N}$-orbits have lower dimension, finishing the proof for $E = \phi$.

If E is arbitrary, G/B is a fiber space over $G/B(E)$. Since the orbit $\overline{N}(\dot{e})$ is open and dense in G/B, the same is true for the $\overline{N}$-orbit of $\dot{e}$ in $G/B(E)$. It is easy to see that $\overline{N} \cap B(E) = \overline{N}^E$; therefore we have $\overline{N}(E)\,(\dot{e}) = \overline{N}(\dot{e})$ in $G/B(E)$, and the proof is finished.

Remark. The proof even shows that the map $\overline{N}(E) \to G/B(E)$ defined by $\bar{n} \to \dot{\bar{n}}$ is injective.

Lemma 2.4. For any $E \subset F$, $G/B(E)$ is a boundary of $X = G/K$.

Proof. Since $B(E) \supset AN$, K is transitive on $G/B(E)$, so this is a weak boundary. Since the mapping in the Remark above is smooth, there exists a smooth function ψ on $\overline{N}(E)$ such that for every integrable f on $G/B(E)$ we have

$$\int_{G/B(E)} f\,dm = \int_{\overline{N}(E)} f_1(\bar{n})\,\psi(\bar{n})\,d\bar{n}$$

where f_1 denotes the function $f_1(\bar{n}) = f(\dot{\bar{n}})$ and $d\bar{n}$ is a Haar measure on $\overline{N}(E)$. In particular, we have $\int\psi = 1$.

Suppose that f is continuous. For $a \in A$, we have, using that $a \in A \subset B$,

$$\int f\,d(am) = \int f \circ a\,dm = \int_{\overline{N}(E)} f_1(\bar{n}^a)\,\psi(\bar{n})\,d\bar{n}$$

where $\bar{n}^a$ stands for $a\bar{n}a^{-1}$. Now we let a run through a sequence such that $\bar{n}^a \to e$ for all $\bar{n} \in \overline{N}(E)$ (e.g., setting $a = \exp tH$ with H in $\mathfrak{a}^+(E)$, the subset of $\mathfrak{a}(E)$ on which all positive roots that do not vanish on $\mathfrak{a}(E)$ assume positive values, and letting $t \to \infty$). The last integral will then tend to $f_1(e) = f(\dot{e})$ by the Lebesgue dominated convergence theorem. This shows that $G/B(E)$ is a boundary.

Theorem 2.1. The boundaries of $X = G/K$ are exactly the spaces $G/B(E)$ $(E \subset F)$.

Proof. By Lemma 2.9 every $G/B(E)$ is a boundary. Now let U be an arbitrary boundary. B acts on the convex weakly compact set of all probability measures on U, and by the Markov–Kakutani theorem keeps such a measure μ fixed. By the Iwasawa decomposition, $G\mu = KB\mu = K\mu$; since K is compact, this set of measures is compact, hence closed. Since U is a boundary, by Remark 2 after Definition 1.3 there exists a sequence $\{g_n\}$ such that $g_n\mu$ tends to the delta measure δ_u based at some point $u \in U$. Since $G\mu$ is closed, it contains δ_u. Since $G\mu = K\mu$, we have $\delta_u = k\mu$, hence $\mu = \delta_{ku}$ with some $k \in K$. It follows that B fixes the point ku, i.e., the isotropy group of G at ku contains B. By Lemma 2.2, this group is then a $B(E)$.

Remark. G/B can be regarded as a fiber space over any other $G/B(E)$. It is called the maximal boundary of Furstenberg. As we shall prove in the last section, every bounded harmonic function on X is the Poisson integral of an L^∞-function on G/B. The Poisson integrals of L^∞-functions on the other $G/B(E)$-s only give subclasses of the bounded harmonic functions. These other boundaries are still of interest, mainly because the Bergman–Šilov boundary of a bounded symmetric domain is in general one of them (cf. Section 5).

Theorem 2.2. The boundary $G/B(E)$ of X is faithful if and only if writing X as a product of irreducible factors $X = X_1 \times \cdots \times X_r$, $X_i = G_i/K_i$, and identifying F with the union of the sets F_i of restricted roots for G_i/K_i, E does not contain any of the sets F_i $(1 \le i \le r)$.

Proof. It follows from Theorem 2.1 that every boundary U of X is equal to a product of boundaries U_i of X_i $(1 \le i \le r)$. As pointed out in Section 1, such a product is faithful if and only if each U_i is faithful for X_i. This reduces the problem to the case of an irreducible X.

So let X be irreducible. If $E = F$, $G/B(E)$ consists of a single point and is clearly not faithful. If $E \neq F$, then $H = \{g \in G \mid gm = m\}$ is a closed subgroup of G containing K and not equal to G since $g_n m \to \delta$ for some $\{g_n\}$. Since K is a maximal subgroup [9, p. 214], it follows that $H = K$ and so $G/B(E)$ is faithful.

3. A Differentiation Theorem

Our goal in this section is to prove Corollaries 1 and 2 of Theorem 3.1; they will be needed in Section 4. Through Theorem 3.1, these results are reduced to the classical Hardy–Littlewood Maximal Theorem. Even in the case $N = \mathbf{R}^n$ this approach, while not the only possible one, seems to be the most efficient [*29*, Chapter XVII, Section 2]. We should also point out that in the case $p = 1$ both Theorem 3.1 and its Corollaries are in general false (even if $N = \mathbf{R}^n$, cf. [*29*]). If, however, A is one-dimensional, it is possible to make useful statements for $p = 1$, cf. the article of N. Weiss in this volume.

Lemma 3.1. Let N be a connected, simply connected nilpotent Lie group, and let A be a commutative group of automorphisms of N such that for each $\alpha \in A$ the induced automophism α_* of the Lie algebra $\mathfrak{n}$ of N is diagonalizable.

Then there exist one-parameter subgroups $N_1, \ldots, N_r$ of N such that

(i) each N_j $(1 \le j \le r)$ is invariant under A,

(ii) $N = N_1 N_2 \ldots N_r$,

(iii) for each $1 \le j \le r$, $N_j N_{j+1} \ldots N_r$ is a normal subgroup of N.

Proof. Since the exponential map is in the present case a diffeomorphism, it is enough to prove the Lemma on the Lie algebra level.

The proof will be by induction on the dimension r of N. Let $\mathfrak{z}$ be the center of $\mathfrak{n}$; it is invariant under A, and therefore contains a one-dimensional subspace $\mathfrak{n}_r$ which is a common eigenspace of all elements of A. Clearly $\mathfrak{n}_r$ is an ideal in $\mathfrak{n}$. Let $\bar{\mathfrak{n}} = \mathfrak{n}/\mathfrak{n}_r$ and let $\pi : \mathfrak{n} \to \bar{\mathfrak{n}}$ be the natural map. By the induction hypothesis, $\bar{\mathfrak{n}} = \bar{\mathfrak{n}}_1 + \cdots + \bar{\mathfrak{n}}_{r-1}$ with each $\bar{\mathfrak{n}}_j$ one-dimensional, A-invariant, and each $\bar{\mathfrak{n}}_j + \cdots + \bar{\mathfrak{n}}_{r-1}$ an ideal in $\bar{\mathfrak{n}}$. Let $\mathfrak{n}'_j = \pi^{-1}(\bar{\mathfrak{n}}_j)$ $(1 \le j \le r-1)$; then each $\mathfrak{n}'_j$ is a two-dimensional A-invariant subspace of $\mathfrak{n}$ containing $\mathfrak{n}_r$. By the diagonalizability of A, each $\mathfrak{n}'_j$ contains an A-invariant subspace $\mathfrak{n}_j$ complementary to $\mathfrak{n}_r$. We have now $\mathfrak{n} = \mathfrak{n}_1 + \cdots + \mathfrak{n}_r$ with each $\mathfrak{n}_j$ one-dimensional, A-invariant, and each $\mathfrak{n}_j + \cdots + \mathfrak{n}_r$ an ideal in $\mathfrak{n}$. This finishes the proof.

Theorem 3.1. Let N and A be as in Lemma 3.1. Let $V \subset N$ be bounded measurable with positive Haar measure $|V|$. For all $f \in L^p(N)$ $(1 < p \leqq \infty)$, define

$$f^*(n) = \sup_{\alpha \in A} \frac{1}{|\alpha V|} \int_{\alpha V} |f(nn')| \, dn'.$$

Then $\|f^*\|_p \leq C_p \|f\|_p$ with a constant C_p independent of f.

Proof. The case $p = \infty$ is trivial; from now on we assume $1 < p < \infty$. We take a decomposition $N = N_1 \ldots N_r$ having the properties of Lemma 3.1. We will first prove the theorem for the case where $V = I_1 \ldots I_r$ with I_j denoting a connected symmetric neighborhood of the identity of N_j. This proof will be by induction on $r - j$; we prove the statement of the theorem for each of the subgroups $N_j \ldots N_r$ $(1 \leq j \leq r)$.

If $r - j = 0$, the statement is just the classical Maximal Theorem, since N_r is isomorphic with $\mathbf{R}$, I_r is a symmetric interval around 0, and A acts on $\mathbf{R}$ by automorphisms. Now assume the statement proved for the group $T = N_j \ldots N_r$, denote $S = N_{j-1}$ for brevity, and prove the statement for the group $Q = ST$.

The main fact to be used is that, for all integrable f on Q,

$$\int_Q f(q)\, dq = \int_{S \times T} f(st)\, ds dt = \int_{S \times T} f(ts)\, ds dt \tag{3.1}$$

(with an appropriate normalization of the Haar measures of Q, S, T). This is a special case of much more general results [*4*, Chapter 6, Section 7, § 2]; it also can easily be checked directly by using the translation-invariance of the measures and the nilpotency of the groups.

We write $I_S = I_{j-1}$, $I_T = I_j \ldots I_r$, and define, for $f \in L^p(Q)$,

$$f^{**}(q) = \sup_{\alpha \in A} \frac{1}{|\alpha I_S|} \int_{\alpha I_S} f_1(qs')\, ds',$$

$$f_1(q) = \sup_{\alpha \in A} \frac{1}{|\alpha I_T|} \int_{\alpha I_T} |f(qt')|\, dt'.$$

It is clear that

$$f^*(q) \leq f^{**}(q) \tag{3.2}$$

for all $q \in Q$. By (3.1) and Fubini's theorem, the function $t \to f(st)$ is in $L^p(T)$ for a.a. fixed $s \in S$. For every such s,

$$\int_T |f_1(st)|^p\, dt \leq C_p' \int_T |f(st)|^p\, dt$$

by the induction hypothesis. Integrating both sides with respect to s, we have, by (3.1),

$$\|f_1\|_p^p \leq C_p^{(r-j)}\|f\|_p^p. \tag{3.3}$$

For a.a. $t \in T$, $s \to f_1(ts)$ is now in $L^p(S)$. For such t, we have

$$\int_S |f^{**}(ts)|^p ds \leq C_p^{(0)} \int_S |f_1(ts)|^p \, ds$$

by definition of f^{**} and by the classical Maximal Theorem. Integrating both sides with respect to t, we obtain

$$\|f^{**}\|_p^p \leq C_p^{(0)}\|f_1\|_p^p. \tag{3.4}$$

By (3.2)–(3.4), we have $\|f^*\|_p^p \leq C_p^{(0)} C_p^{(r-j)}\|f\|_p^p$, concluding the induction.

In the case of an arbitrary V, we note that $V \subset I = I_1 \ldots I_r$ for appropriate symmetric neighborhoods of the identity $I_j \subset N_j$ $(1 \leq j \leq r)$. We have

$$\frac{1}{|\alpha V|} \int_{\alpha V} |f(nn')| \, dn' \leq \frac{|\alpha I|}{|\alpha V|} \frac{1}{|\alpha I|} \int_{\alpha I} |f(nn')| \, dn'$$

and $|\alpha I|/|\alpha V| = |I|/|V| =$ const. Therefore, f^* relative to V is majorized by a constant times f^* relative to I, finishing the proof of the theorem.

Remark. The proof shows that as long as V is of the special form $V = I_1 \ldots I_r$, C_p is even independent of V.

Corollary 1. Assume that A is partly ordered and for every $\alpha_1, \alpha_2 \in A$, there exists $\alpha \in A$ such that $\alpha_1, \alpha_2 \leq \alpha$. Assume further that $\lim \alpha n = e$ for all $n \in N$ as $\alpha \to \infty$. Then for any $f \in L^p(N)$ $(1 < p \leq \infty)$ and any bounded $V \subset N$ of positive measure, we have

$$\lim_{\alpha \to \infty} \frac{1}{|\alpha V|} \int_{\alpha V} f(nn') \, dn' = f(n)$$

for a.a. $n \in N$.

Proof. From the hypothesis about A, it follows that for any neighborhood U of e in N there exists $\alpha_0 \in A$ such that $\alpha \geq \alpha_0$ implies $\alpha V \subset U$. This implies at once that the corollary holds for all continuous functions f.

As for the general statement, it suffices to prove it for $1 < p < \infty$, since every L^∞-function is locally in, say, L^2, and the assertion involves only local properties. Now, given $f \in L^p(N)$ $(1 < p < \infty)$, it can be written as $f = f_1 + f_2$ with f_1 continuous and $\|f_2\|_p$ arbitrarily small. Using the notation

$$(M_\alpha f)(n) = \frac{1}{|\alpha V|} \int_{\alpha V} f(nn')\, dn',$$

we have

$$|(M_\alpha f)(n) - f(n)| \leq |(M_\alpha f_1)(n) - f_1(n)| + |(M_\alpha f_2)(n) - f_2(n)|.$$

Taking lim sup as $\alpha \to \infty$, the first term on the right is zero; the second is majorized in the L^p norm by $(C_p + 1)\|f_2\|_p$. It follows that for any $\varepsilon > 0$, the set of points n where $\limsup |(M_\alpha f)(n) - f(n)| \geq \varepsilon$ has measure at most $\varepsilon^{-p}(C_p + 1)^p \|f_2\|_p^p$. Since $\|f_2\|_p$ can be made arbitrarily small, the proof is finished.

Corollary 2. Under the same hypotheses as in Corollary 1,

$$\lim_{\alpha \to \infty} \frac{1}{|\alpha V|} \int_{\alpha V} |f(nn') - f(n)|\, dn' = 0$$

for a.a. $n \in N$.

Proof. (cf. [29, Vol. I, p. 65].) Applying Corollary 1 to the functions $|f - q|$, we have, for every rational number q and a.a. $n \in N$,

$$\lim_{\alpha \to \infty} \frac{1}{|\alpha V|} \int_{\alpha V} |f(nn') - q|\, dn' = |f(n) - q|.$$

Since $f(n)$ can be approximated by rational q arbitrarily closely, the assertion follows.

4. Boundary Behavior of Poisson Integrals

It was seen in the proof of Lemma 2.4 that if f is continuous on $G/B(E)$ and F is its Poisson integral, then $\lim_{t\to\infty} F((\exp tH)\cdot o) = f(\dot{e})$ for $H \in \mathfrak{a}^+(E)$, which, of course, also implies that $\lim_{t\to\infty} F(k(\exp tH)\cdot o) = f(\dot{k})$ for all $k \in K$. Here we want to improve this result in two ways: First, extend it to the case of $f \in L^\infty(G/B(E))$ and its Poisson integral (in this case, we can only expect convergence a.e. on $G/B(E)$), and second, instead of "radial" convergence along the geodesic lines $k(\exp tH)\cdot o$, we want to generalize "nontangential" convergence, i.e., prove convergence on certain sets with nonempty interior in X which generalize the nontangential domains in the unit disk.

There are several possible generalizations of nontangential convergence even in the case of the polydisk. We proceed with the general definitions, then illustrate them on examples, and then prove a Fatou-type theorem for Poisson integrals of L^∞-functions.

In the following, we always write $k(g)$ to denote the part in K of an element $g \in G$ in the Iwasawa decomposition; so $g = k(g)an$ ($k(g) \in K$, $a \in A$, $n \in N$).

Let $E \subset F$ be fixed. We want to define convergence to the boundary $G/B(E)$.

Definition 4.1. Let $H \in \mathfrak{a}^+(E)$, $g \in G$. A *truncated restricted admissible domain at* $\dot{g} \in G/B(E)$ (*with respect to* H) will be a set

$$\mathscr{R}^{\tau}_{H,C}(\dot{g}) = \{k(g)\,(\exp tH)\cdot x \,|\, t \geq \tau,\ x \in C\}$$

where $\tau \in \mathbf{R}$ and $C \subset X$ is compact with nonempty interior and invariant under $M(E)$ (the last condition is not essential).

Definition 4.2. A *truncated semirestricted admissible domain at* $\dot{g} \in G/B(E)$ will be

$$\mathscr{A}^{T}_{C}(\dot{g}) = \{k(g)a\cdot x \,|\, a \in A(E),\ \log a \geq T,\ x \in C\}$$

where $T \in \mathfrak{a}(E)$, C is as above, and $\log a \geq T$ means $\log a - T \in \bar{\mathfrak{a}}^+$.

Definition 4.3. A *truncated unrestricted admissible domain at* $\dot{g} \in G/B(E)$ will be

$$\{k(g)ma\cdot x \,|\, m \in K^E,\ a \in A,\ \log a \geq T,\ x \in C\}$$

where $T \in \mathfrak{a}$ and C is as before.

Definition 4.4. We say that a function F on X converges to the number r at $\dot{g} \in G/B(E)$ admissibly and restrictedly (with respect to $H \in \mathfrak{a}^+(E)$) if for all compact $M(E)$-invariant $C \subset X$ with nonempty interior and all $\varepsilon > 0$, where exists $\tau \in \mathbf{R}$ such that $x \in \mathscr{R}^{\tau}_{H,C}(\dot{g})$ implies $|F(x) - r| < \varepsilon$.

A function F on X *converges* to a function f on $G/B(E)$ *admissibly and restrictedly* (*with respect to* H) a.e. if F converges to $f(\dot{g})$ in the sense just described at almost all $\dot{g} \in G/B(E)$.

Semirestricted and *unrestricted* convergence are defined analogously.

Remarks. 1. In Theorem 4.1 we will prove semirestricted admissible convergence a.e. for the Poisson integrals of L^∞-functions; this of course also implies restricted admissible convergence with respect to any H. There are however cases (e.g., Poisson integrals of L^1-functions on polydisks [*29*, Chapter XVII]) where *only* restricted convergence holds. As for unrestricted convergence, in general it does not hold even for L^∞-functions, as can be shown by examples [*26*].

2. If $E = \phi$, the notions of unrestricted and semirestricted convergence coincide.

3. If $\mathfrak{a}(E)$ is one-dimensional (as in the case of rank $(X) = 1$ or if X is an irreducible bounded symmetric domain and $G/B(E)$ its Bergman–Šilov boundary, cf. Section 5), then restricted and semirestricted convergence coincide.

Example 1. The unit disk in **C**. The disk is equivalent under a Cayley transformation to the upper halfplane; we shall look at the latter first. As a symmetric space it is equal to

$$SL(2,\mathbf{R})/SO(2) \qquad \text{with} \qquad \begin{pmatrix} a & b \\ c & d \end{pmatrix} \in SL\,(2,\mathbf{R}) = G$$

acting by $z \to (az + b)/(cz + d)$. (The action has kernel $\{I, -I\}$; the identity coset corresponds to the point i.) $\mathfrak{g}$ is the algebra of 2-by-2 real matrices of trace zero, $\mathfrak{k}$ is the subset of skew-symmetric, $\mathfrak{p}$ the subset of symmetric matrices. $\mathfrak{a}$ can be chosen to be the (one-dimensional) subspace of diagonal matrices, the ordering can be chosen so that

$$H = \begin{pmatrix} -\frac{1}{2} & 0 \\ 0 & \frac{1}{2} \end{pmatrix} \in \mathfrak{a}^+.$$

Then $M = \{I, -I\}$, N consists of the lower, $\overline{N}$ of the upper triangular unipotent matrices in $SL(2,\mathbf{R})$. Since F consists of a single element, the only nontrivial boundary is G/B; it can be identified with $\mathbf{R} \cup \{\infty\}$ since G

acting on the extended complex plane in the usual way is transitive on $\mathbf{R}\cup\{\infty\}$ and the isotropy group at 0 is easily seen to be $B = MAN$. The orbit of 0 under $\overline{N}$ is $\mathbf{R}$. A consists of the elements $\exp tH$; these act by $z \to e^{-t}z$.

It is easy to see that a truncated admissible domain (all three kinds coincide now) at $\dot{e} = 0$ is just a nontangential neighborhood of 0, i.e., a small triangle in the upper half-plane with vertex at 0. Transformation to the unit disk maps this domain onto a nontangential domain at 1 (since the Cayley transformation is conformal even on the boundary). On the unit disk, K acts by ordinary rotations, so it is clear that the admissible domains are just the nontangential domains at every boundary point, and we get the usual notion of nontangential convergence.

EXAMPLE 2. A product of l copies of the unit disk. Again, this is equivalent to a product of halfplanes. G is the direct product of l copies of $SL(2,\mathbf{R})$, so $\mathfrak{a}$ is l-dimensional, F consists of l elements. First we consider the case $E = \phi$. Let $H = (H_1, \ldots H_l)$ where each H_j is the element

$$\begin{pmatrix} -\frac{1}{2} & 0 \\ 0 & \frac{1}{2} \end{pmatrix}$$

in the corresponding direct summand of $\mathfrak{a}$. Then $\exp tH$ acts by $(z_1, \ldots, z_l) \to (e^{-t}z_1, \cdots, e^{-t}z_l)$. One sees at once that $\mathscr{R}^{\tau}_{H,C}(\dot{e})$ is a set of points $(z_1, \cdots, z_l)$ such that each z_l is contained in a nontangential triangle at 0 and $\operatorname{Im} z_j \leq \alpha \operatorname{Im} z_k$ for all j, k with some $\alpha > 1$. Transformed to the product of disks, this is a product of nontangential domains with the added conditions $1 - |z_j| \leq \alpha(1 - |z_k|)$ $(1 \leq j, k \leq l)$. The corresponding notion of convergence coincides with *restricted nontangential convergence* in the sense of [*29*, Chapter XVII].

H can also be chosen differently (with not each of the H_j equal). $\mathscr{R}^{\tau}_{H,C}(\dot{e})$ in this case is the subset of a product of nontangential domains determined by $(\operatorname{Im} z_j)^{e_j} \geq \alpha(\operatorname{Im} z_k)^{e_k}$ $(1 \geq j, k \geq l)$ with some positive numbers $e_1, \ldots, e_l$.

The semirestricted and unrestricted domains in the case $E = \phi$ are simply products of nontangential domains in each of the factors.

If $E \neq \phi$, say if E consists of the positive roots of the first s $(0 < s < l)$ factors, then $\mathscr{R}^{\tau}_{H,C}(\dot{g})$ and $\mathscr{A}^{T}_{C}(\dot{g})$ are products of a rotation-invariant compact set in the product of the first s disks and of a restricted (resp. unrestricted) nontangential domain in the product of the last l-s disks. An unrestricted admissible domain is the product of the first s disks and of an unrestricted nontangential domain in the product of the last l-s disks.

Of course, in this case we are talking about convergence to $G/B(E)$, which is the product of the trivial one-point boundaries of the first s disks and of the ordinary boundaries of the last l-s disks.

It appears that here the case $E \neq \phi$ does not give anything essentially new; it merely means that we disregard some of the factors in our product. This is so as long as we are looking at products of symmetric spaces of rank one, not however in general.

EXAMPLE 3. The unit ball in $\mathbf{C}^n$. This domain can be regarded as the symmetric space $SU(n, 1)/S(U(n) \times U(1))$; $G = SU(n, 1)$ acts by holomorphic automorphisms. The rank is 1, so there is only one kind of admissible domain; at the boundary point $(1, 0, \cdots, 0)$ it is described [*18*] by an inequality $|1 - z_1| \leq ((1 + \alpha)/2)(1 - \|z\|^2)$, $\alpha > 0$. If $n > 1$, this is not a nontangential domain in the geometric sense (its intersection with the $(2n - 1)$-dimensional real plane spanned by Re $z_1, \cdots$ Re z_n, Im z_2, $\cdots$, Im z_n is a ball touching the boundary). It is because of this example that we talk about "admissible" instead of "nontangential" convergence.

EXAMPLE 4. There are further details about the case of bounded symmetric domains in Section 5.

Lemma 4.1. Given $C \subset X$, compact, $M(E)$-invariant with nonempty interior and $g' \in G$, there exists $C' \subset X$, compact $M(E)$-invariant with nonempty interior such that $g'\mathscr{R}^{\tau}_{H,C}(\dot{g}) \subset \mathscr{R}^{\tau}_{H,C'}(g'\dot{g})$ for all $\tau > 0$, $H \in \mathfrak{a}^+(E)$, $g \in G$.

Proof. It is clear that $g'k(g) = k(g'g)an$ and, by continuity of the Iwasawa decomposition, a and n stay in compact sets $C_A \subset A$, $C_N \subset N$ as g varies in G. We have

$$\begin{aligned} g'\mathscr{R}^{\tau}_{H,C}(\dot{g}) &= g'k(g)\mathscr{R}^{\tau}_{H,C}(\dot{e}) = k(g'g)an\mathscr{R}^{\tau}_{H,C}(\dot{e}) \\ &= \{k(g'g)an\,(\exp tH)\cdot x \mid t \geq \tau,\ x \in C\} \\ &= \{k(g'g)\,(\exp tH)\ an^{\exp(-tH)}\cdot x \mid t \geq \tau,\ x \in C\} \end{aligned}$$

from which it is clear that

$$C' = \{an^{\exp(-tH)}\cdot x \mid a \in C_A,\ n \in C_N,\ t \geq \tau,\ H \in \mathfrak{a}^+(E),\ x \in C\}$$

has the required properties.

Corollary 1. Given C and g' as above, there exists C' such that $g'\mathscr{A}_C^T(\dot{g}) \subset \mathscr{A}_{C'}^T(g'\dot{g})$ for all $T \in \mathfrak{a}^+(E)$ and all $g \in G$.

Proof. It is immediate from the definitions that

$$\mathscr{A}_C^T(\dot{g}) = \bigcup_{H \geq T} \mathscr{R}_{H,C}^1(\dot{g}).$$

Corollary 2. Restricted admissible convergence (with respect to a fixed H) and semirestricted admissible convergence are G-invariant notions.

(More exactly, if F converges to r at $\dot{g}$ in one of these senses, then $F \circ g'$ does the same at $g'^{-1}\dot{g}$.)

Definition 4.5. For any $\bar{n} \in \bar{N}(E)$, $T \in \mathfrak{a}(E)$ and for any compact sets $U \subset \bar{N}(E)$, $V \subset A^E$ with nonempty interior, let

$$\Gamma_{U,V}^T(\bar{n}) = \{\bar{n}a\bar{n}_1 ma' \cdot o \mid a \in A(E), \log a \geq T, \bar{n}_1 \in U, m \in K^E, a' \in V\}.$$

(Note that this could also be written $\bar{n}a\bar{n}_1 \cdot x^E$, with a, $\bar{n}_1$ as above and $x^E \in X^E$ restricted to a compact set.)

Lemma 4.2. Let $\bar{n} \in \bar{N}(E)$. A function F on X converges to the number r at $\dot{\bar{n}}$ semirestrictedly and admissibly if and only if for all $\varepsilon > 0$ and all compact $U \subset N(E)$, $V \subset A^E$ with nonempty interior, there exists $T \in \mathfrak{a}(E)$ such that $x \in \Gamma_{U,V}^T(\bar{n})$ implies $|F(x) - r| > \varepsilon$.

Proof. By Corollary 2 above and since $\Gamma(\bar{n}) = \bar{n} \cdot \Gamma(e)$, it suffices to consider the case $\bar{n} = e$.

Let $\Gamma_{U,V}^T(e)$ be given. It is clearly equal to (hence contained in) $\mathscr{A}_C^T(\dot{e})$ with

$$C = \{a_1 \bar{n}_1 ma' \cdot o \mid a_1 \in A(E), 0 \leq \log a_1 \leq T_0, \bar{n}_1 \in U, m \in K^E, a' \in V\}$$

where T_0 is an arbitrary element in $\mathfrak{a}^+(E)$.

Conversely, let $\mathscr{A}_C^T(\dot{e})$ be given. By the Corollary of Lemma 2.1, $C \subset C_A UK^E V \cdot o = C'$, where $C_A \subset A(E)$, $U \subset \bar{N}(E)$, $V \subset A^E$ are compact sets. By compactness there exists $T_0 \in \mathfrak{a}(E)$ such that $\log a_1 \geq T_0$ for all $a_1 \in C_A$. From the definitions it follows that $\mathscr{A}_C^T(\dot{e}) \subset \Gamma_{U,V}^{T+T_0}(e)$.

From these observations the lemma follows.

Lemma 4.3. There exists a smooth positive function ψ on $\bar{N}(E)$ such that whenever $f \in L^1(G/B(E))$ and F is the Poisson integral of f we have, denoting $f_1(\bar{n}) = f(\dot{\bar{n}})$ for $\bar{n} \in \bar{N}(E)$,

$$F(\bar{n}_0 ma\cdot o) = \int_{\bar{N}(E)} f_1(\bar{n}_0 \bar{n}^{ma})\, \psi(\bar{n})\, d\bar{n}$$

for all $\bar{n}_0 \in \bar{N}(E)$, $m \in K^E$, $a \in A$.

Proof. As in the proof of Lemma 2.4, let ψ be defined as the function such that

$$\int_{G/B(E)} f = \int_{\bar{N}(E)} f_1 \psi$$

for all $f \in L^1(G/B(E))$; the existence of ψ follows from Lemma 2.3. Using the definition of the Poisson integral and then using that $ma \in B(E)$ we have

$$F(\bar{n}_0 ma\cdot o) = \int_{G/B(E)} f \circ \bar{n}_0 ma = \int_{\bar{N}(E)} f(\bar{n}_0 ma\cdot\dot{\bar{n}})\, \psi(\bar{n})\ d\bar{n}$$
$$= \int_{\bar{N}(E)} f_1(\bar{n}_0 \bar{n}^{ma})\, \psi(\bar{n})\, d\bar{n},$$

finishing the proof.

Theorem 4.1. Let $f \in L^\infty(G/B(E))$ and let F be the Poisson integral of f. Then F converges to f admissibly and semirestrictedly a.e.

Proof. Let $f_1(\bar{n}) = f(\dot{\bar{n}})$ for $\bar{n} \in \bar{N}(E)$. We shall show that F converges to $f(\dot{\bar{n}}_0)$ at every point $\dot{\bar{n}}_0$ having the property that

$$\lim_{\log a \to \infty} \frac{1}{|W^a|} \int_{W^a} |f_1(\bar{n}_0 \bar{n}) - f_1(\bar{n}_0)|\, d\bar{n} = 0$$

for bounded sets $W \subset \bar{N}(E)$ with positive Haar measure. ($\log a \to \infty$ is understood with respect to the partial ordering induced in $\mathfrak{a}(E)$ by $\mathfrak{a}^+(E)$.) By Corollary 2 to Theorem 3.1 and by Lemma 2.3 this will finish the proof.

By Lemma 4.2 it is enough to show that for every $\varepsilon > 0$ and every compact $U \subset \bar{N}(E)$, $V \subset A^E$, there exists $T \in \mathfrak{a}(E)$ such that $x \in \Gamma^T_{U,V}(\bar{n}_0)$ implies $|F(x) - f_1(\bar{n}_0)| > \varepsilon$. Now $x \in \Gamma^T_{U,V}(\bar{n}_0)$ means that $x = \bar{n}_0 a \bar{n}_1 m a' \cdot o$ with $a \in A(E)$, $\log a \geq T$, $\bar{n}_1 \in U$, $m \in K^E$, $a' \in V$. We also have $x = \bar{n}_0 \bar{n}_1^a m a a' \cdot o$, and by Lemma 4.3,

$$|F(x) - f_1(\bar{n}_0)| = \left|\int_{\bar{N}(E)} (f_1(\bar{n}_0\bar{n}_1^a\bar{n}^{maa'}) - f_1(\bar{n}_0))\,\psi(\bar{n})\,d\bar{n}\right|$$

$$= \left(\int_{\bar{N}(E)-U_1} + \int_{U_1}\right)| f_1(\bar{n}_0\bar{n}_1^a\bar{n}^{maa'}) - f_1(\bar{n}_0)|\,\psi(\bar{n})\,dn$$

$$= I_1 + I_2.$$

Choosing U_1 to be a compact neighborhood of the identity such that

$$\int_{\bar{N}(E)-U_1} \psi(\bar{n})\,d\bar{n} < \frac{1}{4\|f\|_\infty},$$

it is clear that $I_1 > \varepsilon/2$. ψ is bounded on U_1 by some constant M_1; therefore

$$I_2 \leqq M_1 \int_{U_1} |f_1(\bar{n}_0\bar{n}_1^a\bar{n}^{maa'}) - f_1(\bar{n}_0)|\,d\bar{n}.$$

We make the change of variable $\bar{n}' = \bar{n}_1^a\bar{n}^{maa'}$ in this integral. m is in the compact group K^E, hence it acts on $\bar{N}(E)$ with Jacobian 1. aa' acts on $\bar{\mathfrak{n}}(E)$ by a linear transformation; the Haar measure of $\bar{\mathfrak{n}}(E)$ goes under exponentiation into the Haar measure of $\bar{N}(E)$, hence the Jacobian of aa' is equal to the constant $|U_1|/|U_1^{aa'}|$. So

$$I_2 \leqq \frac{M_1|U_1|}{|U_1^{aa'}|} \int_{(\bar{n}_1 U_1^{ma'})^a} |f_1(\bar{n}_0\bar{n}') - f_1(\bar{n}_0)|\,dn'.$$

$\bar{n}_1 U_1^{ma'}$ is contained in $UU_1^{K^E V} = W$ which is still compact. Since a acts linearly, $|W^a|/|U_1^{aa'}| = |W|/|U_1^{a'}|$, and this is bounded for $a' \in V$ by compactness of V. It follows that

$$I_2 \leqq \frac{M_2}{|W^a|}\int_{W^a} |f_1(\bar{n}_0\bar{n}') - f_1(\bar{n}_0)|\,d\bar{n}'.$$

By our assumption about $\bar{n}_0$, there exists T such that this is $<\varepsilon/2$ whenever $\log a > T$. This finishes the proof.

Remark: In the case of the upper half-plane in **C** (cf. Example 1 above), if we identify $\bar{N}$ with its orbit **R**, we have $\psi(x) = \pi^{-1}(1 + x^2)^{-1}$; so ψ is essentially the Poisson kernel. For the present proof, all we needed to know about ψ was that it was integrable and continuous.

5. The Bergman-Šilov Boundary of a Symmetric Domain

In this section, $X = G/K$ will be a Hermitian symmetric space of non-compact type. X has a standard realization as a bounded domain in $\mathbf{C}^n$ (cf. the article of J.A. Wolf in this volume; also [*21a*, *20*]). The Bergman–Šilov boundary of this domain is isomorphic with one of the boundaries $G/B(E)$; we are going to describe exactly which one. Furthermore, we will describe in geometric terms the various notions of convergence to the boundary.

As in the preceding section, let $\mathfrak{g} = \mathfrak{k} + \mathfrak{p}$ be the Cartan decomposition (in Wolf's notation $\mathfrak{g}_0 = \mathfrak{k} + \mathfrak{m}_0$). Let $\mathfrak{h}$ be a Cartan subalgebra contained in $\mathfrak{k}$. As in Wolf's article, let Ψ be a maximal system of strongly orthogonal $\mathfrak{h}^C$-roots ($\mathfrak{h}^C$ is the complexification of $\mathfrak{h}$); we number the elements of Ψ as $\psi_1, \ldots \psi_l$. Let $H_j = H_{\psi_j}$ $(1 \leq j \leq l)$ be the corresponding elements of $\mathfrak{h}^C$. Let $\mathfrak{h}^-$ be the real span of $H_1, \ldots, H_l$; then $i\mathfrak{h}^- \subset \mathfrak{h}$. Let $\mathfrak{a}$ be the real span of all $E_\psi + E_{-\psi}$ $(\psi \in \Psi)$, this is a maximal Abelian subspace of $\mathfrak{p}$. Let, finally, $c = \exp(\pi/4)\sum_{\psi\in\Psi}(E_{-\psi} - E_\psi)$ be the Cayley transformation.

$\mathrm{Ad}(c)$ is an automorphism of $\mathfrak{g}^C$ which interchanges $\mathfrak{a}$ and $\mathfrak{h}^-$. Hence the system of restricted roots with respect to $\mathfrak{a}$ is isomorphic with the set of restrictions of the $\mathfrak{h}^C$-roots to $\mathfrak{h}^-$. By a result of Harish-Chandra [*8a*, p. 586] (which is based on the same lemmas that lead to the construction of Ψ), these restrictions are necessarily of the form $\pm\frac{1}{2}(\psi_j \pm \psi_k)$, $\pm\psi_j$, and maybe $\pm\frac{1}{2}\psi_j$. Since they form a root system, it is clear that in the irreducible case *all* $1 \leq j < k \leq l$ occur. The system of restricted roots is therefore BC_l or C_l, depending on whether the $\pm\frac{1}{2}\psi_j$ occur or not (cf., e.g. [*3*]).

The set F of simple roots can be chosen to consist of $\frac{1}{2}(\psi_j - \psi_{j+1})$ $(1 \geq j \geq l - 1)$ and $\frac{1}{2}\psi_l$ or ψ_l, depending on which case we are in. It is known ([*21a*] or [*20*, pp. 156–161]) that the Lie algebra of the isotropy group of the Bergman–Šilov boundary is the $\mathrm{Ad}(c)$-transform of the sum of the nonpositive eigenspaces of $\mathrm{ad}(H_\Psi) = \mathrm{ad}(H_1 + \cdots + H_l)$; the group is the normalizer of its Lie algebra. It is clear therefore, that the Bergman–Šilov boundary is $G/B(E)$ where E is the subset of all elements of F vanishing on H_Ψ, i.e.,

$$E = \{\tfrac{1}{2}(\psi_j - \psi_{j+1}) \mid 1 \leq j \leq l - 1\}.$$

So $\mathfrak{a}(E)$ is one-dimensional and is spanned by H_Ψ.

In the reducible case, of course, we are simply dealing with a product of several factors of the type just described.

We may also note that it is easy to see without going through the details above that the Bergman–Šilov boundary is a homogeneous space of both G and K, and the other defining property of a "boundary" (Definition 1.3) follows from Ref. [*17*, Proposition 2.5]. This already shows that we are dealing with one of the $G/B(E)$, but does not tell exactly which one is the subset E of F which occurs.

We turn to the description of convergence to the boundary. We shall only consider the irreducible case; the generalizations to products are obvious.

It is easiest to start not with the bounded Harish-Chandra realization of X, but with its Cayley transform [*21a*, *20*], i.e., the orbit of $c \cdot o = iE_{-\Psi}$ under $\mathrm{Ad}(c)G$ in $\mathfrak{p}^-$. It is known that this is a generalized halfplane

$$\{(z_1,z_2) \in \mathfrak{p}_1^- + \mathfrak{p}_2^- = \mathfrak{p}^- \mid \operatorname{Im} z_1 - \Phi(z_2,z_2) \in \Omega\}$$

where Ω is an open cone in a real form of $\mathfrak{p}_1^-$ and $\Phi : \mathfrak{p}_2^- \times \mathfrak{p}_2^- \to \mathfrak{p}_1^-$ is a Hermitian bilinear form such that $\Phi(z_2, z_2) \in \overline{\Omega}$. The set $\operatorname{Im} z_1 - \Phi(z_2, z_2) = 0$ is the image of that part of the Bergman–Šilov boundary which is not moved out of $\mathfrak{p}^-$ by c; it is exactly the orbit of 0 under $\overline{N}(E)$.

By group-invariance of the notions of convergence, it suffices to describe admissible domains at the boundary point 0. $\mathfrak{a}(E)$ is spanned by H_Ψ, and it is easy to see that $\exp tH_\Psi$ acts on $\mathfrak{p}_1^-$ by multiplication by e^{-2t}, on $\mathfrak{p}_2^-$ by e^{-t}. It follows from this [*19*] that the restricted (= semirestricted) truncated admissible domains are of the form

$$\{\bar{n}\cdot(iy, 0) \,|\, y \in \omega, |\bar{n}| < \alpha|y|, y < e^{-\tau}\}$$

where $\alpha > 0$, ω is a cone such that $\bar{\omega} \subset \Omega_\cup\{0\}$, $y \to |y|$ is a norm on $\mathfrak{p}_1^-$ and $\bar{n} \to |\bar{n}|$ is a "gauge" on $\overline{N}(E)$ [*19*, p. 399]. If $\mathfrak{p}_2^- = 0$, i.e., the domain is a tube, then the gauge on $\overline{N}(E)$ simply becomes a vector space norm. As N. J. Weiss [*27*] pointed out, in this case our notion of convergence simply means that z approaches 0 in such a way that $d(z, 0)/d(z, \partial)$ remains bounded; here d denotes the Euclidean distance and ∂ is the topological boundary of the domain. Since the inverse Cayley transform is regular at 0, this same characterization is valid for the bounded realization as well. In the nontube case, our notion of convergence is even stronger than this (cf. Example 3 after Definition 4.4).

An *unrestricted* admissible domain at 0 contains the intersection of Ω with a (vector space) neighborhood of 0. It is exactly for this case that the counterexamples to unrestricted convergence are worked out in Ref. [*26*].

Finally we want to ask the question: "Where" is the maximal boundary

G/B of a symmetric domain? It is clear in general that $G/B = K/M$ can be identified (as a K-space) with the set of orbits $k(\exp \mathfrak{a}^+)\cdot o$ $(k \in K)$; the corresponding unrestricted admissible domains are sets of points within a finite Riemannian distance of such orbits. (Exactly analogously $G/B(E) = K/M(E)$ could be regarded as the set of all $k(\exp \mathfrak{a}^+(E))\cdot o$, etc.) K/M can be regarded as a fiber space over $K/M(E)$, the fiber being the set of all $k(\exp \mathfrak{a}^+)\cdot o$ which contain $(\exp \mathfrak{a}^+(E))\cdot o$ For example, in the case of $X = Sp(2, \mathbf{R})/U(2)$ realized as the tube over a circular cone in $\mathbf{R}^3$, the points of the fiber lying over the point 0 of the Bergman–Šilov boundary are the 2-dimensional angular domains spanned by the axis of the cone and a straight line on the cone.

6. Further Results on Boundary Behavior

In the case of the unit disk, it is well known that Fatou's theorem extends to Poisson integrals of L^p-functions ($1 \leq p \leq \infty$) and of regular measures; furthermore, there is a "local" version of this theorem, due to Privalov.

Whether in the case of an arbitrary symmetric space, Theorem 4.1 extends to Poisson integrals of functions more general than L^∞-functions is not known. In certain special cases, however, this question has been answered in the affirmative. Such cases are the bounded symmetric domains with Bergman–Šilov boundary [*26–28*], (cf. the article of N. J. Weiss in this volume) and the products of symmetric spaces of rank one [*29, 15, 19, 21*].

In every case, the following lemma seems to be useful.

Lemma 6.1. Let $1 \leq p \leq \infty$. In order to prove that for all $f \in L^p(G/B(E))$ the Poisson integral of f converges to f admissibly and restrictedly (or semirestrictedly) a.e., it suffices to prove the same statement for all f such that $f_1 \in L^p(\overline{N}(E))$.

Proof. We know that the image of $\overline{N}(E)$ under $\bar{n} \to \dot{\bar{n}}$ is open in $G/B(E)$. Let U be an open set whose closure is contained in this image. By compactness, there exist elements $e = g_1, g_2, \cdots, g_l$ in G such that $g_1 U, \cdots, g_l U$ cover $G/B(E)$. Now any $f \in L^p(G/B(E))$ can be written as $f = \sum f^{(j)}$ with $f^{(j)} \in L^p(G/B(E))$ and $\operatorname{supp}(f) \subset g_j U$. Each $f^{(j)} \circ g_j$, having its support in U, is in $L^p(\overline{N}(E))$. By Corollary 2 to Lemma 4.1, our assertion follows.

In order to prove Fatou-type theorems for Poisson integrals of functions more general than L^∞-functions, one needs an explicit formula, or at any rate a considerable amount of information about the function ψ of Lemma 4.3. The following is universally true but is not explicit enough to be immediately useful.

Lemma 6.2. For the function ψ of Lemma 4.3, we have

$$\psi(\bar{n}^{-1}) = e^{-2\rho_E(H(\bar{n}))}$$

where $2\rho_E$ is the sum of all positive restricted roots which do not vanish on $\mathfrak{a}(E)$.

A simple proof can be found in [*19*, Lemma 1.3], where there is also a related expression for the Poisson kernel $P_E(x, u)$ (Proposition 1.2).

For the case of symmetric spaces of rank one, Helgason [*11*] proved a completely explicit formula for $\psi(\bar{n})$. From this formula, as pointed out in [*21*], one can immediately see that if $f_1 \in L^p(\bar{N})$ $(1 \leq p \leq \infty)$ and f_1^* is its maximal function as in Section 3, then

$$\sup_{x \in \mathscr{A}_C^T(\bar{n})} |F(x)| \leq C' f_1^*(\bar{n})$$

where F is the Poisson integral of f and C' is a constant depending only on C and T. (The same inequality was also proved in [*19*] in a more complicated way.) From this, the admissible convergence a.e. of F to f follows easily by classical methods [*29*, Chapter XVII]; the result also extends immediately to Poisson integrals of measures.

All this can be extended to products of spaces of rank one by the same arguments as in Ref. [*29*, Chapter XVII]. As is apparent from the discussion of Example 2 in Section 4, it is enough to consider the case $E = \phi$ since the other cases reduce trivially to this one. The final result can be summarized as follows.

Theorem 6.1. Let $X = G/K$ be a product of symmetric spaces of rank one. Let f be a function or a measure on G/B and let F be its Poisson integral. If $f \in L^p(G/B)$ $(1 < p \leq \infty)$, then F converges to f admissibly and unrestrictedly a.e. If $f \in L^1(G/B)$ or if f is a measure, then F converges to f or to the derivative of f admissibly and restrictedly a.e.

Examples show [*29*] that this result can not be improved to unrestricted convergence in the last two cases. For the case of radial convergence, the theorem is due to Knapp [*15*].

For bounded symmetric domains with the Bergman–Šilov boundary, Theorem 6.1 remains true if we replace the word "unrestrictedly" by "semirestrictedly." The proof, which is much more difficult than in the previous case, is in the article of N. J. Weiss in this volume. In this case, one finds explicit formulas for ψ (i.e., the Poisson kernel) by a method described in Ref. [*17*] which does not use Lemma 6.2.

As for Privalov's local version of the Fatou theorem, a method of proof due to Calderón in the case of the polydisk [*29*, Chapter XVII] has been extended to the complex unit ball in Ref. [*18*] and to spaces of rank one in Ref. [*19*]. It also extends in an obvious way to products of such spaces. More precisely, we have the following.

Theorem 6.2. Let $X = G/K$ be a product of symmetric spaces of rank one, and let F be a harmonic function on X. Let $S \subset G/B$ be a measurable set, and assume that for each $\dot{g} \in S$ there exists a truncated unrestricted admissible domain $\mathscr{A}_C^T(\dot{g})$ such that F is bounded on $\mathscr{A}_C^T(\dot{g})$. Then for a.a. $\dot{g} \in S$, F converges admissibly and unrestrictedly at $\dot{g}$ to a finite value.

Whether this theorem is true for more general spaces is an open question. For the proof, we refer to the articles cited above. We have to point out, however, an error and a misprint. In the proof [*19*, p. 408], we need a positive harmonic function on X constant on $\bar{N}$-orbits, which we will now denote by Φ. This function in [*18*, p. 515] is equal to $h(z)^n$, not to $h(z)$; in the general case, it is correctly given by

$$\Phi(\bar{n}a\cdot o) = e^{-2\rho(\log a)} \qquad (\bar{n} \in \bar{N},\ a \in A)$$

where 2ρ is the sum of all positive restricted roots (with multiplicities counted). The reference to [*14*] in [*19*] proves that Φ is weakly harmonic; actually also the following is true.

Lemma 6.3. On any symmetric space, the function Φ defined above is harmonic.

Proof. Let Ψ be defined by $\Psi(g\cdot o) = e^{-2\rho(H(g^{-1}))}$. By [*19*, Proposition 1.2], Ψ is harmonic. Denoting by s the symmetry of X, it follows that $\Psi \circ s$ is harmonic. By [*9*, Lemma 3.3, p. 222], $s(\bar{n}\ a) = a^{-1}n$ with $n \in N$. Hence $\Psi \circ s = \Phi$, finishing the proof.

Remarks Added in Proof

1. L. Lindahl (Fatou's theorem for symmetric spaces, to be published) proved that for every boundary of every symmetric space there exist finite values p_0 such that the statement of Theorem 4.1 holds for all $f \in L^p(G/B(E))$ if $p > p_0$. The best possible p_0 is in general not known.

2. The proof of Theorem 6.2 sketched in [*18*] and [*19*] contains an error. A correct proof, at least for the case of a space of rank one, will be given in a forthcoming article of the author. This article will also contain a Fatou-type theorem relative to the boundary components in the Furstenberg-Satake compactifications of arbitrary symmetric spaces; cf. the indications on p. 132 of Ref. [*21*].

7. Representation Theorems

Up to here, with the only exception of Theorem 6.2, we have always talked about functions that were assumed to be Poisson integrals. Now we want to describe what is known about determining whether a given harmonic function is a Poisson integral. Throughout this section, $X = G/K$ will be an arbitrary Riemannian symmetric space of noncompact type, G/B will be its maximal boundary as in Section 2.

Theorem 7.1. If F is harmonic and bounded on X, then F is the Poisson integral of a function $f \in L^\infty(G/B)$.

Proof. [5]. Let Q be the set of all bounded measurable functions $\varphi : G \to \mathbf{R}$ such that $\|\varphi\|_\infty \leqq \|F\|_\infty$ and

$$F(g \cdot o) = \int_K \varphi(gkg')\, dk \tag{7.1}$$

for all $g, g' \in G$. Q is bounded and closed in $L^\infty(G)$, hence compact in the weak star-topology. It is clearly convex and it is also nonempty since the function $g \to F(g \cdot o)$ belongs to Q by Theorem 1.1. Therefore, by the Markov–Kakutani theorem, B acting on Q by right translations has a fixed point φ_0.

Let $f : G/B \to \mathbf{R}$ be defined by $f(\dot{g}) = \varphi_0(g)$.

Setting $\varphi = \varphi_0$ and $g' = e$ in (7.1), we have (m denoting the normalized K-invariant measure on G/B),

$$F(g\cdot o) = \int_K \varphi_0(gk)\, dk = \int_K (f \circ g)(\dot{k}) dk = \int_{G/B} f \circ g\; dm$$

which shows that F is the Poisson integral of f.

Furstenberg [*5*] also proved the following theorem strengthening Theorem 7.1 to a considerable extent.

Theorem 7.2. Every bounded weakly harmonic function on X is harmonic.

The proof [*5*] uses probability theory. A different way [*14*] of proving it is indicated below in the remarks after Theorem 7.5. A proof proposed in [*2*] is incomplete and seems difficult to complete. It would be of interest to find a simple proof.

The following extension of Theorems 7.1 and 7.2 is due to Knapp and Williamson [*16*].

Theorem 7.3. Let $1 \leq p \leq \infty$. If F is weakly harmonic and such that

$$\sup_{a\in A} \int_K |F(ka\cdot o)|^p\, dk < \infty, \tag{7.2}$$

then F is the Poisson integral of a signed measure on G/B if $p = 1$, and of a function in $L^p(G/B)$ if $p > 1$. (In particular, F is harmonic.)

Similarly, if F is weakly harmonic and

$$\sup_{a\in A} \int_{\bar{N}} |F(\bar{n}a\cdot o)|^p\, d\bar{n} < \infty, \tag{7.3}$$

then F is the Poisson integral of a signed measure whose lift to $\bar{N}$ under $\bar{n} \to \dot{\bar{n}}$ is finite, resp. of a function whose lift to $\bar{N}$ belongs to $L^p(\bar{N})$.

The proof [*16*] is by a "regularization" argument; one takes a convolution of F on K (resp. $\bar{N}$) with an approximate identity, thus reducing to the bounded case covered by Theorems 7.1 and 7.2. In the case of $\bar{N}$, there is one difficult point in the proof: In order to make the approximation argument work, one needs to know that the function ψ of Lemma 4.3 vanishes at infinity. This is proved in [*16*] with the aid of some results of Ref. [*15*].

In the following theorem, we have to assume (strong) harmonicity.

Theorem 7.4. If F is nonnegative and harmonic, then it is the Poisson integral of a positive measure on G/B.

This result follows immediately from Theorem 7.3, even from a weaker version of it which assumes F to be (strongly) harmonic. In fact, it is clear from the mean value theorem (Theorem 1.1) that our F satisfies (7.2) with $p = 1$.

Theorem 7.4 also follows easily from the next theorem; related matters are also in [*6*].

Writing out the kernel explicitly (e.g. [*19*]), Theorem 7.4 can be formulated as stating that F can be written in the form

$$F(g \cdot o) = \int_{K/M} e^{-2\rho(H(g^{-1}k))}\, d\mu(\dot{k})$$

with a positive measure μ on K/M (= B/G). For weakly harmonic functions, this result does not hold; we have instead the following.

Theorem 7.5. If F is nonnegative and weakly harmonic, then it can uniquely be written in the form

$$F(g \cdot o) = \int_{K/M \times \Lambda_0} e^{-(\rho+\lambda)(H(g^{-1}k))}\, d\mu(\dot{k},\lambda) \tag{7.4}$$

where Λ_0 is the set $\{\lambda \in \mathfrak{a}^+ \,|\, (\lambda, \lambda) = (\rho, \rho)\}$ and μ is a positive measure on $K/M \times \Lambda_0$.

This result is due to Karpelevič [*14*]; the proof involves the study of the heat equation on X and is quite complicated. The same result was also announced without proof in [*22*]. It would be desirable to have a simple proof.

As Karpelevič [*14*] points out, Theorem 7.5 contains not only Theorem 7.4 but also Theorems 7.1 and 7.2. In fact, it is clear that if $0 \le F_1 \le F_2$ and F_1, F_2 are weakly harmonic, then we have $\mu_1 \le \mu_2$ for the corresponding measures. Now the function 1 is its own Poisson integral, i.e.,

$$1 = \int_{K/M} e^{-2\rho(H(g^{-1}k))}\, d\dot{k}$$

which shows that the measure μ_0 corresponding to 1 in (7.4) is supported on $K/M \times \{\rho\}$ and equals $d\dot{k}$ there. If F is bounded and weakly harmonic, we may assume (by adding a constant) that $0 \leq F \leq c$. It follows that the measure μ corresponding to F in (7.4) is also supported on $K/M \times \{\rho\}$ and majorized by $c\, d\dot{k}$. This proves that F is the Poisson integral of a bounded function.

References

[*1*] S. Araki, On root systems and an infinitesimal classification of irreducible symmetric spaces, *J. Math.* (*Osaka City Univ.*), **13** (1962), 1–34.

[*2*] F. A. Berezin, An analog of Liouville's theorem for symmetric spaces of negative curvature, *Dokl. Akad. Nauk SSSR*, **125** (1959), 1187–1189. (In Russian.)

[*3*] N. Bourbaki, Groupes et algèbres de Lie, Chapters IV–V–VI, Hermann, Paris, 1968.

[*4*] ———, Intégration, Chapter VI, Hermann, Paris, 1963.

[*5*] H. Furstenberg, A Poisson formula for semi-simple Lie groups, *Ann. of Math.*, **77** (1963), 335–386.

[*6*] ———, Translation-invariant cones of functions on semi-simple Lie groups, *Bull. Amer.Math. Soc.*, **71** (1965), 271–326.

[*7*] R. Godement, Une généralisation du théorème de la moyenne pour les fonctions harmoniques, *C. R. Acad. Sci. Paris*, **234** (1952), 2137–2139.

[*8*] Harish-Chandra, On a lemma of F. Bruhat, *J. Math. Pures Appl.*, **35** (1956), 203–210.

[*8*a] ———, Representations of semi-simple Lie groups VI, *Amer. J. Math.*, **78** (1956), 564–628.

[*9*] S. Helgason, "Differential Geometry and Symmetric Spaces," Academic Press, New York, 1962.

[*10*] ———, Lie groups and symmetric spaces "Battelle Rencontres," Benjamin, New York, 1967, pp. 1–71.

[*11*] ———, Application of the Radon transform to representations of semisimple Lie groups, *Proc. Nat. Acad. Sci. U.S.A.*, **63** (1969), 643–647.

[*12*] S. Helgason and A. Korányi, A Fatou-type theorem for harmonic functions on symmetric spaces, *Bull. Amer. Math. Soc.*, **74** (1968), 258–263.

[*13*] R. Hermann, Geometric aspects of potential theory in symmetric spaces I, II, III, *Math. Ann.*, *148* (1962), 349–366; **151** (1963), 143–149; *153* (1964), 384–394.

[*14*] F. I. Karpelevič, Geometry of geodesics and eigenfunctions of the Laplace–Beltrami operator in symmetric spaces, *Trudy Moskov. Mat. Obsv.*, **14** (1965), 48–185. (In Russian.)

[*15*] A. W. Knapp, Fatou's theorem for symmetric spaces I, *Ann. of Math.*, (2)**88** (1968), 106–127.

[*16*] A. W. Knapp and R. E. Williamson, Poisson integrals and semisimple groups. To appear in *J. Analyse Math.*

[*17*] A. Korányi, The Poisson integral for generalized half-planes and bounded symmetric domains, *Ann. of Math.*, (2) **82** (1965), 332–350.

[*18*] ———, Harmonic functions on Hermitian hyperbolic space, *Trans. Amer. Math. Soc.*, **135**(1969), 507–516.

[*19*] ———, Boundary behavior of Poisson integrals on symmetric spaces, *Trans. Amer. Math.Soc.*, **140** (1969), 393–409.

[*20*] ———, Holomorphic and harmonic functions on bounded symmetric domains, C.I.M.E. Summer Course on Geometry of Bounded Homogeneous Domains, Cremonese, Roma, 1968, pp. 125–197.

[*21*] ———, Generalizations of Fatou's theorem to symmetric spaces. *Rice Univ. Studies* **56** (1970), 127–136.

[*21*a] A. Korányi and J. A. Wolf, Realization of Hermitian symmetric spaces as generalized half-planes, *Ann. of Math.*, **81** (1965), 265–288.

[*22*] G. Laha, Nonnegative eigenfunctions of Laplace–Beltrami operators on symmetric spaces, *Bull. Amer. Math. Soc.*, **74** (1968), 167–170.

[*23*] C. C. Moore, Compactifications of symmetric spaces I, II, *Amer. J. Math.*, **86** (1964), 201–218, 358–378.

[*24*] I. Satake, On representations and compactifications of symmetric Riemannian spaces, *Ann. of Math.*, (2) **71** (1960), 77–110.

[*25*] E. M. Stein, The analogues of Fatou's theorem and estimates for maximal functions, C.I.M.E. Summer Course on Geometry of Bounded Homogeneous Domains, Cremonese, Roma, 1968, pp. 291–307.

[*26*] E. M. Stein and N. J. Weiss, On the convergence of Poisson integrals, *Trans. Amer. Math. Soc.*, **140** (1969), 35–54.

[*27*] N. J. Weiss, Almost everywhere convergence of Poisson integrals on tube domains over cones, *Trans. Amer. Math. Soc.*, **129** (1967), 283–307.

[*28*] ———, Convergence of Poisson integrals on generalized upper halfplanes, *Trans. Amer. Math. Soc.*, **136** (1969), 109–123.

[*29*] A. Zygmund, "Trigonometric Series," 2nd ed., Cambridge Univ. Press, New York, 1959.

FATOU'S THEOREM FOR SYMMETRIC SPACES*

NORMAN J. WEISS

Department of Mathematics
Columbia University
New York, New York

*This work has been partially supported by the National Science Foundation under Grant No. NSF-GP-13693.

1. Introduction

An important class of symmetric spaces consists of those equivalent to the bounded symmetric domains in complex n-space. A useful tool in the study of function theory on these spaces is the Poisson integral, which, for instance, reproduces the real part of a well-behaved holomorphic function from the real part of its boundary value. The purpose of this note is the extension to such spaces of the theorem of Fatou which states that the Poisson integral of an L^p function on the line (the boundary of the upper half-plane) has the same function as an almost everywhere boundary value. Also included is an interesting negative result which has no real analog in the classical situation.

In our version, the spaces involved are realized as generalized upper half-planes (also known as Siegel domains of Type II). These are spaces of the form

$$D = \{(z, w) \in V_1 \times V_2 : \operatorname{Im} z - \Phi(w, w) \in \Omega\},$$

where V_1 and V_2 are complex Euclidean spaces, $\Omega \subset \operatorname{Re} V_1$ is an open convex cone, and $\Phi : V_2 \times V_2 \to V_1$ is a Hermitian bilinear form taking the diagonal of $V_2 \times V_2$ into $\overline{\Omega}$. If Ω is homogeneous and symmetric and Φ has certain homogeneity and symmetry properties, the case we consider here, D is holomorphically equivalent to a bounded symmetric domain. When $V_2 = \{0\}$, D is a tube domain over the cone Ω, and the corresponding bounded symmetric domain is of so-called tube type.

The Silov boundary of D is

$$B = \{(z, w) : \operatorname{Im} z - \Phi(w, w) = 0\},$$

and the pairing of $(\operatorname{Re} z + i\Phi(w, w), w) \in B$ with $[\operatorname{Re} z, w]$ allows us to identify B with $\operatorname{Re} V_1 \times V_2$. There is a nilpotent Lie group, $\mathfrak{N}$, of affine transformations of D. As a set, $\mathfrak{N}$ is also $\operatorname{Re} V_1 \times V_2$; the action and group multiplication of $\mathfrak{N}$ are described, respectively, by

$$[x, w]\cdot(z_0, w_0) = (z_0 + x + 2i\Phi(w_0, w) + i\Phi(w, w), w_0 + w),$$

$$[x, w]\cdot[x', w'] = [x + x' + 2\operatorname{Im}\Phi(w, w'), w + w'].$$

$\mathfrak{N}$ acts simply transitively on B, and we will consistently identify $g \in \mathfrak{N}$ with $g\cdot 0 \in B$.

The $\mathfrak{N}$-invariant measure on B is induced from Haar measure on $\mathfrak{N}$, which in turn is induced from Euclidean measure on Re $V_1 \times V_2$. The Poisson kernel is a nonnegative function $P(u, \zeta)$ defined on $B \times D$, and the Poisson integral of a function f on B is

$$F(\zeta) = \int_B P(u, \zeta) f(u) du.$$

Details of the above are in Koranyi's paper [3].

We are interested in the behavior of such F at the Silov boundary. To be precise, for $u = [x, w] \in B$ and $y \in \Omega$, we define

$$u_y = (x + iy + i\Phi(w, w), w) \in D,$$

and say that $u_y \to u$ *restrictedly* if $y \to 0$ in Ω, subject to the restraint dist $(y, \partial\Omega) > \alpha|y|$. If $y \to 0$ in Ω without further conditions, the convergence of u_y to u is said to be *unrestricted.*

Theorem A. Suppose that $f \in L^p(B)$ for some p, $1 \leqq p \leqq \infty$, and that F is the Poisson integral of f.

Then for a.e. $u \in B$, $F(u_y) \to f(u)$ as $u_y \to u$ restrictedly.

Theorem B. Suppose that D is irreducible, i.e., is not the Cartesian product of two generalized half-planes, and that Ω is not one-dimensional.

(a) There exists $f \in L^\infty(B)$ such that for a.e. $u \in B$, $F(u_y)$ has no limit as $u_y \to u$ unrestrictedly.

(b) If $1 \leqq p < \infty$, there exists $f \in L^p(B)$ such that for a.e. $u \in B$, lim sup $F(u_y) = \infty$ as $y \to 0$ unrestrictedly.

In case Ω is one-dimensional, i.e., is a half-line, the distinction between restricted and unrestricted convergence disappears. This, of course, is the case when D is the classical upper half-plane.

For completeness, we state separately a result which is an immediate consequence of the proof of Theorem A in the case $p > 1$.

Theorem C. Suppose that $D = D_1 \times \cdots \times D_N$ and that $y = (y_1, \ldots, y_N) \to 0$ in $\Omega = \Omega_1 \times \cdots \times \Omega_N$ in such a way that the convergence of each y_j to 0 in Ω_j is restricted.

Then if $f \in L^p(B)$, $1 < p \leqq \infty$, $F(u_y) \to f(u)$ a.e. on B.

In order to apply Theorem A to obtain the same result for $p = 1$, one needs the further restraint $|y_j|/|y_k| \leqq \gamma$, $1 \leqq j, k \leqq N$. We note here that Theorems A and C in the case when the D_j are ordinary half-planes are due respectively to Marcinkiewicz and Zygmund [4] and Jessen et al. [2].

The proofs given below for Theorems A–C will be for two special, but typical, kinds of generalized half-planes. For the full proofs, together with any details omitted here, the reader is referred to the work of E. M. Stein and the author, Refs. [8–10].

Section 2 is devoted to several more general preliminary results. In Section 3, we prove Theorem A in the hardest case, $p = 1$. In Section 4, we prove Theorem A for $p > 1$, and, incidentally, Theorem C, including a simplification of the author's original method in the tube domain case due to Professors R. Coifman and G. Weiss. The negative result, Theorem B, is established in Section 5; the proof has been simplified owing to a suggestion by R. Coifman. It should also be pointed out that Theorem B extends a result originally proved by E. M. Stein et al. [7].

2. Preliminary Results

Given a measurable function f on a measure space $(\mathcal{M}, d\mu)$, we define the distribution function $m(f)$ on the positive half-line by

$$m(f)\,(s) = \mu(\{x \in \mathcal{M} : |f(x)| > s\}).$$

We say that f is of weak-type 1 if $m(f)\,(s) \leqq A/s$, $s > 0$. Notice that

$$m(\,(f_1 + f_2)/2)\,(s) \leqq m(f_1)\,(s) + m(f_2)\,(s);$$

this estimate cannot be improved. The following lemma is used in Section 3 in place of the missing subadditivity. (See [6] and [8].)

Lemma 2.1. Suppose that $f_j \geqq 0$ is measurable on $\mathcal{M}$ and that $m(f_j)(s) \leqq 1/s$, $j = 1, \ldots, N$. Let $f = (1/N)\sum_{j=1}^{N} f_j$.
Then $m(f)\,(s) \leqq 2(\log N + 2)/s$.

Proof. Fix s, and for each j, define

$$l_j(x) = \begin{matrix} f_j(x), & f_j(x) \leqq s/2, \\ 0, & \text{otherwise}, \end{matrix} \qquad u_j(x) = \begin{matrix} f_j(x), & f_j(x) \geqq Ns/2, \\ 0, & \text{otherwise}, \end{matrix}$$

$n_j(x) = f_j(x) - [l_j(x) + u_j(x)]$. Letting $l(x)$, $u(x)$, and $n(x)$ be the respective averages, we see immediately that $l(x) \leqq s/2$ for every x and that $u(x) = 0$ except on a set of measure $2/s$. Finally,

$$\int_{\mathcal{M}} n_j(x)\, d\mu(x) = -\int_{s/2}^{Ns/2} \sigma\, dm(f_j)(\sigma)$$

$$= \int_{s/2}^{Ns/2} m(f_j)(\sigma)\, d\sigma - \sigma m(f_j)(\sigma)\Big|_{s/2}^{Ns/2} \leqq \int_{s/2}^{Ns/2} \sigma^{-1} d\sigma + 1,$$

so that $\int_{\mathcal{M}} n(x)\, d\mu(x) \leqq \log N + 1$, and $n(x) \leqq s/2$ except on a set whose measure does not exceed $(\log N + 1)2/s$. This concludes the proof.

Note. It is not hard to see that the estimate in the lemma is the best possible.

The second preliminary result extends to arbitrary locally compact groups, via methods of Wiener and Marcinkiewicz–Zygmund, a well-known result of Hardy and Littlewood.

Lemma 2.2. Let G be a locally compact group with left Haar measure μ. Suppose that $\{\alpha_t : t > 0\}$ is a multiplicative group of automorphisms of G, and that P is a compact neighborhood of the identity such that $\mu(P) > 0$ and $\alpha_t(P) \subset \alpha_{t'}(P)$ if $t < t'$.

For any locally integrable function f on G, define

$$f^*(x_0) = \sup_{t>0} \mu(\alpha_t P)^{-1} \int_{\alpha_t P} |f(x_0 x^{-1})|\, dx.$$

Then $m(f^*)(s) < M_P \|f\|_1/s$, $s > 0$, where

$$M_P = \mu(PP^{-1}P)/\mu(P).$$

We express the conclusion of 2.2 by saying that the (sublinear) map $f \to f^*$ is of weak-type 1–1 with bound M_P. By a standard method [*11*, Chapter 17], one sees that 2.2 is an immediate consequence of

Lemma 2.3 Let G, $\{\alpha_t\}$, P be as in 2.2. Suppose that to every point x of a measurable subset of E of G there is associated $t(x) > 0$ in such a way that $\sup_{x \in E} \mu(\alpha_{t(x)} P) < \infty$.

Then there is a sequence $\{x_j\}$ of points of E satisfying:

(a) the translates $x_j \alpha_{t(x_j)} P$ are disjoint;

(b) $\sum_j \mu(\alpha_{t(x_j)} P) \geqq M_P^{-1} \mu(E)$.

Proof. We choose x_1 by the condition $t(x_1) = \sup_{x\in E} t(x)$, assuming temporarily, here and below, that such supremums are attained. The points $x_2, x_3, \ldots$ are chosen inductively by the conditions

$$x_{j+1} \notin \bar{R}_k, \quad 1 \leqq k \leqq j; \quad t(x_{j+1}) = \sup\{t(x) : x \notin \bar{R}_k, \quad 1 \leqq k \leqq j\},$$

where $\bar{R}_k = \cup\{x\alpha_{t(x_k)}P : x\alpha_{t(x_k)}P \cap x_k\alpha_{t(x_k)}P \neq \emptyset\}$.

Now the sets $R_j = x_j\alpha_{t(x_j)}P$ are clearly disjoint, and $\mu(R_j)$ decreases with j. If $\Sigma\, \mu(R_j) = \infty$, we are done. Otherwise, $\mu(R_j) \to 0$, which implies that $t(x_j) \to 0$, and thus that the $\bar{R}_j$ cover E, since for any x outside their union, we would have $t(x) = 0$, which is impossible.

Finally, it follows easily from the construction of the $\bar{R}_j$ that $\bar{R}_j = x_jP_jP_j^{-1}P_j$, where $P_j = \alpha_{t(x_j)}P$, so that

$$\begin{aligned}\mu(E) &\leqq \textstyle\sum_j \mu(\bar{R}_j) \leqq \sum_j [\mu(P_jP_j^{-1}P_j)/\mu(P_j)]\mu(P_j)\\ &= M_P \textstyle\sum_j \mu(P_j).\end{aligned}$$

The assumption that $\sup\{t(x): \ldots\}$ is attained can be dropped by adding a multiplicative constant, arbitrarily close to 1, to the estimate.

3. Proof of Theorem A

Before beginning the proof, we note that the Poisson integral is actually a convolution in terms of the group structure on $\mathfrak{N}$. In fact, making the identification of $g \in \mathfrak{N}$ with $g \cdot 0 \in B$ and noticing that for $\zeta \in D$, $g \in \mathfrak{N}$, $\partial(g\zeta)/\partial\zeta = 1$, we have from [3, 3.4] that $P(h, \zeta) = P(gh, g\cdot\zeta)$. Moreover, if $\zeta = g_y$, $g \in \mathfrak{N}$, $y \in \Omega$, $h\cdot\zeta = (hg)_y$. Therefore, the Poisson integral F of a function f is given by

$$\begin{aligned}F(\zeta) = F(g_y) &= \int_{\mathfrak{N}} P(h, g_y)f(h)\, dh\\ &= \int_{\mathfrak{N}} P(g^{-1}h, 0_y)f(h)\, dh = \int_{\mathfrak{N}} P_y(h)f(gh)\, dh\\ &= \int_{\mathfrak{N}} P_y(h)f(gh^{-1})\, dh, \qquad (1)\end{aligned}$$

since $P_y(h) = P(h, 0_y) = P_y(h^{-1})$.

The first special class of domains we consider consists of the typical tube domains D_n defined by taking $V_2 = \{0\}$ and letting V_1 be the $\frac{1}{2}n(n+1)$-dimensional complex vector space of symmetric $n \times n$ matrices with complex entries. Re V_1 is the space of real symmetric matrices, and $\Omega \subset \text{Re } V_1$ is the cone of positive definite matrices. $B = \mathfrak{N} = \text{Re } V_1$ is given real Euclidean coordinates by considering the diagonal and upper triangular matrix entries. Note that the group structure of $\mathfrak{N}$ in this case is that of Euclidean space.

We have the usual (non 1–1) map of $E_n \times SO(n)$ onto (almost all of) B given by $(r, k) \to k^{-1} d(r)k$, where $d(r)$ is the diagonal matrix whose entries are the coordinates r_j of the vector r, i.e., the eigenvalues of $k^{-1}d(r)k$. If dx is Lebesgue measure on B, dr is Lebesgue measure on E_n, and dk is Haar measure on $SO(n)$,

$$dx = \Delta(r)\, dr\, dk, \qquad \Delta(r) = c_n \prod_{i<j} |r_i - r_j|.$$

The Poisson kernel is given by

$$P_y(x) = b_n[\det y / |\det(x + iy)|^2]^{(n+1)/2},$$

and, in particular, if $\eta > 0$, $x = k^{-1} d(r)k$,

$$P_{\eta I}(x) = b_n[\eta^n / \prod_{j=1}^{n} (r_j^2 + \eta^2)]^{(n+1)/2}. \tag{2}$$

The first step in the proof of Theorem A is to reduce the case of restricted convergence to the case of convergence along the single ray, $y = \eta I$, $\eta > 0$. As will be seen, this amounts to

Lemma 3.1. Suppose that $y \in \Omega$ and that

$$\text{dist}(y, \partial\Omega) > \alpha|y|.$$

Then for any $x \in B$,

$$P_y(x) \leqq A_\alpha P_{|y|I}(x).$$

The reader is referred to [9] for the proof of 3.1, which is straightforward but involved.

The advantage of using $|y|I$ instead of arbitrary $y \in \Omega$ is that formula (2) becomes available. This in turn enables us to estimate $F(x + iy) = P_y * f(x)$ in terms of maximal averages of f over certain classes of sets, which we proceed to define.

Definition 3.2. Let $(j) = (j_1, \ldots, j_n)$ be an n-tuple of nonnegative integers satisfying $j_1 \geqq j_2 \geqq \cdots \geqq j_n$.

$$R_{(j)} = \{r \in E_n : |r_i| \leqq 2^{j_i}, i = 1, \ldots, n\};$$

$$B_{(j)} = \{x \in B : x = k^{-1} d(r)k, r \in R_{(j)}, k \in SO(n)\}.$$

Also, with the conventions that $2^{-1} = 0$, $(j-1) = (j_1 - 1, \ldots, j_n - 1)$, $\Gamma_{(j)} = B_{(j)} - B_{(j-1)}$.

(It may be helpful here to consider the case $n = 2$. In this case, $B = E_3$, and $B_{(j)}$ is the conical surface obtained by rotating about the line $x_1 = x_2$, $x_3 = 0$ the rectangle $|x_1| \leqq 2^{j_1}$, $|x_2| \leqq 2^{j_2}$, $x_3 = 0$.)

Before turning to the maximal averages whose sums dominate the Poisson integral, we record an estimate on the Poisson kernel which is an immediate consequence of (2) and 3.2:

$$\sup_{t \in \Gamma_{(j)}} P_I(t) \leqq b_n \prod_{i=1}^{n} [1 + 2^{2(j_i - 1)}]^{-(n+1)/2}, \tag{3}$$

where we have kept in force the convention $2^{-1} = 0$.

One more minor estimate is needed. Here and below we write $|E|$ for the Lebesgue measure of a set $E \subset E_n$ and, more generally, the Haar measure of a set $E \subset \mathfrak{N}$. Also, we take $\varphi(\cdot) \sim \psi(\cdot)$ to mean that $A_1|\varphi(\cdot)| \leqq |\psi(\cdot)| \leqq A_2|\varphi(\cdot)|$, where the constants are independent of the argument.

Lemma 3.3. $|B_{(j)}| \sim 2^{[nj_1 + (n-1)j_2 + \cdots + j_n]}$.

Proof. $$|B_{(j)}| = \int_{R_{(j)}} \Delta(r)\, dr = c_n \int_{R_{(j)}} \prod_{i<j} |r_i - r_j|\, dr_1 \ldots dr_n,$$

and the expression on the RHS of 3.3 obviously dominates a multiple of $|R_{(j)}| \sup_{r \in R_{(j)}} \Delta(r)$, and thus a multiple of $|B_{(j)}|$. The reverse inequality is obtained by restricting the integration to those parts of $R_{(j)}$ where the coordinates are close to their maximum values. Details are in [9, 6.7].

The maximal averages to which we have been referring are described in

Definition 3.4. For any set $E \subset B$, $\eta > 0$, $E^\eta = \{\eta x : x \in B\}$ is the η-dilation of E.

For any locally integrable function f on B,

$$f^*_{(j)}(x) = \sup_{\eta > 0} |B^\eta_{(j)}|^{-1} \int_{B^\eta_{(j)}} |f(x-t)|\, dt.$$

The domination of the Poisson integral by sums of the $f^*_{(j)}$ can now be established.

Lemma 3.5. Let $f \in L^p$, $1 \leqq p \leqq \infty$, and let F be its Poisson integral. Let $\alpha > 0$, and define

$$f^{**}_\alpha(x) = \sup\{|F(x + iy)| : \mathrm{dist}(y, \partial\Omega) > \alpha|y|\}.$$

Then $f^{**}_\alpha(x) \leqq B_\alpha \sum_{(j)} 2^{-(j_1 + \cdots + j_n)} f^*_j(x)$.

Proof. The formula (1) is now

$$F(x + iy) = \int_B f(x - t) P_y(t)\, dt.$$

By Lemma 3.1, we may assume y to be of the form ηI, $\eta > 0$. Also, referring to 3.2, we see that for any $\eta > 0$, $B = \bigcup_{(j)} \Gamma^\eta_{(j)}$, and so

$$|F(x + iy)| \leqq \sum_{(j)} \int_{\Gamma^\eta_{(j)}} |f(x - t)| P_y(t)\, dt. \tag{4}$$

Moreover, $\Gamma_{(j)} \subset B_{(j)}$, and so

$$\begin{aligned} \int_{\Gamma^\eta_{(j)}} |f(x-t)|\, P_y(t)\, dt &\leqq [\sup_{t \in \Gamma^\eta_{(j)}} P_y(t)] \int_{\Gamma^\eta_{(j)}} |f(x-t)|\, dt \\ &\leqq [\sup_{t \in \Gamma^\eta_{(j)}} P_y(t)] \int_{B^\eta_{(j)}} |f(x-t)|\, dt \\ &\leqq [\sup_{t \in \Gamma^\eta_{(j)}} P_y(t)]\, |B^\eta_{(j)}| f^*_{(j)}(x). \end{aligned} \tag{5}$$

Finally, since the dimension of B is $n(n + 1)/2$, $|B^\eta_{(j)}| = \eta^{n(n+1)/2} |B_{(j)}|$, while on the other hand it is clear from (2) that $P_{\eta I}(\eta x) = \eta^{-n(n+1)/2} P_I(x)$. Combining (3), 3.3, (5), and (4), the proof is complete.

The proof of Theorem A is routine once we obtain an estimate on f^{**}_α in terms of $\|f\|_p$. In case $p > 1$, we can show that $\|f^{**}_\alpha\|_p \leqq A_{p,\alpha} \|f\|_p$; this will be done in Section 4. As for the case $p = 1$, we will show that the transformation $f \to f^{**}_\alpha$ is of weak-type 1–1 with a bound depending only on α. The method is to use Lemma 3.5 together with weak-type estimates on the $f^*_{(j)}$.

For simplicity, we write $|j| = j_1 + \cdots + j_n$.

Lemma 3.6. $m(f^*_{(j)})(s) \leqq A(1 + |j|)\,\|f\|_1/s,\ s > 0.$

Proof. The proof consists of finding maximal averages ${}^kf^*_{(j)}$, $k = 1, \ldots, N_{(j)}$, satisfying

(i) $m({}^kf^*_{(j)})(s) \leqq A\|f\|_1/s,\ s > 0;$

(ii) $f^*_{(j)}(x) \leqq BN^{-1}_{(j)} \sum_{k=1}^{N_{(j)}} {}^kf^*_{(j)}(x);$

(iii) $N_{(j)} \leqq 2^{\alpha|j|}\beta.$

By Lemma 2.1, it is indeed enough to find such ${}^kf^*_{(j)}$.

Recall that the $f^*_{(j)}$ are maximal averages over dilations $B^\eta_{(j)}$ of the sets $B_{(j)}$. We will find rectangular sets ${}^kP_{(j)}$, $k = 1, \ldots, N_{(j)}$ (with sides not necessarily parallel to the coordinate axes) such that $B_{(j)} \subset \bigcup_{k=1}^{N(j)} {}^kP_{(j)}$, and define

$$ {}^kf^*_{(j)}(x) = \sup_{\eta>0} |{}^kP^\eta_{(j)}|^{-1} \int_{{}^kP^\eta_{(j)}} |f(x-t)|\,dt. $$

The properties (i), (ii), (iii) of the ${}^kf^*_j$ follow from

(i′) $M_{{}^kP_{(j)}} = |{}^kP_{(j)} - {}^kP_{(j)} + {}^kP_{(j)}|/|{}^kP_{(j)}| \leqq A;$

(ii′) $|{}^kP_{(j)}| \leqq BN^{-1}_{(j)}|B_{(j)}|;$

(iii) $N_{(j)} \leqq 2^{\alpha|j|}\beta.$

In fact, (i) follows from (i′) and Lemma 2.2. Moreover, since the ${}^kP_{(j)}$ are rectangular and $\mathfrak{N}$ is Euclidean, (i′) is automatically satisfied with $A = 3^\nu$, $\nu = n(n+1)/2 =$ dimension of $\mathfrak{N}$. To see that (ii) follows from (ii′), we write

$$ \begin{aligned} f^*_{(j)}(x) &= \sup_{\eta>0} |B^\eta_{(j)}|^{-1} \int_{B^\eta_{(j)}} |f(x-t)|\,dt \\ &\leqq \sup_{\eta>0} |B^\eta_{(j)}|^{-1} \sum_{k=1}^{N_{(j)}} \int_{{}^kP^\eta_{(j)}} |f(x-t)|\,dt \leqq BN^{-1}_{(j)} \textstyle\sum {}^kf^*_{(j)}. \end{aligned} $$

The sets ${}^kP_{(j)}$ are constructed as follows. We cover $SO(n)$ with neighborhoods $U_1, \ldots, U_{N_{(j)}}$, set

$$ {}^kB_{(j)} = \{v^{-1}d(r)v : r \in R_{(j)},\ v \in U_k\}, $$

and show that the ${}^kB_{(j)}$ can be covered by rectangular sets ${}^kP_{(j)}$ in such a way that (ii′) and (iii) are satisfied simultaneously. Clearly (cf. 3.2), $B_{(j)} =$

$\bigcup^k B_{(j)}$. (The geometrical picture in the case $n = 2$ is again helpful; here $B_{(j)}$ is a conical surface with some thickness, and the ${}^kB_{(j)}$ are the wedges determined by decomposing the circle into arcs $U_1, \ldots, U_{N(j)}$.)

By homogeneity, it suffices to find a neighborhood U_1 of the origin in $SO(n)$ such that the corresponding ${}^1B_{(j)}$ can be covered by a rectangular ${}^1P_{(j)}$ satisfying (ii′), while $SO(n)$ can be covered by $N_{(j)}$ translates of U_1, with $N_{(j)}$ satisfying (iii).

Let us take for U_1 the image under the exponential map of the spherical neighborhood $\mathscr{U}_{(j)}$ of the origin in $\mathfrak{so}(n)$:

$$\mathscr{U}_{(j)} = \{X \in \mathfrak{so}(n) : |X| \leqq \rho_{(j)}\},$$

where $\rho_{(j)}$ is to be determined. Now

$${}^1B_{(j)} = \{v^{-1}\, d(r)v : |r_i| \leqq 2^{j_i},\ v \in U_1\}.$$

If we approximate $v^{-1}d(r)v$ by $(I - X)d(r)\,(I + X)$ and keep in mind that $j_1 \geqq j_2 \geqq \cdots \geqq j_n$, we see that the greatest possible absolute value of the ith diagonal coordinate, x_{ii}, of any $x \in {}^1B_{(j)}$ is $2^{j_i} + \rho_{(j)}^2\, 2^{j_1}$, while the greatest possible absolute value of x_{ik}, $i < k$, is $\rho_{(j)}(2^{j_i} + \rho_{(j)}2^{j_1})$. (Taking more terms in the expansion of exp X merely adds terms involving higher powers of $\rho_{(j)}$.)

Given these estimates, we can clearly cover ${}^1B_{(j)}$ by a rectangular ${}^1P_{(j)}$ (with sides parallel to the axes) such that

$$|{}^1P_{(j)}| \leqq B' \prod_{i=1}^{n} (2^{j_i} + \rho_{(j)}^2 2^{j_1}) \prod_{i=1}^{n-1} [\rho_{(j)}(2^{j_i} + \rho_{(j)}2^{j_1})]^{n-i}.$$

Taking $\rho_{(j)} = 2^{-|j|} \leqq 2^{-j_1}$, and recalling 3.3, we have

$$\begin{aligned} |{}^1P_{(j)}| &\leqq B'' \rho_{(j)}^{n(n-1)/2} 2^{[nj_1 + (n-1)j_2 + \cdots + j_n]} \\ &\leqq B''' \rho_{(j)}^{n(n-1)/2} |B_{(j)}|. \end{aligned}$$

On the other hand, since dim $SO(n) = n(n - 1)/2$, it is not hard to see [8, 3.4] that $SO(n)$ can be covered by $N_{(j)}$ translates of U_1, where

$$N_{(j)} \sim \rho_{(j)}^{-n(n-1)/2} = 2^{n(n-1)|j|/2}.$$

Thus we have established (ii′) and (iii) and completed the proof of Lemma 3.6.

We turn now to the second typical class of domains to be considered here. There is one of these, denoted $D_{n,m}$, for each pair of integers $n > 0$, $m \geqq 0$. In the language of Section 1, V_1 is now the n^2-dimensional vector space of $n \times n$ matrices with complex entries, Re V_1 is the space of Hermitian symmetric matrices [so that Re $z = \frac{1}{2}(z + z^*)$], and Ω is the cone of positive definite matrices. V_2 is the nm-dimensional space of $n \times m$ matrices with complex entries, and $\Phi(w_1, w_2) = w_1 w_2^*$. The Poisson kernel is given by

$$P_y(u) = P_y([x, w]) = c_{n,m}\{\det y / |\det(x + i[ww^* + y])|^2\}^{n+m}. \tag{6}$$

As before, the program for proving Theorem A begins with a reduction of restricted convergence to radial convergence.

Lemma 3.1′. [*10*, 3.7] Suppose that $y \in \Omega$ and that

$$\operatorname{dist}(y, \partial\Omega) > \alpha|y|.$$

Then for any $u \in B$,

$$P_y(u) \leqq A_\alpha P_{|y|I}(u).$$

The Euclidean coordinates we use for $B = \text{Re } V_1 \times V_2$ are again matrix entries (diagonal and upper triangular ones in the case of Re V_1). There is a diagonalization which depends on whether $m \leqq n$ or vice versa; we assume the former. B is then the image of $E_n \times E_m \times U(n) \times U(n) \times U(m)$ under the map

$$(r, s, k, u, v) \to [k^{-1}d(r)k, u\tilde{d}(s)v],$$

where $d(r)$ is as above and $\tilde{d}(s)$ is the diagonal-type $n \times m$ matrix whose entries are the coordinates s_i and whose last $n - m$ rows consist of zeroes. (In case $n \leqq m$, the role of E_m is played by another copy of E_n; this is the only reason for assuming $m \leqq n$.)

Given n- and m-tuples (j) and (k) of nonnegative integers satisfying $j_1 \geqq \cdots \geqq j_n$, $k_1 \geqq \cdots \geqq k_m$, we define $R_{(j)} \subset E_n$ as in Definition 3.2 and $S_{(k)} \subset E_m$ similarly.

Definition 3.2′.

$$C_{(j)} = \{x = k^{-1}d(r)k \in \text{Re } V_1 : r \in R_{(j)}, k \in U(n)\},$$

$$E_{(k)} = \{w = u\tilde{d}(s)v \in V_2 : s \in S_{(k)}, u \in U(n), v \in U(m)\}.$$

Lemma 3.3′. [*10*, 3.3].

$$|C_{(j)}| = 2^{-[(2n-1)j_1+(2n-3)j_2+\cdots+j_n]},$$

$$|E_{(k)}| = 2^{-[(n+m-1)k_1+(n+m-3)k_2+\cdots+(n-m+1)k_m]}.$$

Recall that we make no distinction between $g \in \mathfrak{N}$ and $g \cdot 0 \in B$. For $g = [x, w] \in \mathfrak{N}$ and $\eta > 0$, we define

$$g^\eta = [\eta x, \eta^{1/2} w],$$

and call g^η the η-dilation of g. If $G \subset \mathfrak{N}$, $G^\eta = \{g^\eta : g \in G\}$. Checking the rule for multiplication in $\mathfrak{N}$ (Section 1), we see that $g \to g^\eta$ is an automorphism of $\mathfrak{N}$. And checking (6), we see that

$$P_{\eta I}(g^\eta) = \eta^{-n(n+m)} P_I(g). \tag{7}$$

On the other hand, if $C \subset \operatorname{Re} V_1$, $E \subset V_2$, then, since the real dimensions of $\operatorname{Re} V_1$ and V_2 are, respectively, n^2 and $2nm$, it follows that

$$|(C \times E)^\eta| = \eta^{(n^2+nm)} |C \times E|. \tag{8}$$

Suppose that we try to continue the analogy with the proof of Theorem A for the domains D_n, i.e., to dominate the Poisson integral by maximal averages. We would define $f^*_{(j)(k)}$ as a maximal average over η-dilations of $C_{(j)} \times E_{(k)}$, set

$$\Gamma_{(j)(k)} = (C_{(j)} - C_{(j-1)}) \times (E_{(k)} - E_{(k-1)}),$$

and attempt to establish a series of inequalities like

$$|F(g_{\eta I})| = \left|\int f(gh^{-1}) P_{\eta I}(h)\, dh\right| \leqq \sum_{(j)(k)} \int_{\Gamma^\eta_{(j)(k)}} |f(gh^{-1})| P_{\eta I}(h)\, dh;$$

$$\begin{aligned}
\int_{\Gamma^\eta_{(j)(k)}} |f(gh^{-1})| P_{\eta I}(h)\, dh &\leqq \Big[\sup_{h \in \Gamma^\eta_{(j)(k)}} P_{\eta I}(h)\Big] \int_{\Gamma^\eta_{(j)(k)}} |f(gh^{-1})|\, dh \\
&\leqq [\sup P_{\eta I}(h)] \int_{(C_{(j)} \times E_{(k)})\eta} |f(gh^{-1})|\, dh \\
&\leqq [\sup P_{\eta I}(h)]\, |(C_{(j)} \times E_{(k)})^\eta|\, f^*_{(j)(k)}(g) \\
&\leqq A 2^{-(|j| + |k|)} f^*_{(j)(k)}(g).
\end{aligned} \tag{9}$$

Unfortunately, the last inequality, the only one that is not trivial, is also, in general, not true. The problem is most easy to analyze in the case when

$$(j) = (j, 0, \ldots, 0), \qquad (k) = (j/2, 0, \ldots, 0),$$

where j is an even integer. It follows from (7) and (8) that η can be assumed to be 1. The condition $g = [x, w] \in \Gamma_{(j)(k)}$ means that both x and ww^* have one eigenvalue between 2^{j-1} and 2^j, and the others between 0 and 1.

Suppose, for instance, that $x = d(2^j, 0, \ldots, 0)$. Considering the two extreme cases, $ww^* = d(2^j, 0, \ldots, 0)$, $ww^* = d(0, 2^j, 0, \ldots, 0)$, we see that

$$P_I([x, w]) = c_{n,m}|\det(x + i[ww^* + I])|^{-2(n+m)}$$

can range between multiples of $[2^{2j} + (1 + 2^j)^2]^{-(n+m)}$ and $[(2^{2j} + 1)(2^j + 1)^2]^{-(n+m)}$ as $[x, w]$ ranges over $\Gamma_{(j)(k)}$. The difference is accounted for by the varying degrees of "interaction" between the largest eigenvalues of x and of ww^*. In particular,

$$\sup_{g \in \Gamma_{(j)(k)}} P_I(g) \geqq C 2^{-2j(n+m)}$$

while it follows from 3.3′ that

$$|C_{(j)} \times E_{(k)}| \geqq B\, 2^{(2n-1)j}\, 2^{2(n+m-1)j/2} = B\, 2^{(3n+m-2)j}$$

and an inequality like (9) cannot hold.

A deeper analysis of the situation begins with the observation that $P_I([x, w])$ takes on its largest value for x and w with fixed eigenvalues, $x = k^{-1}d(r)k$, $w = u\tilde{d}(s)v$, when $k^{-1} = u$ (since $ww^* = u\tilde{d}(s)\tilde{d}(s)^*u^{-1}$), i.e., $ku = I$. A more precise estimate follows.

For $\rho > 0$, $\mathscr{V}_\rho$ is defined to be the neighborhood of 0 in $\mathfrak{u}(n)$ defined by the inequalities $|Z_{1i}| = |Z_{i1}| \leqq \rho$, $2 \leqq j \leqq n$, where $\mathfrak{u}(n)$ is realized as the space of skew Hermitian symmetric $n \times n$ matrices.
$V_\rho \subset U(n)$ is the exponential image of $\mathscr{V}_\rho$.

Lemma 3.7. Let $(j) = (j, 0, \ldots, 0)$, $(k) = (j/2, 0, \ldots, 0)$, and suppose that $x = k^{-1}d(r)k$, $w = u\tilde{d}(s)v$, where $r \in R_{(j)} - R_{(j-1)}$, $s \in S_{(k)} - S_{(k-1)}$, and $ku \notin V_\rho$.
Then

$$P_I([x, w]) \leqq A[2^{2j}(1 + 2^{2j}\rho^4)]^{-(n+m)}. \tag{10}$$

Proof. The inequality (10) is a consequence of (6) and the first order approximation of exp Z by $I + Z$. Details are in [*10*, 3.2].

Using Lemma 3.7 and its variants for other values of (j) and (k), we can dominate the Poisson integral of a function on $\mathfrak{N}$ by sums of maximal averages over more complicated kinds of sets.

Definition 3.8. Let $\{V_{(j)(k)l}\}_{l=1,\ldots,L+1}$ be a decreasing family of neighborhoods of the identity in $U(n)$, where $V_1 = U(n)$, $V_{L+1} = \emptyset$.

$$H_{(j)(k)l} = \{[k^{-1}d(r)k, u\tilde{d}(s)v] : r \in R_{(j)},\ s \in S_{(k)},\ ku \in V_{(j)(k)l}\};$$

$$K_{(j)(k)l} = \{[k^{-1}d(r)k, u\tilde{d}(s)v] : r \in R_{(j)} - R_{(j-1)},\ s \in S_{(k)} - S_{(k-1)},$$
$$ku \in V_{(j)(k)l} - V_{(j)(k)l+1}\}.$$

(Note that the H and K can be considered as subsets of either B or $\mathfrak{N}$.)

$$f^*_{(j)(k)l}(g) = \sup_{\eta>0}\left|H^\eta_{(j)(k)l}\right|^{-1}\int_{H^\eta_{(j)(k)l}} \left|f(gh^{-1})\right| dh.$$

Lemma 3.5′. Let F be the Poisson integral of f and let

$$f^{**}_\alpha(g) = \sup\{\left|F(g_y)\right| : \mathrm{dist}(y, \partial\Omega) > \alpha|y|\}.$$

Then there exists for every (j) and (k) a decreasing family of neighborhoods of the identity in $U(n)$.

$$U(n) = V_{(j)(k)1} \supset \cdots \supset V_{(j)(k)L} \supset V_{(j)(k)L+1} = \emptyset,$$

where L does not depend on (j) or (k), such that

$$f^{**}_\alpha(g) \leqq A_\alpha \sum_{(j)(k)} 2^{-(|j|/2+|k|)} \sum_{l=1}^{L} f^*_{(j)(k)l}(g). \tag{11}$$

Proof. As in 3.5, we can assume y to be of the form $y = \eta I, \eta > 0$. Now

$$F(g_{\eta I}) = \int_{\mathfrak{N}} f(gh^{-1})\, P_{\eta I}(h)\, dh = \sum_{(j)(k)l} \int_{K^\eta_{(j)(k)l}} \cdots \tag{12}$$

We consider only (j) and (k) as before, i.e., $(j) = (j, 0, \ldots, 0)$, $(k) = (j/2, 0, \ldots, 0)$. (This is the worst possible case, in the sense that it allows for the least amount of what we have called "interaction" between the eigenvalues of the x- and w-components of g.) The proof amounts to finding neighborhoods $V_{(j)(k)l}$ such that the following chain of inequalities is valid:

$$\left|\int_{K^{\eta}_{(j)(k)l}} f(gh^{-1})P_{\eta l}(h)\, dh\right| \leqq \Big[\sup_{h\in K^{\eta}_{(j)(k)l}} P_{\eta l}(h)\Big]\int_{H^{\eta}_{(j)(k)l}} |f(gh^{-1})|\, dh$$
$$\leqq \Big[\sup_{h\in K^{\eta}_{(j)(k)l}} P_{\eta l}(h)\Big]\, |H^{\eta}_{(j)(k)l}|\, f^{*}_{(j)(k)l}(g)$$
$$\leqq A2^{-(|j|/2+|k|)} f^{*}_{(j)(k)l}(g). \tag{13}$$

Referring again to (7) and (8), we see that it is enough to establish (13) in the case $\eta = 1$. Motivated by Lemma 3.7, we will find sequences $\{\rho_{(j)(k)l}\}$ of positive numbers, decreasing in l, and set

$$V_{(j)(k)l} = \exp \mathscr{V}_{(j)(k)l} = \{\exp Z : |Z_{1i}| = |Z_{i1}| \leqq \rho_{(j)(k)l},\ i = 2, \ldots, n\}.$$

Since $(j) = (j, 0, \ldots, 0)$, $(k) = (j/2, 0, \ldots, 0)$, it follows from Lemma 3.3′ and the definition of $\mathscr{V}_{(j)(k)l}$ that

$$|H_{(j)(k)l}| = |C_{(j)}|\,|E_{(k)}|\,|V_{(j)(k)l}| \sim 2^{(3n+m-2)j}(\rho_{(j)(k)l})^{2(n-1)}. \tag{14}$$

On the other hand, it follows from Lemma 3.7 that

$$\sup\nolimits_{h\in K_{(j)(k)l}} P_l(h) \leqq A(2^{2j}[1 + 2^{2j}(\rho_{(j)(k)l+1})^4])^{-(n+m)}. \tag{15}$$

Combining (14) and (15), and writing ρ_l for $\rho_{(j)(k)l}$, we see that the last stage of (13) amounts to

$$(2^{2j}[1 + 2^{2j}\rho_{l+1}^4])^{-(n+m)}2^{(3n+m-2)j}\rho_l^{2(n-1)} \leqq B2^{-j}, \tag{16}$$

where ρ_1 is sufficiently large so that $V_1 = U(n)$, and $\rho_{L+1} = 0$. We are thus reduced to finding $\rho_1 \geqq \rho_2 \geqq \cdots \geqq \rho_L \geqq \rho_{L+1} = 0$, where L is independent of j, such that

$$(2^{4j}\rho_2^4)^{-(n+m)}\, 2^{(3n+m-2)j} \leqq B2^{-j}, \tag{17a}$$

$$(2^{4j}\rho_{l+1}^4)^{-(n+m)}\, 2^{(3n+m-2)j}\, \rho_l^{2(n-1)} \leqq B2^{-j}, \quad l = 2, \ldots, L-1, \tag{17b}$$

$$(2^{2j})^{-(n+m)}\, 2^{(3n+m-2)j}\, \rho_L^{2(n-1)} \leqq B2^{-j}. \tag{17c}$$

The main problem is the nondependence of L on j. Neglecting multiplicative constants, we see that (17a)–(17c) are satisfied if

$$\rho_2 = 2^{-[(n+3m+1)/4(n+m)]j},$$

$$\rho_{l+1} = \rho_l^{[(n-1)/2(n+m)]}\rho_2, \qquad l = 2, \ldots, L-1,$$

$$\rho_L \leqq 2^{-j/2} \leqq 2^{-[(n-m-1)/2(n-1)]j}.$$

The numbers $\delta_l = \log_2 \rho_l$ satisfy the recursion

$$\delta_{l+1} = \alpha\delta_l + \delta_2, \qquad \alpha = \frac{n-1}{2(n+m)}, \qquad \delta_2 = -\frac{n+3m+1}{4(n+m)}j,$$

and are thus the partial sums of the geometric series $\delta_2 + \alpha\delta_2 + \alpha^2\delta_2 + \cdots$, which converges to

$$\frac{\delta_2}{1-\alpha} = -\frac{n+3m+1}{2(n+2m+1)}j < -\frac{j}{2}.$$

Therefore we can indeed choose L, independently of j, so that $\rho_L < 2^{-j/2}$. (And, in general, L can be chosen to be independent of (j) and (k).)

To summarize, we have established (16), and thus (13), and have completed the proof of Lemma 3.5′.

The next step in the proof of Theorem A for the domains $D_{n,m}$ is an analog of Lemma 3.6, that is, a weak-type 1 estimate on the $f^*_{(j)(k)l}$ in terms of $\|f\|_1$. The idea is similar. We will cover the $H_{(j)(k)l}$ with sets ${}^iP_{(j)(k)l}$, $i = 1, \ldots, N_{(j)(k)l}$ that are rectangular with respect to the Euclidean structure on $\mathfrak{N}$ in such a way that suitable variants of the conditions (i′), (ii′), and (iii) in the proof of Lemma 3.6 are satisfied. The main new problem will be the verification of (i′), which was automatic in the tube domain case since $\mathfrak{N}$ was then simply a Euclidean space with vector addition.

Lemma 3.6′. $m(f^*_{(j)(k)l})(s) \leqq A(1 + |j| + |k|)\,\|f\|_1/s$, $s > 0$.

Proof. Note that the η-dilations,

$$g = [x, w] \to g^\eta = [\eta x, \eta^{1/2}w], \qquad \eta > 0,$$

form a multiplicative group of automorphisms of $\mathfrak{N}$; in fact, $[x, w]^\eta[x', w']^\eta = [\eta(x + x' + 2\,\mathrm{Im}\, ww'^*), \eta^{1/2}(w + w')] = ([x, w]\,[x', w'])^\eta$. Moreover, these automorphisms together with the sets $P = H_{(j)(k)l}$ (see Definition 3.8) satisfy the hypotheses of Lemma 2.2. The problem is that in this case M_P is on the order of $2^{\gamma(|j|+|k|)}$, $\gamma \geqq 1$. Exactly as in 3.6, we will overcome this difficulty by covering the $H_{(j)(k)l}$ by sets ${}^iP_{(j)(k)l}$, $i = 1, \ldots, N_{(j)(k)l}$ to which 2.2 also applies, subject to the restrictions

(i″) $M_{i(j)(k)l} = |{}^iP_{(j)(k)l}{}^iP_{(j)(k)l}^{-1}{}^iP_{(j)(k)l}|/|{}^iP_{(j)(k)l}| \leqq A$,

(ii″) $|{}^iP_{(j)(k)l}| \leqq BN_{(j)(k)l}^{-1}|H_{(j)(k)l}|$,

(iii″) $N_{(j)(k)l} \leqq 2^{\alpha(|j|+|k|)}\beta$.

The sets ${}^iP_{(j)(k)l}$ will again be rectangular, in terms of the Euclidean structure on $\mathfrak{N}$ inherited from its identification with Re $V_1 \times V_2$. Also, the ${}^iP_{(j)(k)l}$ will be constructed so as to cover subsets of $H_{(j)(k)l}$ of the form

$$ {}^iH_{(j)(k)l} = \{[k^{-1}d(r)k, u\tilde{d}(s)v] \in H_{(j)(k)l} : k \in U_{\lambda(i)}, \\ u \in U'_{\mu(i)}, v \in U''_{\nu(i)}\}, $$

where the U_λ and U'_μ are appropriate subsets of $U(n)$ and the U''_ν of $U(m)$. As before, the U_λ, U'_μ, U''_ν will be translates of neighborhoods of the origin in their respective groups.

The reasoning of Lemma 3.6 can be used to show that (ii″) and (iii″) can be satisfied simultaneously. The U_λ, U'_μ, U''_ν will be exponential images of spherical neighborhoods of 0 in the appropriate Lie algebras whose radii are multiples of $2^{-|j|}$, $2^{-[|j|+2|k|]}$, $2^{-[|j|+2|k|]}$, respectively. Also, it is not hard to see that the U_λ, U'_μ can be chosen in a way that is consistent with the requirement $ku \in V_{(j)(k)l}$ satisfied by $[k^{-1}d(r)k, u\tilde{d}(s)v] \in H_{(j)(k)l}$.

Let us consider now the condition (i″). Recall the formula for group multiplication in $\mathfrak{N}$:

$$ [x, w]\,[\xi, \omega] = [x + \xi + 2 \operatorname{Im} w\omega^*, w + \omega]. $$

Since the ${}^iP_{(j)(k)l}$ are rectangular, $M_{i(j)(k)l}$ would be 3^ν, $\nu = n^2 + 2mn = \dim \mathfrak{N}$, except for the term Im $w\omega^*$ in the formula. The ${}^iP = {}^iP_{(j)(k)l}$ are Cartesian products of rectangular sets ${}^iR \subset$ Re V_1 and ${}^iQ \subset V_2$. We will show that for $w, \omega \in {}^iQ$, Im $w\omega^*$ lies in a cube whose side is one-quarter the smallest side of ${}^iR$; this guarantees that $M_{i(j)(k)l} \leqq 5^\nu$ and thus finishes the proof of the lemma.

Since Re V_1 consists of the Hermitian symmetric $n \times n$ complex matrices, Im $z = (z - z^*)/2i$. Thus

$$ |\operatorname{Im}(uwv)\,(u\omega v)^*| = |\operatorname{Im} uw\omega^*u^{-1}| = |\operatorname{Im} w\omega^*|, $$

and it is sufficient to prove our assertion about $w, \omega \in {}^iQ$ for $i = 1$. The rectangles ${}^1R$ and ${}^1Q$ determined by the neighborhoods of the identity U_1, U'_1, U''_1 have sides parallel to the axes in their respective spaces. As can be seen by referring to the construction in the proof of 3.6 (the tube domain case), the smallest side of ${}^1R$ is of a length which is a fixed multiple of $2^{-|j|}$. Also, if $w = (w_{ij}) = (w'_{ij} + iw''_{ij})$ is the representation of w as an $n \times m$ matrix and $\rho = c2^{-[|j|+2|k|]}$ is the radius of the balls in $\mathfrak{u}(n)$ and $\mathfrak{u}(m)$ which determine U'_1 and U''_1, and thus ${}^1Q$, we have for $w \in {}^1Q$ that

$$|w'_{ii}| \leqq C2^{|k|}, \qquad |w'_{ij}| \leqq C2^{|k|}\rho, \quad i \neq j, \qquad |w''_{ij}| \leqq C2^{|k|}\rho, \quad \text{all } i, j. \quad (18)$$

(${}^1Q$ is a rectangular set covering the "wedge" in V_2 obtained by the action on diagonal-type matrices $\tilde{d}(s)$, $s \in S_{(k)}$, of $u \in U'_1$, $v \in U''_1$.)

The next observation, while not complicated, is at the heart of our argument for the verification of (i″). It is simply that if w is coordinatized as above, and ω similarly, Im $w\omega^* = \eta_{ij}/2i$, where

$$\eta_{ij} = \sum_{k=1}^{m} (w_{ik}\,\bar{\omega}_{jk} - \omega_{ik}\,\bar{w}_{jk}).$$

In particular, η_{ij} contains no terms of the form $w'_{ii}\,\omega'_{jj}$, so that if $w, \omega \in {}^1Q$, it follows from (18) that

$$|\eta_{ij}| \leqq C(2^{|k|}\rho)2^{|k|} = C'2^{-|j|}.$$

Therefore, after multiplication of the radius ρ by another constant, we can conclude that the term Im $w\,\omega^*$ for $w, \omega \in {}^1Q$ indeed lies in a cube in Re V_1 whose side is one-quarter of the smallest side of ${}^1R$ (which we observed is a multiple of $2^{-|j|}$).

The proof of (i″) is now complete. Since (ii″) and (iii″) have also been verified, we can [by (ii″)] dominate $f^*_{(j)(k)l}$ by $N^{-1}_{(j)(k)l}\sum_{i=1}^{N(j)(k)i}{}^if^*_{(j)(k)l}$, conclude [from (i″) and 2.2] that $m({}^if^*_{(j)(k)l})\,(s) \leqq A\|f\|_1/s$, $s > 0$, and finish the proof of 3.6′, by observing [via (iii″) and 2.1] that $m(f^*_{(j)(k)l})\,(s) \leqq C(1 + |j| + |k|)\,\|f\|_1/s$, $s > 0$.

The remainder of the proof of Theorem A is now easy. The penultimate step consists of combining Lemmas 3.5 and 3.6 (or 3.5′ and 3.6′) to show that the map $f \to f^{**}_\alpha$ is of weak-type 1–1. Finally, given $f \in L^1(\mathfrak{N})$, we write $f = f_1 + f_2$, where f_1 is continuous and $\|f_2\|_1$ is small. Then, since the Poisson integral of a continuous function converges to the function everywhere, we use the weak-type estimate on $(f_2)^{**}_\alpha$ to show that f is the limit of its Poisson integral except on a set of (arbitrarily) small measure. Details of the last two steps are well known. See, e.g., Ref. [*11*, Chapter 17].

We close this section with a few remarks concerning our choice of the domains D_n and $D_{n,m}$ as typical. Any tube domain equivalent to an irreducible bounded symmetric domain can be imbedded in the complexification of a real, compact, simple Jordan algebra in such a way that the Poisson kernel is expressible in terms of the structure of the algebra. The features of D_n that we made use of above, primarily the diagonalizability of symmetric matrices and the existence of square roots of positive definite matrices, have their analogs in the case of the more general Jordan algebras and can be used to prove Theorem A. In fact, the proof we have given for D_n can be followed step by step in the more general situation, once the Jordan algebra machinery has been set up.

Similarly, the proof of Theorem A given for $D_{n,m}$ contains the essential ingredients of the proof for any irreducible non-tube domain equivalent to a bounded symmetric domain. Finally, the proof in the case of non-irreducible domains follows from known techniques which will be discussed in Section 4.

4. The Case $p > 1$

In this section we first prove Theorem A for L^p functions, $p > 1$, and then prove Theorem C. For any $p < \infty$, the continuous functions are dense in L^p, and so it follows from the considerations of Section 3 that to establish Theorem A for such p, it is enough to prove that

$$\|f_\alpha^{**}\|_p \leqq A_{\alpha,p}\|f\|_{p'} \qquad 1 < p < \infty, \tag{19}$$

where, as before,

$$f_\alpha^{**}(g) = \sup\{|F(g_y)| : \operatorname{dist}(y, \partial\Omega) > \alpha|y|\}.$$

Given the proof of Theorem A in the case $p = 1$, Eq. (19) is almost immediate. In fact, since $\int_{\mathfrak{N}} P_y(g)\, dg = 1$ for all $y \in \Omega$,

$$\|F(g_y)\|_\infty = \|P_y * f(g)\|_\infty \leqq \|f\|_\infty.$$

An application of the Marcinkiewicz interpolation theorem [*11*, Chapter 12] to this inequality and the estimate $m(f_\alpha^{**})(s) \leqq A\|f\|_1/s$, $s > 0$, which was established in Section 3, indeed yields (19).

After considering the case $p = \infty$, which is easy, we indicate how (19) can be proved without the complicated estimates necessary in the case $p = 1$. In the case of the tube domains, we give a particularly simple proof due to R. Coifman and G. Weiss. In the non-tube case, the simplified method is the author's.

Suppose now that f is essentially bounded. We show that the Poisson integral of f converges to f almost everywhere on any finite cube Q, which is clearly sufficient. Let $\bar{Q}$ be the cube concentric with Q of twice the side of Q, and write $f = \varphi + \psi$, where $\varphi = f$ on $\bar{Q}$, $\varphi = 0$ off $\bar{Q}$. Then $\varphi \in L^p$, $1 \leqq p \leqq \infty$, and so we have only to show that if Ψ is the Poisson integral of ψ, $\Psi(g_y) \to 0$ for all $g \in Q$ as $y \to 0$ in Ω. But if

$$\delta = \sup\{|h| : Qh^{-1} \subset \bar{Q}\} \qquad \text{and} \qquad g \in Q,$$

$$|\Psi(g_y)| = \left|\int_{|h|>\delta} \psi(gh^{-1})\, P_y(h)\, dh\right| \leqq \|\psi\|_\infty \int_{|h|>\delta} P_y(h)\, dh,$$

and the last integral vanishes as $y \to 0$ in Ω, since P_y is an approximation of the identity [3, 2.4]. So much for the case $p = \infty$.

The technique of Coifman and G. Weiss depends on the study of a general class of convolution operators which act on functions of one real variable. Suppose that $d\mu$ is a Borel measure on the line and that $k(\tau)$ is an even, nonnegative function on the line which is nonincreasing for positive values of τ and is integrable with respect to $d\mu$.

Lemma 4.1. Suppose that $f \in L^p((-\infty, \infty), d\mu)$, $1 \leqq p \leqq \infty$. Then

$$\left|\int_{-\infty}^{\infty} f(\sigma - \tau) k(\tau)\, d\mu(\tau)\right| \leqq \left[\int_{-\infty}^{\infty} k(\tau)\, d\mu(\tau)\right] \bar{f}(\sigma),$$

where

$$\bar{f}(\sigma) = \sup_{\varepsilon > 0} \mu([-\varepsilon, \varepsilon])^{-1} \int_{-\varepsilon}^{\varepsilon} |f(\sigma - \tau)|\, d\mu(\tau).$$

Proof. We assume temporarily that k is of the form $k = \sum_{j=1}^{N} a_j \chi_j$, where $a_j > 0$ and χ_j is the characteristic function of the interval $[-b_j, b_j]$, $0 < b_1 < \cdots < b_N$. For k of this form,

$$\left|\int_{-\infty}^{\infty} f(\sigma - \tau) k(\tau)\, d\mu(\tau)\right| \leqq \sum a_j \int_{-b_j}^{b_j} |f(\sigma - \tau)|\, d\mu(\tau)$$

$$= \sum a_j \mu([-b_j, b_j]) \left\{\mu([-b_j, b_j])^{-1} \int_{-b_j}^{b_j} |f(\sigma - \tau)|\, d\mu(\tau)\right\},$$

which implies the desired inequality, since $\int_{-\infty}^{\infty} k(\tau)\, d\mu(\tau) = \sum a_j \mu([-b_j, b_j])$.

The assumption that k is of the special form above can be dropped easily. In fact, any function satisfying the hypotheses on k can be approximated from below by an increasing sequence of such special functions. The proof in general thus follows from the proof above after an application of the monotone convergence theorem.

Now we turn to the proof of Theorem A, i.e., the inequality (19). The sole property of the Poisson kernel that will be needed is the following: let $t = \tau t'$, $\tau \geqq 0$, $|t'| = 1$, be the ordinary polar decomposition of $t \in B$ (which is Euclidean since D is now assumed to be a tube domain); then for fixed t', $P_I(\tau t')$ is a nonincreasing function of τ.

We verify this property of the Poisson kernel for the domains D_n. The verification in the general case, which makes use of the structure of Jordan algebras, is not very different. Referring to (2), we see that it is enough to show that for fixed $t' \in B$, $|\det(\tau t' + iI)|$ is nondecreasing in τ. Diagonalizing t', this is immediate.

We are now ready to prove (19). First, as usual, we appeal to Lemma 3.1 and deduce that it suffices to show that $\|f^{**}\|_p \leqq A_p\|f\|_p$, where

$$f^{**}(u) = \sup_{\eta > 0} |F(u + i\eta I)|.$$

Letting $\nu = n(n+1)/2 = \dim B$, and $u \in B$,

$$\begin{aligned} F(u + \eta I) &= \int_B f(u - t)\, P_{\eta I}(t)\, dt \\ &= \int_{S^{\nu-1}} \int_0^\infty f(u - \tau t')\, P_{\eta I}(\tau t') \tau^{\nu-1}\, d\tau\, dt' \\ &= \tfrac{1}{2} \int_{S^{\nu-1}} [\int_{-\infty}^\infty f(u - \tau t')\, P_{\eta I}(\tau t') |\tau|^{\nu-1}\, d\tau]\, dt', \end{aligned} \tag{20}$$

where the last equality is a consequence of $P_{\eta I}(t)$ being an even function.

We have just seen that $P_I(\tau t')$ and thus $P_{\eta I}(\tau t')$, $\eta > 0$, is nonincreasing in τ for fixed t'. It is thus possible to apply Lemma 4.1 with $d\mu(\tau) = |\tau|^{\nu-1} d\tau$, and obtain from (20) the inequality

$$|F(u + i\eta I)| \leq \tfrac{1}{2} \int_{S^{\nu-1}} [\int_{-\infty}^\infty P_{\eta I}(\tau t') |\tau|^{\nu-1} d\tau] \bar{f}_{t'}(u)\, dt', \tag{21}$$

where

$$\bar{f}_{t'}(u) = \sup_{\varepsilon > 0} [\int_{-\varepsilon}^{\varepsilon} |\tau|^{\nu-1}\, d\tau]^{-1} \int_{-\varepsilon}^{\varepsilon} |f(u - \tau t')|\, |\tau|^{\nu-1}\, d\tau.$$

The function $\bar{f}_{t'}$ is a weighted maximal average of f over one-dimensional intervals pointed in the direction of t'.

Considering again formula (2), we have

$$P_{\eta I}(\tau t') = \eta^{-\nu} P_I(\tau \eta^{-1} t').$$

The right side of (21) is thus independent of η. Since

$$\tfrac{1}{2}\int_{S^{\nu-1}} \left[\int_{-\infty}^{\infty} P_I(\tau t')|\tau|^{\nu-1}\, d\tau\right] dt' = \int_B P_I(t)\, dt = 1,$$

the desired norm estimate on f^{**} will follow from the Minkowski inequality for integrals and the estimate

$$\|\bar{f}_{t'}\|_p \leqq A_p \|f\|_p, \qquad 1 < p \leqq \infty, \tag{22}$$

where A_p is independent of t'.

To establish (22), we notice first that for any function φ on $(-\infty, \infty)$, if $\bar{\varphi}$ is defined as in 4.1 with $d\mu(\tau) = |\tau|^{\nu-1} d\tau$, then

$$\begin{aligned}\bar{\varphi}(\sigma) &= \sup_{\varepsilon>0} \left[\int_{-\varepsilon}^{\varepsilon} |\tau|^{\nu-1}\, d\tau\right]^{-1} \int_{-\varepsilon}^{\varepsilon} |\varphi(\sigma-\tau)||\tau|^{\nu-1}\, d\tau \\ &\leqq \sup_{\varepsilon>0} \left\{(2\nu^{-1}\varepsilon^{\nu})^{-1}\varepsilon^{\nu-1}\int_{-\varepsilon}^{\varepsilon} |\varphi(\sigma-\tau)|\, d\tau\right\} = \nu\varphi^*(\sigma),\end{aligned}$$

where φ^* is the ordinary Hardy–Littlewood maximal function [*11*, Chapter 1] of φ. The Hardy–Littlewood maximal theorem (of which 2.2 is a generalization) states that $\|\varphi^*\|_p \leqq A_p \|\varphi\|_p$.

Taking $\varphi(\sigma) = f(u + t'\sigma)$ for fixed u, t', we have $\bar{\varphi}(\sigma) = \bar{f}_{t'}(u + t'\sigma)$, and so

$$\begin{aligned}\int_{-\infty}^{\infty} |\bar{f}_{t'}(u+\sigma t')|^p\, d\sigma &= \|\bar{\varphi}\|_p^p \leqq \nu^p \|\varphi^*\|_p^p \\ &\leqq A_p \|\varphi\|_p^p = A_p \int_{-\infty}^{\infty} |f(u+\sigma t')|^p\, d\sigma.\end{aligned}$$

Integrating over the space of all lines parallel to $\{u + \sigma t', -\infty < \sigma < \infty\}$, we obtain (22), and thus (19), for tube domains.

Only a bare sketch will be given of the simplified proof of Theorem A, $p > 1$, in case $D = D_{n,m}$. The domination of f_α^{**} by sums of $f_{(j)(k)l}$ (Lemma 3.5′) is still needed; the proof of (19) is reduced to verification of

$$\|f_{(j)(k)l}^*\|_p \leqq A_p \|f\|_p, \qquad 1 < p < \infty. \tag{23}$$

The method for obtaining (23) is similar to the one just given in the tube case. The place of $\{\tau t' : -\infty < \tau < \infty\}$ is taken by the subgroup

$$\mathfrak{N}_{k,u,v} = \{[k^{-1}d(r)k,\ u\tilde{d}(s)v] : r \in E_n,\ s \in E_m\}$$

of $\mathfrak{N}$, which is isomorphic to $E_n \times E_m$. The place of $S^{\nu-1}$ is taken by $U(n) \times U(n) \times U(m)$; in fact $f^*_{(j)(k)l}$ is dominated by the integral over $U(n) \times U(n) \times U(m)$ of functions $\bar{f}_{(j)(k)l;k,u,v}$ which are maximal averages with respect to isomorphic images of rectangular sets in the cosets of $\mathfrak{N}_{k,u,v}$. Continuing the analogy, the place of the one-dimensional Hardy–Littlewood theorem is taken by its variant for $(n + m)$-dimensional Euclidean space. And the final integration over a space of parallel lines becomes integration over the coset space $\mathfrak{N}/\mathfrak{N}_{k,u,v}$. Details of the verification of (23) are in [*10*].

Proof of Theorem C. The idea of the proof is simply to iterate the L^p estimates proved above. For simplicity, assume that $D = D_1 \times D_2$, $\Omega = \Omega_1 \times \Omega_2$. Let $f \in L^p(B) = L^p(B_1 \times B_2)$, and let F be its Poisson integral. Since $P_{y_1,y_2}(g_1, g_2) = {}^1P_{y_1}(g_1)\, {}^2P_{y_2}(g_2)$, we have

$$f^{**}_{\alpha_1,\alpha_2} = \sup \{ |F((g_1)_{y_1}, (g_2)_{y_2})| : \operatorname{dist}(y_i, \Omega_i) \geqq \alpha_i |y_i| \}$$

$$\leqq \sup_{\ldots y_1 \ldots} \int_{\mathfrak{N}_1} P_{y_1}(h_1) \Big[\sup_{\ldots y_2 \ldots} \int_{\mathfrak{N}_2} P_{y_2}(h_2) \, |f(g_1 h_1^{-1}, g_2 h_2^{-1})| \, dh_2 \Big] dh_1,$$

and so $\|f^{**}_{\alpha_1,\alpha_2}\|_p \leqq A_p \|f\|_p$, which, as we have seen, is sufficient.

The proof of Theorem A for such products resembles the proof for products of upper half-planes. Again assuming $D = D_1 \times D_2$, etc., the proof is first reduced to the case $(y_1,y_2) = (\eta I_1, \eta I_2)$, $\eta > 0$. Then $F((g_1)_{\eta I_1}, (g_2)_{\eta I_2})$ is dominated by sums of maximal averages over sets which are Cartesian products of the sets over which we took maximal averages in the irreducible case.

Theorem C is known to be false for $p = 1$, even in the case when D is the product of classical upper half-planes.

5. The Negative Result

We shall prove Theorem B in the case when $D = D_n$ and $n = 2$. The extension to all tube domains will be indicated; the extension to the non-tube domains will be omitted.

The crux of Theorem B is part (a), which we prove by showing that unrestricted a.e. convergence must fail for some L^∞ function, and, in particular, for the characteristic function of some set. Part (b) follows from part (a) and a general result from functional analysis.

More specifically, the idea of the proof of (a) is to show that unrestricted a.e. convergence of Poisson integrals of characteristic functions of sets implies a kind of differentiation of the indefinite integral of such functions and then to show the differentiation to be impossible.

Let us recall the details of the analysis on $D = D_2$. The boundary B is the space of 2×2 real symmetric matrices, and B is identified with Euclidean 3-space by pairing $\begin{pmatrix} x_1 & x_3 \\ x_3 & x_2 \end{pmatrix}$ with (x_1, x_2, x_3). The cone Ω consists of the positive definite matrices. The Poisson kernel is

$$P_y(x) = c_2[\det y/|\det(x + iy)|^2]^{3/2},$$

and so if we define $y \circ x \in B$ to be $y^{1/2}\, xy^{1/2}$ for $y \in \Omega$, $x \in B$, we have $P_y(y \circ x) = (\det y)^{-3/2} P_I(x)$.
Also, the determinant of the linear transformation $x \to y \circ x$ is $(\det y)^{3/2}$, and so

$$F(x + iy) = P_y * f(x) = \int_B f(x - u)P_y(u)\, du = \int_B f(x - y \circ u)P_I(u)\, du.$$

Since $\int_B P_I(u)\, du = 1$, it follows that

$$F(x + iy) - f(x) = \int_B [f(x - y \circ u) - f(x)]P_I(u)\, du.$$

If f is the characteristic function of a set, the last integrand is of the same sign (or 0) for every value of u. Moreover, the Poisson kernel P_I is bounded below on the cube

$$Q = \{x : |x_i| \leqq 1,\ i = 1, 2, 3\}.$$

It follows that if f is a characteristic function,

$$|F(x + iy) - f(x)| \geqq A|\textstyle\int_Q [f(x - y \circ u) - f(x)]\, du|$$
$$= A'|y \circ Q|^{-1} \left| \int_{y \circ Q} [f(x - u) - f(x)]\, du \right|.$$

Convergence of the Poisson integral of f to f at the point x is thus seen to imply differentiation of the indefinite integral of f with respect to the sets $y \circ Q$ at x. The derivative of the indefinite integral of the characteristic function of a set E with respect to a collection of sets $\mathscr{R}$ is usually called the E-density with respect to $\mathscr{R}$. That is, the E-density at a point x is

$$\lim_{\substack{R \in \mathscr{R} \\ \text{diameter } R \to 0}} |(x + R) \cap E|/|R|,$$

if the limit exists. Using this language, and the observations above, we see that to prove the existence of a function for which a.e. unrestricted convergence of the Poisson integral fails, it is enough to establish

Lemma 5.1. There exists a set $H \subset E_3$ having a subset of positive measure consisting of points which do not have H-density 1 with respect to the collection $\{y \circ Q : y \in \Omega\}$.

Proof. The sets $y \circ Q$ are not hard to analyze. Suppose first that y is diagonal, i.e., $y = d(y_1, y_2)$. Then $y \circ (x_1, x_2, x_3) = (y_1x_1, y_2x_2, (y_1y_2)^{1/2}x_3)$, and so $y \circ Q$ is the rectangular set centered at the origin with sides $2y_1$, $2y_2$, $2(y_1y_2)^{1/2}$. Moreover, if $k \in SO(2)$, $(k^{-1}yk)^{1/2} = k^{-1}y^{1/2}k$; therefore $(k^{-1}yk) \circ x = k^{-1}(y \circ kxk^{-1})k$. The sets $y \circ Q$ are thus essentially (modulo the fact that $k^{-1}Qk \neq Q$) the images under the rotations $x \to k^{-1}xk$ of the rectangles of sides $2y_1$, $2y_2$, $2(y_1y_2)^{1/2}$. (Note that if $k = \begin{pmatrix} \cos\theta & \sin\theta \\ -\sin\theta & \cos\theta \end{pmatrix}$, the map $x \to k^{-1}xk$ is given on 3-space in terms of the coordinates we have chosen as rotation by an angle θ about the line $x_1 = x_2$, $x_3 = 0$.)

Let us consider $\mathscr{R}$ defined as the collection of rectangular sets in E_3, centered at the origin, having one side longer than the other two and their long axes along the boundary $\partial\Omega$ and its negative. (In the case at hand,

$$\begin{aligned}\partial\Omega &= \{x : \operatorname{tr} x \geqq 0, \det x = 0\} \\ &= \{(x_1, x_2, x_3) : x_1x_2 = x_3^2,\ x_1 + x_2 \geqq 0\}.)\end{aligned}$$

It is clear from the discussion above that every member R of $\mathscr{R}$, while perhaps not of the form $y \circ Q$ (since $k^{-1}Qk \neq Q$), can be covered by a set of the form $y \circ Q$ in such a way that $|y \circ Q|/|R| < C$. In particular, if H is any set, H-density 1 at a point x with respect to $\{y \circ Q : y \in \Omega\}$ implies H-density 1 at x with respect to $\mathscr{R}$. In fact,

$$\begin{aligned}|(x + R) - H|/|R| &< [|(x + y \circ Q) - H|/|y \circ Q|]\,[|y \circ Q|/|R|] \\ &< C|(x + y \circ Q) - H|/|y \circ Q|,\end{aligned}$$

where R and $y \circ Q$ are related as above, and the last expression is arbitrarily small once y is small enough. The proof of the lemma is thus reduced to showing the existence of a set H for which not a.e. point has H-density 1 with respect to $\mathscr{R}$.

Suppose to the contrary that a.e. point of every $H \subset E_3$ does have H-

density 1 with respect to $\mathscr{R}$. We show that this implies that a.e. point of every plane set $\tilde{H}$ has $\tilde{H}$-density 1 with respect to the collection $\tilde{\mathscr{R}}$ of all plane rectangles centered at the origin (with sides not necessarily parallel to the axes), which is known to be impossible.

Let, then, $\tilde{H}$ be a plane set, and imbed $\tilde{H}$ in the plane through the origin in 3-space perpendicular to the axis of Ω, i.e., the plane Π determined by the equation $\operatorname{tr} x = x_1 + x_2 = 0$. Let H be the (largest) set between Π and the plane $\operatorname{tr} x = 1$ whose orthogonal projection on Π is $\tilde{H}$. We use our assumption about all $H \subset E_3$ to prove our assertion about the plane set $\tilde{H}$.

Let $\tilde{x}$ be a point of $\tilde{H}$ which is the image under projection on Π of a point $x \in H$ of H-density 1 with respect to $\mathscr{R}$, and note that a.e. $\tilde{x} \in \tilde{H}$ is, by assumption, such a point. We claim that $\tilde{x}$ has $\tilde{H}$-density 1 with respect to $\tilde{\mathscr{R}}$. We would like to proceed as follows: given $\tilde{R} \in \tilde{\mathscr{R}}$, find one of the (nonunique) sets $R \in \mathscr{R}$ such that $x + R$ projects onto $\tilde{x} + \tilde{R}$, and then estimate $|(\tilde{x} + \tilde{R}) \cap \tilde{H}|/|\tilde{R}|$ in terms of $|(x + R) \cap H|/|R|$. These quotients would be equal except that the line segments in R which are the inverse images under projection of the points in $\tilde{R}$ are not all of the same lengths, but get smaller near the ends of R.

To get around this technical difficulty, we introduce the collection $\mathscr{R}'$ consisting of the members of $\mathscr{R}$ with their ends sliced off by planes parallel to the plane $x_1 = x_2$. The effect of this construction is that if $R' \in \mathscr{R}'$, the inverse image in R' under projection of the projection $\bar{x} \in \Pi$ of any $x \in R'$ has length L independent of R'. Moreover, just as H-density 1 with respect to $\{y \circ Q\}$ implied the same with respect to $\mathscr{R}$, H-density 1 with respect to $\mathscr{R}$ implies the same with respect to $\mathscr{R}'$.

Now suppose $\tilde{x} \in \tilde{H}$ is the image under projection of $x \in H$, where x has H-density 1 with respect to $\mathscr{R}'$, and let $\tilde{R} \in \tilde{\mathscr{R}}$. Choosing $R' \in \mathscr{R}'$ such that R' projects onto $\tilde{R}$, it follows from the construction of H and the property above of $\mathscr{R}'$ that

$$\begin{aligned} |(\tilde{x} + \tilde{R}) \cap \tilde{H}|/|\tilde{R}| &= L|(x + R') \cap H|/L|R'| \\ &= |(x + R') \cap H|/|R'|. \end{aligned}$$

Finally, we may always assume that the diameter of R' is no more than twice that of $\tilde{R}$, and so it goes to zero when the latter does, in which case the above quotients tend to 1.

To summarize, if a.e. point of every subset of H of 3-space has H-density 1 with respect to $\{y \circ Q : y \in \Omega\}$ (and thus with respect to $\mathscr{R}$ and $\mathscr{R}'$), the same is true for a.e. point of every plane set with respect to the collection $\tilde{\mathscr{R}}$. This assertion for $\tilde{\mathscr{R}}$, however, is known [*1*, *5*] to be false, and so we have established by contradiction the conclusion of 5.1.

The remainder of the proof of Theorem B presents no problems. First we produce a bounded function f whose Poisson integral has no unrestricted limit at a.e. point. If $\{r_j\}$ is an enumeration of the points of 3-space with rational coordinates, $H_j = H + r_j$, where H is the set whose existence is guaranteed by 5.1, and if f_j is the characteristic function of H_j, $f = \sum_j 2^{-j} f_j$ is the required function. By construction, the Poisson integral F of f fails at a.e. point to converge unrestrictedly to f. Since F converges restrictedly to f at a.e. point, F clearly has no unrestricted limit at a.e. point.

As for part (b) of Theorem B, we note first that the set H of 5.1 can be taken to be bounded, so that its characteristic function provides an example for any $p \geqq 1$ of an L^p function whose Poisson integral does not converge unrestrictedly to it a.e. On the other hand, convergence is uniform for functions which are bounded and continuous. These observations, together with a Banach–Steinhaus-type result [*8*, 4.1(b)], suffice to complete the proof of Theorem B in the case $D = D_n$, $n = 2$.

The proof of the negative result for any other tube domain follows the same pattern. In particular, a.e. unrestricted convergence to f of the Poisson integral of the characteristic function f of a set H implies that a.e. point of H has H-density 1 with respect to a class of rectangular sets having one side arbitrarily longer than all the others and their long axes on the boundary of the cone Ω. Again, the latter assertion leads to a contradiction.

The proof of the theorem for the non-tube domains consists of a reduction to the tube case [*8*].

6. Additional Results

Two results which extend Theorem A are recorded here without their proofs, which may be found in [*8*] and [*10*].

The first additional result concerns a type of convergence (first considered by A. Koranyi) that reduces to nontangential convergence in the case of the tube domains. Suppose that $\mathfrak{N}$ and B are as in Section 1, and let $g_0 \in \mathfrak{N}$, $u_0 = g_0 \cdot 0 \in B$. For $\beta > 0$, we set

$$\Gamma_\beta(u_0) = \{g_0 g \cdot [iy, 0] : g = [x, w], \max\{|x|, |w|^2\} < \beta|y|\}.$$

Convergence of $\zeta \in D$ to $u_0 \in B$ is said to be *admissible* if ζ stays within some $\Gamma_\beta(u_0)$. If $\zeta = u_y$ converges to u_0 admissibly in such a way that dist $(y, \partial\Omega) > \alpha|y|$ for some $\alpha > 0$, the convergence is said to be *restricted admissible.*

Theorem D. Let $f \in L^p(B)$, $1 \leqq p \leqq \infty$, and let F be its Poisson integral. Then for a.e. $u \in B$, $F(\zeta) \to f(u)$ as $\zeta \to u$ restrictedly and admissibly.

Theorem E. Let λ be a finite Borel measure on B and let f be its Radon–Nikodym derivative. Let

$$F(\zeta) = \int_B P(u, \zeta)\, d\lambda(u)$$

be the Poisson integral of λ.

Then for a.e. $u \in B$, $F(\zeta) \to f(u)$ as $\zeta \to u$ restrictedly and admissibly.

References

[1] H. Busemann and W. Feller, "Zur differentiation der Lebesguesehn Integrale," *Fund. Math.*, **22** (1934), 226–256.

[2] B. Jessen, J. Marcinkiewicz, and A. Zygmund, "Note on the differentiability of multiple integrals," *Fund. Math.*, **25** (1935), 217–234.

[3] A. Koranyi, "The Poisson integral for generalized half-planes and bounded symmetric domains, *Ann. of Math.*, (2) **82** (1965), 332–350.

[4] J. Marcinkiewicz and A. Zygmund, "On the summability of double Fourier series," *Fund. Math.*, **32**(1939), 112–132.

[5] O. Nikodym, "Sur les ensembles accessibles," *Fund. Math.*, **10** (1927), 116–168.

[6] E. M. Stein, "Maximal functions and Fatou's theorem," Summer course on Homogeneous Bounded Domains, Cremonese (1967).

[7] E. M. Stein, G. Weiss, and M. Weiss, "H^p classes of holomorphic functions in tube domains," *Proc. Nat. Acad. Sci. U.S.A.*, **52** (1964), 1035–1039.

[8] E. M. Stein and N. J. Weiss, "On the convergence of Poisson integrals," *Trans. Amer. Math. Soc.*, **140** (1969), 35–54.

[9] N. J. Weiss, "Almost everywhere convergence of Poisson integrals on tube domains over cones," *Trans. Amer. Math. Soc.*, **129** (1967), 283–307.

[10] N. J. Weiss, "Convergence of Poisson integrals on generalized upper half-planes," *Trans. Amer. Math. Soc.*, **136** (1969), 109–123.

[11] A. Zygmund, *Trigonometric Series*, Cambridge Univ. Press, New York, (1959).

NEW AND OLD RESULTS IN INVARIANT THEORY WITH APPLICATIONS TO ARITHMETIC GROUPS*

STEPHEN J. RALLIS

Department of Mathematics
State University of New York
Stony Brook, New York

*Supported in part by Army Research Office-Durham grant ARO-D-31-124-G892, S2.

1. New and Old Results in Invariant Theory

In this paper, we are going to study the general problem of a group G operating linearly on a vector space V over $\mathfrak{C}$. The general questions asked in the theory of invariants are the following:

(1) Let $S(V)$ be the polynomial algebra of V, and $I = S(V)^G$ the ring of invariant polynomials under the induced action of G. The first and classical question is the structure of the ring I, i.e., is I finitely generated, and if so, how much does I differ from being a polynomial ring?

(2) What is the G orbit structure in V; how does this relate to the ring I? What are the closed orbits, the principal orbits (orbits of maximum dimension in the case that G is a Lie group), and what is the possible quotient structure, i.e., does the map $V \to V/G$ have a cross section and what is the significance of the cross section?

(3) What is the representation theory of G on $S(V)$? This means we already know the structure of I and thus how $S(V)$ behaves as an I module under multiplication, i.e., what is the cohomological dimension of $S(V)$ (is it free, etc.), and then is there a reasonable complement H which is G stable so that $S(V) = I \otimes H$. Thus the problem reduces to determining the G module structure of H.

1.1 Structure of I

We assume that G is a reductive algebraic group. Then by a theorem of Noether, I is finitely generated. Note that given any subgroup $\Gamma \subseteq \mathrm{Aut}(V)$ we can form its Zariski closure $\overline{\Gamma}$ which is an algebraic group and, in fact, that $S(V)^{\Gamma} = S(V)^{\overline{\Gamma}}$. In addition, the theorem of Weitzenbock states if Γ is any one parameter group in $\mathrm{Aut}(V)$, then I is finitely generated (the point here is that one splits $\overline{\Gamma} = \Gamma_s \cdot \Gamma_u$ with Γ_u a one-dimensional unipotent subgroup or equals 1, Γ_s reductive, and the elements of Γ_s and Γ_u commuting. Then by the classical theorem of Weitzenbock, $S(V)^{\Gamma_u}$ is finitely generated and $I = (S(V)^{\Gamma_u})^{\Gamma_s}$ also has this property since Γ_s is reductive.) One knows that the Fourteenth Problem of Hilbert is not true (by an example of Nagata [*14*]).

This is not our main concern; it is to determine when I is a polynomial ring. The following is known.

(A) G any finite group.

Then I is a polynomial ring $\Leftrightarrow$ G is a finite unitary reflection group, i.e., $G \subseteq U(V)$, and G is generated by unitary reflections (i.e., an element r

which takes the value 1 on a $n - 1$ dimensional subspace V_+ and $\exp(2\pi i/s)$ a primitive root of unity on a one-dimensional complement to V_+; this generalizes the usual orthogonal reflection group). This is a result of Shepard, Todd, and Chevalley. Other facts one has in this case are: (i) that the number of algebraically independent generators (homogeneous) of I is exactly dim V, and although these polynomials are not necessarily unique, their degrees m_i are, and (ii) that the order of G is $O(G) = \prod_{i=1}^{\dim V} m_i$ and if q_i is the number of elements in G with eigenvalue 1 with multiplicity i, then $\sum_{i \geqq 0} q_i t^i = \prod(m_i - 1 + t)$.

(B) G any reductive connected Lie group.

The ring I from the comment above is determined by the maximal compact subgroup $G_u \subseteq G$. The first obvious tool is to determine the Poincaré polynomial $P(I, \lambda) = \sum_{k \geqq 0}(\dim I^{(k)})\lambda^k$ as used by Shepard–Todd in Case A. By the formula of Molien and the classical Weyl character formula, one has the integral formula

$$\int_T \frac{|D(t)|}{\det(\pi(t)\cdot\lambda - I_n)}\, dt = P(I, \lambda)$$

where T is the maximal torus in G_u, π the associated representation of G_u, and $D(t)$ the associated weight function on T. One knows that if I is a polynomial ring, then $P(I, \lambda)$ has the form $\prod[1/(1 - \lambda^{a_i})]$, where the a_i are the degrees of the homogeneous generators. The first problem is to get a reasonable way to compute the integral above.

Example. $G_u = SU(2)$ and π_i, the finite dimensional orthogonal representations of G_u.

(1) $P_1(I, \lambda) = 1$, the trivial representation.

(2) $P_3(I, \lambda) = \dfrac{1}{1 - \lambda^2}$, the adjoint representation of G_u on its Lie algebra $\mathscr{T}$.

(3) $P_5(I, \lambda) = \dfrac{1}{(1 - \lambda^2)(1 - \lambda^3)}$, the representation of G_u on the orthocomplement with respect to the Killing form in $SU(3)$ (a special case of K on p is shown in [*11*]).

(4) $P_7(I, \lambda) = \dfrac{1 + \lambda^{15}}{(1 - \lambda^2)(1 - \lambda^4)(1 - \lambda^6)(1 - \lambda^{10})}$, hence this is not a polynomial ring while cases (1)–(3) are.

(These computations are carried out in [*8*].)

Conversely, we ask from purely local ring theory: if $P(I, \lambda)$ has the form $\prod[1/(1 - \lambda^{a_i})]$ then is I a polynomial ring. Also recent work of Igusa on the Siegel formula has given other examples of I as a polynomial ring (however this is case by case checking).

We would also like to mention that there is a connection between Cases (A) and (B). Namely if $\mathcal{T}$ is the Lie algebra of G, and π is the adjoint representation, then one has the following isomorphism:

$$S(\mathcal{T})^G \to S(\mathcal{H})^W$$

where $\mathcal{H}$ is a Cartan subalgebra in $\mathcal{T}$ and W is the Weyl group of G. (The map here is just the restriction of functions.)

1.2 Orbit Structure

We now look at the question of the orbit structure of G in V. The motivation of this study comes from the following case.

(A) If G is any finite group and V is any affine variety on which G acts regularly (in the sense of algebraic geometry), then V/G (the space of orbits) is an affine variety and the associated affine ring is I, i.e., $\mathrm{Spec}(I) \approx V/G$. The correspondence here is between the orbits of G in V and the maximal ideals of the ring I. This leads to the following generalization in Case (B).

(B) If G is reductive connected and algebraic, then we have a map

$$\{\text{set of closed orbits in } V\} \leftrightarrow \mathrm{Spec}(I) \qquad (V, \quad \text{vector space}).$$

The map is given by the restriction of functions in I to a closed orbit. The map is injective by the well-known principle that distinct closed orbits of reductive groups can be separated by invariant polynomials (see [*13*]). It is instructive to see why the map is surjective. We start with a maximal ideal $\mathcal{M}$ in I and then form $\mathcal{M} \cdot S(V)$. By projecting onto I, we can easily see that $\mathcal{M} \cdot S(V) \neq S(V)$. This means that there is a maximal ideal $\mathcal{T}$ in $S(V)$ containing $\mathcal{M} \cdot S(V)$. However every maximal ideal of $S(V)$ is determined by point evaluation in V. Thus there is a point x in V such that the homomorphism φ_x determined by the point has the property that Kernel(φ_x) $\supset \mathcal{M} \cdot S(V)$. Then by the principle above one knows that there is a unique closed orbit $G \cdot y$ in the "closure $G \cdot x$" (for orbits, the Zariski closure is the same as the Euclidean closure). Thus the homomorphism φ_y determined by y has the same property as φ_x.

What we have done here is essentially to define an algebraic morphism ψ from V to Spec(I), where we associate to every element x in V the unique closed orbit associated to it as defined above. We enumerate several important properties to the map ψ.

(i) It is not surjective in general but the image is Zariski dense.

(ii) If G is semisimple, then there exists a G stable open dense subset $V' \subseteq V$ such that the map ψ' induced on quotient $V'/G \to \mathrm{Spec}(I)$ is finite to one (V'/G, space of G orbits in V').

The proof of (ii) is based on two important facts in Invariant Theory that must be mentioned. First we note that the codimension in V of an orbit of maximal dimension is equal to the transcendence degree of the field $Q(V)^G$ of rational invariants, i.e., $Q(V)$ is the field of rational functions on the vector space V and $Q(V)^G$ is the set of G invariants in $Q(V)$. (For this fact, G need not be semisimple.) And second we note that if G is semisimple, then every element of $Q(V)^G$ has the form a/b where $a, b \in I$ (also I is a unique factorization domain).

From above, we see it is important to study the closed orbits in the G action on V. With the same assumptions as in Case (B), we note that if G^x is the isotropy subgroup of the point x, then the induced map $G/G^x \to G\cdot x$ is a biregular map of algebraic varieties. Then if the orbit $G\cdot x = O_x$ is closed, it is affine, and hence the homogeneous space G/G^x is an affine variety. Then it follows from [*2*] that G^x is a reductive subgroup of G. The converse of this is not true; however if G is any connected algebraic group and H is a closed reductive subgroup, then the space G/H is affine (this means that there exists a rational representation $\pi : G \to \mathrm{Aut}(V)$ with a point $x \in V$ such that the isotropy group of the point x is H and the G orbit of x is closed). We note that for G unipotent, the G orbit of any point is always closed. However one has a criterion for closed orbits in Case (B) above. That is, if A is a maximal torus in G, then the set of A fixed points in V has the property that every G orbit determined by a point in this set is closed (use Iwasawa decomposition of G, see [*9*]). Aside from special examples, there seems to be no better way to characterize closed orbits. We close this section by mentioning the adjoint representation of G on its Lie algebra $\mathscr{T}$. Here it is known that a point x lies in a closed orbit if and only if the operator $\mathrm{ad}(x)$ is semisimple (diagonalizable).

In (ii) above we comment that the set V' can be chosen so that all the orbits are principal, i.e., are of maximal dimension. In fact if $\mathscr{P}$ is the set of all principal elements, then it is open and dense in V. As in [*9*] and [*11*], one wants necessary and sufficient conditions to characterize $\mathscr{P}$. For instance, a first "invariant" is the condition of the linear independence of the differentials $df_1, \cdots, df_l$ at $x \in \mathscr{P}$ where the f_i are a maximal set of algebraically independent elements of the ring I. Then we comment upon the relation between the dimension of the vector space and the dimension of the principal orbit. In [*1*], the notion of index of a representation is used; the index i_π is the unique nonzero scalar which is the ratio of two trace forms $i_\pi = \mathrm{Tr}_V(\pi(x)\pi(y))/\mathrm{Tr}_{\mathscr{T}}(\mathrm{ad}\ x\ \mathrm{ad}\ y)$ for all $x, y \in \mathscr{T}$, the Lie algebra of G. Then one has $i_\pi > 1 \Leftrightarrow$ dim. of max. orbit $= \dim G \Leftrightarrow \dim V > \dim G$ (see [*1*]; here G is assumed to be simple and π irreducible).

The point in (ii) above is to see what type of "algebraic geometric" structure can be put on the set $\mathscr{P}$ (see [*13*]). For instance, one idea used in [*11*] is to find a larger group G' containing G with the property that it operates transitively on the principal orbits lying above a fixed closed orbit (i.e., the role of the group K_θ as opposed to K in [*11*]).

In this direction, one asks more generally: what structure does the whole orbit space V/G have? For instance, does V/G have a geometric quotient structure in the sense of Mumford (see [*13*, p. 4])? Here one knows that if V is any affine variety and if G is reductive, then V/G has a geometric quotient if and only if all the orbits of G are closed. Thus it is easy to deduce that if a geometric quotient exists, then all the G orbits have the same dimension. Hence the right candidate for this set is $\mathscr{P}$ defined above. However $\mathscr{P}$ is not an affine set so the criterion above does not apply. But at worst, one arrives at the problem of finding a smooth cross section for the set $\mathscr{P}$, that is, an algebraic set S such that every G orbit in $\mathscr{P}$ meets S exactly once. For instance, in the classical case of the adjoint representation of SL_n (the $n \times n$ matrices of determinant 1) on its Lie algebra (the $n \times n$ matrices of trace zero), one knows that $\mathscr{P} = \{X \mid \text{characteristic polynomial of } X = \text{minimal polynomial of } X\}$ and that the set of companion matrices

$$X = \left\{ \begin{pmatrix} 0 & 1 & & 0 \\ & & 0 & 1 \\ x_1 & \cdots & x_{n-1} & 0 \end{pmatrix} \;\middle|\; x_i \in \mathbb{C} \right\}$$

serves as a cross section for the set of principal orbits. The main point is that the map

$$S(\mathscr{T})^G \to S(X) \qquad \text{(restriction)}$$

is an algebra isomorphism where $S(X)$ is the affine algebra of the "hyperplane" X; thus, in fact, we have linearized the generators of the ring of invariants.

1.3 Representation Theory

We are interested in the question of the representation theory of G on the polynomial ring $S(V)$. As above, we examine Case (A) first.

(A) If G is a finite orthogonal reflection group, then by Chevalley, $S(V)$ is a free I module under multiplication; in particular, if we let $D(V)$ be the ring of constant coefficient differential operators on V, then we have a nonsingular pairing between $D(V)$ and $S(V)$ defined by $\langle \partial, f \rangle = \partial(f)(0)$ for $\partial \in D(V)$ and $f \in S(V)$. Also we know that G has a linear action on the space $D(V)$ as algebra automorphisms. Then let I_* be the G invariants and I_*^+ the ideal of I_* with zero constant term. Define $H = \{f \in S(V) | \langle \partial, f \rangle = 0$ for all $\partial \in I_*^+ D(V)\}$, called the space of harmonic polynomials (generalizing the usual notion). Then H is G stable, and the Separation of Variables Theorem says that $I \otimes H \to S(V)$ under the multiplication is an isomorphism. In fact, one knows that this statement is an alternative characterization of finite orthogonal reflection groups (see [*4*]). Thus the problem of finding the G module structure of $S(V)$ reduces to that of finding the G module structure of H. However, also by Chevalley, H forms a linear base of the quotient field $Q(V)$ of $S(V)$ over the subfield $Q(V)^G$ of G invariants in $Q(V)$ (one knows that every element of $Q(V)^G$ has the form a/b where $a, b \in I$). Then by elementary Galois theory, $Q(V)^G[G]$ and $Q(V)^G \otimes H$ are equivalent G modules over the field $Q(V)^G$. Hence it is easy to deduce that H is isomorphic to $\mathfrak{C}[G]$ with the left regular representation of G. Hence H affords the regular representation of G. Thus the problem of determining the multiplicity of a fixed irreducible representation λ of G reduces to that of the regular representation which is just dim V_λ, where V_λ is the space corresponding to the representation λ. However the problem at hand is to determine the degrees in which λ occurs in H, i.e., if the isotypic component is denoted by H^λ, then for what k do we have $S^k \cap H^\lambda \neq (0)$, where S^k denotes those polynomials homogeneous of degree k?

We give an example of what can be done in this direction. First we assume that G operates irreducibly on the vector space V (this is equivalent to the associated root diagram being connected, see [*4*, p. 127]). Then we form the exterior algebra $\Lambda V = \bigoplus_{k=0}^n \Lambda^k V$ where $\Lambda^k V$ is the kth exterior power. Then G acts irreducibly on $\Lambda^k V$, and if $k \neq j$, then $\Lambda^k V$ and $\Lambda^j V$ afford inequivalent representations of G. Now we form the Jacobian of the basic invariants:

$$\frac{\partial(u_1, \ldots, u_n)}{\partial(x_1, \ldots, x_n)} = \det \begin{vmatrix} \partial_{x_1}(u_1) \cdots \partial_{x_n}(u_1) \\ \vdots \\ \partial_{x_1}(u_n) \cdots \partial_{x_n}(u_n) \end{vmatrix}$$

The representation V of $\Lambda^k V$ described above has multiplicity $\binom{n}{k}$ in H.

To find these spaces in H, we start by taking any increasing sequence $1 \leqq i_1 < i_2 < \cdots < i_k \leqq n$. Then look at rows $i_1, i_2, \ldots, i_k$ in the matrix above and form the linear space spanned by all $k \times k$ minors of this new matrix. Now we take the projection of this space on H; it is easy to see that this space transforms according to the $\Lambda^k V$ representation and that the degree of homogeneity is $(\sum m_{i_j}) - k$ where the m_i are the degrees of the basic invariants (see Section 1.1). The representations of G given this way do not exhaust all the irreducible representations and the problem still remains for the general representation (see [*4*] and [*16*] where the solution as above is given).

To return to Case (B), we note that as above we can again form the space H of harmonics and ask for a separation of variables theorem. It will not be the case that the space H is finite dimensional; the analog in this case is that an isotypic component H^λ is a finite module over I under multiplication. We have a criterion for such a theorem under a certain restriction; we assume that G leaves invariant a nonsingular symmetric bilinear form B on V. For notation, we let I^+ be the ideal generated by invariants with zero constant term. It is straightforward to show that $S(V) = I^+ \oplus H$ and that the multiplication map $I \otimes H \rightarrow S(V)$ is a surjection (see [*7*] and [*9*]). The problem is to show it is an injection. As in [*9*] and [*11*] sufficient conditions are:

(i) I^+ is a radical ideal, and

(ii) there exists an algebraic group G' such that $\mathrm{Aut}(V) \supseteq G' \supseteq G$, $S(V)^{G'} = S(V)^G = I$, and G' has an open dense orbit in $V(I^+) = \{x \mid f(x) = 0 \text{ for all } f \in I^+\}$, the zero fiber (recall G connected in assumptions above).

Thus if (i) and (ii) are satisfied, then $S(V)^\lambda \cong I \otimes H^\lambda$ where λ is the class of an irreducible representation (finite dimensional) of G. Examples showing (i) and (ii) satisfied are given in [*11*]. Another example is $G' = O(B)$ ($G = SO(B)$), the standard complex orthogonal group of the form $B = \sum z_i^2$ (the subgroup of G' of elements of determinant $+1$) and $V = \mathfrak{C}^n \oplus \cdots \oplus \mathfrak{C}^n$ (the standard representation taken k times with $k \leqq [n/2]$). Here $V(I^+)$ is the set of k-tuples $(\xi_1, \ldots, \xi_k)$ such that the subspace of V generated by $(\xi_1, \ldots, \xi_k)$ is totally isotropic with respect to B.

As in Case (A), the problem thus reduces to determining the multiplicity of λ in H. It is easy to see that the map

$$H \rightarrow S(O_x) \qquad \text{(restriction)}$$

is a G module surjection where O_x is any orbit and $S(O_x)$, the restriction of $S(V)$ to O_x. If $R(O_x)$ is the ring of regular algebraic functions on the orbit O_x and if O_x is a closed orbit (hence an affine variety), then one has

$$R(G/G^x) = R(O_x) = S(O_x)$$

(see [9]). Then the map $H \to R(G/G^x)$ is an isomorphism if (i^*) O_x is a closed orbit, (ii*) if φ is the homomorphism of I determined by O_x, then the ideal in $S(V)$ generated by Kernel(φ) is radical, and (iii*) $O_x = \{y \in V \mid f(y) = 0$ for all $f \in \text{Kernel}(\varphi) \cdot S(V)\}$, "$\varphi$ fiber" (we assume here that $S(V) = I \otimes H$ which implies that $S(V) = (\text{Kernel}(\varphi) \cdot S(V)) \oplus H)$. Also note that (iii*) implies that the orbit O_x is principal.

Thus H is equivalent to the G module $R(G/G^x)$ under left translation. It follows from a simple analog to Frobenius reciprocity that the multiplicity of λ in H is $\dim(V_\lambda)^{G^x}$ (fixed point set of G^x in V_λ). The problem of determining the degree of homogeneity of the component λ can be solved in special cases (see [9] and [10]).

From the criterion above, it appears that the zero fiber must be studied very carefully. For this, we have the following useful test (Hilbert–Mumford Criterion): if G is reductive algebraic and $V(I^+) = \{x \mid f(x) = 0$ for all $f \in I^+\}$, "zero fiber," then a point $x \in V$ belongs to $V(I^+)$ if and only if there exists a one-parameter subgroup of G (a homomorphism χ of $\mathfrak{C}^*$ into G in the category of algebraic groups) such that 0 lies in the closure of $\chi(\lambda)(x)$, $\lambda \in \mathfrak{C}^*$. In fact, the last part of the statement can be replaced by the statement that there exists $X \in \mathscr{T}$, the Lie algebra of G, diagonalizable with real eigenvalues such that 0 lies in the Euclidean closure of $\exp(\lambda X)(x)$, $(\lambda \in \mathfrak{C})$. We observe that in the adjoint case (G on $\mathscr{T}$), the Jacobson–Morosov Theorem is essentially a reformulation of this criterion: an element $x \in \mathscr{T}$ belongs to the zero fiber if and only if ad(x) operating on $[\mathscr{T}, \mathscr{T}]$ is nilpotent (if and only if there exists an element $y \in \mathscr{T}$ so that $\text{ad}(y)(x) = x$; then to apply the criterion above one uses the one-parameter group $\exp(\lambda\, \text{ad}(y))$, $\lambda \in \mathfrak{C}$).

2. Applications to Arithmetic Groups

One important application of the theory of invariants is to arithmetic groups. One proves the Godemont Conjecture by very simple tools in invariant theory.* As motivation, we give the following example.

*The proof given here follows that in [2].

We let A_Q be a quaternion algebra over Q, the field of rational numbers. Then A_Q is a four dimensional algebra over Q, i.e., A_Q is defined by two nonzero rational numbers a, b. We can choose a linear base of A_Q as 1, x_1, x_2, x_3 where the multiplication table becomes:

	1	x_1	x_2	x_3
1	1	x_1	x_2	x_3
x_1	x_1	$a\cdot 1$	x_3	ax_2
x_2	x_2	$-x_3$	$b\cdot 1$	$-bx_1$
x_3	x_3	$-ax_2$	bx_1	$-ab\cdot 1$

Then the algebra A_Q has a conjugation operation $\overline{}$ given by $\overline{(\xi_0\cdot 1 + \xi_1 x_1 + \xi_2 x_2 + \xi_3 x_3)} = \xi_0\cdot 1 - \xi_1\cdot x_1 - \xi_2\cdot x_2 - \xi_3 x_3$. $\overline{}$ is a linear map and an antiautomorphism of the algebra structure, i.e., $\overline{(x + y)} = \bar{x} + \bar{y}$, $\overline{(xy)} = \bar{y}\bar{x}$. The norm function on A_Q is $N(x) = x\cdot\bar{x}$. Then $N(x) = \xi_0^2 - \xi_1^2 a - \xi_2^2 b + \xi_3^2 ab$; this means that N defines a quadratic form F on A_Q. The criterion for A_Q to be a division algebra is then easy to state: (i) A_Q is a division algebra $\Leftrightarrow$, (ii) if $N(x) = 0$, then $x = 0$ (the quadratic form F defined by N has no nonzero isotropic vectors)$\Leftrightarrow$, (iii) a is not a norm in $Q(\sqrt{b})$, i.e., a cannot be expressed as $x^2 - by^2$ with $x, y \in Q \Leftrightarrow$, (iv) the form F restricted to the subspace spanned by x_1, x_2, x_3 does not have a nonzero isotropic vector (see [*15*, p. 146]). We now observe that if K is any field containing Q, then the tensor product $A_K = K \otimes_Q A_Q$ is a quaternion algebra over K with the same multiplication table, conjugation operation, norm function, and criterion to be a division ring (where Q is replaced by the field K). For instance, if $K = \mathfrak{C}$, the complex numbers, then $A_{\mathfrak{C}}$ is isomorphic to $M_2(\mathfrak{C})$, the 2×2 complex matrices, where multiplication in $A_{\mathfrak{C}}$ becomes the usual multiplication of matrices (this result comes from Wedderburn's Theorem). However one knows that every element of $M_2(\mathfrak{C})$ has a Jordan decomposition, $X = X_s + X_n$, X_s and X_n its semisimple and nilpotent parts. It is easy to check that if $X \in A_Q$, then X_s and $X_n \in A_Q$.

We let G be the group of invertible elements in $A_{\mathfrak{C}}$; let $G' = \{x \in G \mid N(x) = 1\} = \{x \in M_2(\mathfrak{C}) \mid \text{Det}\,(x) = 1\}$. Then G' is an algebraic group and operates faithfully on $A_{\mathfrak{C}}$ by left multiplication. Let A_Q be as above and form $G'_Q = \{x \in G' \mid x(A_Q) \subseteq A_Q\} = \{x \in A_Q \mid N(x) = 1\}$. Then we choose a lattice $L \subseteq A_Q$, a $\mathfrak{Z}$-module of maximal rank in A_Q (for instance, L can be all elements of the form $\xi_0\cdot 1 + \xi_1 x_1 + \xi_2 x_2 + \xi_3 x_3$ where $\xi_0, \xi_1, \xi_2, \xi_3 \in \mathfrak{Z}$). Then let $G_L = \{g \in G_Q \mid g(L) \subseteq L\}$.

In the case that $K = \mathfrak{R}$, the real numbers, then $A_{\mathfrak{R}}$ is isomorphic to either $M_2(\mathfrak{R})$, the 2×2 real matrices, or the unique division quaternion algebra D over $\mathfrak{R}$ (by (iii) it follows that $A_{\mathfrak{R}} = D$ only when $a < 0$ and $b < 0$). Also we can form $G'_{\mathfrak{R}} = \{g \in G' | g(A_{\mathfrak{R}}) \subseteq A_{\mathfrak{R}}\}$ as above.

Then the Godement Conjecture is formulated as follows: (i) A_Q is a division algebra $\Leftrightarrow$ (ii) G'_Q consists only of semisimple elements $\Leftrightarrow$ (iii) the homogeneous quotient $G'_{\mathfrak{R}}/G'_L$ is compact.

To begin the proof, we note that the important feature is not the algebra structure of A_Q but rather the presence of the quadratic form F. For instance, if A_Q is a division algebra, then F does not represent zero rationally, i.e., it has no nonzero isotropic vectors in A_Q. It would then suffice to prove that $O(F)_Q$, the rational points of the orthogonal group $O(F)$ of the form F, has only semisimple elements (since it is clear that $G' \subseteq O(F)$). Conversely if A_Q is not a division algebra, then there exist three rational numbers ξ_1, ξ_2, ξ_3, not all zero, such that $-\xi_1^2 a - \xi_2^2 b + \xi_3^2 ab = 0$. Then the element $x = 1 + \xi_1 x_1 + \xi_2 x_2 + \xi_3 x_3$ operating in A_Q by multiplication has characteristic polynomial $(\lambda - 1)^4$; this implies that x is unipotent or that $1 - x$ is nilpotent.

Thus we can try to prove a more general version of the Godement Conjecture. We assume that F is a nondegenerate quadratic form defined over V_Q, the rational points of a complex vector space V, i.e., $V = V_Q \otimes_Q \mathfrak{C}$. In case $\dim V = 2$, the group $O(F)$ is a torus (a one-dimensional algebraic group of semisimple elements) and if $\dim V \geqq 3$, then $O(F)$ is semisimple. As above we let L be a lattice in V_Q and $\Gamma = \{g \in O(F)_Q | g(L) \subseteq L\}$. Then in case $\dim V \geqq 3$, we want the equivalence of the following: (i) F has no nonzero isotropic vectors in V, (ii) $O(F)_Q$ has only semisimple elements, (iii) the quotient $O(F)_{\mathfrak{R}}/\Gamma$ is compact. A suitable modification in the general version of the Godement Conjecture is introduced to take care of the case when $\dim V = 2$.

It is instructive to first prove the equivalence of (i) and (ii). If $O(F)_Q$ has a unipotent element g, then g has a nonzero fixed vector $v \in V_Q$. Since F is nondegenerate and assuming V is nonisotropic, then $V = v \oplus v^\perp$ where $v^\perp$ is the orthogonal complement of (v) in V. However $v^\perp$ is stable under g and so by the same reasoning as above we find that g must be identity operator. Conversely if F has a nonzero isotropic vector v, it follows that there exists an isotropic vector $w \neq 0$ such that the space spanned by (v, w) is a hyperbolic plane $\mathscr{H}$, i.e., $\mathscr{H} = v \oplus w$, $F(v, w) = 1$. Since $\dim V \geqq 3$, the operator T which maps any $z \in \mathscr{H}^\perp$ to $z + w$ and is identity on $\mathscr{H}$ is clearly unipotent and belongs to $O(F)_Q$.

Thus we are at the point of showing the equivalence of (ii) and (iii). However it is easier now to state the general form of the conjecture and give a proof using invariant theory. First we must recall some definitions.

We are given an algebraic group $G \subseteq \mathrm{Aut}(V)$ defined over Q. We let L be a lattice in V_Q and G_L be the subgroup of G_Q where $g \in G_L$ if and only if $g(L) \subseteq L$. Then a subgroup Γ of G_Q is called *arithmetic* if Γ is commensurable with G_L ($\Gamma \cap G_L$ has finite index in Γ and G_L). This definition is not dependent on the choice of L or the embedding (invariance under Q-isomorphisms of G) of G in $\mathrm{Aut}(V)$. We note that commensurability is an equivalence relation.

We now recall that every algebraic group G is the semidirect product of a reductive group H and a normal connected unipotent group N. In fact if G is defined over k (any field of characteristic zero), then H and N have the same property. The group N is uniquely determined; it is the unipotent radical of G, i.e., the largest normal connected subgroup of G formed of unipotent elements. We also note that every reductive subgroup of G is conjugate by an element of N to a subgroup of H. And if G^0 denotes the identity component of G, one has that G is connected if and only if H is connected, i.e., $G^0 = H^0 \cdot N$.

We also recall that a character of the algebraic group is a homomorphism of G into $\mathfrak{C}^*$, the multiplicative complex numbers (homomorphism in the category of algebraic groups). Let $X(G)$ be the Abelian group of characters of G and $X(G)_Q$ the subgroup of those characters defined over Q; the rational numbers.

Then the version of the Godement Conjecture which we will prove is the following: (i) *the homogeneous space $G_{\mathfrak{R}}/\Gamma$ is compact if and only if* (ii) $X(G^0)_Q = 1$ *and all the elements of G_Q which are unipotent belong to the unipotent radical N.*

We note in passing here that when the last part of (ii) is omitted, then we get the following statement: (i) $G_{\mathfrak{R}}/\Gamma$ *has finite volume if and only if* (ii) $X(G^0)_Q = 1$ (see [2]).

We let $\mathscr{R}$ be the set of all lattices in $V_{\mathfrak{R}}$. If we fix the lattice L as the standard lattice, then we have a choice of base of V given by L. This base also determines the $n \times n$ complex matrices of nonzero determinant, $GL_n(\mathfrak{C})$ so that $\mathrm{Aut}(V)$ is Q-isomorphic to $GL_n(\mathfrak{C})$. Then the set $\mathscr{R}$ is a single orbit under $GL_n(\mathfrak{R})$; the isotropy subgroup of L is $GL_n(\mathfrak{Z})$, the $n \times n$ matrices with integral coefficients and determinant ± 1. Then $GL_n(\mathfrak{Z})$ is an arithmetic group and hence discrete; if we form the quotient, then it clearly has the structure of a locally compact Hausdorff space. It is this topology we put on the set $\mathscr{R}$ under the identification map $GL_n(\mathfrak{R})/GL_n(\mathfrak{Z}) \to \mathscr{R}$. Then from above the Lie group $G_{\mathfrak{R}} = G \cap GL_n(\mathfrak{R})$ acts on the space $\mathscr{R}$. As above, we want information about the homogeneous space $G_{\mathfrak{R}}/\Gamma$. However it is easy to see that $G_{\mathfrak{R}}/\Gamma$ is compact if and only if $G_{\mathfrak{R}}/G_{\mathfrak{Z}}$ is compact where $G_{\mathfrak{Z}}$ is the isotropy subgroup of the lattice L ($G_{\mathfrak{Z}} = G \cap GL_n(\mathfrak{Z})$). If we know that the $G_{\mathfrak{R}}$ orbit of L is closed in $\mathscr{R}$, then the map $G_{\mathfrak{R}}/G_{\mathfrak{Z}} \to G_{\mathfrak{R}}(L)$ is a homeomorphism where the orbit $G_{\mathfrak{R}}(L)$ carries the induced topology of $\mathscr{R}$ (this statement follows from a result of Arens cited in [3]: if H is a locally compact group operating continuously on a locally compact space M with H countable at infinity and if the H orbit of a point v is closed in M, then the map $H/H^v \to H(v)$ is a homeomorphism where $H(v)$ carries the induced topology of M). Thus assuming that the orbit is closed to show that $G_{\mathfrak{R}}/G_{\mathfrak{Z}}$ is compact, it suffices to show that $G_{\mathfrak{R}}(L)$ is a compact set in $\mathscr{R}$.

Now the Mahler criterion of compactness of subsets in $\mathscr{R}$ can be applied: *let $\mathscr{M} \subseteq \mathscr{R}$, then $\mathscr{M}$ is relatively compact in $\mathscr{R}$ if and only if $|det(g)|$ is bounded for all $g \in GL_n(\mathfrak{R})$ such that $g(L) \in \mathscr{M}$ and there exists a neighborhood U of the origin in $V_{\mathfrak{R}}$ such that $U \cap L' = (0)$ for all $L' \in \mathscr{M}$.*

We must interpret this statement in terms of invariant theory. We observe first that since the algebraic group G is defined over Q, then the ring I has a Q form, i.e., a subspace I_Q so that $I = I_Q \otimes \mathfrak{C}$ (in fact if $S^k(V)$ is the space of homogeneous polynomials of degree k and I, the G invariants, then $I^k = S^{k'}(V) \cap I$ has a Q form). And since a base of V can be chosen from the lattice L, then for a given P in I_Q there exists a number $c > 0$ so that $|P(\xi)| \geqq c$ for all $\xi \in L$ so that $P(\xi) \neq 0$. Hence it follows that if $|\det(g)|$ is bounded for all $g \in G_{\mathfrak{R}}$ and $V(I^+) \cap L = 0$ where $V(I^+)$ is the zero fiber defined in Section 1, then $G_{\mathfrak{R}}/G_{\mathfrak{Z}}$ is compact (assuming that $G_{\mathfrak{R}}(L)$ is closed in $\mathscr{R}$). The point of the proof of this fact is that if $g_j \in G_{\mathfrak{R}}$, $v_j \in L$ ($j \in \mathfrak{Z}$) so that $g_j(v_j) \to 0$ as $j \to \infty$, then for j sufficiently large the generators F_i, $i = 1, \ldots, k$, of the ideal determined by I^+ vanish, i.e., $F_i(v_j) = 0, j \gg 0$. Conversely if G is reductive, it is an easy exercise to show that compactness of $G_{\mathfrak{R}}/G_{\mathfrak{Z}}$ implies that $V(I^+) \cap L = 0$ (use Hilbert-Mumford criterion).

With this machinery, we can now begin the proof of the Godement Conjecture. The proof has three steps; first we reduce to the reductive case, then prove the statement in this case, and finally we clear up loose ends about closed orbits in the space $\mathscr{R}$.

Recalling the decomposition of G above, we note that there exist arithmetic subgroups Γ_1, Γ_2 of H, N, respectively, such that $G_{\mathfrak{Z}}$ is commensurable to $\Gamma_1 \cdot \Gamma_2$. Then if $N_{\mathfrak{R}}/\Gamma_2$ is compact, $H_{\mathfrak{R}}/\Gamma_1$ is compact if and only if $G_{\mathfrak{R}}/G_{\mathfrak{Z}}$ is compact. And since every character of a unipotent group is trivial, then $X(G^0) = X(H^0)$. And since $G_Q = H_Q \cdot N_Q$, then all the elements of H_Q are semisimple if and only if every unipotent element of G_Q is in the radical of G_Q, i.e., belongs to N_Q [one uses here the fact that $x \in G_Q$ if and only if x_s and x_u (the semisimple and unipotent components) $\in G_Q$].

We comment here that an important point in Mahler's criterion is that $G \subseteq \mathrm{Aut}(V)$, i.e., the group G has a faithful matrix realization in $\mathrm{Aut}(V)$. Then the compactness of the homogeneous space $N_{\mathfrak{R}}/\Gamma_2$ follows from comments above. In fact, we will prove a stronger result: let U be unipotent and $\pi : U \to \mathrm{Aut}(W)$ any rational representation of U defined over Q, W_Q a rational form of W, L a lattice in W_Q; then if $\pi(g_j)(v_j) \to 0$ as $j \to \infty$ (where $g_j \in U_{\mathfrak{R}}$, $v_j \in L$), one has $v_j = 0$ for j sufficiently large. The proof goes by induction on dimension of W. It is clear for $\dim W = 1$ since every rational representation of U has a nonzero fixed vector. Thus we can assume that $\dim W > 1$. Then we know that the contragredient representation of U on W^*, the complex dual space of W, is defined over Q and hence there exists a nonzero U invariant linear form λ in W_Q^*. From above it is clear that for j sufficiently large $\lambda(v_j) = 0$; hence $v_j \in \mathrm{Kernel}(\lambda)$ for $j \gg 0$. However the space $\mathrm{Kernel}(\lambda)$ is stable under U and so defines a rational representation of U. Then by the induction assumption $v_j = 0$ for $j \gg 0$. The proof of the compactness of $N_{\mathfrak{R}}/\Gamma_2$ is concluded by noting that $\det(g) = 1$ for any element $g \in U$ since g is unipotent (we also note that the condition $V(I_W^+) \cap L = (0)$ is not true in general where I_W is the ring of U invariants in $S(W)$ but $U_{\mathfrak{R}}/U_3$ is always compact).

Hence we are at the point of proving for G reductive algebraic and $G \subseteq \mathrm{Aut}(V)$ (faithful representation), the equivalence of the following: (i) $G_{\mathfrak{R}}/G_3$ *is compact* and (ii) $X(G^0)_Q = 1$ and *every element of* G_Q *is semisimple*.

We show that (i) implies (ii). We let H be a locally compact group operating continuously on a locally compact space M and D a closed subgroup of H such that H/D is compact. Then if $D \cdot m$ (for some m) is a closed orbit, then $H \cdot m$ is also closed (use the fact that there exists compact set C so that $H = C \cdot D$). We apply this criterion to the following case: Let G be an algebraic group, not necessarily reductive, Γ an arithmetic subgroup, π a rational representation of G ($\pi : G \to \mathrm{Aut}(W)$ defined over Q), $G_{\mathfrak{R}}/\Gamma$ compact, and $v \in V_Q$, nonzero; then the orbit $\pi(G_{\mathfrak{R}})(v)$ is closed in $V_{\mathfrak{R}}$. Indeed there is a lattice L' in V_Q stable under $\pi(\Gamma)$ and containing the point v. Then the orbit $\pi(\Gamma)(v)$ is closed and hence we can apply our criterion.

We let $\chi : G^0 \to \mathrm{Aut}(\mathfrak{C})$ be a character of G^0 defined over Q. Then if $G_{\mathfrak{R}}/G_3$ is compact, there exists $v \in Q^*$, so that $\chi(G_{\mathfrak{R}}) \cdot v$ is closed. Now if $\chi \neq 1$, then $\chi(G_{\mathfrak{R}})(v)$ contains at least the positive real numbers (χ nontrivial on $G_{\mathfrak{R}}$ since $G_{\mathfrak{R}}$ has a finite number of components and is Zariski dense in G). However $0 \notin \chi(G_{\mathfrak{R}})v$, which is a contradiction. Hence $\chi(G^0)_Q = 1$.

Next we let $\pi : G \to \mathrm{Aut}(M_n)$ be the adjoint representation of G on M_n, the $n \times n$ complex matrices. The representation is clearly defined over Q and in fact we have $G_Q \subseteq M_n(Q)$. Above we noted that if $g \in G_Q$, then $g_s, g_u \in G_Q$ (the semisimple and unipotent components of g). But also by the previous remarks, every orbit of $G_{\mathfrak{R}}$ in $M_n(Q)$ must be closed. But by Jacobson–Morosov Theorem, we know that there exists a one parameter group $x(t)$ such that $x(t)g_u x(t)^{-1} \underset{(t \to \infty)}{\longrightarrow} 1$. Hence $g = g_s$ and all the elements of G_Q are semisimple.

Thus we are now at the point of proving (ii) implies (i) in the Godement Conjecture for G reductive. We first assume that G is connected semisimple with Center $(G) = 1$. Then we have $G \subseteq \mathrm{Aut}(\mathscr{T})$ where $\mathscr{T}$ is the Lie algebra of G and we can use the comments mentioned above. We recall that $G_{\mathfrak{R}}/G_{\mathfrak{Z}}$ is compact if and only if $\mathscr{N} \cap L = (0)$ where $\mathscr{N}$ is the set of nilpotent elements of $\mathscr{T}$ (see [9]). However the map $\exp : \mathscr{N} \to \hat{\mathscr{N}}$ is a biregular rational map from $\mathscr{N}$ to the variety $\hat{\mathscr{N}}$ of unipotent elements of G. And since G is defined over Q, we know that $\mathscr{T}$ has a Q-form, $\mathscr{T}_Q$; hence all the elements of G_Q are diagonalizable if and only if all the elements of $\mathscr{T}_Q$ have the same property. Thus (ii) $\Rightarrow$ (i) becomes a triviality. (We note that $|\det g| = 1$ for all $g \in G_{\mathfrak{R}}$ when $G \subseteq \mathrm{Aut}(\mathscr{T})$ as above.) To pass now to the general case when G is reductive is rather technical and we refer the reader to [2, pp. 55–56]. As in [2], we also comment that with the assumption (ii) about G one would like to have a faithful matrix representation $\pi : G \to \mathrm{Aut}(W)$ so that $V(I_W^+) \cap L = 0$ with L a lattice in W_Q and I_W the usual G invariants (also require that $|\det \pi(g)|$ be bounded for all $g \in G_{\mathfrak{R}}$). Then from our comments above $G_{\mathfrak{R}}/G_{\mathfrak{Z}}$ is compact.

To clear up loose ends, we must now determine when the $G_{\mathfrak{R}}$ orbit of L, the standard lattice in $V_{\mathfrak{R}}$, is closed in $\mathscr{R}$ (see the definition of lattices above). Here we are assuming that G is reductive; then it follows by the remarks in Section 1 that we have a representation $\pi : GL_n(\mathfrak{C}) \to \mathrm{Aut}(W)$ defined over Q and a point $v \in W_Q$, $v \neq 0$ so that the isotropy group of v is G and the $GL_n(\mathfrak{C})$ orbit of v is closed. Then we can embed the vector v in a lattice L' stable under $GL_n(\mathfrak{Z})$. Thus the set $GL_n(\mathfrak{Z})\cdot v$ is closed and, noting that the map $\varphi : GL_n(\mathfrak{R}) \to GL_n(\mathfrak{R})\,(v)$ is continuous, we have that the inverse image of $GL_n(\mathfrak{Z})\cdot v$ $(= GL_n(\mathfrak{Z})\cdot G_{\mathfrak{R}})$ is closed in $GL_n(\mathfrak{R})$. Hence $G_{\mathfrak{R}}\cdot GL_n(\mathfrak{Z})$ is closed in $GL_n(\mathfrak{R})$; thus the orbit $G_{\mathfrak{R}}(L)$ is closed in $\mathscr{R}$. Similarly if G is unipotent, one shows that $G_{\mathfrak{R}}(L)$ is closed; this is left as an exercise to the reader.

Notes

1. In Section 1.1 for a proof of Weitzenbock theorem we refer reader to [*14*].
2. In Section 1.3 a general discussion of the Hilbert-Mumford criterion is given in [5]. Also see chapter II in [*13*].
3. We note that the proof of the Godemont conjecture as given in [*2*] is based on [*12*].

References

[*1*] E. M. Andreev, E. B. Vinberg, and A. G. Elashvili, Orbits of greatest dimension in semisimple linear Lie groups, *Functional Analysis and Its Applications*, Vol. 1, No. 4, pp. 257–262. (English translation.)

[*2*] A. Borel, Introduction aux groupes arithmetiques, *Publ. Inst. Math. Univ. Strasbourg*, **15** (1969).

[*3*] A. Borel, and Harish-Chandra, Arithmetic subgroups of algebraic groups, *Ann. Math.*, (2) **75** (1962), 485–535.

[*4*] N.Bourbaki, Groupes et algebres de Lie, *Elements de Mathematique*, Vol. XXXIV, Hermann, Paris, 1968, Chapters 4–6.

[*5*] J. Dieudonnè and J. B. Carrell, "Invariant Theory, Old and New," *Advances in Math.*, (No. 1) **4** (1970).

[*6*] J. Fogarty, *Invariant Theory*, Benjamin, New York, 1969.

[*7*] S. Helgason, Invariants and fundamental functions, *Acta. Math.*, **109** (1963), 241–258.

[*8*] J. Igusa, Modular forms and projective invariants, *Amer. J. Math.*, **89** (1967), 817–855.

[*9*] B. Kostant, Lie group representations on polynomial rings, *Amer. J. Math.*, **85** (1963), 327–402.

[*10*] B. Kostant and S. Rallis, On representations associated with symmetric spaces, *Bull. Amer. Math. Soc.*, (No. 4) **75** (1969), 884–888.

[*11*] ———, On orbits associated with symmetric spaces, *Bull. Amer. Math. Soc.*, (No. 4) *75* (1969), 879–883.

[*12*] G. D. Mostow and T. Tamagawa, On the compactness of arithmetically defined homogeneous spaces, *Ann. of Math.*, (2) **76** (1962), 446–463.

[*13*] D. Mumford, Geometric invariant theory, *Ergeb. Math.*, **34** (1965).

[*14*] M. Nagata, Lectures on the Fourteenth Problem of Hilbert, *Tata Inst. Notes*, (No. 31) 1965.

[*15*] O. T. O'Meara, *Introduction to Quadratic Forms*, Academic Press, New York, 1963.

[*16*] L. Solomon, Invariants of Euclidean reflection groups, *Trans. Amer. Math. Soc.*, **113**, (1964), 274–286.

[*17*] H. Weyl, *Classical Groups*, Princeton Univ. Press, Princeton (1964), New Jersey, 1946.

TOPICS ON TOTALLY DISCONTINUOUS GROUPS*

HSIEN-CHUNG WANG

Department of Mathematics
Cornell University
Ithaca, New York

*Supported in part by the National Science Foundation under Grant GP 11320.

This is an elaboration of the author's lectures in the "Symmetric Spaces Program" at Washington University. The main purpose is to give a report of the recent developments on *some* aspects of the theory of totally discontinuous groups. It is not intended to be a general survey. We leave untouched many important topics such as the theory of automorphic functions and compactification of locally symmetric spaces. At the end we include an appendix to serve as a handy glossary for the reader. Most of the definitions concerning algebraic groups used in this article can be found there.

1. Introduction

Let G be a Lie group (not necessarily connected), and Γ a discrete subgroup of G. Each right invariant measure μ of G induces a measure $\bar{\mu}$ on the coset space G/Γ such that the projection $\pi : G \to G/\Gamma$ is, locally, measure-preserving. This measure $\bar{\mu}$ is relatively G-invariant in the sense that, to each element g of G, there corresponds a positive number $\Delta(g)$ such that $\bar{\mu}(gE) = \Delta(g)\bar{\mu}(E)$ for all measurable subsets E of G/Γ. Since G is transitive on G/Γ, any two such relatively G-invariant measures differ only by a positive constant multiple. If $\bar{\mu}(G/\Gamma) < \infty$, then $\Delta(g) = 1$ for all g, and $\bar{\mu}$ is invariant under the action of G, whence it follows that μ is two-sided invariant.

Definition. A discrete subgroup Γ of G will be called a *lattice* if G/Γ has a finite measure invariant under the action of G. If, moreover, G/Γ is compact, then we shall call Γ a *uniform lattice* or a *C-lattice*.

Discrete subgroups are very closely related with the discontinuous transformation groups of coset spaces of G. Let C be a compact subgroup of G such that G acts on $C\backslash G$ almost effectively (or what is the same, C does not contain any connected normal subgroup of G except the identity). Put $M = C\backslash G$. Siegel [*33*] observed: a subgroup D of G is discrete if and only if D acts on M in the discontinuous manner (i.e., each orbit of D is closed in M and is discrete with respect to the induced topology). Now suppose Γ to be a discrete subgroup of G, and consider the double coset space $M/\Gamma = C\backslash G/\Gamma$. Since the projection $p : G/\Gamma \to M/\Gamma$ is a proper map, the measure $\bar{\mu}$ on G/Γ induces a Borel measure μ' on M/Γ. In fact, for each Borel set A of M/Γ, $\mu'(A) = \bar{\mu}(p^{-1}(A))$. Obviously, (a) Γ is a lattice of G if and only if $\mu'(M/\Gamma) < \infty$, and (b) Γ is a uniform lattice if and only if M/Γ is compact.

Let B be a symmetric bounded domain, and $I(B)$ the group of all holomorphic automorphisms of B. Then $I(B)$ is a semisimple Lie group with only a finite number of connected components. We can identify B with the coset space $K\backslash I(B)$ where K is a maximal compact subgroup of $I(B)$. The class of discontinuous groups of holomorphic automorphisms of B coincides with the class of discrete subgroups of $I(B)$. In fact, this is one of the motivations to study the lattices of semisimple Lie groups.

2. Existence of Lattices

A Lie group G may not have any lattice at all. We have seen, in the preceding section, that if a lattice exists in G, then G must be unimodular. The following is another necessary condition for the existence of lattices:

(2.1) (Garland–Goto [*15*]) If a connected Lie group G has a lattice, then the group of all inner automorphisms of G is closed in the group Aut G of all continuous automorphisms of G. Here, Aut G is given the compact-open topology.

Concerning the existence problem of lattices, we have quite a bit of information. Nevertheless, a *simple* criterion for a general Lie group to possess a lattice is still lacking.

In 1949, Malcev [*20*] proved that a connected nilpotent Lie group has a lattice if and only if the Lie algebra $\mathfrak{G}$ of G possesses a basis referred to which all the structure constants are rational numbers (i.e., the Lie algebra $\mathfrak{G}$ has a $\mathbf{Q}$-structure, $\mathbf{Q}$ denoting the field of rational numbers).

It was Siegel who first discussed the lattices of a semisimple Lie group other than the special linear group. By using the arithmetic method, the lattices of some simple Lie groups are constructed. In 1958, Weil [*44*] proved the existence of lattices in all the "classical groups." Along this line, the most general results were proved by Borel–Harish-Chandra in 1962 [*9*]. They can be stated as follows:

(2.2) Let G be an algebraic group defined over $\mathbf{Q}$, $G_{\mathbf{R}} = G \cap GL(n, \mathbf{R})$ the set of real matrices in G, and $G_{\mathbf{Z}}$ the set of matrices in G with integer entries and with determinant equal to ± 1. Then $G_{\mathbf{Z}}$ is a lattice of $G_{\mathbf{R}}$ if and only if G has no character defined over $\mathbf{Q}$ except the trivial one. In particular, if G is semisimple, then $G_{\mathbf{Z}}$ is always a lattice of $G_{\mathbf{R}}$.

(2.3) Suppose that G, $G_{\mathbf{R}}$, $G_{\mathbf{Z}}$ have the same meaning as before. Then $G_{\mathbf{R}}/G_{\mathbf{Z}}$ is compact if and only if (a) G has no character defined over $\mathbf{Q}$ other than the trivial one, and (b) every unipotent matrix in $G_{\mathbf{Z}}$ lies in the radical of $G_{\mathbf{R}}$. (By radical, we mean the maximal connected, normal solvable subgroup.)

By using a theorem of Mahler, Mostow–Tamagawa [*28*] gave another proof of (2.3). This criterion for the uniformity of lattices has been generalized to the case by which the lattice may not be arithmetic. In fact, we have

(2.4) Let Γ be a lattice of a connected, semisimple linear Lie group L without compact factor. Then L/Γ is not compact if and only if Γ contains a unipotent matrix other than the identity.

This was first proved by A. Selberg when G is the product of a finite number of copies of $SL(2, \mathbf{R})$ and $SL(2, \mathbf{C})$. Then H. Garland proved it when G is simple and has $\mathbf{R}$-rank one. The general case was due to Kazdan-Margulis [*18*, 8].

In a solvable Lie group, Mostow [*25*] has shown that each lattice is uniform.

Now we come back to the existence problem of lattices. Let L be a connected, noncompact, semisimple Lie group, $\mathfrak{L}$ its Lie algebra, and Aut $\mathfrak{L}$ the group of automorphisms of $\mathfrak{L}$. Using the Weyl–Chevalley basis of the complexification of $\mathfrak{L}$, one [*5*] can construct a basis $\{X_1, X_2, \ldots, X_n\}$ of the real Lie algebra $\mathfrak{L}$ such that (a) the set $\mathfrak{B} = \mathbf{Q}X_1 + \mathbf{Q}X_2 + \cdots + \mathbf{Q}X_n$ forms a Lie algebra over $\mathbf{Q}$, and (b) there exists an element X in $\mathfrak{B}$ with $X \neq 0$ and ad X nilpotent. We identify Aut $\mathfrak{L}$ with a group of matrices by means of the basis $\{X_1, X_2, \ldots, X_n\}$. Then, from (a), Aut $\mathfrak{L}$ is the set of $\mathbf{R}$-points in an algebraic group defined over $\mathbf{Q}$. Because of (b), the set of $\mathbf{Z}$-points in Aut $\mathfrak{L}$ is a nonuniform lattice of Aut $\mathfrak{L}$.

Using arithmetic methods, Siegel [*32*] constructed uniform lattices of the symplectic groups. This result has been extended by Borel [*5*] to the group of automorphisms of an arbitrary real semisimple Lie algebra. Therefore the group Aut $\mathfrak{L}$, and hence its identity component Ad L, has both uniform and nonuniform lattices. Since L is a covering group of Ad L, we have

(2.5) Theorem. Let L be a connected, noncompact, semisimple Lie group. Then L has both uniform and nonuniform lattices.

All the lattices constructed above belong to a class of subgroups called "arithmetic subgroups" which is defined below.

Definition. Two groups A and B are called *commensurable* if their intersection $A \cap B$ is of finite index both in A and in B.

Definition. A subgroup D of a Lie group L is called *arithmetic* if there exist a locally one-to-one representation $f : L \to GL(n, \mathbf{R})$ and a subgroup L' of $GL(n, \mathbf{R})$ with the following properties: (a) $L' = K \times f(L)$ is the direct product of $f(L)$ with a compact subgroup K, and (b) the image $\pi(L' \cap SL(n, \mathbf{Z}))$ is commensurable with $f(D)$ where $\pi : K \times f(L) \to f(L)$ denotes the natural projection.

Definition. A subgroup D of a Lie group L is called *decomposable* if L can be written as a local direct product $L = L_1 \cdot L_2$ such that (a) both L_1 and L_2 are of positive dimension, and (b) D is commensurable with $(D \cap L_1) \cdot (D \cap L_2)$.

It is natural to ask for the Lie groups which have only arithmetic lattices. Selberg conjectured that, in most semisimple Lie groups, all nonuniform lattices are arithmetic. Concerning this question, we have the following:

(i) (G. D. Mostow.) In a connected, simply connected solvable Lie group, each lattice is arithmetic.

(ii) (Makarov [*19*], Vinberg [*34*]) In the hyperbolic groups $SO(1, 3)$, $SO(1, 4)$, and $SO(1, 5)$, there exist nonarithmetic lattices.

(iii) (Margulis [*21*].) Let $G \subset GL(n, \mathbf{R})$ be a connected semisimple linear Lie group without compact factor, and Γ a nonuniform, non-decomposable lattice of G. If G has $\mathbf{R}$-rank greater than one, there exists an element $g \in GL(n, \mathbf{R})$ such that $g\Gamma g^{-1} \cap GL(n, \mathbf{Q})$ is commensurable with $g\Gamma g^{-1}$.

In (iii), we note that, up to an inner automorphism of $GL(n, \mathbf{R})$, and up to commensurability, the lattice Γ contains only matrices with entries in $\mathbf{Q}$.

3. Density Property

Let F be a closed subgroup of a locally compact group G such that the coset space G/F has a finite G-invariant measure. Denote by $\pi : G \to G/F$ the projection. Suppose we are given an element g of G and a neighborhood U of the identity in G. Choose another neighborhood V of the identity with $V = V^{-1}$ and $V \subset U$. The subsets $\pi(V), \pi(gV), \pi(g^2V), \ldots$ of G/F have the same measure and so they cannot be disjoint as G/F has finite measure. Hence we can find integers m, n with $n \neq 0$ such that $\pi(g^m V)$ intersects $\pi(g^{m+n}V)$. It follows that $g^n \in VFV^{-1} \subset UFU$. This property was first observed by Selberg [*31*]. For this reason, let us say a subgroup H (not necessarily closed) of G has "*the Selberg property*" or simply has "*the property* S" if, given an element g of G and a neighborhood U of the identity, there exists an integer n different from zero such that $g^n \in UHU$. When G is abelian, this property is similar to the Kronecker approximation.

As we have just seen, a lattice of G must have the property S. It is an open question whether a discrete subgroup of G with property S is necessarily a lattice. S. P. Wang [*42*] has proved that, if G is a connected solvable Lie group, the answer is affirmative.

The following theorem was proved by Selberg [*31*] when $G = SL(n, \mathbf{R})$ and by Borel [*4*] in general:

(3.1) Density Theorem. Let G be a connected semisimple Lie group without compact factor, and H a subgroup of G having the property S. For any linear representation $f : G \to GL(n,\mathbf{R})$, the closure of $f(H)$ in $GL(n,\mathbf{R})$ with respect to the Zariski topology contains $f(G)$. In other words, for any finite-dimensional representation f of G, each $f(H)$-invariant subspace must be $f(G)$-invariant.

Now suppose that G is a connected semisimple Lie group without compact factor, and Γ a lattice of G. Denote by N the normalizer of Γ in G, and by N_e the identity component of N. Then N_e centralizes Γ and thus $(\mathrm{Ad}\ \Gamma)X = X$ for every vector X in the Lie algebra $\mathfrak{N}$ of N_e where Ad denotes linear adjoint representation of G on its Lie algebra. By the density theorem, we have $(\mathrm{Ad}\ G)X = X$. It follows then from the semisimplicity of G that $X = 0$. Hence $\mathfrak{N} = 0$, and N is discrete. But G/Γ has a finite measure and so N/Γ must be finite. We have thus proved

(3.2) Corollary. Let Γ be a lattice of a connected semisimple Lie group G without compact factor, and N the normalizer of Γ in G. Then N/Γ is finite.

Generalizations of (3.2) in different directions have been made. As an example, let us state a result of S. P. Wang [*43*].

(3.3) Let Γ be a lattice of a connected semisimple Lie group G, and Λ a subgroup of Γ not centralizing any nontrivial connected abelian subgroup of G. Then $N_G(\Lambda)/N_\Gamma(\Lambda)$ is finite where $N_G(\Lambda)$ and $N_\Gamma(\Lambda)$ denote the normalizers of Λ in G and in Γ, respectively.

Depending essentially on the density theorem is the following:

(3.4) (Maximality Property of Lattices [37]). Let Γ be a lattice of a connected semisimple Lie group G without compact factor. Then Γ is contained in only a finite number of discrete subgroups of G.

By using the same arguments, one can extend (3.4) to

(3.5) [*37*, p. 132.] Let Γ be a lattice of a connected semisimple Lie group G. Then Γ is contained in only a finite number of discrete subgroups of G if and only if Γ does not centralize any nontrivial connected Abelian subgroup of the maximal connected, compact normal subgroup of G.

4. Intersection of Lattices with Certain Subgroups

In this section, we shall list some theorems which assert that, for certain subgroups L of a Lie group G, the intersection of L with every lattice of G is a lattice of L. The classical Bieberbach theorem and some recent results are of this nature.

Definition. A normal subgroup L of G will be said to have "*the Bieberbach property*" or simply to have "*the property* B" if, for every lattice Γ of G, the subgroup ΓL is closed in G.

Suppose that L is a normal subgroup having the property B in G, and Γ is a lattice of G. Then ΓL is a closed subgroup of G. Now we have three locally compact groups Γ, ΓL, G with $\Gamma \subset \Gamma L \subset G$. Since G/Γ has a finite invariant measure, we can prove by measure-theoretical arguments

[*25*, Lemma 2.5] that $\Gamma L/\Gamma$ also has a finite invariant measure. But $\Gamma L/\Gamma = L/(L \cap \Gamma)$ and so $\Gamma \cap L$ is a lattice of L. Therefore, *the closedness of* ΓL *implies that* $\Gamma \cap L$ *is a lattice of* L.

(4.1) (Bieberbach [*3*]) Let G be the group of all motions of the Euclidean space $\mathbf{R}^n$, and T the subgroup consisting of all the translations. Then T has the property B in G.

(4.2) (Malcev [*20*]) Let N be a connected, simply connected nilpotent Lie group. Then the derived group $[N, N]$ has the property B in N.

(4.3) (Mostow [*24*]) Let R be a connected solvable Lie group. Then its nil-radical (i.e., the maximal connected nilpotent normal subgroup) has the property B in R.

(4.4) (Auslander [*1*]) Let N be a connected, simply connected nilpotent Lie group, and G an extension of N by a compact group. If the centralizer of N in G has no connected compact subgroup other than the identity, then N has the property B in G.

(4.5) (Auslander [*2*]) Let G be a connected Lie group, R its radical (not necessarily simply connected), S a maximal connected semisimple subgroup of G, and C the maximal connected compact normal subgroup of S. Then CR has the property B in G. Moreover, each lattice of G intersects CR in a uniform lattice of CR.

Auslander's arguments essentially go as follows: let Γ be a lattice of G. He showed that the identity component $Q = (\overline{\Gamma R})_0$ of the closure of ΓR is a solvable subgroup of CR. Hence ΓQ coincides with $\overline{\Gamma R}$, whence ΓQ is closed and $\Gamma \cap Q$ is a lattice of Q. But Q is solvable, and so $\Gamma \cap Q$ is a uniform lattice of Q. From the compactness of CR/Q, it follows that $\Gamma \cap CR$ is a uniform lattice of CR. There exists then a compact subset K of CR such that $CR \subset \Gamma K$. We have therefore $\Gamma CR = \Gamma K$. The subgroup ΓCR is therefore closed.

This result of Auslander has many applications. Suppose that G, S, C, and R have the same meaning as above and let S' be the product of the noncompact factors of S. Then (4.4) reduces the problem of lattices in the general Lie group G to that of CR and that of S'. It also enables us to prove the following:

(4.6) Let G be a connected Lie group, L a connected closed normal subgroup of G, and D a discrete subgroup of G. If $D \cap L$ is a lattice of L, then DL is closed.

Proof. Let us first assume that L is semisimple without compact factor. Then G takes the form of a local direct product $H \cdot L$ where H is a connected normal subgroup of G. Denote by N the normalizer of $D \cap L$ in L. From (3.2), $N/(N \cap D)$ is finite, and so $D^* = N \cdot D$ is also a discrete subgroup of G. Let D' be the totality of elements g in L such that there exists h in H with $hg \in D$. Then $D' \subset N \subset D^* \cap L$ and

$$D \subset (D^* \cap H) \cdot (D^* \cap L)$$

from which it follows that

$$D^* = (D^* \cap H) \cdot (D^* \cap L), \qquad D^*L = (D^* \cap H)L.$$

Let $\pi : G \to G/L$ be the projection and $\sigma : H \to G/L$ denote the restriction of π over H. We have $\ker \sigma = H \cap L \subset D^* \cap H$. Therefore $\sigma(D^* \cap H)$ is a closed subgroup of G/L. Since $\pi(D^*) = \pi(D^*L) = \sigma(H \cap D^*)$, D^*L must be a closed subgroup of G. DL, being a subgroup of D^*L with finite index, is therefore a closed subgroup of G.

Now let us consider the general case. Let R be the radical of L, S a maximal connected semisimple subgroup of L, and C the maximal connected compact normal subgroup of S. Then CR is a characteristic subgroup of L, and hence a normal subgroup of G. Let $p : G \to G/CR$ be the projection. From (4.5), the intersection $D \cap CR = (D \cap L) \cap CR$ is a uniform lattice of CR. It follows that DCR is closed in G, or what is the same, $p(D)$ is discrete. Since $D \cap L$ is a lattice of L, the discreteness implies that $p(D) \cap p(L)$ is a lattice of $p(L)$. Here, $p(L)$ is semisimple without compact factor. From the preceding paragraph, we know that $p(D)p(L)$ is closed in $p(G)$. Hence DL is closed in G. This completes the proof.

Definition. Let H be a reductive algebraic group defined over $\mathbf{Q}$, T a maximal $\mathbf{Q}$-split torus in H, and N a maximal unipotent subgroup such that T belongs to the normalizer of N. Denote by $\mathfrak{T}$, $\mathfrak{N}$ the Lie algebras of T, N, respectively. Then $\mathfrak{N}$ can be written as a sum of root spaces:

$$\mathfrak{N} = \textstyle\sum_{\alpha \in \Delta} \mathfrak{N}_\alpha ,$$

where the α's are linear forms over $\mathfrak{T}$, and $\mathfrak{N}_\alpha = \{X \in \mathfrak{N} : [A, X] = \alpha(A)X, A \in \mathfrak{T}\}$. Let $\Phi = \{\alpha_1, \alpha_2, \ldots, \alpha_s\}$ be the set of simple roots in Δ, and $\{\alpha_{i_1}, \alpha_{i_2}, \ldots, \alpha_{i_a}\}$ a subset of Φ. Denote by $\mathfrak{M}$ the sum of all $\mathfrak{N}_\alpha$ as α runs through the linear combinations

$$\alpha = C_1\alpha_1 + C_2\alpha_2 + \cdots + C_s\alpha_s \in \Delta$$

with $C_{i_1} + C_{i_2} + \cdots + C_{i_a} > 0$, and denote by M the subgroup corresponding to the subalgebra $\mathfrak{M}$. Then M is called a *horospherical subgroup* of H. The group $M_{\mathbf{R}}$ of real points in M is called a *horospherical subgroup* of the group $H_{\mathbf{R}}$ of real points in H.

(4.7) (Selberg, 1965, unpublished) Let G be the product of a finite number of copies of $SL(2, \mathbf{R})$ and $SL(2, \mathbf{C})$, and Γ a nondecomposable lattice of G. If Γ is not uniform, then Γ contains a unipotent matrix g such that (a) the centralizer $Z_G(g)$ of g in G is a horospherical subgroup, and (b) Γ intersects $Z_G(g)$ in a lattice of $Z_G(g)$.

This has been generalized. In fact we have

(4.8) (Garland–Raghunathan [*16*]) Let G be a connected simple Lie group of **R**-rank one, and Γ a nonuniform lattice of G. There exist horospherical subgroups H of G such that Γ intersects H in a lattice of H.

(4.9) (Margulis [*21*]) Let G be a semisimple linear group without compact factor, Γ a nonuniform, nondecomposable lattice of G, and U_Γ a maximal unipotent subgroup of Γ. Then the least connected subgroup U_G of G which contains U_Γ is a horospherical subgroup of G.

We note here that U_Γ is automatically a lattice of U_G.

5. Volume of Lattices

In this section, G always stands for a connected semisimple Lie group without compact factor. Let $\mathfrak{G}$ denote the Lie algebra of G, and k the Killing form. Let $\sigma : \mathfrak{G} \to \mathfrak{G}$ be a Cartan involution. The inner product over $\mathfrak{G}$ defined by $\langle X, Y\rangle = k(\sigma X, Y)$ gives a right invariant Riemannian metric over the underlying manifold of G. We shall call it a *canonical Riemannian metric*. This metric depends on the choice of σ. Nevertheless, any two such metrics differ only by an inner automorphism. It follows that all these metrics give the same volume element (in fact, this is the same as the volume element induced by the Killing form). The measure defined by this volume element will be called the *canonical measure* or *normalized*

Haar measure. To avoid ambiguity, we shall always use this measure in what follows.

Let D be a discrete subgroup of G. Then the canonical measure induces a G-invariant measure $\bar{\mu}$ over G/D. We define $v(D)$ to be $\bar{\mu}(G/D)$ and call it *the volume of* D.

C. L. Siegel has proved that the set $\{v(D) : D = \text{discrete subgroup of } SL(2, \mathbf{R})\}$ has a positive lower bound. Recently, Kazhdan–Margulis showed that this is valid for all semisimple Lie groups G without compact factor. To study this lower-bound problem, let us first consider some properties of the discrete nilpotent subgroups of G, and then give a proof of the theorem of Kazdan–Margulis.

Let $\mathfrak{G}_\pi$ be the totality of elements X in $\mathfrak{G}$ such that the imaginary part of every eigenvalue of Ad X lies in the open interval $(-\pi, \pi)$. Restricted to $\mathfrak{G}_\pi$, the exponential map exp is a diffeomorphism [37]. Denote by G_π the image of $\mathfrak{G}_\pi$ under exponential map, and denote by $\log : G_\pi \to \mathfrak{G}_\pi$ the inverse of the restriction of exp on $\mathfrak{G}_\pi$. The set G_π is a neighborhood of the identity in G invariant under all automorphisms. We can see easily that a subset $\mathfrak{J}$ of $\mathfrak{G}_\pi$ generates a nilpotent subalgebra if and only if its image exp $\mathfrak{J}$ generates a nilpotent subgroup.

Let us fix a Cartan involution of $\mathfrak{G}$, and thus we have an inner product $\langle , \rangle$ on $\mathfrak{G}$ and a canonical Riemannian metric over G. Use $\|X\|$ to denote the length of X in $\mathfrak{G}$. For each subset $\mathfrak{F}$ of $\mathfrak{G}$, we put $r(\mathfrak{F}) = \inf\{\|X\| : X \neq 0, X \in \mathfrak{F}\}$.

(5.1) [*38*] Let $\mathfrak{F}$ be a closed discrete subset of a nilpotent subalgebra of $\mathfrak{G}$ containing at least a nonzero element. Then there exists $g \in G$ such that $r(\mathfrak{F}) < r((\text{Ad } g)\mathfrak{F})$. Moreover, this g can be chosen arbitrarily close to the identity.

For each $x \in G_\pi$, there is a unique element X of $\mathfrak{G}_\pi$ with $x = \exp X$. We define $\|x\|$ to be $\|X\|$. For any subset F of G_π, set

$$r(F) = \inf\{\|x\| : x \neq e, x \in F\}$$

Then from (5.1), we have

(5.1′) Let N be a discrete nilpotent subgroup of G, and $F = N \cap G_\pi$. If F contains more than one element, then there exists g in G such that $r(F) < r(gFg^{-1})$. Moreover, g can be chosen arbitrarily close to the identity e.

This leads to the following:

(5.2) Theorem [*38*]. Let Q be a compact subset of G_π. Given any discrete nilpotent subgroup D of G, there exists h in G such that $hDh^{-1} \cap Q$ is either the identity or void.

Proof. Assuming (5.1), we can prove (5.2) quite easily. In fact, for each g in G, let $F(g) = G_\pi \cap gDg^{-1}$, and let $b = \sup\{r(F(g)) : g \in G\}$. We are going to show that b is infinite. Suppose b to be finite. We can find elements $g_1, g_2, \ldots$ in G such that

$$r(F(g_1)) \leqq r(F(g_2)) \leqq \cdots \leqq b < \infty.$$

Let $U = \{x \in G_\pi : \|x\| < r(F(g_1))\}$. This U is a neighborhood of the identity, and $U \cap g_i D g_i^{-1} = e$ for all i. By a theorem of Chabauty [*12*], the sequence $\{g_i D g_i^{-1}\}$ of subgroups contains a convergent subsequence. Therefore, we can assume that $\{g_i D g_i^{-1}\}$ is already convergent and approaches a discrete subgroup D^* as its limit. D^* is obviously nilpotent, and $r(D^* \cap G_\pi) = b$. By (5.1′), we can find k in G such that $r(D^* \cap G_\pi) < r(kD^*k^{-1} \cap G_\pi)$. It follows that

$$\lim r(F(kg_i)) = r(kD^*k^{-1} \cap G_\pi) > b.$$

This contradicts the definition of b. Hence the set $\{r(F(g)) : g \in G\}$ is not bounded.

Since Q is a compact subset of G_π, we can find a large number q such that $\|x\| \leqq q$ for all x in Q. By the preceding discussions, we can find h in G with $r(F(h)) > q$. It follows then that the intersection $Q \cap hDh^{-1}$ is either the identity or void. Theorem (5.2) is thus proved.

Let L be a subset of G. We put $C_0(L) = L$, and define inductively $C_n(L) = \{xyx^{-1}y^{-1} : x \in L, y \in C_{n-1}(L)\}$, $n = 1, 2, \ldots$. A set M is said *to have property* F if $\lim C_n(M) = \{e\}$ in the sense that, given any neighborhood V of e, there exists n_0 such that $C_n(L) \subset V$ for all $n \geqq n_0$. It is well known that a small neighborhood of the identity has property F (cf. Pontrjagin [*28a*, p. 237]).

Now suppose that M is a subset with property F. Let D be a discrete subgroup of G. From the discreteness, it follows that we can find an integer m with $C_m(D \cap M) = \{e\}$. By a theorem of Zassenhaus [*48*], $D \cap M$ generates a nilpotent subgroup.

Let r be a positive real number. We use $\mathfrak{S}_r$ to denote the set $\{X \in \mathfrak{G} : \|X\| \leqq r\}$, and put $S_r = \exp \mathfrak{S}_r$. They are compact sets. Now we can prove the following version of the theorem of Kazhdan–Margulis:

(5.3) Theorem. Suppose that $\mathfrak{S}_a$ belongs to $\mathfrak{G}_\pi$ and S_a has property F (these two conditions are satisfied when a is small). Then given any discrete subgroup D of G, there exists an element h in G such that $hDh^{-1} \cap S_a = \{e\}$.

Proof. The proof is similar to that of (5.2). For each subset H of G_π, we use $r(H)$ to denote the greatest lower bound of $\{\|x\| : x \neq e, x \in H\}$. Let $b = \sup\{r(gDg^{-1} \cap S_a) : g \in G\}$. Since either $r(gDg^{-1} \cap S_a) \leqq a$ or $r(gDg^{-1} \cap S_a) = \infty$, we have that either $b \leqq a$ or $b = \infty$. Suppose that b is finite. We can find elements $g_1, g_2, \ldots$ in G such that $r(g_iDg_i^{-1} \cap S_a) \geqq r(g_{i-1}Dg_{i-1}^{-1} \cap S_a)$ for all i, and $\lim r(g_iDg_i^{-1} \cap S_a) = b$. Let $U = \{x \in G_\pi : \|x\| < r(g_1Dg_1^{-1} \cap S_a)\}$. Then $g_iDg_i^{-1} \cap U = \{e\}$ for all i, and so by the Chabauty theorem, we can assume that the sequence $\{g_iDg_i^{-1}\}$ of subgroups is convergent and has a discrete subgroup D^* as its limit. We see immediately that $r(D^* \cap S_a) \leqq b$. Now let us show that they are equal. For this purpose, choose $u \in D^* \cap S_a$ such that $\|u\| = r(D^* \cap S_a)$. If u is a boundary point of S_a, then $r(D^* \cap S_a) = \|u\| = a \geqq b$ whence $r(D^* \cap S_a) = a = b$. If u is an interior point of S_a, then there exist $x_i \in g_iDg_i^{-1} \cap S_a$ for sufficiently large i's with $\lim x_i = u$, whence $\|u\| = \lim\|x_i\| \geqq b$ and $r(D^* \cap S_a) = b$. Therefore, in both alternatives, we have shown $r(D^* \cap S_a) = b = \lim r(g_iDg_i^{-1} \cap S_a)$. By assumption, S_a has property F, and so the intersection $D^* \cap S_a$ generates a nilpotent subgroup. By (5.1′), we can find k in G so that $r(k(D^* \cap S_a)k^{-1}) > b$. The complement $D^* - S_a$ is a closed set and does not intersect the compact set S_a. Therefore, we can choose k sufficiently close to the identity such that $k(D^* - S_a)k^{-1}$ is disjoint from S_a. This tells us that $r(kD^*k^{-1} \cap S_a) \geqq r(k(D^* \cap S_a)k^{-1} > b$. On the other hand, $\lim kg_iDg_i^{-1}k^{-1} = kD^*k^{-1}$. Just as above, we can prove that $r(kD^*k^{-1} \cap S_a) = \lim r(kg_i Dg_i^{-1}k^{-1} \cap S_a)$. Hence $r(kg_iDg_i^{-1}k^{-1} \cap S_a) > b$ for large i which contradicts our definition of b. Therefore b is infinite, and (5.3) is proved.

Suppose that S_a has the same meaning as in (5.3). Choose a neighborhood W of the identity so that $WW^{-1} \subset S_a$. Let Γ be a lattice of G. Find h in G with the property that $h\Gamma h^{-1} \cap S_a = e$. It follows that $v(\Gamma) = v(h\Gamma h^{-1}) \geqq \mu(W)$. Hence

(5.4) Corollary. Let Φ_G denote the totality of volume of all the lattices of G. Then Φ_G has a positive lower bound.

As we shall see in Section 8, this set Φ_G is a closed and discrete subset of **R**. For some G, the ratio of any two members of Φ_G is a rational number. However, it is an open question whether this holds in general.

Remark. The number a in (5.3) has been discussed in [*38*]. For example, we can take a to be $288\sqrt{n}/1000$ when $G = SL(n, \mathbf{R})$, and take a to be $288\sqrt{n+1}/1000$ when $G = Sp(n, \mathbf{R})$.

6. Space of Homomorphisms and Rigidity

Let G be a Lie group, and D a discrete subgroup of G. Denote by $R(D, G)$ the totality of homomorphisms of D into G with the compact-open topology. Assume that D is finitely generated. Choose a set of generators $h_1, h_2, \ldots, h_s$ subject to the relations

$$w_j(h_1, h_2, \ldots, h_s) = e, \qquad j \in J,$$

where each w_j is a word in $2s$ symbols, the h's and their inverses. Consider the map

$$R(D, G) \to \underbrace{G \times G \times \cdots \times G}_{s}$$

defined by $r \to (r(h_1), r(h_2), \ldots, r(h_s))$, $r \in R(D, G)$. We see immediately that this map is a homeomorphism. Its image

$$\{(x_1, x_2, \ldots, x_s) \in G \times G \times \cdots \times G : w_j(x_1, x_2, \ldots, x_s) = e, j \in J\}$$

is a closed subset of the analytic manifold $G \times G \times \cdots \times G$. From now on, we identify $R(D, G)$ with this image set. If J is finite, then $R(D, G)$ is an analytic subset of the analytic manifold $G \times \cdots \times G$, and hence is locally connected.

Suppose that G is a real algebraic group defined over a subfield k of **R** (i.e., G is the set of **R**-points of some algebraic linear group defined over k). Then the product $G^{(s)} = G \times \cdots \times G$ of s copies of G is also a real algebraic group defined over k. For each j in J, define $F_j : G^{(s)} \to G$ by $F_j(x_1, x_2, \ldots, x_s) = w_j(x_1, x_2, \ldots, x_s)$, $x_i \in G$. This map is rational and defined over k. Therefore, the intersection $R(D, G) = \cap F_j^{-1}(e)$, as j runs over J, is a real algebraic set defined over k. (It does not matter whether J is finite or infinite.)

Definition. Let D be a finitely generated discrete subgroup of a Lie group G. Denote by I the identity map of D into G, and by $R'(D, G)$ the totality of members r in $R(D, G)$ such that $r(D)$ is closed in G. Then D is called *strongly rigid* (*rigid*) if there exists a neighborhood W of I in $R(D, G)$ such that, given any $r \in W$ ($r \in W \cap R'(D, G)$), we can find an automorphism σ of G with $\sigma(u) = r(u)$ for all u in D.

The rigidity problem of lattices is of considerable importance. In this section, we shall describe the relation between rigidity and the first cohomology as observed by Weil [*46*].

Let H be a topological group, V a vector space, and ρ a representation of H over V. By a 1-cocycle of the $\rho(H)$-module V, we mean a continuous map $z : H \to V$ such that $z(hh') = z(h) + \rho(h)z(h')$ for all h, h' in H. A map $z : H \to V$ is called a 1-coboundary if there exists v in V with $z(h) = v - \rho(h)v$ for all h in H. Obviously, 1-coboundaries are 1-cocycles. The quotient of the space of 1-cocycles by the space of 1-coboundaries is called the first cohomology space $H^1(H, \rho, V)$. The space of 1-cocycles will be denoted by $Z^1(H, \rho, V)$.

Denote by Aff(V) the group of all affine transformations of V. Each map $z: H \to V$ induces a map $z^{\#} : H \to \text{Aff}(V)$ defined by $z^{\#}(h)v = \rho(h)v + z(h)$. We see immediately that z is a 1-cocycle if and only if $z^{\#}$ is a homomorphism of the topological group H into Aff(V). *If H is a Lie group, then the continuity of the 1-cocycle z implies real analyticity.*

Now let us come back to the discrete subgroup D of the Lie group G. Choose generators $h_1, h_2, \ldots, h_s$ subject to the relations $w_j(h_1, h_2, \ldots, h_s) = e$, $j \in J$. Take V to be the Lie algebra $\mathfrak{G}$ of G, and ρ to be the linear adjoint representation Ad. Each 1-cocycle of the (Ad D)-module $\mathfrak{G}$ is uniquely determined by its value on $h_1, h_2, \ldots, h_s$. We want to find the condition under which a map $z : \{h_1, h_2, \ldots, h_s\} \to \mathfrak{G}$ can be extended to a cocycle. This condition, obviously, is the same as the condition under which the corresponding map $z^{\#} : \{h_1, \ldots, h_s\} \to \text{Aff}(\mathfrak{G})$ can be extended to a homomorphism of D into Aff($\mathfrak{G}$). The latter condition can be written as

$$\textbf{(6.1)} \qquad w_j(z^{\#}(h_1), \ldots, z^{\#}(h_s)) = e, \qquad j \in J.$$

We write each word $w_j(h_1, \ldots, h_s)$ as the product $q_1 q_2 \ldots q_{n(j)}$ where each of the q's is either of the form h_i or h_i^{-1}, and put

$$\textbf{(6.2)} \quad u_{jk}(X_1, \ldots, X_s) = \begin{cases} \text{Ad}(q_1 \ldots q_{k-1})X_i, & \text{if } q_k = h_i \\ -\text{Ad}(q_1 \ldots q_k)X_i, & \text{if } q_k = h_i^{-1} \end{cases}$$

where $X_1, \ldots, X_s$ are elements of $\mathfrak{G}$. With these notations, it can be seen without difficulty that (6.1) is equivalent to

$$\textbf{(6.3)} \qquad \textstyle\sum_{k=1}^{n(j)} u_{jk}(z(h_1), \ldots, z(h_s)) = 0, \qquad j \in J.$$

Thus the space $Z^1(D, \mathrm{Ad}, \mathfrak{G})$ of 1-cocycles is in one-to-one correspondence with the set of s-tuples $(z(h_1), \ldots, z(h_s))$ of elements of $\mathfrak{G}$ satisfying (6.3).

The following lemma will be used.

(6.4) Lemma. (Weil [*46*]) Let A, B, C_j $(j \in J)$ be analytic manifolds, and $f : A \to B$, $F_j : B \to C_j$ analytic maps such that, for each j, the iteration $F_j \circ f : A \to C_j$ is a constant map. Suppose a to be a point in A, $b = f(a)$, and $c_j = F_j(b)$. Denote by $df : T_a(A) \to T_b(B)$, $dF_j : T_b(B) \to T_{c_j}(C_j)$, respectively, the differential of f at a and the differential of F_j at b, where the T's signify the tangent spaces. Let $M = \cap F_j^{-1}(c_j)$ be the intersection as j runs through the set J. If the image of df coincides with the intersection $\mathfrak{K} = \cap (dF_j)^{-1}(0)$ of the kernels of dF_j, then a sufficiently small neighborhood V of b in B has the properties that $M \cap V$ is a submanifold and $f(A) \cap V = M \cap V$.

Proof. For each $j \in J$, take local coordinate functions $h_j^1, h_j^2, \ldots, h_j^{m(j)}$ of the manifold C_j around the point c_j where $m(j) = \dim C_j$. For convenience, we assume that $h_j^i(c_j) = 0$. The iteration $h_j^i \circ F_j$ is an analytic function defined in a neighborhood of the point b in B. In the set $\Theta = \{h_j^i \circ F_j : j \in J, i = 1, 2, \ldots, m(j)\}$ of local functions, we can find finitely many functions $\varphi_1, \varphi_2, \ldots, \varphi_t$ such that (a) the differentials $(d\varphi_1)_b, \ldots, (d\varphi_t)_b$ at b are linearly independent, and (b) for any function $\varphi \in \Theta$, $(d\varphi)_b$ is a linear combination of $(d\varphi_1)_b, \ldots, (d\varphi_t)_b$. It follows that $\mathfrak{K}$ is defined by

$$(d\varphi_1)_b = (d\varphi_2)_b = \cdots = (d\varphi_t)_b = 0,$$

and hence $\dim \mathfrak{K} = n - t$ where $n = \dim B$. From the linear independence of the differentials at b, we can find a neighborhood V_1 of b in B such that $\varphi_1, \varphi_2, \ldots, \varphi_t$ are all defined in V_1 and that the set $Q = \{p \in V_1 : \varphi_1(p) = \varphi_2(p) = \cdots = \varphi_t(p) = 0\}$ is an $(n-t)$-dimensional submanifold of B. Obviously, $f(A) \cap V_1 \subset M \cap V_1 \subset Q$.

On the other hand, choose a linear subspace $\mathfrak{P}$ of the tangent space $T_a(A)$ of A at the point a such that df carries $\mathfrak{P}$ isomorphically onto $(df)(T_a(A))$ which coincides with $\mathfrak{K}$. Take a submanifold P of A passing through the point a with $T_a(P) = \mathfrak{P}$. Moreover, we can take P so small that (i) the restriction $f|_P$ of f on P is a diffeomorphism, and (ii) $f(P) \subset V_1$. We have then $f(P) \subset V_1 \cap M \subset Q$. Now $f(P)$ and Q are manifolds of the same dimension, and so we can find a neighborhood V_2 of b in B such that $f(P) \cap V_2 = Q \cap V_2$. Therefore, any neighborhood V of b in B which is contained in $V_1 \cap V_2$ has the required properties. (6.4) is thus proved.

Let $\beta : G \to \mathfrak{G}$ be a real analytic map. Define a vector field X over G by putting $X(g) = (dR_g)\beta(g)$, $g \in G$ where R_g denotes the right translation $x \to xg$ and $dR_g : T_e(G) \to T_g(G)$ the differential of R_g. We shall say that β and X are *associated.* Obviously, this gives a one-to-one correspondence between the set of all real analytic maps from G to $\mathfrak{G}$ and the set of all real analytic vector fields over G. The following can be proved by a direct computation:

(6.5) Suppose that a map $\beta : G \to \mathfrak{G}$ and a vector field X over G are associated. Then the following two conditions are equivalent: (a) β is a 1-cocycle of the (Ad G)-module $\mathfrak{G}$ (i.e., $\beta(xy) = \beta(x) + (\mathrm{Ad}\ x)\beta(y)$; $x,y \in G$). (b) Let I be an open interval, and $k_1 : I \to G$, $k_2 : I \to G$ be integral curves of X. Then the map of I into G defined by $t \to k_1(t)k_2(t)$, $t \in I$ is also an integral curve of X.

From (6.5), we have:

(6.6) Let X be the infinitesimal transformation of a 1-parameter group of automorphisms of a Lie group G. Then the map $\beta : G \to \mathfrak{G}$ associated with X is a 1-cocycle of the (Ad G)-module $\mathfrak{G}$. Conversely, if G can be generated by a compact subset (or what is the same, the group of connected components of G is finitely generated), then every 1-cocycle can be so obtained.

Proof. The first part is an immediate consequence of (6.5). Now let us prove the second part. Suppose that $\beta : G \to \mathfrak{G}$ is a 1-cocycle and X the vector field over G associated with β. Let C be a compact subset of G. We denote, by $r(C)$, the largest real number r such that, for each $g \in C$, there exists an integral curve $k : (-r, r) \to G$ of X with $k(0) = g$. For an arbitrary subset H of G, we put

$$r(H) = \inf\{r(C) : C \quad \text{is a compact subset of } H\}.$$

From our assumption, G is generated by a compact subset, say K, of G.

By (6.5), we have $r(K^2) = r(K)$, $r(K^3) = r(K)$, . . . and so $r(G) = r(K) > 0$. Therefore, X is the infinitesimal transformation of a global 1-parameter group α_t $(-\infty < t < \infty)$ of diffeomorphisms of the manifold G. From (6.5), α_t must be an automorphism, and (6.6) is thus proved.

Now we are in a position to prove the following result of Weil [*46*].

(6.7) Theorem. Let D be a finitely generated discrete subgroup of a Lie group G. Suppose that G can be generated by a compact subset and that the restriction map $H^1(G, \mathrm{Ad}, \mathfrak{G}) \to H^1(D, \mathrm{Ad}, \mathfrak{G})$ is surjective. Then D is strongly rigid.

Proof. Choose a set of generators $h_1, h_2, \ldots, h_s$ of D, and denote the relations by $w_j(h_1, \ldots, h_s) = e, j \in J$. Let us apply (6.4). Take $A = \mathrm{Aut}\, G$ to be the group of all automorphisms of G, $B = G^{(s)} = \underbrace{G \times \cdots \times G}_{s}$, $C_j = G$, $b = (h_1, h_2, \ldots, h_s) \in B$, and take a, c_j to be the identities of A and G, respectively. Define $f : A \to G^{(s)}$ and $F_j : G^{(s)} \to G$ by $f(\alpha) = (\alpha(h_1)), \ldots, \alpha(h_s))$, and $F_j(g_1, \ldots, g_s) = w_j(g_1, \ldots, g_s)$, $\alpha \in A$, $g_i \in G$, $j \in J$. Then $b = f(a)$, $e = c_j = F_j(b)$ and the iteration $F_j f$ is a constant map.

For each x in G, we identify the tangent space $T_x(G)$ of G at x with $T_e(G)$ by means of the right translation, and thus with the Lie algebra $\mathfrak{G}$. Hence the tangent space $T_{(h_1, \ldots, h_s)}(G^{(s)})$ is identified with the direct sum $\mathfrak{G} \oplus \mathfrak{G} \oplus \cdots \oplus \mathfrak{G}$ of s copies of $\mathfrak{G}$. The differential $dF_j \colon T_{(h_1, \ldots, h_2)}(G^{(s)}) \to T_e(G)$ is therefore a linear map of $\mathfrak{G} \oplus \cdots \oplus \mathfrak{G}$ into $\mathfrak{G}$. Now let us verify the hypothesis of (6.3) that $df(T_a(A))$ coincides with the intersection $\cap\, (dF_j)^{-1}(0)$ of the kernels of dF_j as j runs through J. Suppose that $X_i \in G$ $(i = 1, 2, \ldots, s)$ and $dF_j(X_1, X_2, \ldots, X_s) = 0$ for all j in J. By a direct computation, we can show that these imply

$$\textstyle\sum_{k=1}^{n(j)} u_{jk}(X_1, X_2, \ldots, X_s) = 0, \qquad j \in J\,,$$

where the u's are given by (6.2). From the discussion just before Lemma (6.4), we know that there exists a cocycle $z \in Z^1(D, \mathrm{Ad}, \mathfrak{G})$ such that $z(h_i) = X_i$ for all i. By assumption, the restriction map $H^1(G, \mathrm{Ad}, G) \to H^1(D, \mathrm{Ad}, \mathfrak{G})$ is surjective. Therefore, the map $z \colon D \to \mathfrak{G}$ can be extended to a cocycle $\beta \in Z^1(G, \mathrm{Ad}, \mathfrak{G})$. By (6.6), there is a 1-parameter group $\alpha_t(-\infty < t < \infty)$ of automorphisms of G whose infinitesimal transformation is associated with β. We can thus find an element Λ in the Lie algebra $\mathfrak{A}$ of A with $\alpha_t = \exp t\Lambda$. From the definition of our map f, we have $(df)(\Lambda) = (X(h_1), \ldots, X(h_s))$. But $T_g(G)$ is identified with $\mathfrak{G}$ by right translation for every g. Hence

$$(df)(\Lambda) = (\beta(h_1), \ldots, \beta(h_s)) = (z(h_1), \ldots, z(h_s)) = (X_1, \ldots, X_s).$$

We have thus shown that the image of df contains the intersection $\cap (dF_j)^{-1}(0)$ of the kernels of dF_j. Since $F_j \circ f$ is a constant map for each j, it follows that the image of df coincides with the intersection $\cap (dF_j)^{-1}(0)$. All hypotheses in (6.4) are satisfied. Therefore, $f(A)$ contains a neighborhood of $(h_1, \ldots, h_s)$ in the intersection $\cap F_j^{-1}(e)$. According to our identification, $\cap F_j^{-1}(e)$ is precisely the space $R(D, G)$ of homomorphisms and the point $(h_1, \ldots, h_s)$ is the identity map l. Thus D is strongly rigid in G.

Remark. If G is semisimple, then $H^1(G, \mathrm{Ad}, \mathfrak{G}) = 0$, and then the condition in (6.7) on the cohomology is the same as the vanishing of $H^1(D, \mathrm{Ad}, \mathfrak{G})$.

7. Finite Presentability and Rigidity of Lattices

In this section, the Lie group G is always assumed to be connected. There are various ways to see that each uniform lattice of G is finitely presentable. Suppose D to be a uniform lattice of G. We consider the universal covering group $\tilde{G}$ of G, and denote by $p : \tilde{G} \to G$ the projection. Then $\tilde{G}/p^{-1}(D) = G/D$ is a compact analytic manifold and hence a finite polyhedron whence $\pi_1(G/D)$ is finitely presentable. We know that $p^{-1}(D) = \pi_1(G/D)$, $D \approx p^{-1}(D)/(\text{kernel of } p)$, and that kernel of p is finitely generated. Therefore D is finitely presentable.

The finite presentability of nonuniform lattices is a much deeper problem. We have the following:

(a) Let G be a noncompact simple Lie group with **R**-rank r greater than one. Then each lattice in G is finitely generated. (This was proved for $r > 2$ by Kazdan [*17*] and for $r = 2$ by S. P. Wang [*40*].)

(b) Let G be a simple Lie group with **R**-rank one. Then each lattice of G is finitely presentable (Garland–Raghunathan [*16*].)

(c) Let G be a semisimple Lie group without compact factor with **R**-rank greater than one. Then each nondecomposable lattice in G is finitely presentable (Margulis [*21*]).

Now let G be a general connected Lie group, R its radical, S a maximal connected semisimple subgroup of G, and C the maximal connected, compact, normal subgroup of S. Then CR is a normal subgroup of G. We denote by $\pi : G \to G/CR$ the projection. Suppose D to be a lattice of G. By Auslander's theorem (4.4), $D \cap CR$ is a uniform lattice of CR, and so $\pi(D)$ is a lattice of G/CR. Since G/CR is semisimple and has no compact factor, we have from the above proposition (b) and (c) that $\pi(D)$ is finitely presentable. The intersection $D \cap CR$, being a uniform lattice, is also finitely presentable. D is an extension of $D \cap CR$ by $\pi(D)$, and so D is finitely presentable. Therefore

(7.1) The lattices of connected Lie groups are finitely presentable.

Now let us come to the rigidity problem. It is a classical result that lattices in $SL(2, \mathbf{R})$ may not be rigid. But this is an exceptional case. For historical interest, we list some particular results about rigidity before stating the most general theorems.

"Let G be a connected Lie group. The uniform lattices of G are rigid in the following cases: (i) G *is compact* (Montgomery–Zippin, 1942), (ii) G is simply connected and nilpotent (Malcev, 1949 [*20*]), (iii) $G = SL(n, \mathbf{R})$, $n > 2$ (Selberg, 1959 [*31*]), (iv) $G = SO(n, 1)$, $n \geqq 4$ (Calabi, 1960 [*10*]), (v) G is the group of all automorphisms of an irreducible symmetric bounded domain of complex dimension $\geqq 2$ (Calabi–Vesentini, 1960 [*11*])."

A. Weil gave a systematic study of uniform lattices. The following two theorems were proved:

(7.2) [*45*] Let Γ be a uniform lattice of a connected Lie group G, and $l : \Gamma \to G$ denote the identity map. There exists a neighborhood W of l in $R(\Gamma, G)$ such that (a) each r in W is an injective map of Γ into G, and (b) we can find a neighborhood U of e in G and a compact subset K in G so that $U \cap r(\Gamma) = \{e\}$, $G = K \cdot r(\Gamma)$ for every r in W. In particular, the set $R_0(\Gamma, G) = \{r \in R(\Gamma, G)\colon r(\Gamma)$ is a uniform lattice and r is injective$\}$ is an open subset of $R(\Gamma, G)$. Moreover, given any two homomorphisms r_1, $r_2 : \Gamma \to G$ belonging to the same connected component of $R_0(\Gamma, G)$, there exists an analytic homeomorphism $\sigma : G \to G$ of the manifold G such that $\sigma(x \cdot r_1(\gamma)) = \sigma(x) \cdot r_2(\gamma)$ for all x in G and $\gamma \in \Gamma$.

This tells us that, for uniform lattices, rigidity coincides with strong rigidity.

(7.3) [*45*] Let G be a connected semisimple Lie group without compact factor, and Γ a nondecomposable uniform lattice of G. If G is not three dimensional, then $H^1(\Gamma, \mathrm{Ad}, \mathfrak{G}) = 0$ and Γ is strongly rigid.

The uniform lattices of a connected and simply connected solvable Lie group L have been thoroughly studied. In this case, lattices are automatically uniform. Take a lattice Γ. Let $R_0(\Gamma, L) = \{r \in R(\Gamma, L): r \text{ is injective and } r(\Gamma) \text{ is a lattice}\}$. Denote by $R_1(\Gamma, L)$ the connected component of $R_0(\Gamma, L)$ which contains the identity map I, and by N the nil-radical of L. Then the space $R_1(\Gamma, L)$ (usually called the space of deformations of Γ) can be described in a very precise manner by means of the nonconnected group ΓN [*36*]. In particular, we know that Γ may not be rigid and that the space $R_1(\Gamma, L)$ is a manifold.

The rigidity problem of nonuniform lattices Γ in a simple Lie group G was settled at first only when Γ is arithmetic (A. Borel and then M. S. Raghunathan, in general; Bass and Garland independently when G is split over **Q**).

In 1969, by studying the fundamental domains of lattices, Garland–Raghunathan proved:

(7.4) [*16*] Let G be a connected simple Lie group of **R**-rank one, and Γ a nonuniform lattice of G. Then (a) G/Γ is diffeomorphic with the interior of a compact manifold with boundary; (b) if G is locally isomorphic with neither $SL(2, \mathbf{R})$ nor $SL(2, \mathbf{C})$, then $H^1(\Gamma, \mathrm{Ad}, \mathfrak{G}) = 0$ and Γ is strongly rigid; and (c) if G is locally isomorphic with $SL(2, \mathbf{C})$, then Γ is rigid, though not necessarily strongly rigid.

A little later, G. A. Margulis, among other results, announced the following:

(7.5) [*21*] Let G be a connected semisimple Lie group without compact factor and with **R**-rank greater than one. Then each nondecomposable nonuniform lattice in G is strongly rigid.

8. Some Applications of the Rigidity Theorem

In this section we shall discuss three consequences of the rigidity theorem.

(8.1) Theorem. Let G be a connected semisimple Lie group with neither compact nor three dimensional factor. Given any positive number c, there are a finite number of lattices $D_1, D_2, \ldots, D_p$ such that every discrete subgroup Φ of G with $v(\Phi) \leqq c$ is conjugate with one of the D's.

[See A Borel "Commensurability classes and volumes of hyperbolic 3-manifolds", Ann. Sc. Norm. Sup. Pisa, 8 no. 1 (1981) 1-33. , See §8.3
He claims this is not correct.] (Oct. 2011)

Proof. Taking account of (5.3), (7.1), and the rigidity theorem, we can prove this by the same kind of arguments as the author used in [*37*, p. 131]. In fact, suppose that (8.1) is false. Then we can find a sequence $\Gamma_1, \Gamma_2, \ldots$ of lattices such that $v(\Gamma_{i-1}) \leqq v(\Gamma_i)$ for all i, $\lim v(\Gamma_i) = b \leqq c$, and no two of the Γ's are conjugate to each other. From (5.3), there exists a neighborhood U_0 of e in G such that every discrete subgroup of G is conjugate to one which intersects U_0 only at the identity. Therefore, we can choose Γ_i in the very beginning so that $\Gamma_i \cap U_0 = e$. In other words, the sequence $\{\Gamma_i\}$ is uniformly discrete, and so by the Chabauty theorem [*12*], it has a convergent subsequence. To simplify the notations, we assume that the sequence $\{\Gamma_i\}$ is already convergent. Let $\Gamma = \lim \Gamma_n$ be the limit. Since $v(\Gamma) \leqq \lim\inf\{v(\Gamma_i): i = 1, 2, \ldots\} = b$, Γ is a lattice of G.

Let $h_1, h_2, \ldots, h_s$ be a set of generators of Γ with relations $w_j(h_1, \ldots, h_s) = e$, $j = 1, 2, \ldots, t$. Since the number of relations is finite, we can find a neighborhood V of e in G such that $w_j(h_1v_1, h_2v_2, \ldots, h_sv_s) \subset U_0$ for all j and all elements $v_1, v_2, \ldots, v_s$ in V. There exists an integer n_0 so that, whenever $n \geqq n_0$, Γ_n contains elements $h_{n1}, h_{n2}, \ldots, h_{ns}$ with $h_{nj} \in h_jV$ for all j. It follows that

$$w_j(h_{n1}, h_{n2}, \ldots, h_{ns}) \subset U_0 \cap \Gamma_n = \{e\},$$

and therefore we can find a homomorphism $r_n : \Gamma \to \Gamma_n$ which carries h_j to h_{nj} for all j. As elements in $R(\Gamma, G)$, the sequence $\{r_n\}$ tends to the identity map l as the limit. The image $r_n(\Gamma)$ is a discrete subgroup of G. Since Γ is rigid, for large n, we can find an inner automorphism σ_n of G so that $\sigma_n(\gamma) = r_n(\gamma)$, $\gamma \in \Gamma$. The ratio $v(\Gamma)/v(\Gamma_n) = v(r_n(\Gamma))/v(\Gamma_n)$ is a positive integer, and the limit $\lim v(\Gamma)/v(\Gamma_n) = v(\Gamma)/b$ actually exists. It follows that $v(\Gamma)/b \geqq 1$ and that we can find an integer n' such that $v(\Gamma_n) = v(\Gamma_{n'})$ for all $n \geq n'$. But we already know that $v(\Gamma) \leq b$ and so $v(\Gamma) = b$ and $V(\Gamma) = V(r_n(\Gamma)) = V(\Gamma_n)$ for $n \geq n'$. In other words, $r_n(\Gamma) = \Gamma_n$ and each Γ_n is conjugate to Γ for $n \geqq n'$. A contradiction is thus obtained, and (8.1) is thus proved.

As an immediate consequence of (8.1), we have

(8.2) Suppose that G has the same meaning as in (8.1), and $\Theta = \{v(D): D = \text{lattices of } G\}$. Then Θ is a closed and discrete subset of **R**.

Now let us consider another property which can be derived from rigidity. G has the same meaning as before. We shall prove that any group Γ_0 can be imbedded in G as a lattice in only a finite number of ways up to inner automorphisms.

In $\mathbf{R}^n$, we have two topologies—the usual metric topology of Euclidean space and the Zariski topology. To avoid ambiguity, topological concepts referred to Zariski topology will be given the prefix "Zariski." If no prefix is given, we refer to the ordinary metric topology. An algebraic set in $\mathbf{R}^n$ is the same as a Zariski-closed set. Let m be an integer smaller than n. Following H. Whitney, by an *algebraic partial manifold* of dimension $n - m$, we mean a subset P of $\mathbf{R}^n$ such that, for each point p of P, there exists a neighborhood U of p in $\mathbf{R}^n$, and polynomials $f_1, \ldots, f_m$ such that their differentials $df_1, \ldots, df_m$ are linearly independent at p and $U \cap M = \{x \in U: f_1(x) = f_2(x) = \cdots = f_m(x) = 0\}$. Therefore, algebraic partial manifolds are analytic submanifolds of $\mathbf{R}^n$, in general not closed and not connected. H. Whitney proved the following [*47*]:

(a) Let V be an algebraic set in $\mathbf{R}^n$. Any Zariski-open subset of V has only a finite number of connected components.

(b) Each algebraic set V in $\mathbf{R}^n$ can be written as a disjoint union $V = M_1^* \cup M_2^* \cup \cdots \cup M_s^*$ where each M_i^* is an algebraic partial manifold of dimension r_i with $r_1 > r_2 > \cdots > r_s$. Moreover, for each i, the union $M_i^* \cup M_{i+1}^* \cup \cdots \cup M_s^*$ is Zariski-closed.

(8.3) Lemma. Let V be an algebraic set of $\mathbf{R}^n$, and L a Lie group (not necessarily connected) of automorphisms of the vector space $\mathbf{R}^n$ leaving V invariant. Let q be a given integer. Then L has only a finite number of orbits O such that (i) O is an open subset of V, and (ii) O is q-dimensional.

Proof. Let O be an orbit of L satisfying (i) and (ii), and take $x \in O$. Let $V = M_1^* \cup \cdots \cup M_s^*$ be the Whitney decomposition of V as mentioned above. Then $x \in M_i^*$ for a fixed i. Each element g of L carries M_i^* onto itself, and so $O = L(x) \subset M_i^*$. It follows that the dimension r_i of M_i^* is equal to q. Hence all the orbits with properties (i) and (ii) belong to M_i^*. Put $V' = M_i^* \cup M_{i+1}^* \cup \cdots \cup M_s^*$.

For each $y \in \mathbf{R}^n$, let $f_y: L \to \mathbf{R}^n$ be defined by $f_y(g) = g(y)$, $g \in L$. Denote by $df_y: T_e(L) \to T_y(\mathbf{R}^n)$ the differential of the map f_y at the point e, and put $r(y) = \dim (df_y)(T_e(L))$, or what is the same, $r(y)$ denotes the rank of the differentiable map f_y at the identity e. We note that the orbit $L(y) = f_y(L)$ contains a manifold of dimension $r(y)$. Now consider the set $Q = \{y \in \mathbf{R}^n: r(y) \geqq q\}$. It is Zariski-open in $\mathbf{R}^n$. Since V' is the union of manifolds of dimension $\leqq q$ and $L(M_j^*) = M_j^*$ for all j, we have $Q \cap V' = \{y \in V': r(y) = q\} = \{y \in M_i^*: r(y) = q\}$. Since L satisfies the second countability axiom, and M_i^* is a q-dimensional manifold, the following three conditions are equivalent for any point y in M_i^*: (a) $L(y)$ is open in M_i^*, (b) $r(y) = q$, and (c) $L(y)$ is q-dimensional. Let $B = \{O_1, O_2, \ldots\}$ be the totality of q-dimensional orbits under L in M_i^*. Then $Q \cap M_i^*$ is the disjoint union $\cup O_j$ of members of B. Each orbit O_j is open in $Q \cap M_i^*$. Therefore, O_j is both open and closed in $Q \cap M_i^*$. Since $Q \cap M_i^*$ is a Zariski-open subset of the algebraic set V', it has only a finite number of connected components. B is therefore a finite set. The family of orbits under L in V satisfying (i) and (ii), being a subset of B, must be also finite. This proves the Lemma.

(8.4) Theorem. Let G be a semisimple real algebraic group in which every lattice is strongly rigid and Zariski-dense (or what is the same, G has neither compact factor nor factor locally isomorphic with $SL(2, \mathbf{R})$ or $SL(2, \mathbf{C})$). Suppose Γ to be a lattice of G. Then, up to inner automorphisms of G, there are only a finite number of isomorphisms $r : \Gamma \to G$ such that the images $r(\Gamma)$ are lattices.

Proof. Take a set of generators $h_1, h_2, \ldots, h_s$ of Γ with relations $w_j(h_1, h_2, \ldots, h_s) = e$, $j = 1, 2, \ldots, t$. Just as before, we regard $R(\Gamma, G)$ as a subset of the product $G^{(s)}$ of s copies of G. Since G is algebraic, $G^{(s)}$ is also a real algebraic group. For each j, consider the map $F_j : G^{(s)} \to G$ defined by

$$F_j(x_1, x_2, \ldots, x_s) = w_j(x_1, x_2, \ldots, x_s).$$

Obviously, this is a rational map. $R(\Gamma, G)$ coincides with the intersection of the kernels of F_j as j runs through $1, 2, \ldots, t$. Therefore, $R(\Gamma, G)$ is an algebraic subset of the algebraic group $G^{(s)}$.

For large m, we can imbed G in the algebra $M(m, \mathbf{R})$ of all real matrices of order m as an affine variety. Let $n = sm^2$, and let us identify $\mathbf{R}^n$ with the direct sum $M(m, \mathbf{R}) \oplus \cdots \oplus M(m, \mathbf{R})$ of s copies of $M(m, \mathbf{R})$. Thus $R(\Gamma, G) \subset G^{(s)} \subset \mathbf{R}^n$, and $R(\Gamma, G)$ is an algebraic set in $\mathbf{R}^n$. Denote by ρ the representation of G on $\mathbf{R}^n$ given by $\rho(g)(u_1, u_2, \ldots, u_s) = (gu_1g^{-1}, gu_2g^{-1} \ldots, gu_sg^{-1})$ where $g \in G$, $u_i \in M(m, \mathbf{R})$. Then $R(\Gamma, G)$ is $\rho(G)$-invariant. Now let us put $V = R(\Gamma, G)$, $L = \rho(G)$, $p = \dim G$, and apply Lemma (8.3). We know that L has only a finite number of orbits O such that O is p-dimensional and is an open subset of $R(\Gamma, G)$. Suppose that $r \in R(\Gamma, G)$ is an isomorphism and $r(\Gamma)$ is a lattice of G. By assumption, $r(\Gamma)$ is strongly rigid, and hence the orbit $L(r) = \rho(G)r$ is an open subset of $R(\Gamma, G)$. On the other hand, $r(\Gamma)$ is Zariski-dense in G. It follows that $L(r) = G/Z$ where Z denotes the center of G, whence $\dim L(r) = \dim G = p$. We know that these kinds of orbits under L are finite in number. In other words, up to inner automorphisms of G, there are only a finite number of isomorphisms r which carry Γ into lattices of G. Theorem (8.4) is thus proved.

(8.5) Corollary. Suppose G and Γ have the same meaning as in (8.4). Then the group of outer automorphisms of Γ is finite.

Proof. Let N be the normalizer of Γ in G, A the group of all automorphisms of Γ, A'' the group of inner automorphisms of Γ, and A' the group of automorphisms of Γ induced by elements in N. Since Γ is Zariski-dense in G, N/Γ is finite and so A'/A'' is finite. Each automorphism of Γ can be regarded as an isomorphism of Γ into G with its image being a lattice. Therefore, from (8.4), we can find $\alpha_1, \ldots, \alpha_p \in A$ such that every member of A is equivalent to one of the α's under Ad G, and hence under Ad N. This implies that A/A' is finite. The finiteness of the group A/A'' of outer automorphisms of Γ then follows.

(8.6.) Theorem (Selberg, Garland–Raghunathan [*16*]). Let $G \subset GL(m, \mathbf{R})$ be a real algebraic group defined over a subfield k of $\mathbf{R}$, and D a finitely generated discrete subgroup of G. If $H^1(D, \mathrm{Ad}, \mathfrak{G}) = 0$, then there exist an element g in G and a finite extension k' of k such that $gDg^{-1} \subset GL(m, k')$.

9. Appendix

We shall always use $\mathbf{C}$ to denote the complex field, $\mathbf{R}$ the real field, $\mathbf{Q}$ the field of rational numbers, k a subfield of $\mathbf{C}$, and k^* the multiplicative group of nonzero elements in k. $M(n, k)$ stands for the algebra of all $(n \times n)$-matrices with entries in k, and $GL(n, k)$ the group of all the invertible elements in $M(n, k)$.

$M(n, \mathbf{C})$ can be identified with $\mathbf{C}^{n^2}$ in the obvious manner. Thus, in addition to the usual topology, it has the Zarisky topology. A subset F of $M(n, \mathbf{C})$ is called *Zariski-closed* if F consists of all the matrices g whose entries g_{ij} annihilate a system of polynomials $P_\alpha \in \mathbf{C}[X_{11}, X_{12}, \ldots, X_{nn}]$, $\alpha = 1, 2, \ldots$ with complex coefficients in n^2 variables. The P's are called the defining polynomials of F. $GL(n, \mathbf{C})$ is a Zariski-open subset of $M(n, C)$.

The topology on $GL(n, \mathbf{C})$ induced by the Zariski topology of $M(n, \mathbf{C})$ is also called the Zariski topology. A topological term without the prefix "Zariski" is always referred to the Lie group topology.

By an *algebraic group* in this note, we shall always mean a subgroup G of $GL(n, \mathbf{C})$ which is Zariski-closed. G is said *to be defined over a subfield* k of $\mathbf{C}$ if it has a system of defining polynomials whose coefficients are all contained in k. Suppose that G is an algebraic group defined over a subfield of $\mathbf{R}$. Then the set $G \cap GL(n, \mathbf{R})$ of real points in G is called *a real algebraic group* and denoted by $G_{\mathbf{R}}$. We note that a real algebraic group can be regarded as a subgroup of $GL(n, \mathbf{R})$ which consists of all matrices whose entries annihilate a system of polynomials with real coefficients.

Each algebraic group is a complex Lie group. *By its dimension, we always mean the complex dimension.*

Let G be an algebraic group. By a *character of* G, we mean a function $\chi : G \to \mathbf{C}^*$ such that (1) χ preserves multiplication, and (2) χ is a regular function over G, or more precisely, there exists a polynomial $P(X_{ij}, Z)$ in $n^2 + 1$ variables with complex coefficients such that $\chi(g) = P(g_{ij}, (\det g)^{-1})$ for all g in G where g_{ij} denotes the (i, j)th entry of g. Suppose that G is, moreover, defined over a subfield k of $\mathbf{C}$. Then a character χ of G is said *to define over* k if the polynomial P has all its coefficients in k.

An *algebraic torus* is an algebraic group which is Zariski-connected and conjugate to a diagonal group in $GL(n, \mathbf{C})$. It is isomorphic with the product of a finite number of copies of $\mathbf{C}^*$. Now let T be an algebraic torus defined over k. It is called *k-split* if every character of T is defined over k.

Suppose G to be a Zariski-connected algebraic group, defined over k. Then any two maximal k-split tori of G are conjugate by elements of $G \subset GL(n, k)$. Their common dimension is called the *k-rank* of G. If $k = \mathbf{C}$, we simply call it the *rank of* G.

By a *parabolic subgroup* of an algebraic group G, we mean a Zariski-connected and Zariski-closed subgroup P of G such that G/P is compact (with respect to the ordinary topology of coset space of Lie groups).

Note added in proof: Recently, C. D. Mostow proved the following: let E, E' be simply connected symmetric Riemannian spaces of negative curvature having no factor of rank one, and G, G' the groups of isometries of E, E' respectively. Let D be a uniform subgroup of G, and D' a uniform subgroup of G'. Then each isomorphism $f: D \to D'$ can be extended to an isomorphism of G onto G'.

Acknowledgment

The author has had many profitable conversations regarding subjects treated here with Professors Howard Garland and Oscar Rothaus. To both of them, the author is indebted.

References

[1] L. Auslander, Bieberbach's theorem on space groups and discrete uniform subgroups of Lie groups, II, *Amer. J. Math.*, **83** (1961), 279–280.

[2] L. Auslander, On radicals of discrete subgroups of Lie groups, *Amer. J. Math.*, **85** (1963), 145–150.

[3] L. Bieberbach, Ueber die Bewegungsgruppen der Euklidischen Räume I, II, *Math. Ann.*, *70* (1911), 297–336; **72** (1912), 400–412.

[4] A. Borel, Density properties of certain subgroups of semisimple groups without compact components, *Ann. of Math.*, (2) **72** (1960), 179–188.

[5] A. Borel, Compact Clifford–Klein forms of symmetric spaces, *Topology*, **2** (1963), 111–122.

[6] A. Borel, Cohomologie et rigidité d'espaces compacts localement symétriques, Séminaire Bourbaki, 16e année, 1963–1964, Exposé 265, Secrétariat Mathématique, Paris, 1964.

[7] A. Borel, Introduction aux groupes arithmétiques, Publications de Institut Mathématique de l'Université de Strasbourg XV, Hermann, Paris, 1969.

[8] A. Borel, Sous-groupes discrets de groupes semisimple, Séminaire Bourbaki, 21e année, 1968–1969, Exposé 358, Secrétariat Mathématique, Paris, 1969.

[9] A. Borel and Harish-Chandra, Arithmetic subgroups of algebraic groups, *Ann. of Math.*, **75** (1962), 485–535.

[10] E. Calabi, On compact Riemannian manifolds with constant curvature, I., *Proc. Symp. Pure Math.* III (*Differential Geometry*), Providence, Rhode Island, 1961, 155–180.

[11] E. Calabi and E. Vesentini, On compact locally symmetric Kähler manifolds, *Ann. of Math.*, **71** (1960), 472–507.

[12] C. Chabauty, Limite d'ensembles et géométrie des nombres, *Bull. Soc. Math. France*, **78** (1950), 143–151.

[13] H. Garland, On the cohomology of lattices in solvable Lie groups, *Annals of Math.*, **84** (1966), 175–196.

[14] H. Garland, A rigidity theorem for discrete subgroups, *Trans. Amer. Math. Soc.*, **129** (1967), 1–25.

[15] H. Garland and M. Goto, Lattices and the adjoint groups of a Lie group, *Trans. Amer. Math. Soc.*, **124** (1966), 450–460.

[16] H. Garland and M. S. Raghunathan, Fundamental domains for lattices in rank one semisimple Lie groups, *Ann. of Math.*, to appear.

[17] D. A. Kazdan, Connection of the dual space of a group with the structure of its closed subgroups, *Functional × Analysis and Applications*, **1** (1967), 63–65.

[18] D. A. Kazdan and G. A. Margulis, A proof of Selberg's hypothesis, *Mat. Sb.*, (117) **75** (1968), 163–168.

[*19*] V. S. Makarov, Lobachevsky space groups, *Soviet Math. Dokl.* **7** (1966), 328–331.
[*20*] A. I. Malcev, On a class of homogeneous spaces, *Amer. Math. Soc. Transl.*, No. 39, 1951.
[*21*] G. A. Margulis, On the arithmeticity of discrete groups, *Soviet Math. Dokl.*, **10** (1969), 900–902.
[*22*] Y. Matsushima, On the first Betti number of compact quotient spaces of higher dimensional symmetric spaces, *Ann. of Math.*, **75** (1962), 312–330.
[*23*] Y. Matsushima and S. Murakami, On vector-bundle valued harmonic forms and automorphic forms on symmetric spaces, *Ann. of Math.*, **78** (1963), 365–416.
[*24*] G. D. Mostow, Factor spaces of solvable groups, *Ann. of Math.*, **60** (1954), 1–27.
[*25*] G. D. Mostow, Homogeneous spaces with finite invariant measure, *Ann. of Math.*, **75** (1962), 17–37.
[*26*] G. D. Mostow, Quasi-conformal mappings in *n*-spaces and the rigidity of hyperbolic space form, *Inst. Hautes Études Sci. Publ. Math.*, **34** (1968), 53–104.
[*27*] G. D. Mostow, Representative functions on discrete groups and solvable arithemetic subgroups, *Amer. J. Math.*, **92** (1970), 1–32.
[*28*] G. D. Mostow and T. Tamagawa, On the compactness of arithmetically defined homogeneous spaces, *Ann. of Math.*, **76** (1962), 446–463.
[*28*a]Pontrjagin, *Topological Groups*, Princeton Univ. Press, Princeton, New Jersey, 1946.
[*29*] M. S. Raghunathan, On the first cohomology of discrete subgroups of semisimple Lie groups, *Amer. J. Math.*, **87** (1965), 103–139.
[*30*] M. S. Raghunathan, Cohomology of arithmetic subgroups of algebraic groups, *Ann. of Math.*, **86** (1967), 409–424; **87** (1968), 279–304.
[*31*] A. Selberg, On discontinuous groups in higher-dimensional symmetric spaces, *Contribution to Function Theory*, Tata Institute, Bombay, 1960, pp. 147–164.
[*32*] C. L. Siegel, Symplectic geometry, *Amer. J. Math.*, **65** (1943), 1–86.
[*33*] C. L. Siegel, Discontinuous groups, *Ann. Math.*, **44** (1943), 674–689.
[*34*] E. R. Vinberg, Discrete groups generated by reflections in Lobachvsky spaces, *Mat. Sb.*, **72** (1967), 471–486; Correction, ibid., **73** (1967), 303.
[*35*] H. C. Wang, Discrete subgroups of solvable Lie groups I, *Ann. of Math.*, **64** (1965), 1–19.
[*36*] H.-C. Wang, On the deformation of lattices in a Lie group, *Amer. J. Math.*, **85** (1963), 189–212.
[*37*] H.-C. Wang, On a maximality property of discrete subgroups with fundamental domain of finite measure, *Amer. J. Math.*, **89** (1967), 124–132.
[*38*] H.-C. Wang, On discrete nilpotent subgroups of Lie groups, *J. Differential Geometry*, **3** (1969), 481–492.
[*39*] S. P. Wang, Limit of lattices in a Lie group, *Trans. Amer. Math. Soc.*, **133** (1968), 519–526.
[*40*] S. P. Wang, The dual space of semisimple Lie groups, *Amer. J. Math* , **91** (1969), 921–937.
[*41*] S. P. Wang, On a conjecture of Chabauty, *Proc. Amer. Math. Soc.*, **23** (1969), 569–572.
[*42*] S. P. Wang, On *S*-subgroups of solvable Lie groups, *Amer. J. Math.*, **92** (1970), 389–397.
[*43*] S. P. Wang, On a lemma of Mahler, to appear.

[*44*] A. Weil, Discontinuous subgroups of classical groups, Notes, Chicago University, 1958.

[*45*] A. Weil, On discrete subgroups of Lie groups, *Ann. of Math.*, **72** (1960), 369–384; **75** (1962), 578–602.

[*46*] A. Weil, Remarks on the cohomology of groups, *Ann. of Math.*, **80** (1964), 149–157.

[*47*] H. Whitney, Elementary structure of real algebraic varieties, *Ann. of Math. 66* (1957), 545–556.

[*48*] H. Zassenhaus, Beweis eines Satzes ueber diskrete Gruppen, *Abh. Math. Sem. Hansischer Univ.*, **19** (1938), 289–312.